Edible Food Packaging

Amrita Poonia • Tejpal Dhewa
Editors

Edible Food Packaging

Applications, Innovations and Sustainability

Springer

Editors
Amrita Poonia
Department of Dairy Science and Food Technology
Institute of Agricultural Sciences, Banaras Hindu University
Varanasi, Uttar Pradesh, India

Tejpal Dhewa
Department of Nutrition Biology
Central University of Haryana
Mahendragarh, Haryana, India

ISBN 978-981-16-2382-0 ISBN 978-981-16-2383-7 (eBook)
https://doi.org/10.1007/978-981-16-2383-7

This Springer imprint is published by the registered company Springer Nature Singapore Pte Ltd.
The registered company address is: 152 Beach Road, #21-01/04 Gateway East, Singapore 189721, Singapore

This book is dedicated to ***Shri Kashi Vishwanath Ji*** *located on the Western bank of the holy river* ***Ganga****, Varanasi, Uttar Pradesh, India, and is one of the twelve* Jyotirlingas.
The main deity is known by the names Shri Vishwanath and Vishweshwara, literally meaning **Lord of the Universe**. *Varanasi city was called* Kashi *in ancient times, and hence the temple is popularly called Kashi Vishwanath Temple.*

&

Bharat Ratna, ***Mahamana Pandit Madan Mohan Maiaviya Ji***, *Founder of Banaras Hindu University, the largest residential university in Asia and one of the largest in the world.*

Preface

Increasing industrialization and population has put forward two major challenges, namely waste management and continual surge in the global energy demand. Biofuels have recently emerged as ideal fuel carriers to meet these energy requirements in a sustainable manner. Bio-wastes can serve as a valuable renewable source for conversion into biofuels. Food and agro-industrial substrates, such as starch-based wastes, due to their high availability, biodegradability, and rich nutritional composition, can be effectively utilized for packaging edibles as well as eliminate waste disposal problems. Fruits and vegetable peels are suitable candidates to serve this purpose as they have a high fermentable carbohydrate yield. Packaging edibles have generated considerable attention over years as an alternative to synthetic plastics, considering their renewable and biodegradable characteristics. Packaging edibles could be prepared using fruits and vegetable peels as plant-based and biodegradable constituent that may result in high-value commercial use of the by-product. The packaging edibles prepared from these materials could be the best coatings having improved barrier and mechanical properties as compared to individual biopolymers. According to Food and Agriculture Organization (FAO), about three billion tons of food is wasted every year worldwide. Agro-based industries are producing large amount of by-products and waste. Various fruit wastes, i.e., peels, stems, shells, seeds, and trimmings, represent more than 50 % of the fresh produce and these by-products and wastes contain more functional compounds than the fruit. The by-products and wastes also effect the environment, social sector, and economy of a country, and management of these by-products is a big challenge for the society. Due to their functional and nutritional qualities, the utilization of these by-products for packaging edibles is a good solution for preservation of foods. The use of agro-industry and food industry wastes and by-products in preparation of packaging edibles, addition of functional and bioactive compounds with antioxidant and antimicrobial properties against the microbes, is the latest trend in packaging industry and finding a sustainable solution for the plastics. This type of value addition with variation of the productive chain revolutionizes the food industries. New research in the area of packaging edibles is a vital and exclusive field of study which comprises a lot of commercial and environmental potential in near future. In the area of packaging edibles, new research and development is the main area to explore its possibilities for commercialization as well as environmental potential.

The use of sustainable alternatives and innovative and new sources for the development of packaging edibles justifies the publication of this book.

This book is divided into five main parts:

- Sources and origin of packaging edibles
- Sustainable alternatives for packaging edibles
- Shelf-life and safety aspects
- Regulatory aspects
- Innovations and recent trends in edibles

The first part of the book addresses details of edible films and coatings, cups, spoons, and cutlery and various sources like fruit and vegetable industry by-products, grain and oilseed wastes, marine industry by-products, dairy by-products and wastes, and meat industry waste. It also covers animal- and plant-based packaging edibles, scope, and novel microbial sources. Part II addresses on different sustainable alternatives, i.e., seed gums, fruits and vegetable peels, sea weeds, fruits wastes, purees, extracts, juices, dairy by-products, black edible packaging using defatted oil cake, and antioxidant edible packaging. Other food industry by-products like apple pomace, citrus peels, cassava by-products, potato peels, fish skin, and algae have already been proposed as possible environment friendly alternatives for packaging edibles. These by-products are the potential source of biodegradable packaging edibles. They also contain polyphenols and play important role as antioxidants and antimicrobials to the packaging edibles. This may also provide the additional nutritional benefits to the consumers. Utilization of these products as a source of packaging edibles is not only safe for human consumption but also biodegradable in nature and environmental friendly.

Part III of the book addresses about the shelf-life extension foods by using packaging edibles. The packaged food products can be prevented from oxidation and spoilage by microorganisms during storage. Nanolaminates are more advanced and more workable technology that offers maximum utilization potential. The layer-by-layer deposition (LbL) is one of the most powerful techniques whereby charged surfaces are coated on the food with interfacial films consisting of multiple nanolayers of different materials. Nanocoatings have been also studied to incorporate bioactive or functional compounds owing to their ability to control the release of such molecules by the manipulation of coatings' properties. Part IV covers the safety of these packaging systems that is also very crucial. Safety issues related to the microbial spoilage of packaged product had been discussed. Marketing of packaging edibles is affected by many factors such as cost of production, consumer acceptance of these materials, and consuming of these packaging edibles. Part V of the book addresses the innovations in packaging edibles in food as well as other sectors, applications of packaging edibles in liquid foods and pharma sector, and use of sea weeds, seed gums, and agro wastes such as bagasse, cellulose, cutin, molasses, lignin, and paddy straw.

Nowadays, sustainable packaging edibles are gaining considerable interest due to their potential to replace the plastics with food wastes and agro wastes. No book is available about the manufacture of packaging edibles using these types of by-products till date. All the contents of the book are unique and has vast commercial applications and environmental potential.

Varanasi, Uttar Pradesh, India Amrita Poonia
Mahendragarh, Haryana, India Tejpal Dhewa

Contents

Editors and Contributors

About the Editors

Amrita Poonia is working as an assistant professor in the Department of Dairy Science and Food Technology, Institute of Agricultural Sciences, Banaras Hindu University, Varanasi, Uttar Pradesh, India. She has published two laboratory manuals and more than 15 book chapters at national and international level. Dr. Poonia has conducted/offered a Massive Open Online Course (MOOC) under agMOOCs on Functional Foods: Concept, Technology and Health Benefits from (8 October, 2018 to 19 November, 2018) and was awarded with the Certificate of Appreciation for offering a Massive Open Online Course (MOOC) under agMOOCs, by Commonwealth of Learning (COL) Canada. She has published more than 50 publications in esteemed national and international journals. She has successfully completed a project on "Identification of Food Borne Pathogen (Listeria monocytogenes) in Shrikhand Using PCR (Polymerase Chain Reaction)" at BHU and supervised 2 PhD scholars and 28 MSc students and 2 PhD scholars are perusing their research under her supervision.

Tejpal Dhewa is a faculty in the School of Life Sciences, Central University of Haryana, Mahendragarh. He is the coordinator, Food Safety Training and Certification (FoSTaC) Centre, Food Safety and Standards Authority of India (FSSAI), and University SWAYAM Coordinator- Central University of Haryana. Dr. Dhewa is a course coordinator of SWAYAM UGC MOOC on "Food Microbiology and Food Safety." Dr. Dhewa has a diverse industrial, teaching, and research experience. He has published his research works in national and international journals. Dr. Dhewa also supervised several master's theses/dissertations. He has successfully completed DU innovation project (2013-2015) and has conducted short-term courses on "Food Safety, Food Security, and Food Regulations: A Primer" from September 16 to 20, 2018; "Metabolomics in Food and Nutrition Science Research: From Concepts to Applications" from February 15 to 19, 2019; and "Food Fortification and Human Health: From Concepts to Public Health Applications" from September 9 to 13, 2019, under the Ministry of Human Resource Development (MHRD).

Contributors

Partibha Ahlawat Centre of Food Science and Technology, CCS HAU, Hisar, Haryana, India

Usman Ali Department of Food & Nutritional Biotechnology, National Agri-Food Biotechnology Institute, Mohali, Punjab, India

B. N. Ashwija Division of Food Science and Postharvest Technology, Indian Agricultural Research Institute, New Delhi, India

Khalid Bashir Department of Food Technology, New Delhi, India

Aastha Bhardwaj Centre for Rural Development and Technology, Indian Institute of Technology, Hauz Khas, New Delhi, India
Department of Food Technology, New Delhi, India

Suman Bishnoi Department of Livestock Product Technology, College of Dairy Science and Technology, Lala Lajpat Rai University of Veterinary and Animal Sciences, Hisar, Haryana, India

Padma Sangmu Bomzon Department of Nutrition, Raja Narendra Lal Khan Women's College, Paschim Medinipur, Phulpahari, West Bengal, India

Montserrat Calderón-Santoyo Laboratorio Integral de Investigación en Alimentos, Tecnológico Nacional de México/Instituto Tecnológico de Tepic, Tepic, Nayarit, Mexico

Nisha Chaudhary Department of Food Science and Technology, College of Agriculture, Agriculture University, Nagaur, Jodhpur, India

Vandana Chaudhary College of Dairy Science and Technology, Lala Lajpat Rai University of Veterinary and Animal Sciences, Hisar, Haryana, India

Rajni Chopra Department of Food Technology, Institute of Home Economics, University of Delhi, New Delhi, India

Priya Dangi Department of Food Technology and Foods and Nutrition, Institute of Home Economics, University of Delhi, New Delhi, India

Kalyan Das Department of Basic and Applied Science, National Institute of Food Technology Entrepreneurship & Management, Kundli, Sonipat, Haryana, India

Tejpal Dhewa Department of Nutrition Biology, School of Interdisciplinary and Applied Sciences, Central University of Haryana, Mahendragarh, Haryana, India

Para Dholakia Department of Food Technology, Shaheed Rajguru College of Applied Sciences for Women, University of Delhi, New Delhi, New Delhi, India

Vandana Department of Food Technology, Shaheed Rajguru College of Applied Sciences for Women, University of Delhi, New Delhi, New Delhi, India

Himanshi Garg Amity Institute of Food Technology, Amity University, Noida, Uttar Pradesh, India

Meenakshi Garg Department of Food Technology, Bhaskaracharya College of Applied Sciences, University of Delhi, Delhi, India

Neeraj Ghanghas Department of Food Science and Technology, National Institute of Food Technology Entrepreneurship & Management, Kundli, Sonipat, Haryana, India
Department of Food and Human Nutritional Sciences, University of Manitoba, Winnipeg, MB, Canada

Mahendra Gunjal Department of Food Technology and Nutrition, School of Agriculture, Lovely Professional University, Phagwara, Punjab, India

Prerna Gupta Department of Food Technology and Nutrition, School of Agriculture, Lovely Professional University, Phagwara, Punjab, India

T. S. Hanumesh Gowda Division of Food Science and Postharvest Technology, Indian Agricultural Research Institute, New Delhi, India

Himani Department of Food Business Management and Entrepreneurship Development, National Institute of Food Technology Entrepreneurship & Management, Kundli, HR, India

Ankita Hooda Department of Dairy Science and Food Technology, Institute of Agricultural Sciences, Banaras Hindu University, Varanasi, Uttar Pradesh, India

Maricarmen Iñiguez-Moreno Laboratorio Integral de Investigación en Alimentos, Tecnológico Nacional de México/Instituto Tecnológico de Tepic, Tepic, Nayarit, Mexico

Kulsum Jan Department of Food Technology, New Delhi, India

Shumaila Jan Department of Food Science and Technology, National Institute of Food Technology Entrepreneurship and Management, Sonipat, Haryana, India

Sukirti Joshi Centre for Rural Development and Technology, Indian Institute of Technology, Hauz Khas, New Delhi, India

Jaspreet Kaur Department of Food Technology and Nutrition, School of Agriculture, Lovely Professional University, Phagwara, Punjab, India

Sandeep Kaur Department of Food Science, Mehr Chand Mahajan DAV College for Women, Chandigarh, India

Sawinder Kaur Department of Food Technology and Nutrition, School of Agriculture, Lovely Professional University, Phagwara, Punjab, India

Sidra Kazmi Department of Food Technology, New Delhi, India

Ab Lateef Khan School of Biomedical Engineering, Institute for Personalized Medicine, Med-X Shanghai Jiao Tong University, Shanghai, China

Anand Kishore Defence Food Research Laboratory, Grain Science & Technology (GST) Division, DRDO, Ministry of Defence, Government of India, Mysore, India

Anu Kumar University Institute of Biotechnology, Chandigarh University, Mohali, Punjab, India

Anila Kumari Department of Food Science, Mehr Chand Mahajan DAV College for Women, Chandigarh, India

Anju Kumari Centre of Food Science and Technology, CCS HAU, Hisar, Haryana, India

Nitin Kumar Department of Food Engineering, National Institute of Food Technology Entrepreneurship & Management Kundli, Sonipat, Haryana, India

Rakesh Kumar Department of Microbiology, CCS HAU, Hisar, Haryana, India

Sunil Kumar ICMR-RMRC-Regional Medical Research Center, Gorakhpur BRD Medical College, Gorakhpur, UP, India

Yogesh Kumar Department of Agronomy, Food, Natural Resources, Animals and Environment (DAFNAE), University of Padova, Legnaro (PD), Padua, Italy

Bimlesh Mann Department of Dairy Chemistry, National Dairy Research Institute, Karnal, Haryana, India

Murlidhar Meghwal Department of Food Science and Technology, National Institute of Food Technology Entrepreneurship & Management, Kundli, HR, India

Alok Mishra Department of Dairy Science and Food Technology, Institute of Agricultural Sciences, Banaras Hindu University, Varanasi, Uttar Pradesh, India

Pragya Mishra Food Processing and Management, DDU Kaushal Kendra, RGSC, Banaras Hindu University, Varanasi, Uttar Pradesh, India

Aayasee Nanda Department of Food Technology, New Delhi, India

Priya Pal Division of Food Science and Postharvest Technology, Indian Agricultural Research Institute, New Delhi, India

Neeraj Panchal Department of Basic and Applied Science, National Institute of Food Technology Entrepreneurship & Management, Kundli, Sonipat, Haryana, India

Amrita Poonia Department of Dairy Science and Food Technology, Institute of Agricultural Sciences, Banaras Hindu University, Varanasi, Uttar Pradesh, India

A. Poovazhahi Cranfield University, Cranfield, UK

Pramod K. Prabhakar Department of Food Science and Technology, National Institute of Food Technology Entrepreneurship & Management, Kundli, Sonipat, Haryana, India

Mamta Prajapati Sector Expert (Food), National Consumer Helpline, Indian Institute of Public Administration, Delhi, India

Soumya Ranjan Purohit Amity Institute of Food Technology, Amity University, Noida, Uttar Pradesh, India

Juan Arturo Ragazzo-Sanchéz Laboratorio Integral de Investigación en Alimentos, Tecnológico Nacional de México/Instituto Tecnológico de Tepic, Tepic, Nayarit, Mexico

Dinesh Chandra Rai Department of Dairy Science and Food Technology, Institute of Agricultural Sciences, Banaras Hindu University, Varanasi, Uttar Pradesh, India

Prasad Rasane Department of Food Technology and Nutrition, School of Agriculture, Lovely Professional University, Phagwara, Punjab, India

H. Roopa Centre of Food Science and Technology, CCS HAU, Hisar, Haryana, India

Soumen Roy Department of Food Science & Technology, National Institute of Food Technology Entrepreneurship & Management Kundli, Sonipat, Haryana, India

Shalini Gaur Rudra Division of Food Science and Postharvest Technology, Indian Agricultural Research Institute, New Delhi, India

Prabhjot Kaur Sabharwal Department of Food Technology, Shaheed Rajguru College of Applied Sciences for Women, University of Delhi, New Delhi, New Delhi, India

Susmita Dey Sadhu Department of Polymer Science, Bhaskaracharya College of Applied Sciences, University of Delhi, New Delhi, India

Jatindra Kumar Sahu Centre for Rural Development and Technology, IIT Delhi, Hauz Khas, New Delhi, India

Prashant Said College of Agricultural Engineering and Post Harvest Technology, CAU Imphal Gangtok, Sikkim, India

Mrinal Samtiya Department of Nutrition Biology, School of Interdisciplinary and Applied Sciences, Central University of Haryana, Mahendragarh, Haryana, India

Ekta Sehgal Centre of Food Science and Technology, CCS HAU, Hisar, Haryana, India

Anjali Sharma IGNOU Study Centre, National Institute of Food Technology Entrepreneurship & Management Kundli, Sonipat, Haryana, India

Arun Sharma Department of Food Engineering, National Institute of Food Technology Entrepreneurship & Management, Kundli, India

Bhasha Sharma Department of Chemistry, Netaji Subhas University of Technology, Delhi, India

Nitya Sharma Centre for Rural Development and Technology, Indian Institute of Technology, Hauz Khas, New Delhi, India

Sujata Pandit Sharma Department of Life-Sciences, Sharda University Greater Noida, Greater Noida, India

Vandana Sharma Department of Food Science, Mehr Chand Mahajan DAV College for Women, Chandigarh, India

Vasudha Sharma Department of Food Technology, New Delhi, India

Vatsala Sharma Amity Institute of Food Technology, Amity University, Noida, Uttar Pradesh, India

Ritu Sindhu Centre of Food Science and Technology, CCS HAU, Hisar, Haryana, India

Sangeeta C. Sindhu Department of Food and Nutrition, CCS HAU, Hisar, Haryana, India

Jyoti Singh Department of Food Technology and Nutrition, School of Agriculture, Lovely Professional University, Phagwara, Punjab, India

Mohit Singla Department of Food engineering and Technology, Tezpur University, Assam, India

Oscar Alberto Solis-Velazquez Laboratorio de Microbiología, Departamento de Ciencias Médicas y de la Vida, Centro Universitario de la Ciénega, Universidad de Guadalajara, Ocotlán, Jalisco, Mexico

Monika Thakur Amity Institute of Food Technology, Amity University, Noida, Uttar Pradesh, India

Neha Thakur Department of Livestock Product Technology, Lala Lajpat Rai University of Veterinary and Animal Sciences, Hisar, Haryana, India

Pratibha Tiwari Maharshi Markandeshwar University, Maulana, Ambala, Haryana, India

Vandana Department of Food Technology, Shaheed Rajguru College of Applied Sciences for Women, University of Delhi, New Delhi, New Delhi, India

S. Vijay Rakesh Reddy ICAR-Indian Agricultural Research Institute, New Delhi, India

Dev Kumar Yadav Defence Food Research Laboratory, Grain Science & Technology (GST) Division, DRDO, Ministry of Defence, Government of India, Mysore, India

Vedpal Yadav Department of Food Technology, Government Polytechnic, Mandi Adampur, Hisar, Haryana, India

Abbreviations

AG	Arabic gum
Ag	Silver
AgNPs	Silver nanoparticles
Au-MSN	Gold mesoporous silica nanoparticles
BAW	Bulk acoustic wave
BSA	Bovine serum albumin
Ca	Calcium
CAP	Controlled atmospheric packaging
CAS	Controlled atmospheric storage
CFB	Corrugated fiberboard
CFB	Corrugated fiberboard
CFU	Colony-forming units
CMC	Carboxymethylcellulose
CMC	Carboxymethyl cellulose
CNC	Cellulose nanocrystals
CP	Conducting polymer
Cu	Copper
CW	Cellulose whiskers
DHS	Dynamic headspace
DNA	Deoxyribonucleic acid
EC	Edible coatings
ECs	Edible coatings
EOs	Essential oils
EVOH	Ethylene-vinyl alcohol
FAO	Food and Agriculture Organization of the United Nations
FDA	Food and Drug Administration
FLW	Food loss or waste
FSC	Food supply chain
FSSAI	Food Safety and Standards Authority of India
FTIR	Fourier transform infrared spectroscopy
G	A-l-guluronic acid
GRAS	Generally recognized as safe
HCA	Hierarchical cluster analysis

HDPE	High-density and low-density polyethylene
HGA	Homogalacturonan
HM	High methoxyl
HPMC	Hydroxypropyl methylcellulose
LM	Low methoxyl
M	Β-d-mannuronic acid
MAP	Modified atmospheric packaging
MMT	Million metric tonnes
MOSFET	Metal oxide semiconductor field effect transistors
NFC	Nanofibrillated cellulose
OMLs	Overall migration limits
P&T	Purge and trap
PCA	Principal component analyses
PEG	Poly-ethylene glycol
PET	Polyethylene terephthalate
PHA	Polyhydroxyalkanoate
PP	Polypropylene
PPP	Public-private-partnership
PVC	Polyvinyl chloride
ROS	Reactive oxygen species
RSM	Response surface methodology
SEM	Scanning electron microscopy
SH	Sulfhydryl
SHS	Static headspace
SiO_2	Silicon dioxide
SnO_2	Stannic oxide
SPC	Soy protein concentrate
SPI	Soy protein isolate
SS	Disulfide
TDS	Thiamine di-lauryl Sulfide
TiO_2	Titanium oxide
UV	Ultraviolet
WO_3	Tungsten trioxide
WP	Whey protein
WPC	Whey protein concentrate
WPC	Whey protein concentrates
WPI	Whey protein isolate
WVP	Water vapor permeability
ZnO	Zinc oxide
ZnS	Zinc sulfide

List of Figures

List of Tables

Part I

Sources and Origin of Edible Packaging

Edible Packaging: An Overview

1

Jaspreet Kaur, Mahendra Gunjal, Prasad Rasane, Jyoti Singh, Sawinder Kaur, Amrita Poonia, and Prerna Gupta

Abstract

Food packaging exhibits a significant role in chain supply and also is considered one amongst the essential steps in final process. The increased demand of customers for high-quality products with natural ingredients has forced the food and packaging industry to introduce the concept of edible packaging in the market. Till date, a number of techniques have been optimized to preserve the food either by means of adding preservatives or by changing the nature of packaging material. Edible packaging aims to conserve the food quality along with increased shelf life. These are produced either from edible biopolymers, which can be proteins, lipids, polysaccharides (gums and carbohydrates), plasticizers or from food-grade additives. Edible packaging materials include edible coatings, films, pouches and sheets. Depending on the type of final edible packaging material, these can be used either alone or in combination as per the requirement of the food product to be stored in it. For instance, lipids or resins can be combined with polysaccharides or proteins to obtain the edible packaging material having properties that resist water penetration. The best edible film must be a good oxygen barrier, moisture barrier and aroma barrier. The edible film made of protein can be derived from various plant and animal sources such as grains, oilseeds, milk, eggs and other animal tissues. The mechanism of film forming includes drying and extensive interaction of polymer network either with dry or wet casting. At present, various active compounds like antioxidants,

J. Kaur · M. Gunjal · P. Rasane (✉) · J. Singh · S. Kaur · P. Gupta
Department of Food Technology and Nutrition, School of Agriculture, Lovely Professional University, Phagwara, Punjab, India

A. Poonia
Department of Dairy Science and Food Technology, Institute of Agricultural Sciences, Banaras Hindu University, Varanasi, Uttar Pradesh, India

A. Poonia, T. Dhewa (eds.), *Edible Food Packaging*,
https://doi.org/10.1007/978-981-16-2383-7_1

colouring agents, antimicrobials, flavours and nutraceuticals are being incorporated into the films to enhance its quality and stability. The concept of edible packaging seems to be one of the best alternatives to old synthetic plastics that cause severe environmental pollution. The present chapter explores the fundamental understanding about edible packaging and its construction. It highlights the kind of material employed for the construction of edible packaging and the considerations involved in its fabrication. Further, the chapter enlightens the current advancements of edible packaging research and industrial approach towards it.

Keywords

Edible packaging · Preservation · Environmental sustainability · Antioxidants · Food packaging

1.1 Introduction

Utilization of edible packaging is increasing rapidly by using various edible compounds like lipids, polysaccharides, resins and proteins originating from various renewable sources. These packaging are considered as a fundamental fragment of food and can be eaten along with food; therefore, these must be biodegradable in nature (Krochta 2002). The edible packaging usually comprises of sheets or edible films, pouches and coatings. The thickness of edible films is usually less than 254 μm, whereas that of sheets is more than 254 μm. These structures (sheets and films) are manufactured separately from food and are later on placed either on food components or are sealed as edible pouches. The edible coatings are usually the reedy layers made from edible material that are placed directly over food products (Krochta and De Mulder-Johnston 1997). Various examples of edible packaging may include tablet coatings, microcapsules made of edible material, soft and hard gel capsules, etc. There is an increased interest and various researches are in existence because of high purchaser demand for better quality and safe food along with increased storage life. Edible packaging due to their unique properties such as its ability to protect food due to its mechanical and barrier properties, increased sensory characteristics, controlled mass transfer among heterogeneous foods and controlled release of active ingredients are considered to be one amongst the best alternatives for various applications. Despite the fact that conventional packaging cannot be entirely replaced by edible packaging, the edible packaging can, however, improve the efficacy of preservation of food, when combined with other non-edible packaging that acts as additional secondary packaging and protects the edible packaging along with food from the outer environment containing foreign particles. Its use results in enhanced recyclability of packaging along with lesser use of packaging material that too with increased protective functions (Krochta 2002).

Apart from this, these edible packaging can be employed for non-edible packaging, where it acts as an oxygen or gas barrier and, thus, enhances the protective ability along with degradability of the packaging material (Han and Krochta 2001; Hong and Krochta 2003; Lee et al. 2008). The benefits of edible packaging material stand unique and versatile such as providing product stability, variety, better quality and more safety along with convenience to consumers.

1.2 Concept of Edible Films and Coatings

Various coatings and edible films, for instance, wax on fruits were being used since centuries in order to prevent the moisture loss and to increase the aesthetic appeal of the product by glossy appearance. Such practices were carried out since long, even before the chemistries associated with them came into existence, and are still under use at present. The term, 'edible-film' came into existence for food products since past half century. Mostly, the terms coatings and films can be used in either means, which indicates the coated surface of the food by some composition. However, coating and film can be sometimes differentiated by notion that film is a wrapping material, whereas coating can be applied directly onto the surface of the food product. During 1967, the commercialization of edible films was restricted and limited only to use as wax on fruits. In 1986, few companies started growing and their numbers increased up to 10; however, there was drastic increase in companies manufacturing edible packaging from 10 to 600 in 1996. At present, the use of edible film is increasing due to its wide range of benefits such as retaining quality of product and many others, with annual revenue that exceeds by $100 million (Janjarasskul and Krochta 2010).

Any material that enrobes (wraps or coats) different food to increase the shelf life of product, which can be eaten along with the product without removing any of its layer, is considered as coating or edible film. The edible films not only act as coating but also fortify the natural coatings or layers of product, which aid in preventing the loss of moisture from the food and allow only selective gases such as ethylene, carbon dioxide and oxygen to pass in and out through it that participates in the process of respiration. The coating or film with thickness of less than 0.3 mm can also act as surface sterilizer, thereby preventing the loss of various other important components (Janjarasskul and Krochta 2010).

1.3 Properties of Edible Packaging Materials

Packaging serves various functions such as preservation, distribution, marketing of foods, etc. Its dumping and disposal leads to various ecological problems, especially the non-biodegradable ones. Also, recycling of the same requires additional cost. Therefore, in this fast-moving era, edible packaging proves to be a wonderful natural alternative that not only act as packaging but, at the same time, it extends the storage stability of the product as well as hails various other benefits such as acting as barrier against transmission of gases (Álvarez 2000), reduces the process of maturation, increases aesthetic appeal, etc. (Souza et al. 2010).

1.3.1 Functional Properties

Edible packaging serves various functional properties such as environmental barrier, water barrier, oxygen barrier, aroma barrier, oil barrier, etc.

1.3.2 Environmental Barrier

The coatings and edible films serve as environmental barrier and are responsible for controlling the mass transfer between ambient atmosphere and the food. Permeability is considered to be an important factor while selecting the edible material for packaging. The effect of moisture and temperature on film reflects its condition of intended use. Therefore, conducting the measurement of permeability must be done under specific conditions before using it as a packaging material.

1.3.3 Water Barrier

The application of edible packaging has been widely used for food packaging and in various conditions where moisture exchange should be inhibited. The inhibition of moisture exchange results in curbing microbial growth owing to lowered waster activity, averting any change in texture or adverse chemical reactions. Films produced from edible waxes and plastics have fairly less WVP as compared to the hydrocolloid-based films due to substantial polarity, water vapour permeability of hydrophilic films increase at high relative humidity (RH) and high plasticizer concentration. Therefore, for short period duration and in low moisture foods, these films can solely be utilized as shielding barriers against moisture exchange. On the other hand, lipid based hydrophobic compounds are often used to enhance the water barrier properties of the films through low polarity and low affinity densely structured molecular matrices.

1.3.4 Oxygen Barrier

The major factors contributing to food spoilage or food deterioration include discoloration of myoglobin pigment in freshly cut meat pieces, oxidation of lipids and food ingredients or enzymatic browning of the fresh-cut produce. For foods that are sensitive to oxygen, an edible packaging can be used with low oxygen permeability (OP), which aids to preserve quality and extends the shelf life of oxygen-sensitive foods. Moreover, it also helps to reduce the utilization of various expensive and affluent, non-recyclable, oxygen barrier plastics. At the same time, a modified atmosphere can be created by the development of various advancements involved in the formulation of edible films. Such advanced techniques can also help in suppressing the respiration rate of horticultural products and the ethylene production of physiologically active climacteric produce during storage as well as distribution (Rahman 2007).

At low RH value, hydrocolloid-based films generally exhibit remarkable gas barrier properties. Whereas, at high RH conditions, plasticized hydrocolloid based films and hydrophilic EVOH films are used to have good barrier properties.

1.3.5 Aroma Barrier

In food packaging, aroma or volatile flavour and migration of flavours from outside create a problematic situation for any of the food packaging, which must be prevented during storage and distribution. In general, the barrier efficiency of packaging material is improved when a migrating compound has low affinity to film materials and low diffusivity across the polymer matrix. To identify some suitable barriers, include hydrophilicity of protein- and polysaccharide-based edible films for non-polar aroma compounds. Encapsulation of flavour and aroma has been proposed by carbohydrate and protein emulsion-based films (Pegg and Shahidi 2007; Hambleton et al. 2009; Fabra et al. 2009). The chief objective of using this technology is to preserve the hydrophobic organic aroma compounds in a non-polar lipid dispersed phase, whereas the matrix made of hydrophilic polymer prevents aroma loss to the environment or due to oxidation. Further research is still required in order to reveal the various aspects related to the aroma permeability of the edible films (Debeaufort and Voilley 1995).

1.3.6 Oil Barrier Properties

Any sort of lipid-containing food products can be provided resistance against grease by using edible packaging. Zein and whey protein based films have been reported to exhibit excellent resistance against grease owing to the inherent hydrophilicity of protein-based films.

1.4 Food Surface Properties

1.4.1 Wettability

The ability of a given liquid to spread over a solid surface is known as spreading coefficient or wettability (Ws). Higher the value of wettability, more is the coating material suitable for use. The major factors taken into consideration while determining spreading coefficient are the coefficient of adhesion (Wa) and cohesion coefficient (Wc). The former represents force promoting expansion of liquid over a solid surface, while the latter represents force promoting contraction of liquid over a solid surface (Lima et al. 2010).

1.4.1.1 Contact Angle of Coatings in the Surface of Foods

The association between water molecules offers a good understanding of the hydrophobic effect. In order to determine the hydrophobic or hydrophilic property of edible films, contact angle is measured. Studies have revealed that with increase in the number of lipid molecules, the contact angle is also increased (Tang and Jiang 2007; Ramos et al. 2013; Galus and Kadzińska 2016). Similarly, the hydrophilic nature of edible films is related to the solubility of film surface that might come in contact with water. This property is used to validate the ability of outer surface of fruits (like apple and mango) to either participate or not participate in interactions (non-polar) (Medeiros et al. 2012; Ramírez et al. 2012).

1.5 Comparison Between Edible Packaging Films and Synthetic Polymers

The various changes in the properties of various food products result due to the interaction or contact of food and materials used for packaging. Recent reports generated by the government have revealed that food is susceptible to get contaminated due to the possible leaching of chemical components of the cardboard, used for packaging, into the food products like boxes used for packing cereals and pizza. Earlier, phthalates were used while manufacturing cling film but these phthalates contain certain toxic chemicals that could lead to health hazards. But taking this point into consideration, the cling films are now manufactured without using phthalates and, hence, are found to be much safer for use in food industry. The quality of the food commodities is no longer compromised by the use of cling films and reaches the consumer in a very safe form as it aids in protecting as well as preserving the food product till distribution process. A number of factors are considered while choosing the appropriate material for food packaging so that the food is safely delivered to the consumer and any sort of contamination (internal or external) avoided. The most frequently used synthetic films in food industry are low- and high-density polyethylene, polyethylene terephthalate and polyvinylidene chloride, which provide strength, toughness, flexibility, sealability, stiffness and barrier properties against moisture, gas and humidity. Such synthetic packaging materials are extensively used in food industries as plastic bags, container lids, squeezing and juice bottles, etc. (Malhotra et al. 2015).

Synthetic polymers generally being used for packaging material are non-biodegradable and, moreover, their production cost is also very high. The need of the hour is to develop natural polymer-based cling films or edible films from the materials that are biodegradable and get readily converted into carbon dioxide, water and biomass on microbial action. These edible films do not lead to any sort of environmental hazard. Other than these advantages, the shelf life and stability of these natural polymer-based films can be increased by preventing food spoilage caused by microorganisms. Different materials used for manufacturing edible films are polysaccharides, proteins, lipids, nucleic acids and composite materials. Out of all these natural polymers, protein-based films have been accepted wholeheartedly owing to their advantages and use as compared to the synthetic polymer-based films (Malhotra et al. 2015).

1.6 Various Types of Edible Packaging Materials Used in Food and Pharmaceutical Industries

1.6.1 Protein-Based Edible Packaging Materials

Proteins have been widely used for the formation of edible films that involves its denaturation by means of heat treatment, pH change or one or the other solvents. Intermolecular interactions are established between the extended peptide chains, followed by the formation of protein film matrices. The formation of these protein matrices majorly depends on the type of protein chain, denaturation treatment as well as the fabrication conditions, which further leads to the formation of protein films having modified properties. Protein films are reported to have very good optical as well as mechanical properties. These protein-based films are also potent against the transport of aroma, carbon dioxide, oxygen and lipid, but it is not effective to serve as a barrier for moisture due to its innate hydrophilic nature.

For the sake of upgrading the quality and properties of protein-based edible films, many methods have been investigated based on various treatments including chemical, physical and enzymatic. To bring about a change in the physical, structural and chemical properties, different varieties of proteins have been reported to be modified, namely formation of intermolecular cross linking by breaking intramolecular disulphide bonds using heat denaturation; adjustment of pH of protein film-forming solution and modification of protein side chains by either changing solvent or adding salt (Gontard et al. 1992; Avena-Bustillos and Krochta 1993; Brandenburg et al. 1993; Pérez-Gago et al. 1999). And employing physical modifications like lamination, composite formation, ageing, heat curing, annealing and addition of emulsions and nanoparticles (Gennadios et al. 1996; Kim et al. 2002). Usage of enzymatic treatments and irradiation technique is also feasible as it helps in the formation of covalent cross linkage between aromatic amino acids (Gennadios et al. 1998; Vachon et al. 2000). Another research study has reported the use of microfluidization and ultrasound techniques to improve molecular order and enhance intermolecular interactions at the same time (Banerjee et al. 1996).

1.6.1.1 Wheat Gluten

These films are formed by preparing casting solutions with acidic or basic conditions in the presence of reducing agents and alcohol (Cuq et al. 1998). The change in pH caused by acidic and basic conditions of the medium leads to the disruption of hydrophobic, hydrogen and ionic interactions. By the dispersion of wheat gluten in alkaline or acidic environments (addition of reducing agents), the intra- and intermolecular disulphide bridges are cleaved and are reduced to thiol (functional groups). Wheat gluten becomes concentrated after drying of film-forming solution and evaporation of volatile solvents. This results in the formation of new intermolecular interactions as the active sites become free for the sake of bond formations. The cohesive network formed during film formation is contributed by the hydrogen and disulphide bonds as well as hydrophobic interactions.

1.6.1.2 Zein

Zein films are formed by preparing warm casting solutions by dissolving it in aqueous ethyl alcohol or isopropanol. Ethyl alcohol from the film surface is evaporated and it results in the formation of hydrogen, hydrophobic interactions and disulphide bonds within film matrix (Padua and Wang 2002). Zein film acts as a potent barrier against oxygen, moisture and lipid and finds its application in foods like confectioneries, nuts and candies. It is used commercially as a finishing agent that tends to impart surface gloss to the food products. Zein coating is also extensively used in pharmaceutical industry for the controlled release of the active drug ingredients and also to guise the taste of orally administered drugs (Meyer and Mazer 1997). It is also reported that the oil uptake by fried foods and moisture loss is reduced by using zein-based coating materials (Park et al. 1994; Mallikarjunan et al. 1997).

1.6.1.3 Soy Protein Isolate

It is also used for the preparation of film-forming solutions that are formed via hydrophobic interactions as well as intermolecular disulphide when the casting solution is denatured by heat. These films can be used to control lipid oxidation in precooked meat products and also helps to limit moisture loss from the external surface of food commodities (Wu et al. 2000). These films also act as potent carriers for the transport of antioxidants, antimicrobial compounds as well as flavouring agents (Kunte et al. 1997). Soy protein isolate films are used on products like meat pies, cakes containing high moisture content, cheese and coating of fruits and vegetables; where these protective films are used as microencapsulating agents and are highly permeable to water vapours (Petersen et al. 1999).

1.6.1.4 Collagen

Collagen is reported to have high solubility as it tends to swell in polar solvents due to its hydrophilic proteinaceous nature owing to the occurrence of amino acids such as proline, hydroxyproline, glycine, etc. Studies have reported that overwrapping collagen films on refrigerated and thawed beef significantly reduced the colour changes and oxidative changes followed by exudation (Farouk et al. 1990). In order to increase juiciness, reduce shrink loss, absorption of fluids exudates and for the easy removal of nets post-cooking, these collagen-based films have been used for processed meats and meat products.

1.6.1.5 Gelatin

Gelatin coatings are widely used for encapsulating low moisture food products or oil-based foods. It is also used for dietary supplements as it acts as a carrier of bioactive ingredients and oil, and also acts as a barrier against oxygen and moisture.

Gelatin films are used in pharmaceutical industry as an encapsulating agent in forming either soft or hard capsules (Rahman 2007).

1.6.1.6 Casein

Casein films are known to be excellent barriers of oxygen as well as moisture and find its application in retarding lipid oxidation. It also acts as a potent carrier of bioactive ingredients, flavouring agents and nutrients present in food products. Its property of low water activity has been shown to act as a water barrier for fresh produce, frozen foods, water-soluble pouches and dry fruits (Gennadios 1994).

1.7 Carbohydrate-Based Packaging Materials

Carbohydrate-based or polysaccharide coatings are known to act as excellent barriers against oxygen, oil and odour and, at the same time, it provides a little resistance against water migration. For the formation of polysaccharide films, the interactions among long-chain polymeric segments are disrupted and then new bonds are established such as hydrogen bonds and hydrophilic bonds. The solvent is then evaporated in order to obtain a film matrix. These films are known to provide good structural integrity as well as strength. Polysaccharide films help to increase the shelf life of the coated food products as it play a major role in delaying the ripening process of fresh produce. Its hydrogen bonded network contributes to the exceptional oxygen barrier property of these films (Saklani et al. 2019).

1.7.1 Cellulose

Various derivatives of cellulose such as carboxymethyl cellulose, hydroxypropyl methyl cellulose, hydroxypropyl cellulose and methyl cellulose are used as edible packaging materials owing to their excellent film-forming properties. The permeability, solubility, structural and mechanical properties depend upon various factors like the type and degree of substitution for functional group, chain length of the polymer, etc. Cellulose-based coatings are effective barriers against oxygen, moisture as well as oil and are, therefore, applied to a number of food products (Balasubramaniam et al. 1997).

1.7.2 Starch

Starch is composed of two components, namely amylose and amylopectin. These components (amylose and hydroxypropyl amylose) are used for the formation of edible food coatings as well as encapsulating agents. These starch-based films are most commonly used for coating confectionery products, batters, bakery items and meat products owing to its property to act as a barrier against oxygen and lipids. They also help to improve the texture, appearance as well as handling practices.

1.7.3 Chitosan

Properties and viscosity of chitosan films are majorly affected by the type of organic solvent used as these films are formed by casting aqueous acidic solutions. The post-harvest life of fresh fruits and vegetables can be increased by using such chitosan films that are semipermeable in nature. The cationic group (contributed by the amino group) present on chitosan films can easily react with negatively charged groups like ions, cholesterol, fats and proteins. These reactions bring about a good chance of chemical modification. These films exhibit properties like antioxidant, antimicrobial and act as a carrier of bioactive ingredients. It has also been reported in some studies that coatings of chitosan help in delaying enzymatic browning in fresh fruits (Zhang and Quantick 1997; Coma et al. 2002).

1.7.4 Pectin

Water from the pectin gel is evaporated for the formation of edible pectin films. Pectin films act as sacrificing agents by retarding water loss from the food product and prevents it from getting dehydrated; instead the moisture gets lost from the gel matrix. These coatings aid to improve the texture, appearance and handling of food products by retarding water loss and oil migration (Kester and Fennema 1986).

1.7.5 Alginate

Solvent from alginate gels is evaporated to form alginate films. There is another procedure in which the alginate solution is dried and then treated with solution of calcium salt in order to establish cross linkage. Different factors like temperature, pH, time of exposure, concentration of cations and presence of composite constituents play a vital role in the alteration of permeability as well as the strength of alginate films. In similar line with pectin films, the gelatinous alginate films also act as sacrificing agent and, thus, prevent the desiccation or dehydration of enrobed meat. These films are used as encapsulating agents in food and pharmaceutical industries. It also helps to prevent the process of oxidation in foods due to their potent oxygen barrier properties (Hambleton et al. 2009).

1.7.6 Gums

The most widely used gums as coating material for encapsulation are heteropolysaccharides, which have a complex structure including gum ghatti, gum arabic, gum tragacanth and gum karaya. Among them, guar gum is used as a good coating material owing to its properties like solubility (in both hot as well as cold water) and viscosity. It forms a gel-like consistency on binding with calcium. Another example of gum (polysaccharide) is locust bean gum that is neutral in nature and is quite soluble in hot water; thereby producing highly viscous solutions ideal for forming coating material and are also stable at variable temperature and pH (Pegg and Shahidi 2007).

1.8 Lipid-Based Packaging Materials

For many years, lipid-based compounds have been used as wrapping material for protection of food products. Lipid-based coatings are highly fragile, lack cohesiveness and are not self-supporting due to the inability of lipids to form polymers. Lipids lack covalent bonds and do not form long chains as they lack repeating units. These coatings act as a protective barrier against moisture and oxygen. It has been reported that wax films are much more resistant to the migration of water as compared to lipid or non-lipid films used as edible packaging materials. There are various applications of different types of lipid-based coatings. In order to impart a glossy and attractive appearance to food commodities, edible resins like wood resin, terpene resin and shellac are widely used. Shellac possesses low polarity and is used for confectionery and freshly produced food commodities (Saklani et al. 2019). Lipids are used to form composite films as they act as an excellent barrier against water migration. Besides having various advantages as discussed above, the use of lipid-based edible packaging materials is sometimes disadvantageous also due to the greasy surface, possible rancidity as well as their waxy texture and taste (Janjarasskul and Krochta 2010).

1.8.1 Glycol Esters

The properties of edible ingredients for coating food commodities depend upon the degree of saturation, physical state as well as length of chain of fatty acids. Coating films based on glycol esters are either used alone or in combination with other ingredients like fatty alcohols, fatty acids, etc. There are some problems associated with glycol ester coatings like bitter aftertaste, acidic mouthfeel, cracking and flaking during storage (Bourlieu et al. 2008).

1.8.2 Waxes

Wax coatings are much more resistant to water migration as compared to lipid or non-lipid edible films. These are either used alone or in combination with other ingredients. Examples of natural or synthetic edible wax-based coatings include candelilla wax, beeswax, carnauba wax, rice bran wax, paraffin wax, petroleum wax, etc. (Kester and Fennema 1986; Rahman 2007).

1.8.3 Resin

Such coatings like shellac, wood resin and terpene resin are used to give a glossy appearance to food commodities. Shellac consists of aleuritic acid and shellolic acid that are complex polymers of alicyclic and aliphatic -OH acids. Resin coatings are extensively employed in food and pharmaceutical industry as a coating material for fresh produce as well as confectionery food items. They are soluble in alkaline and organic solvents.

1.8.4 Composite Packaging Materials

In general, the fabrication of composite films is done by either of the two methods including lipid emulsion or hydrocolloid bilayer, which is made of a lipid component. The lipid globules are homogenously dispersed by using an emulsifier followed by entrapping into a matrix of hydrocolloid component to provide support. There are two different techniques to form bilayer film, namely emulsion and coating technique. The former comprises of lamination or casting of the molten lipid film-forming solution over a preformed protein-based or polysaccharide-based film for the preparation of final bilayer film, while the latter (emulsion technique) involves the solubilization of lipid in film-forming solution prior to film casting. Post-drying, the bilayer film is formed as a result of phase separation (Pérez-Gago et al. 1999).

Studies have revealed that bilayer films are more efficient as compared to stable emulsion films in terms of exhibiting water vapour barrier properties due to the presence of continuous hydrophilic layer in the film (Debeaufort and Voilley 1995). Besides having advantages, there are certain disadvantages associated with coating technique that the bilayer structure is highly susceptible to get cracked and delaminated. Moreover, two types of castings are required for the coating technique. On the other hand, there are multiple factors that need to be taken into account while fabricating stable emulsion films including melting temperature of the liquid, cross linkage and volatilization of the solvent (Pérez-Gago et al. 1999).

Significant advances have been made in the field of composites and it seems to be correlated with the evolution of nanotechnology. For improving the mechanical barrier properties of edible packaging materials, bio-nanocomposite materials are known for their promising properties (Azeredo et al. 2009; de Moura et al. 2009). Nanoparticles are used in the form of a homogenous dispersion to form a large matrix that further contributes to the thermal and mechanical properties of the material by changing their molecular mobility, nanostructure and relaxation behaviour. Moreover, nanocomponents also aid in refining the barrier properties of the composite films (Dalmas et al. 2007).

1.9 Utilization of Edible Packaging as a Carrier of Bioactive Compounds

The edible packaging is applied on to the food materials by layers that have good impact on preservation, marketing and distribution of foodstuffs by protecting it from microorganisms and physical and chemical damage. Different types of compounds are used to making edible packaging materials including antimicrobials, antioxidants, additives, nutraceuticals and flavouring compounds, which helps to improve the quality of packaging materials and handling (Tavassoli-Kafrani et al. 2016). The use of certain types of antioxidants in edible packaging helps to prevent oxidation of food containing lipids and proteins, which cause a change in flavour and colour (Robertson 2012). The incorporation of antimicrobial compounds into edible packaging materials can help to increase the shelf life of food products and prevents the growth of microorganisms (Bagheripoor et al. 2018; Lappa et al. 2019). The additives also help to improve the bioactive properties of packaging materials; it includes herbs and spices (essential oils, plant extracts and organic compounds) as well as natural antioxidants (Silva-Weiss et al. 2013). In Table 1.1, the role of different bioactive compounds used in edible packaging is shown.

1.10 Formulation of Edible Packaging

The edible packaging formulations involve at least one natural polymer, which is able to formulate a structural matrix that is sufficiently cohesive. The application and functional properties of edible packaging mostly depend on certain cohesion forces, including covalent bonds, ionic bonds and hydrogen bonding, between film-forming polymer molecules. The biopolymer cohesive strengthens it, depending upon the film-making process and parameters, structure and chemistry, plasticizers and additives concentration. In edible coatings, solid fats, resins and waxes are formed by melting and solidification process; for the solubilization, evaporate the solvent with organic solvent followed by the preparation of emulsion by adding water and again evaporating it. The wet and dry processes technology is used to achieve the formation of hydrocolloid-based biopolymer. The wet process also known as solution casting is based on coacervation mechanism. It can also result from the coagulation or thermal gelation process after cooling it in case of gelatin or agar. Based upon the number of biopolymers, the coacervation could be either simple or complex. The fatty substance material and possibly a surfactant is incorporated into the biopolymer solution required if the film formulation is based on an emulsion. This solvent mixture is heated above the melting temperature of fats, homogenized and cast or directly applied while melted or after cooling. For developing film, the different types of solution casting techniques are developed including microwave, air drying, hot surface and infrared. The thermal properties of the film's biopolymer are based on dry thermoplastic extrusion process (Janjarasskul and Krochta 2010).

The several important characteristics of edible packaging are as follows: it does not contain toxic, non-digestible allergic component; it should prevent structural

Table 1.1 The role of different bioactive compounds used into edible packaging

Bioactive compound	Foods	Role of bioactive compound	References
Semi-refined κ-carrageenan and water extract of germinated fenugreek seeds	Chicken breast	• Increase in phenolics content and antioxidant activity • Control the growth of microorganisms	Farhan and Hani (2020)
Lactic acid bacteria and Sodium carboxymethyl cellulose	Banana	• Reduce water vapour and light transmission rates • Prevent lipid oxidation and improve shelf life	Li et al. (2020)
Mango leaf extract	Cashew nuts	• More resistance to oxidation	Rambabu et al. (2019)
Vitamin C and citric acid	Fresh-cut mangoes	• Prevent browning • Increase antioxidant property	Robles-Sánchez et al. (2013)
Clove and oregano oils	Iceberg lettuce	• Effective against coliform bacteria • Reduce discoloration • Improve overall freshness and odour	Wieczyńska and Cavoski (2018)
Limonene	Cucumber	• Possesses weight loss properties, firmness, colour, pH, Tg and organoleptic properties • Reduces fungal growth	Maleki et al. (2018)
Grape seed extract	Rainbow trout fillet	• Maintains the sensorial quality • Extend the shelf life	Hassanzadeh et al. (2018)
Nisin, thymol and lauric arginate	Chicken	• Reduce food-borne pathogens *E. coli, Salmonella spp., L. monocytogenes* and *S. aureus* over 28 days refrigeration condition.	Hassan and Cutter (2020)
Garlic essential oil	Roasted peanuts	• Inhibited growth of *Aspergillus flavus* • Prevents lipid oxidation	Orsuwan and Sothornvit (2018)

stability during handling and display and transportation; the material should uniformly cover the whole surface of food materials; the food materials maintain moisture content and, hence, prevent water transmission rate; the internal environment of foods is maintained during storage like aerobic and anaerobic gases, in addition to the sensory properties, nutritional and organoleptic properties; it should prevent the growth of microbes in food and packaging materials and stop the degradation of compounds; it should provide the micronutrients, flavour and also the colour; the antioxidants and antimicrobial compounds should prevent oxidation of fats and proteins as well as stop the enzymatic activity; the cost of packaging materials should be less and environment friendly as well as the process of manufacturing should be easy (Dinika et al. 2020).

1.11 Applications of Edible Packaging

The selection of appropriate material for fabrication of edible packaging is influence by the storgae conditions of the food that is packaged. The different types of materials used for film and coating of food products include carbohydrates, lipids, proteins and combination of two or more mixture of these. Edible packaging plays important role in extending the shelf life of food products (Falguera et al. 2011). The applications of edible packaging on various food products are explained below.

1.11.1 Fruits and Vegetables

The major cause of spoilage caused in fruits and vegetables is exchange of gas during storage and microbial growth (Bakar et al. 2020). For fruits and vegetables, mineral oils, paraffin, beeswax, candelilla shellac and waxes are used for edible packaging (Baldwin 1994). The main aim of edible coating is to improve the natural vegetable and fruit barrier. This coating material is eaten as food, it helps to increase the shelf life of foods as well as it is eco-friendly (Ali et al. 2010). The different types of processing technologies involved in the processing of fruits and vegetables increase the chances of nutrient loss. The 30% of spoilage of fruits and vegetables are due to action of microorganisms and insects during harvesting time, transportation and maintenance process, which lead to tissue damage and, hence, the integrity of the food products. Fruits and vegetables are becoming more susceptible to contamination, development of undesirable volatile compounds and changes in texture; all these reduce their health benefits (Ribeiro et al. 2020). These changes in fruits and vegetables are controlled by using artificial materials in the form of edible films or coatings (Senturk Parreidt et al. 2018). The application of edible packaging on fruits and vegetables and their effects are shown in Table 1.2.

1.11.2 Dairy Products

The dairy products are described as products that get through by processing of milk; it may contain ingredients or food additives. The demand of milk products has been observed to increase particularly in developing countries due to the increasing population and also change in dietary patterns (FAO 2019). The dairy products are good source of essential nutrients that are important for children's growth and also maintaining good health of adults (Cardador and Gallego 2016). The use of edible packaging material for cheese helps to minimize quality losses. These changes mostly happened due to action of microbes and enzymes, development of undesirable compounds, changes in sensory properties and moisture loss (Senturk Parreidt et al. 2018; Guitián et al. 2019). The application of edible packaging on dairy products and their effects are shown in Table 1.2.

Table 1.2 Application of edible packaging on different foods

Foods	Edible packaging materials	Outcomes and results obtained	References
Fruits and vegetable applications			
Strawberries	Pectin, lemongrass essential oil and cellulose nanocrystals	• The edible coating was effective to minimize weight loss without affecting the chemical parameters	Da Silva et al. (2019)
Orange	Salicylic acid and aloe vera gel	• It help to maintain the quality during cold storage and inhibits growth of microorganisms	Rasouli et al. (2019)
Tomatoes	Almond gum and gum arabic	• Increase the shelf life and affect changes in weight, firmness, titratable acidity, TSS, colour, vitamin C content and decay percentage of tomatoes	Mahfoudhi et al. (2014)
Soybean oil	Pomelo peel flour and tea polyphenol	• Oil oxidation reduces during storage	Wu et al. (2019)
Potato chips	Whey protein and rosemary extract	• The acrylamide and oil content during frying reduce, improve firmness and texture as well as sensory properties	Trujillo-Agudelo et al. (2020)
Dairy products applications			
Paneer	Casein and clove bud essential oil	• Maintain the sensory quality of paneer	Archana et al. (2020)
Cheddar cheese	Linalool, carvacrol and thymol	• Reduce the growth of *A. niger* on the surface of cheddar cheese after 35 days of storage at 15 °C	Kuorwel et al. (2014)
Cheeses	Agar and glycerol	• Control the growth of *L. monocytogenes* in cheeses	Guitián et al. (2019)
Kashar cheese	Sorbitol, whey protein isolate, alginate and ginger essential oil	• Decrease the growth of *E. coli* O157:H7 and *S. aureus*	Kavas et al. (2016)
Meats, poultry and seafood applications			
Beef	Heracleum lasiopetalum essential oil and *Lepidium sativum* seed mucilage	• Maintain the overall acceptance • Prevent microorganism growth and oxidation	Barzegar et al. (2020)
Fresh pork	Grape seed extract, nisin and chitosan alginate	• Inhibit pork oxidation and microbial spoilage	Xiong et al. (2020)
Beef	Basil extract and alginate	• Improve the antioxidant properties and reduced the lipid oxidation during storage	Alexandre et al. (2020)
Fish sausage	Papaya seed extract and sago starch-gelatine	• No colour changes during storage • Maintain overall quality	Bakar et al. (2020)

(continued)

Table 1.2 (continued)

Fruits and vegetable applications			
Foods	Edible packaging materials	Outcomes and results obtained	References
Smoked salmon	Gulfweed seed	• No effects on aroma	Kim et al. (2018)
Chicken breast fillet	Ginger essential oil and sodium caseinate	• Increase the shelf life • Decrease the level of aerobic bacteria under refrigeration storage for 12 days and no significant effect on TBARS value	Noori et al. (2018)
Chicken breast fillets	Black pepper seed extract, turmeric extract and carboxymethyl cellulose	• Good antimicrobial activity against total aerobic mesophilic bacteria and total aerobic psycrotrophic bacteria • Maintain colour, appearance, odour and overall acceptability during storage	Dalvandi et al. (2020)

1.11.3 Meats, Poultry and Seafoods

The quality of meat, poultry and seafood products is improved by using edible packaging as it helps to minimize the moisture loss, prevents oxidation, improves the overall appearance, improves sensory properties and increases the shelf life (Gennadios et al. 1997). The quality of meat is known to increase during storage, when packed with carbohydrate based edible films. The weak water barrier properties of carrageenan, when it is applied on fresh and frozen meat, poultry and fish compared to polysaccharides, help to prevent moisture loss (Debeaufort et al. 1998). The use of this packaging method involves different processes like spraying, dipping, foaming, casting, individual wrapping and rolling on to meats, poultry and seafood products. The application of edible packaging on meats, poultry and seafood products and their effects are shown in Table 1.2.

1.12 Regulations

The edible packaging is an integral part of the edible portion of the food items; therefore, it is necessary to follow all the regulations relevant to food ingredients (Guilbert and Gontard 1995). The edible packaging, depending on its role, could be classified as food contact substances, food packaging materials, food products or food ingredients (Debeaufort et al. 1998). While making edible packaging films or coating, ingredients must be generally recognized as safe (GRAS) for the intended use or approved by the FDA or U.S. Pharmacopoeia/National Formulary. One has to follow the good manufacturing practices (GMP) while using edible packaging materials and additives (Janjarasskul and Krochta 2010). The different types of edible packaging materials are made up of eggs, fish, milk (casein and whey), wheat, peanuts and soy proteins; therefore, taking it into account that some of these are allergens, it must provide all information to consumers according to Food Allergen Labelling and Consumer Protection Act of 2004 (Franssen and Krochta 2003; Janjarasskul and Krochta 2010).

1.13 Advantages and Disadvantages of Edible Packaging

Edible packaging is biodegradable, it prevents moisture loss and helps to stop chemical reaction in foods like one of the enzymatic reactions (Osorio et al. 2011). The edible packaging materials help to increase the shelf life of foods by minimizing loss of oxidation, gas permeation rate, inhibit the growth of microorganisms and oxidative rancidity (Dhanapal et al. 2012). Edible packaging materials do not contain any harmful chemical substances, instead it contain natural and biodegradable substances that are obtained from agricultural sources and, thus, help to protect the environment (Debeaufort et al. 1998). These packaging materials can be directly consumed with food products and do not cause any harm to the animals (Shit and Shah 2014). Edible packaging acts as a carrier with the addition of some naturally available bioactive compounds and agents in accordance with internationally induced dose release of antimicrobial, antioxidants and vitamins to the specific alimentary matrix. It also helps to increase the quality and reduce microbial load of foods (Suhag et al. 2020).

The thick coating is applied onto any type of food products, which prohibit oxygen exchange and development of off flavour in the food products. The edible coating plays an important role as a gas barrier so it causes anaerobic respiration in fruits and vegetables, which results in causing a delay in the ripening process. In addition, some edible packaging materials are present into hygroscopic form so it helps to hinder microorganism growth (Raghav et al. 2016).

1.14 Consumer Acceptance and Commercialization

The consumer acceptance includes properties like sensory properties, marketing, safety as well as cultural and religious restrictions associated with the use of new materials and applications, and is also significantly affected by potential uses of edible materials. The cost of edible packaging materials is another factor that influences consumer acceptance. It has been revealed that the cost of edible packaging is high as compared to the plastic packaging materials. During the manufacturing of edible packaging materials, major focus is kept on reducing the cost of packaging materials so that the consumers are attracted more towards edible packaging materials.

1.15 Conclusion

The demand to use sustainable packaging solution has forced all manufacturing industries to amend their ways and thus, have significantly changed the packaging industry aswell. In the other side of packaging sectors, the edible packaging is recognized as healthy, biodegradable, low cost and sustainable as compared to the plastic packaging materials. The edible packaging (coatings, sheets, pouches and films) provides a good oxygen barrier, moisture barrier and aroma barrier properties. Different types of compounds are used to make edible packaging materials including antimicrobials, antioxidants, additives, nutraceuticals and flavouring compounds, which helps to upgrade the quality of packaging materials and handling. The edible packaging is used for different foods like fruits and vegetables, dairy products, meat, poultry, seafoods, nuts, cereals and confectionary products. The advantages of edible packaging material are unique and versatile such as providing product stability, variety, better quality, more safety along with convenience to consumers. The future research is required on reducing the cost of packaging materials, how to implement it on industrial scale, improve its functional properties and to evaluate the compatibility between packaging, final product along with human consumption.

References

Alexandre S, Vital AC, Mottin C, do Prado RM, Ornaghi MG, Ramos TR, Guerrero A, Pilau EJ, do Prado IN (2020) Use of alginate edible coating and basil (Ocimum spp) extracts on beef characteristics during storage. J Food Sci Technol 22:1–9

Ali A, Maqbool M, Ramachandran S, Alderson PG (2010) Gum arabic as a novel edible coating for enhancing shelf-life and improving postharvest quality of tomato (*Solanum lycopersicum* L.) fruit. Postharvest Biol Technol 58(1):42–47

Álvarez MF (2000) Revisión: Envasado activo de los alimentos/review: active food packaging. Food Sci Technol Int 6(2):97–108

Archana S, Divya KB, Rajakumar SN, Aswin SW (2020) Evaluation of sensory quality of paneer with casein based edible coating incorporated with essential oil. Pharma Innovation 9(11):26–28

Avena-Bustillos RJ, Krochta JM (1993) Water vapor permeability of caseinate-based edible films as affected by pH, calcium crosslinking and lipid content. J Food Sci 58(4):904–907

Azeredo HM, Mattoso LH, Wood D, Williams TG, Avena-Bustillos RJ, McHugh TH (2009) Nanocomposite edible films from mango puree reinforced with cellulose nanofibers. J Food Sci 74(5):31–35

Bagheripoor N, Khoshgozaran-Abras S, Sohrabvandi S, Khorshidian N, Mortazavian AM, MollaKhalili N, Jazaeri S (2018) Application of active edible coatings to improve the shelf-life of cheese. Food Sci Technol Res 24(6):949–962

Bakar J, Kadir NA, Mazlan AA, Ismail-Fitry MR (2020) Effect of edible coating of sago starch-gelatine incorporated with papaya seed extract on the storage stability of Malaysian fish sausage (keropok lekor). Int Food Res J 27(4):618–624

Balasubramaniam VM, Chinnan MS, Mallikarjunan P, Phillips RD (1997) The effect of edible film on oil uptake and moisture retention of a deep-fat fried poultry product. J Food Process Eng 20 (1):17–29

Baldwin EA (1994) Edible coatings for fresh fruits and vegetables: past, present, and future. Edible Coat Films Improve Food Qual 1:25

Banerjee R, Chen H, Wu J (1996) Milk protein-based edible film mechanical strength changes due to ultrasound process. J Food Sci 61(4):824–828

Barzegar H, Behbahani BA, Mehrnia MA (2020) Quality retention and shelf life extension of fresh beef using *Lepidium sativum* seed mucilage-based edible coating containing Heracleum lasiopetalum essential oil: an experimental and modeling study. Food Sci Biotechnol 29 (5):717–728

Bourlieu C, Guillard V, Vallès-Pamiès B, Gontard N (2008) Edible moisture barriers for food product stabilization. In: Food materials science. Springer, New York, pp 547–575

Brandenburg AH, Weller CL, Testin RF (1993) Edible films and coatings from soy protein. J Food Sci 58(5):1086–1089

Cardador MJ, Gallego M (2016) Origin of haloacetic acids in milk and dairy products. Food Chem 196:750–756

Coma V, Martial-Gros A, Garreau S, Copinet A, Salin F, Deschamps A (2002) Edible antimicrobial films based on chitosan matrix. J Food Sci 67(3):1162–1169

Cuq B, Gontard N, Guilbert S (1998) Proteins as agricultural polymers for packaging production. Cereal Chem 75(1):1–9

Da Silva ISV, Prado NS, De Melo PG, Arantes DC, Andrade MZ, Otaguro H, Pasquini D (2019) Edible coatings based on apple pectin, cellulose nanocrystals, and essential oil of lemongrass: improving the quality and shelf life of strawberries (Fragaria Ananassa). J Renew Mat 7 (1):73–87

Dalmas F, Cavaillé JY, Gauthier C, Chazeau L, Dendievel R (2007) Viscoelastic behavior and electrical properties of flexible nanofiber filled polymer nanocomposites. Influence of processing conditions. Compos Sci Technol 67(5):829–839

Dalvandi F, Almasi H, Ghanbarzadeh B, Hosseini H, Karimian Khosroshahi N (2020) Effect of vacuum packaging and edible coating containing black pepper seeds and turmeric extracts on shelf life extension of chicken breast fillets. J Food Bioprocess Eng 1(3):69–78

de Moura MR, Aouada FA, Avena-Bustillos RJ, McHugh TH, Krochta JM, Mattoso LH (2009) Improved barrier and mechanical properties of novel hydroxypropyl methylcellulose edible films with chitosan/tripolyphosphate nanoparticles. J Food Eng 92(4):448–453

Debeaufort F, Voilley A (1995) Effect of surfactants and drying rate on barrier properties of emulsified edible films. Int J Food Sci Technol 30(2):183–190

Debeaufort F, Quezada-Gallo JA, Voilley A (1998) Edible films and coatings: tomorrow's packaging: a review. Crit Rev Food Sci 38(4):299–313

Dhanapal A, Sasikala P, Rajamani L, Kavitha V, Yazhini G, Banu MS (2012) Edible films from polysaccharides. Food Sci Quality Manag 3(0):9

Dinika I, Verma DK, Balia R, Utama GL, Patel AR (2020) Potential of cheese whey bioactive proteins and peptides in the development of antimicrobial edible film composite: a review of recent trends. Trends Food Sci Technol. https://doi.org/10.1016/j.tifs.2020.06.017

Fabra MJ, Hambleton A, Talens P, Debeaufort F, Chiralt A, Voilley A (2009) Influence of interactions on water and aroma permeabilities of ι-carrageenan–oleic acid–beeswax films used for flavour encapsulation. Carbohyd Polym 76(2):325–332

Falguera V, Quintero JP, Jiménez A, Muñoz JA, Ibarz A (2011) Edible films and coatings: Structures, active functions and trends in their use. Trends Food Sci Technol 22(6):292–303

Farhan A, Hani NM (2020) Active edible films based on semi-refined κ-carrageenan: antioxidant and color properties and application in chicken breast packaging. Food Packag Shelf Life 24:100476

Farouk MM, Price JF, Salih AM (1990) Effect of an edible collagen film overwrap on exudation and lipid oxidation in beef round steak. J Food Sci 55(6):1510–1512

Food and Agricultural Organization of the United Nations (2019) Milk and milk products. http://www.fao.org/economic/est/est-commodities/dairy/milk-and-milk-products/en/. Accessed 25 Dec 2020

Franssen LR, Krochta JM (2003) Edible coatings containing natural antimicrobials for processed foods. In: Natural antimicrobials for the minimal processing of foods. Woodhead Publishing, pp 250–262
Galus S, Kadzińska J (2016) Whey protein edible films modified with almond and walnut oils. Food Hydrocol 52:78–86
Gennadios A (1994) Edible and coating and films to improved food quality. In: Edible coating and film based on proteins, pp 201–277
Gennadios A, Ghorpade VM, Weller CL, Hanna MA (1996) Heat curing of soy protein films. Trans ASAE 39(2):575–579
Gennadios A, Hanna MA, Kurth LB (1997) Application of edible coatings on meats, poultry and seafoods: a review. LWT-Food Sci Technol 30(4):337–350
Gennadios A, Rhim JW, Handa A, Weller CL, Hanna MA (1998) Ultraviolet radiation affects physical and molecular properties of soy protein films. J Food Sci 63(2):225–228
Gontard N, Guilbert S, Jean-Louis CUQ (1992) Edible wheat gluten films: influence of the main process variables on film properties using response surface methodology. J Food Sci 57 (1):190–195
Guilbert S, Gontard N (1995) Edible and biodegradable food packaging. In: Ackermann P, Jägerstad M, Ohlsson T (eds) Foods and packaging materials—chemical interactions. The Royal Society of Chemistry, England, UK, pp 159–168
Guitián MV, Ibarguren C, Soria MC, Hovanyecz P, Banchio C, Audisio MC (2019) Anti-*Listeria monocytogenes* effect of bacteriocin-incorporated agar edible coatings applied on cheese. Int Dairy J 97:92–98
Hambleton A, Debeaufort F, Bonnotte A, Voilley A (2009) Influence of alginate emulsion-based films structure on its barrier properties and on the protection of microencapsulated aroma compound. Food Hydrocol 23(8):2116–2124
Han JH, Krochta JM (2001) Physical properties and oil absorption of whey-protein-coated paper. J Food Sci 66(2):294–299
Hassan AH, Cutter CN (2020) Development and evaluation of pullulan-based composite antimicrobial films (CAF) incorporated with nisin, thymol and lauric arginate to reduce foodborne pathogens associated with muscle foods. Int J Food Microbiol 320:108519
Hassanzadeh P, Moradi M, Vaezi N, Moosavy MH, Mahmoudi R (2018) Effects of chitosan edible coating containing grape seed extract on the shelf-life of refrigerated rainbow trout fillet. Vet Res Forum 9(1):73
Hong SI, Krochta JM (2003) Oxygen barrier properties of whey protein isolate coatings on polypropylene films. J Food Sci 68(1):224–228
Janjarasskul T, Krochta JM (2010) Edible packaging materials. Ann Rev Food Sci Technol 1:415–448
Kavas N, Kavas G, Saygili D (2016) Use of ginger essential oil-fortified edible coatings in Kashar cheese and its effects on *Escherichia coli* O157: H7 and *Staphylococcus aureus*. CyTA-J Food 14(2):317–323
Kester JJ, Fennema OR (1986) Edible films and coatings: a review. Food Technol (USA). https://agris.fao.org/agris-search/search.do?recordID=US8836195
Kim KW, Ko CJ, Park HJ (2002) Mechanical properties, water vapor permeabilities and solubilities of highly carboxymethylated starch-based edible films. J Food Sci 67(1):218–222
Kim SY, Kang JH, Jo JH, Min SC (2018) Development of a gulfweed-based edible coating using high-pressure homogenization and its application to smoked Salmon. J Food Sci 83 (12):3027–3034
Krochta JM (2002) Proteins as raw materials for films and coatings: definitions, current status, and opportunities. Protein-Based Films Coat 1:1–40
Krochta JM, De Mulder-Johnston C (1997) Edible and biodegradable polymer films: challenges and opportunities. Food Technol 51(2):61–73
Kunte LA, Gennadios A, Cuppett SL, Hanna MA, Weller CL (1997) Cast films from soy protein isolates and fractions. Cereal Chem 74(2):115–118

Kuorwel KK, Cran MJ, Sonneveld K, Miltz J, Bigger SW (2014) Evaluation of antifungal activity of antimicrobial agents on cheddar cheese. Packaging Technol Sci 27(1):49–58
Lappa IK, Papadaki A, Kachrimanidou V, Terpou A, Koulougliotis D, Eriotou E, Kopsahelis N (2019) Cheese whey processing: integrated biorefinery concepts and emerging food applications. Foods 8(8):347
Lee DS, Yam KL, Piergiovanni L (2008) Food packaging science and technology. CRC Press
Li S, Ma Y, Ji T, Sameen DE, Ahmed S, Qin W, Liu Y (2020) Cassava starch/carboxymethylcellulose edible films embedded with lactic acid bacteria to extend the shelf life of banana. Carbohyd Polym 248:116805
Lima ÁM, Cerqueira MA, Souza BW, Santos ECM, Teixeira JA, Moreira RA, Vicente AA (2010) New edible coatings composed of galactomannans and collagen blends to improve the postharvest quality of fruits–Influence on fruits gas transfer rate. J Food Eng 97(1):101–109
Mahfoudhi N, Chouaibi M, Hamdi S (2014) Effectiveness of almond gum trees exudate as a novel edible coating for improving postharvest quality of tomato (*Solanum lycopersicum* L.) fruits. Food Sci Technol Int 20(1):33–43
Maleki G, Sedaghat N, Woltering EJ, Farhoodi M, Mohebbi M (2018) Chitosan-limonene coating in combination with modified atmosphere packaging preserve postharvest quality of cucumber during storage. J Food Measure Charact 12(3):1610–1621
Malhotra B, Keshwani A, Kharkwal H (2015) Natural polymer based cling films for food packaging. Int J Pharm Pharm Sci 7(4):10–18
Mallikarjunan P, Chinnan MS, Balasubramaniam VM, Phillips RD (1997) Edible coatings for deep-fat frying of starchy products. LWT-Food Sci Technol 30(7):709–714
Medeiros BGDS, Pinheiro AC, Carneiro-da-Cunha MG, Vicente AA (2012) Development and characterization of a nanomultilayer coating of pectin and chitosan–evaluation of its gas barrier properties and application on 'Tommy Atkins' mangoes. J Food Eng 110(3):457–464
Meyer GA, Mazer TB (1997) U.S. Patent No. 5,599,556. U.S. Patent and Trademark Office, Washington, DC
Noori S, Zeynali F, Almasi H (2018) Antimicrobial and antioxidant efficiency of nanoemulsion-based edible coating containing ginger (*Zingiber officinale*) essential oil and its effect on safety and quality attributes of chicken breast fillets. Food Control 84:312–320
Orsuwan A, Sothornvit R (2018) Active banana flour nanocomposite films incorporated with garlic essential oil as multifunctional packaging material for food application. Food Bioprocess Technol 11(6):1199–1210
Osorio FA, Molina P, Matiacevich S, Enrione J, Skurtys O (2011) Characteristics of hydroxy propyl methyl cellulose (HPMC) based edible film developed for blueberry coatings. Procedia Food Sci 1:287–293
Padua GW, Wang Q (2002) Formation and properties of corn zein films and coatings. In: Protein-based films and coatings. CRC Press, pp 43–67
Park HJ, Chinnan MS, Shewfelt RL (1994) Edible corn-zein film coatings to extend storage life of tomatoes. J Food Process Preserv 18(4):317–331
Pegg RB, Shahidi F (2007) Encapsulation, stabilization, and controlled release of food ingredients and bioactives. In: Handbook of food preservation. CRC Press, pp 527–586
Pérez-Gago MB, Nadaud P, Krochta JM (1999) Water vapor permeability, solubility, and tensile properties of heat-denatured versus native whey protein films. J Food Sci 64(6):1034–1037
Petersen K, Nielsen PV, Bertelsen G, Lawther M, Olsen MB, Nilsson NH, Mortensen G (1999) Potential of biobased materials for food packaging. Trend Food Sci Technol 10(2):52–68
Raghav PK, Agarwal N, Saini M (2016) Edible coating of fruits and vegetables: a review. Int J Sci Res Modern Edu 1(1):188–204
Rahman MS (ed) (2007) Handbook of food preservation. CRC Press
Rambabu K, Bharath G, Banat F, Show PL, Cocoletzi HH (2019) Mango leaf extract incorporated chitosan antioxidant film for active food packaging. Int J Biol Macromol 126:1234–1243
Ramírez C, Gallegos I, Ihl M, Bifani V (2012) Study of contact angle, wettability and water vapor permeability in carboxymethylcellulose (CMC) based film with murta leaves (*Ugni molinae* Turcz) extract. J Food Eng 109(3):424–429

Ramos ÓL, Reinas I, Silva SI, Fernandes JC, Cerqueira MA, Pereira RN, Malcata FX (2013) Effect of whey protein purity and glycerol content upon physical properties of edible films manufactured therefrom. Food Hydrocol 30(1):110–122

Rasouli M, Saba MK, Ramezanian A (2019) Inhibitory effect of salicylic acid and *Aloe vera* gel edible coating on microbial load and chilling injury of orange fruit. Sci Hortic 247:27–34

Ribeiro AM, Estevinho BN, Rocha F (2020) Preparation and incorporation of functional ingredients in edible films and coatings. Food Bioprocess Technol:1–23

Robertson GL (2012) Food packaging: principles and practice, 3rd edn. CRC Press, United States of America

Robles-Sánchez RM, Rojas-Graü MA, Odriozola-Serrano I, González-Aguilar G, Martin-Belloso O (2013) Influence of alginate-based edible coating as carrier of antibrowning agents on bioactive compounds and antioxidant activity in fresh-cut Kent mangoes. LWT-Food Sci Technol 50 (1):240–246

Saklani PS, Nath S, Kishor Das S, Singh SM (2019) A review of edible packaging for foods. Int J Curr Microbiol Appl Sci 8(7):2885–2895

Senturk Parreidt T, Müller K, Schmid M (2018) Alginate-based edible films and coatings for food packaging applications. Foods 7(10):170

Shit SC, Shah PM (2014) Edible polymers: challenges and opportunities. J Polym. https://doi.org/10.1155/2014/427259

Silva-Weiss A, Ihl M, Sobral PDA, Gómez-Guillén MC, Bifani V (2013) Natural additives in bioactive edible films and coatings: functionality and applications in foods. Food Eng Rev 5 (4):200–216

Souza BW, Cerqueira MA, Ruiz HA, Martins JT, Casariego A, Teixeira JA, Vicente AA (2010) Effect of chitosan-based coatings on the shelf life of salmon (*Salmo salar*). J Agric Food Chem 58(21):11456–11462

Suhag R, Kumar N, Petkoska AT, Upadhyay A (2020) Film formation and deposition methods of edible coating on food products: a review. Food Res Int 136:109582

Tang CH, Jiang Y (2007) Modulation of mechanical and surface hydrophobic properties of food protein films by transglutaminase treatment. Food Res Int 40(4):504–509

Tavassoli-Kafrani E, Shekarchizadeh H, Masoudpour-Behabadi M (2016) Development of edible films and coatings from alginates and carrageenans. Carbohyd Polym 137:360–374

Trujillo-Agudelo S, Osorio A, Gómez F, Contreras-Calderón J, Mesías-Garcia M, Delgado-Andrade C, Vega-Castro O (2020) Evaluation of the application of an edible coating and different frying temperatures on acrylamide and fat content in potato chips. J Food Process Eng 43(5):e13198

Vachon C, Yu HL, Yefsah R, Alain R, St-Gelais D, Lacroix M (2000) Mechanical and structural properties of milk protein edible films cross-linked by heating and γ-irradiation. J Agric Food Chem 48(8):3202–3209

Wieczyńska J, Cavoski I (2018) Antimicrobial, antioxidant and sensory features of eugenol, carvacrol and trans-anethole in active packaging for organic ready-to-eat iceberg lettuce. Food Chem 259:251–260

Wu Y, Rhim JW, Weller CL, Hamouz F, Cuppett S, Schnepf M (2000) Moisture loss and lipid oxidation for precooked beef patties stored in edible coatings and films. J Food Sci 65 (2):300–304

Wu H, Lei Y, Zhu R, Zhao M, Lu J, Xiao D, Li S (2019) Preparation and characterization of bioactive edible packaging films based on pomelo peel flours incorporating tea polyphenol. Food Hydrocol 90:41–49

Xiong Y, Chen M, Warner RD, Fang Z (2020) Incorporating nisin and grape seed extract in chitosan-gelatine edible coating and its effect on cold storage of fresh pork. Food Control 110:107018

Zhang D, Quantick PC (1997) Effects of chitosan coating on enzymatic browning and decay during postharvest storage of litchi (*Litchi chinensis* Sonn.) fruit. Postharvest Biol Technol 12 (2):195–202

Sources, Origin and Characterization of Edible Packaging

2

Padma Sangmu Bomzon

Abstract

Edible packaging is generally available commercially as edible films or edible coatings depending upon their structural composition and end utilization or function towards food preservation. Edible coatings comprise those materials that are applied directly on food surfaces, either on the outer surface or between layers, and are generally consumed directly as a part of the food that they protect. On the other hand, edible films are usually manufactured separately and applied on the food surface for packaging, and can be removed from or peeled off the food before consumption. Edible packaging should be, as the name suggests, edible and easy to digest. They should be non-toxic to human beings while also being biodegradable. Depending on the function of the end product, a wide range of biodegradable components such as hydrocolloids/polysaccharides, lipids and proteins are extracted from plant sources, animal sources and microorganisms for the process of manufacturing various edible coatings and films. The materials, thus, sourced can be categorized into three broad types as those originating from natural sources, such as agro sources, meaning the biopolymers are extracted directly from the natural biomass; those extracted from biomass developed by action of microorganisms or fermentation or those that are chemically synthesized from biomass. Apart from the characteristics that ensure edible packaging to be used as a food component, the packaging material should also possess various other properties that will allow it to effectively protect food materials from various external factors, while also preserving the structural integrity of the food materials. Edible biopolymers that make up edible coats and films inherently possess the property to form multiple layers or films, which is the most vital

P. S. Bomzon (✉)
Department of Nutrition (U.G.), Raja Narendra Lal Khan Women's College (Autonomous), Paschim Medinipur, West Bengal, India

A. Poonia, T. Dhewa (eds.), *Edible Food Packaging*,
https://doi.org/10.1007/978-981-16-2383-7_2

property for the development of any packaging material. This is in most part due to their ability to form a structural matrix that is both continuous and adequately cohesive. This property of forming a continuous structural matrix enables the resulting packaging film or coating to exhibit a myriad of essential characteristics like the ability to limit or prevent the transfer of moisture, gases (viz. oxygen and carbon dioxide), aroma, lipids, etc. that may affect the shelf life or important organoleptic properties of the food material. Such functionalities aid in the preservation of mechanical integrity of the food being packed. In order to optimize the functionality of edible packaging, their quality is inspected and characterized with regard to various properties including water vapour permeability (WVP), thickness, water solubility, oxygen permeability (OP) and mechanical properties (elongation, tensile strength, etc.). Sensory properties such as appearance (whether transparent/coloured/opaque), odour, taste and texture of the packaging material are also some of the important determinants of quality. As edible packaging is, in its essence, organic and biodegradable, care needs to be taken to ensure that the resulting edible films or coatings do not degrade easily when exposed to various external factors such as moisture, heat and light, among others. The various sources of edible biopolymers required for the development of edible packaging, their origin and characterization have been discussed in detail in this chapter.

Keywords

Edible packaging · Sources · Polysaccharides · Proteins · Wax

2.1 Introduction

Edible packaging is generally available commercially as edible films or edible coatings. Edible coatings comprise of those materials that are formed directly on food surfaces, while edible films are manufactured separately and placed on the food surface for packaging. The materials required to manufacture edible packaging comprise of a wide range of biodegradable components found in plants, animals and microorganisms.

Edible biopolymers that make up edible coats and films inherently possess the property to form multiple layers or films, which is the most vital property for the development of any packaging material. This is in most part due to their ability to form a structural matrix that is both continuous and adequately cohesive. This property of forming a continuous structural matrix enables the resulting packaging film or coating to exhibit a myriad of essential characteristics like the ability to limit or prevent the transfer of moisture, gases (viz. oxygen and carbon dioxide), aroma, lipids, etc. that may affect the shelf life or important organoleptic properties of the food material. Such functionalities aid in the preservation of mechanical integrity of the food being packed.

In order to optimize the functionality of edible packaging, their quality is inspected and characterized with regard to various properties including water vapour permeability (WVP), thickness, water solubility, oxygen permeability (OP) and mechanical properties (elongation Eb%, tensile strength [TS], etc.). Sensory properties such as appearance (whether transparent/coloured/opaque), odour, taste and texture of the packaging material are also some of the important determinants of quality. As edible packaging is, in its essence, organic and biodegradable, care needs to be taken to ensure that the resulting edible films or coatings do not degrade easily when exposed to various external factors such as moisture, heat and light, among others.

2.2 Sources and Characterization

Edible biopolymers, which are utilized in the production of edible packaging, are broadly classified according to their structural framework into polysaccharides, proteins and lipids and composites (Suput et al. 2015; Cerqueira 2015). The biopolymer is usually supplemented with a plasticizer (e.g. glycerol). This is done to optimize the elasticity and flexibility of the end product. Additives, such as antimicrobial agents, artificial food-grade colours and flavour enhancers, can also be added to modify and improve functionality of the packaging material. Various sources of edible packaging are given in Fig. 2.1.

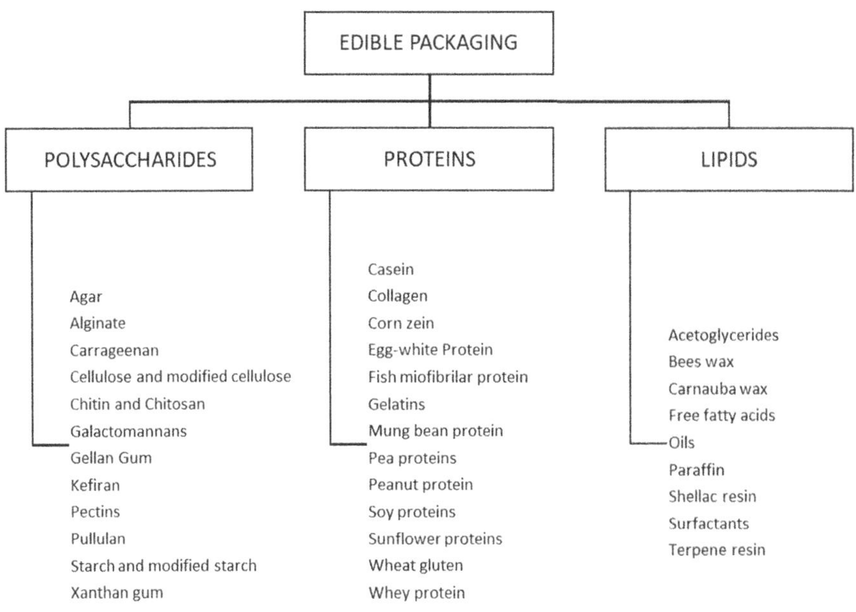

Fig. 2.1 Sources of edible packaging

The sources and characterizations of some commonly used sources of edible packaging are discussed in detail below.

2.3 Edible Polysaccharide Packaging

Polysaccharides are long-chain polymeric carbohydrates formed by the glycosidic linkages between monosaccharide units. The H-bonds play significant roles in the formation of films and their characteristics.

Polysaccharide packaging provides strength and mechanical integrity to the food by creating a barrier against oxygen, aroma and lipid transfer. However, they fall short in terms of protection against moisture.

2.3.1 Agar

Agar is found in the cell walls of algae of the *Rhodophyta* phylum (red algae), also known as agarophytes. It comprises of agarose, which makes up the gelling portion, and agaropectin, which makes up the non-gelling fraction. Usually, agaropectin is removed during commercial processing in order to manufacture agar powder with a higher gelling strength. Apart from being biodegradable, Agar possesses a continuous film-forming ability, has a high retraction ratio, is transparent, heat sealable and, most importantly, it is biologically inert. All these properties are highly favourable for edible food packaging. However, pure agar film has low elasticity, high WVP, high water sensitivity and poor thermal stability because of which it requires to be used in combination with additional substances such as plasticizers, biopolymers, nanoparticles, etc. (Mostafavi and Zaeim 2020).

2.3.2 Alginate

Alginate is a salt of algin (alginic acid), which is a polysaccharide found in the cell walls of brown algae (*Phaeophyceae* sp.). Alginate is a natural hydrophilic polysaccharide biopolymer and is highly favoured for food packaging. In addition to important film-forming properties, they also exhibit characteristics such as moisture retention, maintenance of structural integrity and overall enhancement of various organoleptic properties of the food items being packed. Alginate is known to exhibit antimicrobial properties. These have also been shown to protect food from lipid oxidations (Theagarajan et al. 2019).

2.3.3 Carrageenan

Carrageenan is a polysaccharide found in seaweeds and red algae. Carrageenan biopolymer is a complex hybrid chemical structure comprising of λ-carrageenan, ι-carrageenan and κ-carrageenan, together with non-gelling biological precursors like monomers of ν-carrageenan or μ-carrageenan. These enable the biopolymer to exhibit a myriad of gel-forming properties (Sanchez-García 2011). Carrageenan coatings are mostly used for their anti-microbial properties in gel matrices containing antimicrobial agents. Carrageenans are good lipid and oxygen barriers, and are generally used for packaging of sensitive food items such as meat and its products. These edible films are water soluble.

2.3.4 Cellulose

Cellulose is a polysaccharide that exists in the cell walls of all plants. Its structure comprises a linear chain made up of β-1,4-linked D-glucose units. Owing to its complex crystalline structure of tightly packed polymer chains, ranging from hundreds to several thousands, cellulose possesses high strength, is insoluble in water and exhibits many attractive mechanical properties. Derivatives of cellulose are formed from etherification of cellulose which enhances water solubility. Carboxymethyl cellulose, methyl cellulose, hydroxypropyl cellulose and hydroxypropyl methyl cellulose are some common water-soluble cellulose derivatives used commercially for making edible films.

Edible packaging made from cellulose derivatives provides effective barriers against moisture, oxygen and lipids. Such packaging is generally tasteless, odourless, transparent, water soluble, flexible and resistant to oxygen and carbon dioxide. Bacterial cellulose, such as of *Acetobacter xylinum,* have also been utilized in the production of edible packaging. Bacterial cellulose exhibits high structural strength, high purity and a high moldability factor (Brigham 2018).

2.3.5 Chitosan

Chitosan is a polysaccharide made up of a linear chain comprising of β-1,4-linked D-glucosamine and N-acetyl-D-glucosamine. When Chitin, which is derived from exoskeletons of crustaceans and cell walls of fungi, undergoes deacetylation, it leads to the formation of Chitosan. Edible packaging made from chitosan exhibits antimicrobial activity. Besides that, pure chitosan films are transparent, non-toxic and protect lipid oxidations in food (Cazón and Vázquez 2019).

2.3.6 Galactomannans

Galactomannans are polysaccharides composed of galactose and mannose. These gums are extracted from the endosperm of dicotyledonous seeds of various plants. Guar gum, tara gum and locust bean gum are the most commercially important forms of galactomannans in the food industry. Galactomannans need only water for preparation and have been known to form highly viscous solutions even with very little amounts of the polysaccharide. These are used in combination with other substances such as plasticizers, biopolymers, nanoparticles, etc. in order to decrease WVP and OP, while increasing their TS and Eb% properties (Cerqueira et al. 2011).

2.3.7 Pectin

Pectins are complex polysaccharides with D-galacturonic acid as a principal constituent. It is found in the cell walls of many plants and comprises of 1,4-α-D-galacturonic acid molecules, which are linked to rhamnose residues (main chain) and arabinose, galactose and xylose (side chains). Pectin may be high methoxyl or low methoxyl. Low methoxyl pectins are the ones generally used in food coatings. Pectin forms gels when dissolved in water under suitable conditions, which is the most important property for its use as food packaging. They also provide effective barrier against oxygen, lipids and aroma. They also possess good mechanical properties, but fall short in terms of moisture protection (Valdés et al. 2015).

2.3.8 Starch and Derivatives

Starch $(C_6H_{10}O_5)n$ is a polysaccharide made up of glucose units joined by α-1,4-glycosidic linkages. It is composed of a linear amylose polymer and a branched amylopectin polymer. Starch films are tasteless, odourless, transparent and considerably less permeable to moisture, O_2, CO_2, lipids and aroma. Starch as an edible packaging is particularly promising because of its low cost and easy availability (Pelissari et al. 2019; Versino et al. 2016).

2.4 Edible Protein Packaging

Proteins are large polymers made up of one or more linear chains of amino acids. Edible packaging made from proteins can be classified as those sourced from plant proteins, such as soy protein and wheat gluten, and those sourced from animal proteins, such as casein and whey protein. Edible packaging based on proteins possesses excellent gas barrier and lipid barrier properties, which helps in the prevention of loss of flavours while also controlling the exchange of gases such as oxygen and carbon dioxide. However, protein-based packaging had weak water barrier characteristics, which requires them to be used in combination with polysaccharide derivatives, plasticizers and other substances in order to improve their packaging properties.

2.4.1 Casein

Casein is an animal protein (found in mammalian milk), and is comprised of four phosphoproteins – kappa-casein, beta-casein, alpha s1-casein and alpha s2-casein. Casein possesses excellent film-forming properties. Apart from being biodegradable, it has a high thermal stability, and is non-toxic while also being highly nutritious. However, casein-based packaging is highly sensitive to moisture and, thus, needs to be combined with substances such as plasticizers to improve WVP properties and properties like TS (Chen et al. 2019).

2.4.2 Collagen and Gelatin

Collagen is another animal protein which is popularly used to manufacture commercial edible protein film. It is the most abundant protein found in the connective tissues of mammals. Collagen is a fibrous protein and comprises of a triple helix formed by amino acids bundled together (Shoulders and Raines 2009). Collagen-based packaging exhibits exceptional barrier against oxygen at 0% relative humidity, but an increased relative humidity (RH) leads to weaker barrier properties. Their resistance to moisture is also relatively low. In order to overcome these shortcomings, collagens are combined with polysaccharide derivatives or plasticizers.

Gelatin is produced when collagen is hydrolysed. Gelatin-based packaging is transparent and has excellent oxygen barrier properties. The melting point of gelatin is 35 °C, which makes it highly desirable as an edible food packaging material. Gelatin-based films may also be incorporated with antimicrobial or antioxidant agents to improve their functionality (Chen et al. 2019).

2.4.3 Corn Zein

Zein makes up the major protein component in the endosperm of maize or corn. Alpha-, β-, γ- and δ-zein are the four molecular parts of zein. It is not soluble in water but forms a solution in alcohol. Zein has been widely used as a component of chewing gum and in the coating of various candies. Zein has excellent film-forming properties, is hydrophobic but soluble in organic solvents (e.g. ethanol) and possesses suitable gas barrier properties, which makes it an excellent contender for preparing edible food coatings and films. It is also compatible with most natural antioxidants and antimicrobial agents. However, zein-based films are brittle and do not provide a good barrier against WVP. Combination with other substances such as plasticizers and other biopolymers helps zein-based films and coatings to achieve better mechanical properties (Arcan et al. 2017).

2.4.4 Soy Proteins

Soy proteins are extracted from soybeans. They are available as soy protein isolates, soy flour and soy protein concentrates. Soy protein isolates (SPI) serve as a major source in the production of soy protein-based films. The film-forming characteristics of soy proteins have been extensively researched upon through the years. Soy proteins are usually combined with plasticizers to create edible films. The concentration of SPI and plasticizer is a major determinant of the mechanical properties of the resulting edible packaging material. A higher SPI concentration shows increased thickness with greater tensile strength, while decreasing the Eb%. On the other hand, a higher plasticizer concentration leads to a decreased thickness and lower tensile strength but results in an increase in Eb% (Nandane and Jain 2014).

2.4.5 Wheat Gluten

Gluten is the major protein found in cereal grains like barley, rye and wheat. It is a protein complex comprised of two fractions – gliadin (water soluble) and glutenin (water insoluble). Gluten-based films are transparent, strong and have excellent gas barrier properties at low RH. The WVP property of gluten-based films can be improved drastically by combination with lipidic compounds (Gontard and Guilbert 1998).

2.4.6 Whey Protein

Whey protein is another protein found in mammalian milk. Whey protein comprises of α-lactalbumin, β-lactoglobulin, serum albumin and immunoglobulins. Edible films based on whey protein are transparent, have no distinct taste or smell, are flexible and possess excellent gas and lipid barrier characteristics at low RH. However, whey protein films have poor water barrier properties so they have to be used in combination with other hydrophobic substances such as lipidic compounds to reduce the WVP (Javanmard 2009).

2.5 Edible Lipid Packaging

Lipids are large biopolymers which are soluble in non-polar solvents. Lipids can be classified into fatty acids, glycerolipids, glycerophospholipids, sphignolipids, sterols, prenols, saccharolipids and polyketides. Lipids are hydrophobic in nature.

Lipid-based edible films and coatings have excellent WVP properties, which is its most desirable property in the food packaging industry. However, lipid-based films are usually opaque, inflexible and significantly brittle; thus, lipids are often used in combination with polysaccharide or protein derivatives to achieve better functionality.

2.5.1 Acetoglycerides/Acetylated Monoglycerides

Acetoglycerides constitute a class of compounds comprising of a mixture of carboxylic acid esters of glycerol. Acetylated monoglycerides are emulsifiers and are generally used for their plasticizer properties and incorporated into edible coating formulations.

2.5.2 Resins

Naturally occurring resins are secreted by plants and insects as a protective response to injury. Resins are usually mixtures of organic compounds. Shellac resin is secreted by *Laccifer lacca,* which is an insect in the *Kerriidae* family. It is soluble in alcohols and alkaline solutions and is composed of a mixture of aliphatic alicyclic hydroxyl acid polymers. It is the most commonly used resin in edible food coatings. Terpene resin is derived from wood (Hall 2011).

2.5.3 Waxes and Paraffins

Beeswax is naturally produced by honeybees, which belong to the genus *Apis.* Beeswax is composed of esters of fatty acids and long-chain alcohols. Paraffin wax is derived from crude petroleum. Candelilla is obtained from candelilla plant. Carnauba wax is extracted from the leaves of *C. cerifera* (palm tree). Mineral oil is comprised of a mixture of liquid paraffinic and naphthenic hydrocarbons.

Waxes and paraffins have been used as natural glazing agents on food items. In terms of edible packaging, these are often used in combination with polysaccharide or protein derivatives for its excellent hydrophobic properties to reduce the WVP of the edible packaging. They also possesses excellent gas barrier properties. Paraffin wax is not permitted for use on food items other than raw fruits and vegetables and cheese. If the wax or paraffin coatings on food items are thick, they are generally required to be removed before consumption of the packaged food (Trevisani et al. 2017; Bourtoom 2008).

2.5.3.1 Origin

The categorization of edible biopolymers based on their origin is shown in Fig. 2.2.

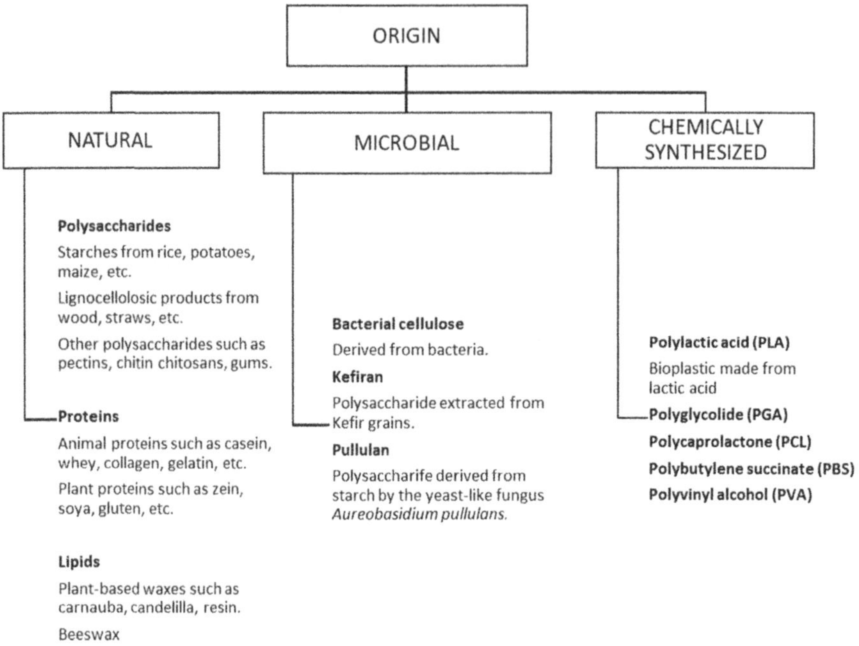

Fig. 2.2 Origin of edible biopolymers (Shankar and Rhim 2018)

References

Arcan İ, Boyacı D, Yemenicioğlu A (2017) The use of Zein and its edible films for the development of food packaging materials. Encyclopedia. In: Reference module in food sciences, 1st edn. Elsevier, pp 1–11

Bourtoom T (2008) Edible films and coatings: characteristics and properties. Int Food Res J 15: 237–248

Brigham C (2018) Biopolymers. In: Green chemistry. Elsevier, pp 753–770

Cazón P, Vázquez M (2019) Applications of chitosan as food packaging materials. In: Sustainable agriculture reviews, vol 36. Springer, Cham

Cerqueira M (2015) Edible food packaging: materials and processing technologies. CRC Press, p 11

Cerqueira MA, Bourbon AI, Pinheiro AC, Martins JT, Souza BWS, Teixeira JA, Vicente AA (2011) Galactomannans use in the development of edible films/coatings for food applications. Trends Food Sci Technol 22(12):662–671

Chen H, Wang J, Cheng Y, Wang C, Liu H, Bian H, Pan Y, Sun J, Han W (2019) Application of protein-based films and coatings for food packaging: a review. Polymers 11(12):2039

Gontard N, Guilbert S (1998) Edible and/or biodegradable wheat gluten films and coatings. In: Guéguen J, Popineau Y (eds) Plant proteins from European Crops. Springer, Berlin, Heidelberg

Hall D (2011) Edible coatings from lipids, waxes and resins. In: Baldwin EA, Hagenmaier R, Bai J (eds) Edible coatings and films to improve food quality, 2nd edn. CRC Press, Boca Raton, pp 79–101

Javanmard M (2009) Biodegradable whey protein edible films as a new biomaterials for food and drug packaging. Iran J Pharm Sci 5(3):129–134

Mostafavi FS, Zaeim D (2020) Agar-based edible films for food packaging applications: a review. Int J Biol Macromol 159:1165–1176
Nandane AS, Jain R (2014) Study of mechanical properties of soy protein based edible film as affected by its composition and process parameters by using RSM. J Food Sci Technol 52(6):3645–3650
Pelissari FM, Ferreira DC, Louzada LB, dos Santos F, Corrêa AC, Moreira FKV, Mattoso LH (2019) Starch-based edible films and coatings: an eco-friendly alternative for food packaging. In: Starches for food application. Elsevier, pp 359–420
Sanchez-García MD (2011) Carrageenan polysaccharides for food packaging. In: Multifunctional and nanoreinforced polymers for food packaging. Woodhead Publishing, pp 594–609
Shankar S, Rhim J-W (2018) Bionanocomposite films for food packaging applications. Ref Module Food Sci 1:1–10
Shoulders MD, Raines RT (2009) Collagen structure and stability. Annu Rev Biochem 78:929–958
Suput D, Lazic V, Popovic S, Hromis N (2015) Edible films and coatings: sources, properties and application. Food Feed Res 42(1):11–22
Theagarajan R, Dutta S, Moses JA, Anandharamakrishnan C (2019) Alginates for food packaging applications. In: Alginates: applications in the biomedical and food industries. Wiley, p 207
Trevisani M, Cecchini M, Siconolfi D, Mancusi R, Rosmini R (2017) Effects of Beeswax coating on the oxidative stability of long-ripened Italian Salami. J Food Qual 2017:1–5
Valdés A, Burgos N, Jiménez A, Garrigós MC (2015) Natural pectin polysaccharides as edible coatings. Coatings 5:865–886. https://doi.org/10.3390/coatings5040865
Versino F, Lopez O, Garcia M, Zaritzky N (2016) Starch-based films and food coatings: an overview. Starch – Stärke 68(11–12):1026–1037

Animal- and Plant-Based Edible Food Packaging for Perishable Foodstuff

3

Montserrat Calderón-Santoyo, Oscar Alberto Solis-Velazquez, Juan Arturo Ragazzo-Sánchez, and Maricarmen Iñiguez-Moreno

Abstract

Nowadays, global awareness of the pollution associated with large amounts of wastes of synthetic polymers from nonrenewable sources and the negative impact of chemical compounds used to preserve perishable foods in the environment and human health has been conducted to the progress of active and comestible food packaging. Furthermore, the global demand for fresh and minimally processed foods such as fruits, vegetables, eggs, meat, and fish is increasing in parallel with population growth and with the tendency to acquire a healthy lifestyle. These foods are crucial in the human diet because of their high nutrients content and the related profits with their consumption. A disadvantage of these foods is their relatively brief shelf life varying from some hours to weeks depending on storage conditions due to the physiological, biochemical, and microbiological changes associated with them. Hence, their preservation is a big challenge to the food manufacturing. Over the last decades, the use of edible packaging based on polysaccharides, proteins, and lipids from animal or plant origin represents an alternative that contributes to mitigate the negative impact of conventional films and coatings. Indeed, a substantial number of works have been oriented on the formulation of biodegradable food packaging made with biopolymers due to their high abundance, obtention from renewable sources, and low cost. Biopolymers from animals and plants matrices are employed to elaborate pertinent packaging for each type of decaying food without altering their sensorial and nutritive

M. Calderón-Santoyo · J. A. Ragazzo-Sánchez · M. Iñiguez-Moreno (✉)
Laboratorio Integral de Investigación en Alimentos, Tecnológico Nacional de México/Instituto Tecnológico de Tepic, Tepic, Nayarit, Mexico

O. A. Solis-Velazquez
Laboratorio de Microbiología, Departamento de Ciencias Médicas y de la Vida, Centro Universitario de la Ciénega, Universidad de Guadalajara, Ocotlán, Jalisco, Mexico

A. Poonia, T. Dhewa (eds.), *Edible Food Packaging*,
https://doi.org/10.1007/978-981-16-2383-7_3

properties. Besides, the functionalization of these edible packaging is possible to transport different active substances such as antimicrobials, antioxidants, antioxidant agents, volatile precursors, nutriments, flavoring, and coloring compounds to perform and enhance the stability, quality, and safety of coated perishable foods. Some of these active compounds can be obtained from agroindustrial residues, which enhance the sustained character of this industry. Furthermore, the enhancement of mechanical, structural, and water vapor and gas barrier qualities of edible films and coatings is achieved through mixing two or more biopolymers to form complex coatings, and through the addition of plasticizers and/or cross-linking agents. This chapter presents the latest updates, tendencies, and challenges in the food industry to develop eco-friendly food packaging from diverse natural sources with different compositions, added bioactive compounds, and their effect on perishable foods. Also, it includes information about their mechanical and physicochemical traits.

Keywords

Shelf life extension · Edible coatings · Biopolymers

3.1 Introduction

Global requirement for fresh, minimally processed and ready-to-eat foods is increasing in parallel with population growth (Pilon et al. 2016; Elik et al. 2019). Foods such as fruits, vegetables, cereals, eggs, meat, chicken, fish, and dairy products are necessary in the human diet. Nevertheless, these foods have a relatively short shelf life between their production and consumption, ranging from few hours to few weeks at room conditions. Some perishable foods such as fruits, vegetables, and eggs remain as living tissues because they carry on physiological and biochemical processes after their obtention until the consumption time, likewise, they can suffer physical, pathological, and/or microbiological damage (Palou et al. 2015; Elik et al. 2019). Food and Agriculture Organization of the United Nations (FAO) reported that about one-third of generated foods for human uptake get lost every year worldwide throughout the food provision chain (FSC) (Gustavsson et al. 2011). The monetary value of this amount of food loss or waste (FLW) is estimated at $936 billion US dollars, regardless of the social and environmental costs paid by society as a whole (FAO 2014). The amount of FLW is enough to relieve one-eighth of the world's population from famine (Gustavsson et al. 2011) and meant to satisfy the global demand for food, which is estimated to increase up to 50–70% at 2050 in comparison with the current demand (Food and Agriculture Organization of the United Nations 2009). The amount of FLW diverges considerably among countries, primarily determined by the amounts of earnings, urbanization, and economic growth (Chalak et al. 2016). In low-developed countries, food is mainly lost during the early and middle stages of the food supply chain. In developing countries, more than 40% of the food losses occur at postharvest and processing levels, while in industrialized countries, over 40% of the food losses occur at retail and consumer levels (Gustavsson et al. 2011). The FLW in low- and medium-developed countries is mostly attributed to poor agricultural practices, technological, financial, and labor restrictions, together with poor infrastructure for storage, processing, and transport (Chalak et al. 2016).

Perishable foods losses caused by microorganisms, mainly fungi and bacteria, along with FSC can reach more than 25% of the total production in industrialized countries, and up to 50% in low- and medium-developed countries if handling and storage conditions during the FSC are not optimal (Sapper and Chiralt 2018; Iñiguez-Moreno et al. 2020a). In fruits and vegetables, weight loss during the postharvest stage is one of the most important characteristics associated with quality maintenance. Changes in this physiological parameter result in loss of texture and development of bruises, shortening the shelf life of the food product. Fruit softening is also attributed to the deterioration of the cell wall components, mainly pectin, associated with the enzymatic activity (Sapper and Chiralt 2018). On the other hand, fish, beef, and chicken are the most perishable foods, due to their short freshness period after death. After obtainment, microbiological spoilage and biochemical reactions occur, such as changes in protein and lipid content and formation of biogenic amines and hypoxanthine; the latter is particularly in fish flesh (Matak et al. 2015). Hence, appropriate food management during the FSC is very important to reduce worldwide food losses (Elik et al. 2019; Yahaya and Mardiyya 2019).

The use of synthetic packaging materials such as polypropylene and polyethylene and/or chemical compounds such as preservatives and fungicides has been used for many years to extend the shelf life of perishable foods. However, the use of packaging obtained from petroleum can remain in the environment as long as 450 years producing negative effects on human health and the environment (Hossain et al. 2020). Regarding the use of chemical compounds, about 23 million kg of fungicides are applied to fruits and vegetables annually (Tripathi and Dubey 2004). Otherwise, nitrates are widely used in the preservation of meat products (Efenberger-Szmechtyk et al. 2020). The residues of these compounds in the foods can affect human health and increase the resistance of pathogenic strains (Tripathi and Dubey 2004; Efenberger-Szmechtyk et al. 2020). The increasing environmental awareness, the restrictions on the use of agrochemicals and food preservatives in many countries, and the increasing consumer demand for high-quality, minimally processed fresh food products have intensified the exploration and application of methods to keep perishable foods safe. Some technologies include refrigeration, controlled atmosphere storage, and sterilization by UV, gamma radiation, or the combination of various preservation methods, which is called barrier technologies. Nevertheless, for many foods, the use of edible films and coatings is one of the most cost-effective ways to maintain their quality and safety. Hence, the food industry is looking for the use of preservation methods and technologies based on abundant, low cost, renewable, and biodegradable materials (Chen and Smith 2017; Dehghani et al. 2018; Sapper and Chiralt 2018). In line with this, some food industry wastes or by-products can be used in food packaging preparation, contributing to a supplementary reduction in overall waste (Avramescu et al. 2020).

Fig. 3.1 Methods used to develop films and coatings

Edible film or coating is any type of thin layer of material with filmogenic capacity used for enrobing or coating foods to extend their shelf life and which can be consumed together with the food (Valdés et al. 2015; Dehghani et al. 2018). Although edible films and coatings have a similar meaning, there exists a difference. Edible films are separately prepared and then applied to the food surface, whereas coatings are applied directly onto the food surface in its liquid form (Fig. 3.1) (Valdés et al. 2015; Bharti et al. 2020). Biodegradable edible films and coatings have been used for centuries to prevent or retard deterioration and improve the appearance of packaged foods. The use of films in foods dates back to the twelfth century in China, where waxes were used to delay the dehydration of citric fruits. The first edible film used for food preservation was made in the fifteenth century from soymilk (Yuba) in Japan (Sánchez-Ortega et al. 2014). In 1930, emulsions and waxes were developed to coat fruits, with the purpose to improve their appearance, control the ripening process, and decrease the water loss (Debeaufort et al. 1998). However, until 1967, edible films and coatings had low commercial use and were limited mostly to the applications of wax coatings on fruits and vegetables. During the 1980s, important businesses grew out based on this concept. Nowadays, edible films are used in a wide variety of foods, with total annual incomes exceeding $100 million US dollars (Huber and Embuscado 2009). The use of edible coatings (ECs) emerged as an effective eco-friendly alternative to extend the shelf life of fresh foods. Their use is not new in the food industry but notorious results in terms of efficiency and versatility to extend the shelf life of perishable foods have been recently demonstrated (Valdés et al. 2015; Marín et al. 2017; González-Saucedo et al. 2019; Avramescu et al. 2020). Edible films and coatings cannot entirely replace synthetic packaging; the efficacy of food protection can be improved by combining primary edible and biodegradable packaging, and secondary nonedible packaging. Usually, secondary packaging is necessary for handling and hygienic practices. In this sense, it is important to apply eco-friendly food preservatives to control the loss of the nutritional value of the perishable foods and to reduce the requirements and waste of conventional packaging, improving the economic efficiency of packaging materials (Cordeiro de Azeredo 2012).

3.2 Characteristics of Edible Coatings to Prolong the Shelf Life of Perishable Foodstuff

Edible films and coatings have similar features as conventional packagings, such as acting as a barrier to gases, water vapor, and flavor compounds, providing physical, chemical, and microbiological protection to the packed foods. Moreover, natural coatings implemented to vegetables and fruits contribute to decreasing the activity of some enzymes responsible for breaking down the polysaccharides of the cell wall, contributing to maintain the firmness (Defilippi et al. 2018). Also, they act as transporters of food additions, improving the nutritional, sensorial, structure integrity, and mechanical properties of coated foodstuffs (Cordeiro de Azeredo 2012; Dehghani et al. 2018). Concerning mechanical as well as barrier properties, natural materials used as edible films and coatings are usually less efficient than food packaging from petroleum (Cordeiro de Azeredo 2012). In edible films and coatings, the most important properties are their restriction to water vapor and gases properties, molecules migration, their aptitude to provide physic-mechanical protection, as well as the appearance of the product such as color and glossiness (Palou et al. 2015). However, the desired properties from an edible film or coating depend on the requirements of the food to be covered. Perishables foods have complex requirements because, in some of these products, the metabolism is still active. Therefore, some substantial characteristics of edible films and coatings are as follows:

1. Formulated with generally recognized as safe (GRAS) compounds.
2. Moderately low permeability to O_2 and CO_2. In perishable foods such as fruits and vegetables, the coatings should create semipermeable barriers to gases. When the concentration of O_2 increases, the respiration process, and ethylene production also increase, increasing the consumption of sugars and other compounds, accelerating the senescence (Bonilla et al. 2012). Hence, the ripening process can be delayed by reducing the penetration of O_2, rather than decreasing CO_2 and ethylene inside the fruit; that is to say, the CO_2/O_2 permeability ratio should be as high as possible (Cordeiro de Azeredo 2012). Films and coatings based on proteins and/or polysaccharides have ratios from 10 to 25, while in the synthetic packaging the ratios are lower than 5.73 (Debeaufort et al. 1998). However, the metabolic activity of the coated food must not create anaerobic conditions, which produce physiological disorders and accelerate the loss of food quality (Kester and Fennema 1986).

3. Low water vapor permeability values contribute to delay dehydration (Garcia and Barrett 2002). In minimally processed foods, it is highly complicated due to the high water contents in food products, which affects the properties of hydrophilic coatings (Hagenmaier and Shaw 1992).
4. Regulate the mass transfer in the food products to prevent moisture loss and changes in flavor, texture, or appearance (Lin and Zhao 2007).
5. Wettability and the effectiveness of an edible coating to protect foods greatly depend on the uniformity of the coating. This behavior is affected by properties of the food surface as well as by the composition of the film- or coating-forming solution: the polymeric substance, the plasticizers, surfactants, cross-linker agents, functional compounds, and others (Falguera et al. 2011). Moreover, this property affects the thickness and permeability of the coating (Cazón et al. 2017).
6. Mechanical properties, particularly resistance are important to avoid the coating fissure and to protect the food product from mechanical and physical damage provoked by collision or pressure during the food storage (Sapper and Chiralt 2018). The addition of plasticizers, emulsifiers, or surfactants to the coating solutions improves the extensibility and flexibility of the structure of the edible food packaging. Adequate adherence is an important behavior to consider because a coating should improve the appearance of the coated food (Pilon et al. 2016).
7. Food compatibility, that is to say, that edible films or coatings must not modify the sensorial properties of the coated food (Cordeiro de Azeredo 2012).
8. Manufacturing process should be easy and economically viable. Low cost, as well as compatibility of the mode of application with current equipment, are important features in ECs (Avramescu et al. 2020).

The characteristics, efficiency, and stability of food packaging depend on chemical and physical factors, such as the chemical structure of the polymeric matrix, pH, viscosity, thickness, cross-linking degree, and processing conditions, as well (Bastarrachea et al. 2015). Besides, the materials used to form edible packaging must be capable to be dissolved or dispersed in a safe solvent (Dehghani et al. 2018). Biodegradable edible films and coatings may be categorized with respect to the source and the biopolymer origin (Dehghani et al. 2018). Lipids, proteins, and polysaccharides from renewable sources are widely used alone or in combination with the coating-forming solutions; each class has its advantages and limitations (Cazón et al. 2017; Dehghani et al. 2018). The efficacy of edible films and coatings to keep the quality and freshness of perishable foods is dependent on the use of the appropriate material to develop the food packaging that will preserve adequate internal gas concentrations according to the gas transfer and transpiration rates of the foods (Casariego et al. 2008).

3.3 Polysaccharide Films and Coatings

Polysaccharides are the most common materials used to develop edible food packaging for perishable foods; this could be associated with their microbial and physical stability over time (Sapper and Chiralt 2018). Furthermore, polysaccharides have high availability, low cost, and great film-forming properties. The main biopolymers used to develop food packing are starch, cellulose, pectin, gums, chitosan, and alginates (Table 3.1) (Cazón et al. 2017; Marín et al. 2017). Polymers are neutral, but some gums are negatively charged because they have a big number of hydroxyl and other polar groups. H-bonds are crucial in the formation and final characteristics of food packaging (Dehghani et al. 2018). Polysaccharides are compatible with several functional compounds and additives that improve their filmogenic properties (Sapper and Chiralt 2018). This is very important because the polysaccharides are hydrophilic (Bourtoom 2008), the water vapor barrier character of polysaccharide coatings is depreciable, but they are selectively permeable for O_2 and CO_2 and avoid lipid migration. Even if polysaccharide coatings do not a have good water vapor barrier character (Dehghani et al. 2018), these coatings could act as "sacrificing agents," retarding water loss from the food by addition of additional moisture on the surface (Kester and Fennema 1986).

3.3.1 Plant Polysaccharides

3.3.1.1 Cellulose and Cellulose Derivates

Cellulose is a homopolymer synthesized by the polymerization of glucose residues linked by β-1,4-O-glycosidic bonds forming a β-1,4-D-glucan (Fig. 3.2a). Cellulose is the most abundant and renewable biopolymer (Delmer 1987; Delmer and Amor 1995). This biopolymer is the main constituent of higher plants, being wood and cotton the principal sources for industrial processes as it constitute 40–50% and nearly 90% of its structure, respectively. Also, nano-, microcrystalline, and nanofibrillar cellulose can be obtained through a sustainable process from sugarcane (Katakojwala and Mohan 2020), cassava (Travalini et al. 2018), and agave bagasse (Palacios et al. 2019), respectively. As well as it is produced by acetic acid bacterium and is found in fungal and green algae cell walls (Delmer 1983). Purity, polymerization degree, and diameter of cellulose microfibrils vary according to the source of obtention. For example, bacterial cellulose has higher purity, mechanical strength, crystallinity, and hydrophilicity than cellulose obtained from other sources (Rozenberga et al. 2016). Cellulose chains have a strong tendency to self-associate by H-bonds to form insoluble, and often highly crystalline fibrils, making them insoluble in aqueous solutions and polar solvents. Hence, the etherification of the hydrogen atom in the hydroxyl groups has been used to obtain water-soluble cellulose derivates, which have great film-forming properties and high biocompatibility (Doelker 1993).

The main cellulose derivates used for coatings are methylcellulose, hydroxypropyl methylcellulose, hydroxyethylcellulose, and carboxymethyl

Table 3.1 Polysaccharides used to develop food packaging for perishable foods

Polymer matrix	Additives	Bioactive compound	Method of application	Perishable food	Significant function in coated food	References
Cellulose nanocrystals from pea hull and carboxymethyl cellulose	Glycerol	–	Film packaged	Red chilies	Reduced weight loss and maintained vitamin C compared with uncoated red chilies after 7 days of storage at 20 °C	Li et al. (2020)
Cellulose nanofibers	Sucrose ester fatty acid	Oleic acid	Brushing	Bananas	Delayed the ethylene and CO_2 biosynthesis, reduced chlorophyll degradation of banana peels and weight loss, and maintaining the firmness of fruit for 10 days at 20 °C	Deng et al. (2017)
Banana starch and chitosan	Sorbitol	*Aloe vera* gel	Dipping	Strawberries	Reduced fungal decay and increased the shelf life up to 15 days of storage, decreasing the water vapor loss from fruit and the structural decay	Pinzon et al. (2020)
Pectin	Tween 80 and ethanol	Oregano essential oil and resveratrol	Nanoemulsions applied by dipping	Fresh pork loins	Shelf life extension by minimizing pH change, retarding lipid and protein oxidation, maintaining meat tenderness, and inhibiting microbial growth during the 20 days of packaging at 4 °C. The oils have synergetic effects on meat color, tenderness, and protein oxidation	Xiong et al. (2020)

Pectin citrus, fish gelatin, and beeswax	Glycerol	Hydroxytyrosol from olives	Packaged with dried films	Beef meat pieces	Reduced lipid oxidation by 100% during 7 days in comparison to packed meat with pectin and gelatin or polypropylene film	Bermúdez-Oria et al. (2019)
Sodium alginate and cellulose nanocrystals	Glycerol	–	Covered with the film	Fresh chicken	Decreased the UV effect and oxygen permeability (25%), also decreased the lipid peroxide value and TBARS (Thiobarbituric acid-reactive substances) after 3 days of storage; besides no oxidative changes were indicated by chicken color after 8 days of storage	Criado et al. (2020)
Sodium alginate	Calcium chloride	Cyclolipopeptides produced by *Bacillus subtilis*	Dipping	Blueberries	Antifungal activity and maintained the freshness properties of vulnerable berries	Xu et al. (2020)
Arabic gum	–	Garlic extract	Dipping	Gola guava fruit	Coating reduced the weight loss, skin browning, and postharvest diseases. Maintained the titratable acidity, ascorbic acid content, and lower total sugars, and reduced the increase in total soluble solids. The antioxidant activity and antioxidant capacity of the fruits were not affected and extended their shelf life compared with	Anjum et al. (2020)

(continued)

Table 3.1 (continued)

Polymer matrix	Additives	Bioactive compound	Method of application	Perishable food	Significant function in coated food	References
					control after for 15 days at 25 °C	
Chitosan	Lactic acid, glycerol, and tween 80	Bacteriocins of *Pediococcus pentosaceus* 147	Dipping	Fresh cheese	Delayed the formation of organic acids and reduced the total color change with respect to the uncoated control, improving the physicochemical properties. Besides, reduced 6.21 log CFU/g of *Listeria monocytogenes*	Jutinico-Shubach et al. (2020)
Chitosan	Acetic acid, glycerol, and tween 20	Orange (*Citrus sinensis* (L.) Osbeck) peel essential oil	Shrimps were collocated between films, and then were put into separate vacuum packages	Pink shrimp	The coating inhibited the lipid oxidation and microbial growth, extending the shelf life of shrimps for nearly 8 days more when compared to the noncoated group	Alparslan and Baygar (2017)
Chitosan and canola oil	Glycerol	*Byrsonima crassifolia* extract (L.) Kunth	Spraying	Bell pepper	The coating reduced the weight loss (30%) and the change in color (15%). After 21 days of storage, the phenolic content, carotenoids, and reducing capacity were increased by 18% and the microbiological activity was reduced by 85%	González-Saucedo et al. (2019)

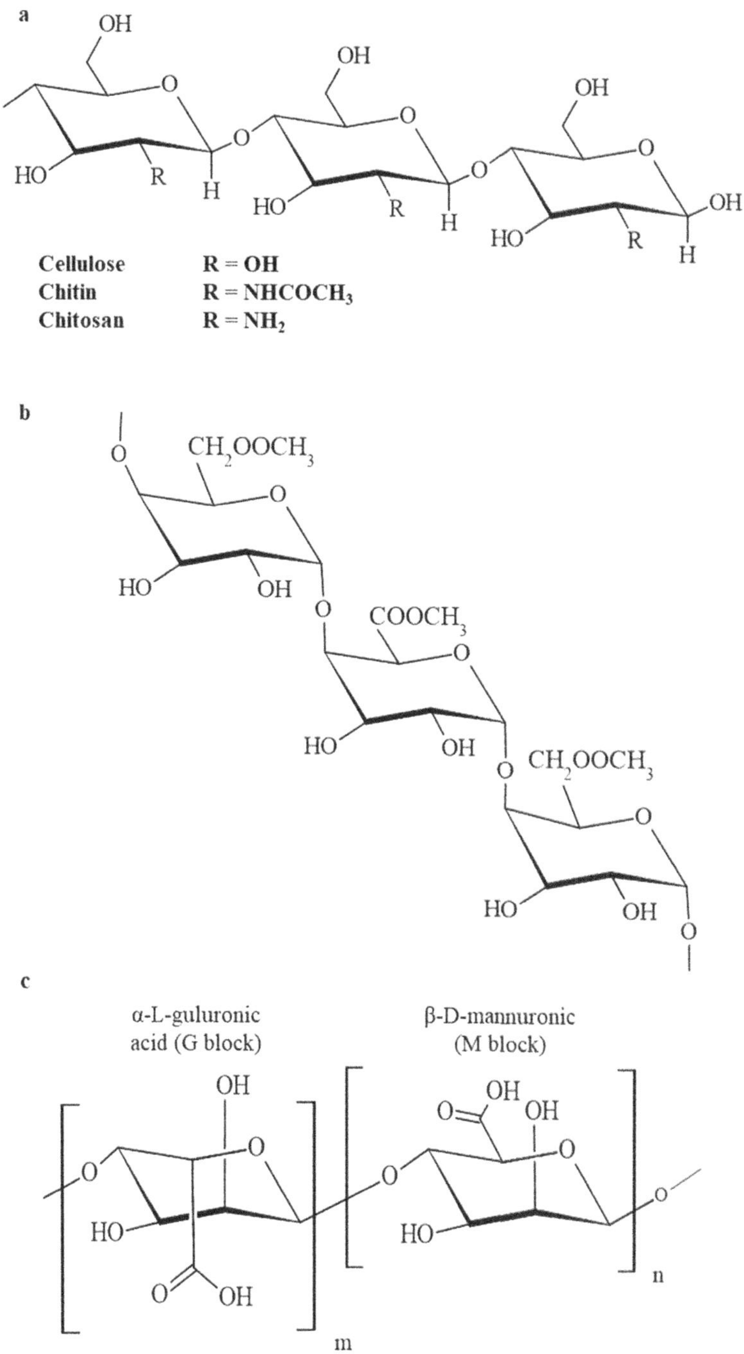

Fig. 3.2 (**a**) Cellulose, chitin, and chitosan (Avramescu et al. 2020), (**b**) pectin, and (**c**) alginate

cellulose (Shahbazi 2018; Francisco et al. 2020). Hydroxypropyl cellulose and methylcellulose films and coatings are beneficial barriers to oxygen, carbon dioxide, and lipids; but with poor resistance to water vapor transfer. Nonetheless, the water vapor barrier properties and water solubility can be increased by the addition of hydrophobic compounds such as lipids into the film- or coating-forming solution or mixing it with other polysaccharides such as starch (Villalobos et al. 2005; Francisco et al. 2020).

3.3.1.2 Starch

Starch is a biodegradable, renewable, and natural origin biopolymer obtained from many plants and represents the biggest carbon reserve (Martin and Smith 1995; Cazón et al. 2017). It is the second-most abundant material in nature. Many plant roots, stalks, crop seeds, and staple crops such as rice, corn, wheat, tapioca, and potato contain starch (Buléon et al. 1998; Corre et al. 2010). Starch is composed of two D-glucose homopolymers, amylose, and amylopectin; however, starch granules may have residues of proteins and lipids (Buléon et al. 1998; Bonilla et al. 2013). Amylose is a linear molecule with glucosyl monomers joined by α-1,4-glycosidic bonds. Amylopectin is the more abundant and highly branched component of the starch. It is composed by chains of α-D-glucopyranosyl residues linked mainly by α-1,4-glycosidic bonds and around 5–6% are joined by α-1,6-glycosidic bonds, which induce the formation of chained branches (Fig. 3.3) (Buléon et al. 1998; James et al. 2003). The amylose molecule gives the film-forming properties of the starch (Bonilla et al. 2013). The different film-forming properties of the starch molecules are determined by the ratio, structure, and molecular weight of amylose and amylopectin composition (Cazón et al. 2017).

The "native starch," used as extracted from the plants, has limited industrial applications due to its insolubility in cold water, hygroscopicity, undesirable texture, retrogradation, etc. Therefore, the starch is chemical, enzymatical, and physically modified to have more adequate properties to obtain "modified starch" (Tharanathan 2005; Corre et al. 2010). An excess of water (>90%, w/w) is required to prepare homogeneous film-forming solutions. The breakage of the amylopectin matrix is necessary to release the amylose by water diffusion through the granules, and to promote the melting of the starch crystallites (Jackson and Ratnayake 2006). The main process to obtain starch films are dry and wet processes; however, starch extrusion is possible due to its thermoplastic properties. In the extrusion, the starch is heated above its glass transition temperature in the presence of low water content. Otherwise, in the wet process, the polymer is solubilized, then the film-forming solution is dried. Usually, the wet process is chosen to form edible preformed films, or to apply coatings by dipping, brushing, or spraying onto food products (Peressini et al. 2003). However, dry methods are more easily implemented at the industrial level (Jiménez et al. 2012). The use of food packaging based on starch is extensive in the food industry because they have no smell or taste, is transparent, and has an oil-free appearance (Chi et al. 2020; Pinzon et al. 2020).

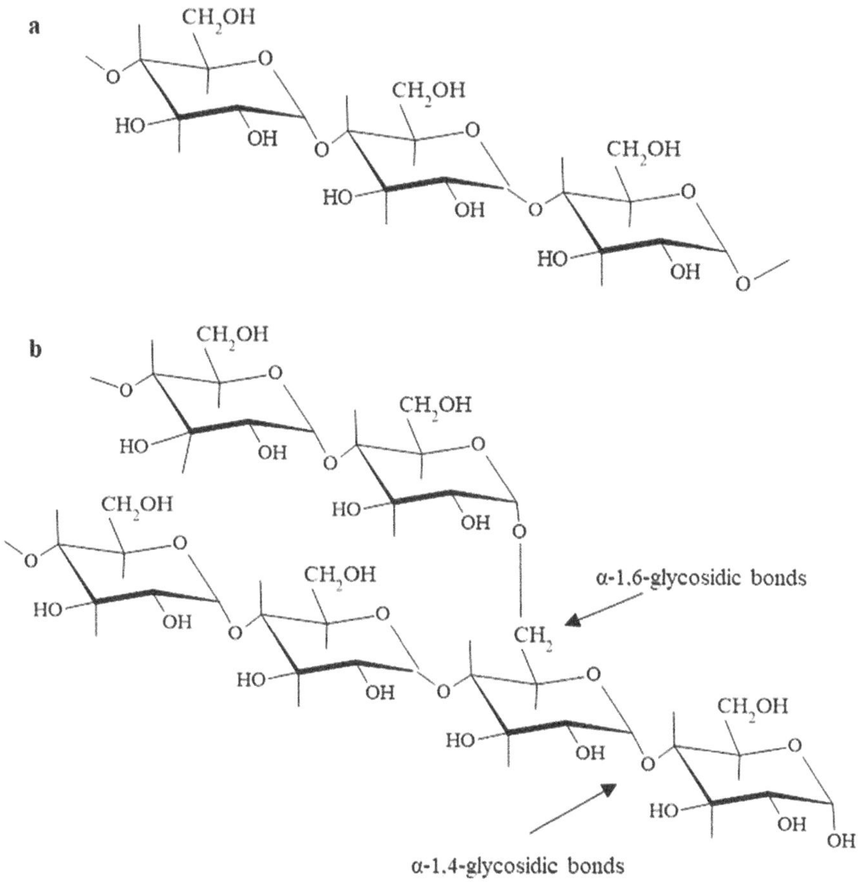

Fig. 3.3 Starch structure. (**a**) Amylose and (**b**) Amylopectin

3.3.1.3 Pectin

Pectin is the most abundant polysaccharide in the cell wall and middle lamella of plants; it represents ~30% of the dry basis (Mohnen 2008; Harholt et al. 2010). However, only a few plants are used to obtain commercial pectin in the food industry; citrus and apple peels are the main materials used for its extraction (Cazón et al. 2017). Pectin is composed of homogalacturonan (HGA), rhamnogalacturonan-I, and rhamnogalacturonan-II. HGA is the major component of the pectic, which are linear chains of (1 → 4)-α-D-galactopyranosyluronic acid units (Fig. 3.2b) (Mohnen 2008). The carboxyl groups of the galacturonic acid units are esterified with methanol and, occasionally, partially acetyl-esterified. According to esterification degree, pectin is classified as high methoxyl (HM, >50% esterified carboxyl groups) or low methoxyl (LM, <50% esterified carboxyl groups) pectin (De Cindio et al. 2015).

Some parameters that should be considered in the pectin gelling process are temperature, pH, and solutes. HM pectin forms gels at pH < 3.5 with more than 55% (w/v) cosolutes as sucrose due to the formation of entanglements, hydrophobic interactions, and hydrogen bonds. The formation of water-insoluble gels from LM pectin occurs in the presence of positive di- or trivalent cations over a wide range of pH values with cosolutes or not (Löfgren et al. 2005). The interaction of carboxyl groups of LM pectin with Ca^{2+} develops a structure similar to the egg box described for alginate (Sikorski et al. 2007). The gelling and mechanical properties, stability, and response to chemical and physical conditions of pectin depending on the source and method applied for its extraction (Munarin et al. 2012).

3.3.1.4 Alginate

Alginate is extracted from the marine brown algae (Phaeophyceae, mainly *Laminaria*) (Rhim 2004; Parreidt et al. 2018), and it is produced by two genera of bacteria, *Pseudomonas* and *Azotobacter* (Hay et al. 2013). Alginate is the alginic acid salt, which is a linear copolymer of (1 → 4) β-D-mannuronic (M) and α-L-guluronic (G) acid. These acid residues are organized in sections of M or G residues, referred to as M or G blocks (Fig. 3.2c). The proportion and distribution pattern of M: G residues vary according to the source of obtention (Fang et al. 2007). Alginate packaging has low moisture barrier properties (Cazón et al. 2017); however, is extensively used in food packaging because of its facility to join at di- or trivalent cations and form gels (Kohn 1975). Divalent ions can establish interactions with M and G blocks, resulting in a three-dimensional stable structure called "egg-box" model is resulted (Grant et al. 1973). The association with Ca^{2+} forms calcium alginate due to electrostatic ionic interaction between negatively charged carboxylate group of sodium alginate and positively charged calcium ions present in the cross-linking solution (Khotimchenko et al. 2001). The cross-linking of sodium alginate with polyvalent cations improves the water barrier properties, mechanical resistance, cohesiveness, and stiffness of alginate films and coatings (Cazón et al. 2017). Alginate-based food packagings are barriers to gases; therefore, reduces the rate of enzymatic browning (Kester and Fennema 1986; Pilon et al. 2016). Moreover, a 1-2% the alginate can encapsulate volatile compounds such as essential oils, as increases its concentration increases the viscosity difficulting the volatilization of the molecules during the film storage (Aloui et al. 2014; Liakos et al. 2014).

3.3.1.5 Arabic Gum

Arabic or acacia gum (AG) is extracted from the exudate of the stems and branches of the *A. senegal* and *A. seyal* trees. It is a highly branched, neutral, or slightly acidic blended polysaccharide composed of arabinogalactan oligosaccharides, polysaccharides, and glycoproteins. The chemical structure of AG is complex; the polymer consists of 1,3-linked β-D-galactopyranosyl units. The side chains are composed of two to five 1,3-linked β-D-galactopyranosyl units; these units are joined to the main chain by 1,6-linkages. The main and the side chains contain units of α-L-arabinofuranosyl, α-L-rhamnopyranosyl, β-D-glucopyranosyl, and 4-*O*-methyl-β- D-glucuronopyranosyl; the latter two mostly as end units (Fig. 3.4)

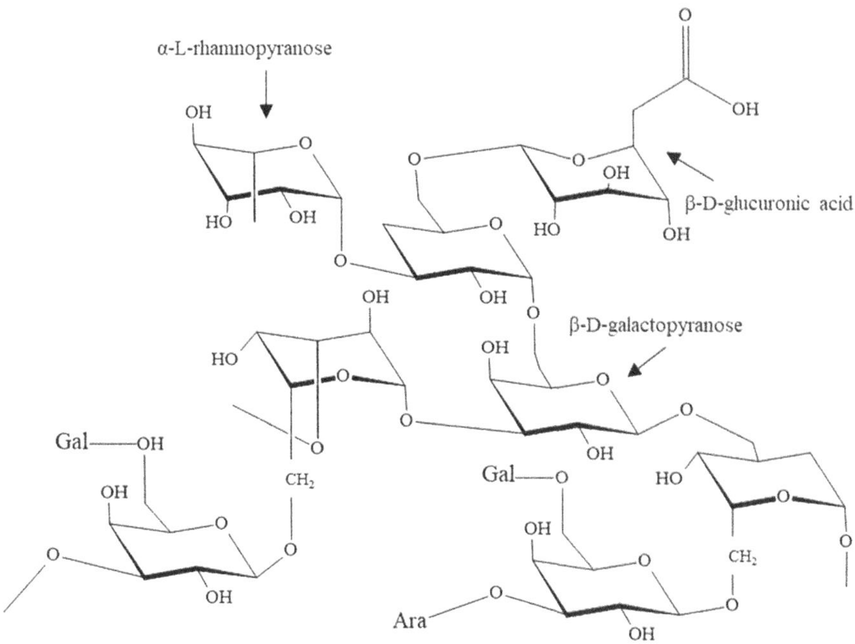

Fig. 3.4 Chemical structure of arabic gum

(Karlton-Senaye and Ibrahim 2013; de la Cruz et al. 2020). AG is covalently associated with glycoproteins rich in amino acid residues of hydroxyproline, proline, and serine. AG is soluble in water (up to 50%, w/v) and has low viscosity. The highly branched structure and low molecular weight confer it with these particular properties. Besides, the proteins provide surface activity, foaming, and emulsifying properties to this polysaccharide (Izydorczyk et al. 2005; de la Cruz et al. 2020). In solution, AG has few sites for intermolecular H bonding because of its structured, highly branch, resulting in low thickening property compared to linear gums and polysaccharides. However, these properties vary in function of the AG source (Huber and Embuscado 2009).

3.3.2 Animal Polysaccharide

3.3.2.1 Chitosan

Chitosan is the *N*-deacetylated derivate of chitin obtained in the presence of concentrated alkali. It consists of 2-acetamido-2-deoxy-β-D-glucose linked by β-bonds (1 → 4) (Fig. 3.2a). Chitin, poly-β-(1–4)-N-acetyl-D-glucosamine, is a natural and abundant mucopolysaccharide; a structural component of crustaceans and insects exoskeleton, and fungal mycelia (Kumar 2000). Chitin and chitosan are very important due to their high nitrogen content (6.89%) as increases in the nitrogen content the film hardness reduces (Lin et al. 2004). Moreover, these molecules have high biocompatibility, biodegradability, nontoxicity, adsorption, chelator, antimicrobial, and antifungal properties. For these reasons, chitosan has been widely used in the biomedical, pharmaceutical, and food industries (Kumar 2000; Hassan et al. 2018). However, chitosan properties strongly depend on the source of obtention, *N*-acetylation degree (highly related to the source and the process used for its obtention), molecular weight, and the conditions of the biological matrix where this molecule is applied. Through the influence of pH and ionic strength, the solutes can react with chitosan by means of electrostatic interaction and/or covalent binding to diminish the reactivity of the amine groups (No et al. 2007; Kaur and Dhillon 2013). Chitosan is similar to cellulose in its high insolubility degree and low chemical reactivity; besides, acid solutions (pH < 6.3) can dissolve the chitosan (Kaur and Dhillon 2013). The good barrier properties to control gases exchange (O_2 and CO_2) and the antagonistic activity of chitosan against several fungi, yeasts, and bacteria make this polysaccharide very important in extending the shelf life of fresh and minimally processed foods (Orgaz et al. 2011; Cazón et al. 2017; Obianom et al. 2019).

3.4 Protein Films and Coatings

Proteins have two forms, fibrous or globular structure. Fibrous proteins are insoluble in water; their main purpose is structural in animal tissues. Otherwise, globular proteins are soluble in water, as well as in acids, bases, or salts aqueous solutions (Huber and Embuscado 2009). Proteins have great filmogenic and barrier properties to O_2 and CO_2 without stopping water diffusion, and good adhesiveness to various materials, particularly hydrophilic surfaces. A chemical potential is created as a function of the distribution and amount of charged, polar, and nonpolar amino acids along the protein chain. The properties of proteins can be adjusted for the specific requirements of various applications through modifications in the different chemical groups linked to the central carbon in every amino acid. These modifications can be performed via chemical, enzymatic, and/or mechanical (Kolster et al. 1998; Sánchez-Ortega et al. 2014).

The principal mechanism to make protein-based films and coatings is the protein denaturation by changing the pH, solvents addition or heating, and followed by the formation of novel intermolecular interactions among the peptides (Fig. 3.5) (Cordeiro de Azeredo 2012). The interactive forces between the polar and nonpolar domains produce an adhesive film. Fibrous proteins associate among them through H-bonds to form films and coatings, while globular proteins do through ionic, covalent, and H-bonds, which allows folding into complex round structures (Jones and Thornton 1996). Also, hydrophobic interactions, van der Waals forces, and disulfide (SS) bonds may occur within the proteins (Wei et al. 2014; Dehghani et al. 2018). The occurrence of these interactions depends on the kind and arrangement of amino acids and on the protein structure. Moreover, extrinsic factors such as process temperature, drying temperature, salt type, pH, ionic strength, and relative humidity

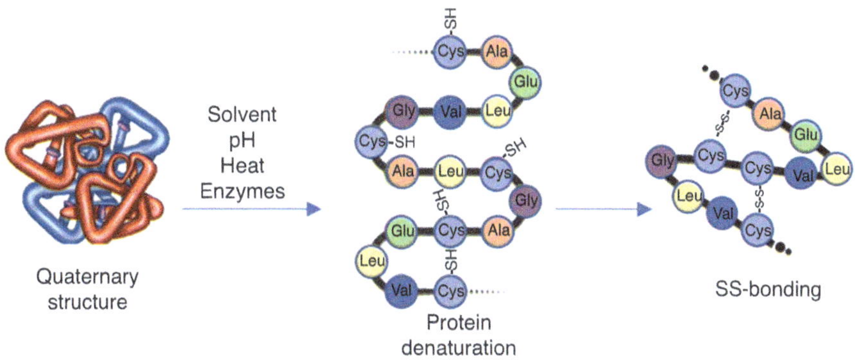

Fig. 3.5 Protein denaturation and new intermolecular associations

during processing and storage affect the interactions within films and coatings. The interactions between chains determine the film strength; higher interactions yield stronger films but these films have lower permeability to vapors, liquids, and gases (Han 2014). Films made with proteins are more stable than films made with polysaccharides or lipids and have a longer duration (Barone and Schmidt 2006; Wei et al. 2014). Generally, protein as polysaccharide food packaging has a poor water vapor barrier, ranging from two to four times more loss of moisture than synthetic polymeric packaging materials such as polyethylene, polypropylene, polyester, and polyvinylidene chloride (Dehghani et al. 2018). Hence, the conventional packaging materials are not suitable for vegetables and fruits that require gas interchange during the ripening process. In order to perform the replacement of films and coatings from petroleum sources, proteins from animal (casein, whey, collagen, gelatin, keratin, and egg albumen) and vegetal (soy, corn zein, and wheat) sources are widely used (Table 3.2). Furthermore, it is believed that ECs developed from proteins provide additional dietary value to the coated food (Huber and Embuscado 2009), particularly due to the presence of essential amino acids (Poppe 1992). However, food packaging based on proteins could be degraded by proteases from meat, fish, and chicken products; besides, some proteins may cause allergenic reactions to the sensitive population (Gennadios et al. 1997).

3.4.1 Plant Proteins

3.4.1.1 Wheat Gluten

Wheat gluten proteins are an inexpensive, cohesive, and viscoelastic proteinaceous material with filmogenic properties. This is a by-product obtained from the starch extraction from wheat flour (Day et al. 2006; Lagrain et al. 2010). Gluten is the storage protein of wheat (Shewry and Halford 2002). Gliadins and glutenins are the water-insoluble fractions of wheat proteins. Gliadin is the principal fraction responsible for the viscosity of wheat gluten. Otherwise, glutenin is the fibrous fraction in

Table 3.2 Proteins used to develop food packaging for perishable foods

Polymer matrix	Additives	Bioactive compound	Method of application	Perishable food	Significant function in the coated food	References
Bovine casein	–	Eugenol	Nanoparticles applied by spraying	Pear fruits	Suppress anthracnose disease in pears after 8 days of storage at 25 °C	Xue et al. (2019)
Whey protein concentrate	Glycerol	*Salvia officinalis* L. extract	Dipping	Pistachio kernels	Decreased *Aspergillus flavus* growth and aflatoxins production in pistachio kernels after 9 days of storage at 20 °C	Tavakolipour et al. (2020)
Whey protein isolate and inulin	Tween 80	Cumin seed essential oil	Filter paper loaded with the microemulsion	Raw beef hamburger	Controlled release and increased stability of the essential oil. The bioactive films decreased TBARS, total mesophilic and psychrophilic bacteria, and keeping water-holding capacity. Moreover, the sensorial attributes of hamburgers were not affected	Hemmatkhah et al. (2020)
Gelatin from porcine skin and zein corn	Glycerol	Propolis extract	Nanoparticles applied by dipping	Raspberries	The encapsulation controlled the release of the extract and extended its efficiency. Inhibited the growth of *Penicillium digitatum* and *Botrytis cinerea* in the fruit after the storage at 5 °C for 11 days	Moreno et al. (2020)
Gelatin from fish	–	Tea polyphenol	Dipping	Golden Pomfret fillets	Reduced significantly the aerobic mesophilic and psychrotrophic count, yeasts, and molds, weight loss, and maintained the pH and the nanostructure of fish fillet, and reducing the oxidative spoilage during cold storage during 17 days	Feng et al. (2017)
Gelatin and chitosan	Acetic acid and ethanol	Eugenol	Dipping	Chilled pork	The coating inhibited the growth of spoilage bacteria in meat and delayed the pH increment. Eugenol shows antioxidant activity, preventing the cross-linking denaturation of myosin in meat and inhibiting its water-holding capacity. Also, the coating	Wang et al. (2020)

					maintains the color properties and inhibited the increase in total volatile basic nitrogen and TBARS values due to its antioxidant activity, contributing to extend the shelf life of coated pork pieces 7 days more than uncoated meat	
Egg albumen and cellulose nanocrystals	Glycerol and egg yolk	Curcumin powder	Dipping	Banana, avocado, and papaya	Decreased microbial growth on the fruit surface while also decreasing O_2 and increasing CO_2 in the microenvironment, which helped to maintain the fruit's freshness. Also reduced the water and gas permeability; hence, maintaining the firmness and reducing the weight loss	Jung et al. (2020)
Wheat gluten	Glycerol and ethanol	–	Dipping	Fresh-cut pineapples	The coated slices have firmer texture, less juice leakage, and lower counts of psychrotrophic, mesophiles, molds, and yeasts than uncoated slices. Besides, *salmonella*, *Escherichia coli*, and total coliform were not detected after 12 days of storage at 5 °C	Pilon et al. (2016)
Soy protein isolated	Glycerol	Thyme or oregano EOs	Dipping	Beef fillets	Decreased the population of *Escherichia coli*, *Listeria monocytogenes*, and *Staphylococcus aureus* in beef fillets from 6 until 3.11 Log_{10} UFC/mL during 14 days at 4 °C. Also improved the color of the beef with acceptable sensory properties	Yemiş and Candoğan (2017)
Soy protein isolated	–	Lemon extract	Dipping	Fresh-cut melon	Reduced the total plate count and yeast and molds amounts. Effective to preserve vitamin C and color, and retarded the respiration. Limited the water loss, thereby protecting the cut tissue of melon samples from wilting, and	Yousuf et al. (2020)

(continued)

Table 3.2 (continued)

Polymer matrix	Additives	Bioactive compound	Method of application	Perishable food	Significant function in the coated food	References
					sensory attributes were preserved to a remarkable extent after 12 d of storage at 4 °C	
Corn zein	Ethanol and titanium oxide	–	Electrospinning	Cherry tomatoes	The coatings exhibited photocatalytic activity against ethylene during the storage (22 days), contributing to delay in the ripening process of this fruit	Böhmer-Maas et al. (2020)

TBARS Thiobarbituric acid-reactive substances

wheat gluten that provides elasticity (Shewry and Halford 2002; Lagrain et al. 2010). For this, wheat gluten films have higher elasticity in comparison to films made from other proteins, also they offer good barrier properties to O_2, comparatively high hydrophobic surface, and adequate thermal stability (Milani and Tirgarian 2020). However, gluten purity affects the appearance and the mechanical properties of the films and coatings, and wheat gluten with a high-purity degree forms more stable and clearer films (Gennadios et al. 1993). Moreover, the stability of these films is affected by the method used for its obtention; flash-dried films are less stable than spray-dried ones (Heralp et al. 1995). The most important disadvantage related to the use of films made with wheat gluten is their water sensitivity and their low mechanical strength. These properties are highly affected in wet conditions due to water sorption and subsequent plasticization (Lens et al. 2003; Milani and Tirgarian 2020).

3.4.1.2 Soy Protein

Soy protein is the principal legume protein; it is constituted by 37% β-conglycinin (7S globulin) and 31% glycinin (11S globulin) of globular proteins. Conglycinin (140–170 kDa) contains various combinations of three subunits highly glycosylated. Otherwise, glycinin (340–375 kDa) consists of six subunits linked via SS-bonds, and is responsible for the gelling, emulsifying, and foaming properties of soy protein (Kunte et al. 1997; Kumar et al. 2002). Soy can be denaturized by heat and alkalis, affecting film formation. Similar to β-lactoglobulin in whey protein, glycinin forms intermolecular SS-bonds when denatured by heat above 80 °C (Fig. 3.5); this affects the tensile properties of films. Soy protein association and stability are ionic strength and pH-dependent (Petruccelli and Añón 1994; Renkema and Vliet 2002). Soy protein is found in three different forms, namely according to the soy protein content, soy flour (54% protein), soy protein concentrate (SPC, 65–72% protein), and soy protein isolate (SPI, ≥90% protein) (Li et al. 2008). The high content of protein in SPI provides a higher film-forming ability in comparison to soy flour and SPC (Netravali et al. 2007). Films made with SPI are clearer, smoother, and more flexible than films made with other proteins. Moreover, they have better gas barrier properties than lipids- or polysaccharides-based films (Huber and Embuscado 2009; Yousuf et al. 2020). SPI contains a large number of polar amino acids which provides hydrophilicity to food packaging, increasing the fragility in wet states, and resulting in low water vapor barrier and mechanical properties (González et al. 2011). Different alternatives such as pH modification, fractionation by molecular weight, protein denaturation by temperature in alkali solutions, and blending with other compounds have been used to meliorate the handicaps of SPI coatings and films (Song et al. 2011).

3.4.1.3 Corn Zein

The global production of corn reached 1147.62 million tons during 2018 (FAOSTAT 2020). The protein content in corn ranges from 6 to 12% (dry basis) according to the variety. Zein is a protein that belongs to prolamines; its content determines the hardness of corn endosperm. Zein has poor nutritional quality due to its deficiency in essential amino acids. However, the greater content of nonpolar amino acids provides a hydrophobic character (Shukla and Cheryan 2001). Hence, it is water-insoluble but can be easily dissolved in organic solvents, such as ethanol (70–80%) or acetone. The solvent used, the temperature of solubilization and the roughness of the coated surface affect strongly the morphology of the zein-based films (Tillekeratne and Easteal 2000; Khalil et al. 2015; Bharathi et al. 2020). The stabilization of zein films and coatings occurs by SS- and H-bonding (Gennadios et al. 1994).

3.4.2 Animal Proteins

3.4.2.1 Casein

Caseins constitute ~80% of milk proteins in bovine milk. This is an abundant, low cost, and highly available food-grade additive (Dalgleish 2011). Alpha-, β-, and κ-casein form colloidal micelles in milk stabilized by calcium-phosphate bridging, with a diameter between 10 and 400 nm (Kinsella and Morr 1984; Semo et al. 2007). Caseins aqueous solutions produce films due to diverse interactions such as H-bonds, and electrostatic and hydrophobic interactions, which increase the inter-chain cohesion. The amphiphilic nature of the caseins makes them worthy for encapsulating low-water-soluble compounds in the hydrophobic core of the micelle (Horne 2006). Usually, casein-based coatings with bioactive compounds consist of nanoparticles applied by casting or spraying methods (Xue et al. 2019).

Caseins are stables at 100 °C and 100 MPa (Kinsella and Morr 1984); however, the heat promotes the formation of the intermolecular SS-bonds (Fig. 3.5). This property is used to improve the properties of casein-based coatings and films. Moreover, gamma-irradiation produces bityrosine bridges between protein chains, this is favored by the addition of calcium chloride, enhancing the mechanical, structural, and barrier properties of casein-based coatings and films (Amirou and Pizzi 2020).

3.4.2.2 Whey Protein

Whey is a by-product of cheese processing composed by proteins issue from the milk serum after caseins coagulation at 20 °C and pH 4.6 (Morr and Ha 1993). The whey protein is constituted by β-lactoglobulin, α-lactalbumin, immunoglobulins, and bovine serum albumin (Jovanović et al. 2005). Depending on the protein content, the powder is called either whey protein concentrate (25–80%) or whey protein isolate if contains more than 90% of protein on a dry weight basis (Morr and Ha 1993). Films made from whey protein are flexible, transparent, soft, with high accurate mechanical hindrance, and great gas barrier properties at low relative humidity (Pérez-Gago et al. 1999). The interactions between the native whey proteins in films or coatings are van der Waals and H-bonding forces. However, films developed from thermally denatured whey protein are more cohesive, stronger, and have better barrier properties than those made from native protein because β-lactoglobulin is thermolabile. After thermal denaturation, the interactions mentioned occurring in the new exposed groups. Moreover, the thiol group in cysteine

121 is available for intermolecular SS-bond formation (Fig. 3.5) (Pérez-Gago et al. 1999; Jooyandeh 2011). An advantage of the use of whey protein is the obtention of bioactive peptides. For example, its hydrolysis with trypsin produces peptides with antibacterial activity against *Listeria monocytogenes* and *Staphylococcus aureus* (MIC 10–20 mg/mL) (Demers-Mathieu et al. 2013).

3.4.2.3 Collagen and Gelatin

Collagen is a complex, fibrous, and insoluble in water protein. It is the major structural component of white connective tissue fibers in animals. Collagen constitutes about 30% of the total protein in vertebrates and invertebrates (Poppe 1992). The incomplete hydrolysis of collagen issue from bones, skin, and tendons of pigs, cows, and lambs produces gelatin, a high molecular polypeptide. Nowadays, gelatin is also obtained from fish skins and bones in order to protect the foods with by-products or waste products from the same food that will be coated (Table 3.2) (Feng et al. 2017). Films and coatings of collagen or gelatin based exhibit good transparency (slightly opaque), adherence to the food products, mechanical, and barrier properties against O_2 and CO_2; besides, they can be manufactured by extrusion or dipping processes. The flexibility of gelatin films increases as a function of the hydration degree; which can be an issue due to this polymer is capable of absorbing water in 5–10 times its volume, producing film instability (Jiang et al. 2010; Hassan et al. 2018; Moreno et al. 2020). Furthermore, films and coatings of gelatin melt at high relative humidity conditions or temperatures ranging from 27 to 34 °C; hence, these coatings are not appropriated for foods that will be stored under similar conditions (Poppe 1992). Otherwise, these coatings are an excellent carrier of active compounds (Scartazzini et al. 2019) or as encapsulating matrices to organic compounds or probiotic cells (Guevara-Avendaño et al. 2018; Zhao et al. 2020b).

3.4.2.4 Egg Protein

Egg albumen is a globular protein and the second abundant component of egg white (water is the first component) and represents approximately 10.5% of the egg white total weight. Egg albumen comprises five proteins: ovomucin, ovalbumin, ovotransferrin, lysozyme, and ovomucoid. Ovalbumin has 44.5 kDa of molecular weight, which is more than 50% of egg white proteins. This protein has four free sulfhydryl (SH) groups available for cross-linking; furthermore, it contains many SS-bonds (Guérin-Dubiard and Audic 2007; Huber and Embuscado 2009). The formation of these bonds improves the mechanical properties and the stability of the egg albumen packaging. Moreover, lysozyme has been widely studied due to its antimicrobial activity, particularly against Gram-positive bacteria (Saito et al. 2019). Egg albumen proteins are sensible to heat and develop strong heat-set gels. Hence, methods to prepare coatings and films with egg white include the denaturation of proteins by increasing the pH from 10.5 to 11.5 at 40 °C for 30 min (Gennadios et al. 1996). The increment of SH- groups permits the formation of inter- and intramolecular SS bonding by oxidation and sulfhydryl-disulfide interchange reactions, which makes films more stretchable (Gennadios et al. 1996). Egg white proteins form stable intermolecular β-sheets structures among ovalbumin, ovotransferrin, and lysozyme during heat denaturation. Ovalbumin is a polypeptide with mostly random coil conferring with good film feasibility (Guérin-Dubiard and Audic 2007; Huber and Embuscado 2009). Egg albumen proteins denaturation is affected by salt concentration, pH, sucrose, and temperature pretreatment. Changes in the conditions of these parameters have been used to improve the mechanical and water resistance properties of egg albumen edible food packaging (Sothornvit 2005). Increases in the drying process temperature induce protein denaturation; consequently, the networks constrain and the film permeability is reduced (Sothornvit 2005). Casting and/or extrusion are methods preferred to obtain films or coatings; besides, albumen has great emulsifying and gelling properties, making it an ideal material for microencapsulation through the coacervation process (Guérin-Dubiard and Audic 2007; Yan and Zhang 2014; de la Cruz et al. 2020). Furthermore, the incorporation of egg yolk into egg albumen films can improve its barrier properties due to the lipids content in the egg yolk, this alternative contributes to reduce the egg waste that reaches approximately 2% in the USA (Jung et al. 2020).

3.5 Lipid Films and Coatings

Lipids are apolar compounds that have low water vapor permeability, making them great barriers to moisture migration and useful to control food desiccation (Vargas et al. 2008; Hassan et al. 2018). The behavior of coatings and films prepared with lipids regarding moisture transfer depends on the scattering of the chemical groups, the length and degree of unsaturation of aliphatic chains, and influences lipid polarity (Cordeiro de Azeredo 2012). Lipids used to coat perishable are obtained from animals, insects, and plant sources. These compounds are grouped into fats and oils fractionated, concentrated, and/or reconstituted (milk butter, lard, tallow, peanut, coconut, palm, palm kernels, and cocoa), waxes, resins, and essential oils. Waxes are the most widely used (Huber and Embuscado 2009) because their hydrophobicity produces extra brittle and thicker films and coatings (Perez-Gago et al. 2002). Some lipids have permeability settings near plastic films, such as polyvinyl chloride and low-density polyethylene. Solid lipids are usually less permeable than liquid lipids. Specificity of physicochemical properties for hydrophobic substances makes edible films based on lipids with variable water barrier properties. The lipid constituents' polarity in the coating formulation depends on the electrostatic potential, chemical groups, aliphatic chain length, and unsaturation (Morillon et al. 2002; Huber and Embuscado 2009).

Lipids are not biopolymers, thereupon cannot form cohesive films (Dehghani et al. 2018). Hence, in most cases, lipids are used to form composite coatings by mixing them with hydrocolloids (Tables 3.1 and 3.2), either by incorporation of lipids into the hydrocolloid coating or by film-forming solution (emulsion technique) (Ahmed et al. 2020) or to obtain a bilayer by lipid deposition of layers onto the surface of the hydrocolloid film (Phan The et al. 2008). These techniques form coatings with improved mechanical and water vapor barriers because of the low polarity obtained (Huber and Embuscado 2009). The hydrocolloid incorporation produces an increment in the moisture permeability compared to the pure lipid (Bravin et al. 2004). Mistakes in the formulations of films and coatings containing lipids can cause damage to the visual aspect of the coated food products (Perez-Gago et al. 2002; Valencia-Chamorro et al. 2010).

3.5.1 Oil and Fat to Films and Coatings

Fats and oils are mixtures where the major components are triglycerides; these molecules are obtained from animals and plants (Lin and Zhao 2007). Oils and fats physicochemical characteristics are influenced by the fatty acids substituted on the triacylglycerol, being fats solids and oils liquids. A significant attribute common in oils and fats from plants is the high proportion of unsaturated fatty acids in the triacylglycerols. A great level of unsaturation of fatty acids in oils of vegetable origin makes them more susceptible to oxidative deterioration (Bradley and Min 1992; Karupaiah and Sundram 2007). It is important to take account of the composition of the fatty acids in fats and oils to adequately characterize them in terms of stability and physicochemical properties (Bockisch 1998; Kostik et al. 2013). For example, oleic acid has been widely used as a hydrophobic and antioxidant compound; when added into cellulose nanofiber coatings, it enhanced the hydrophobicity, stability, and wettability of coatings onto the banana surface (Deng et al. 2017). Moreover, vegetable oils can transport fat-soluble compounds such as vitamins A, D, E, and K, as well as fatty acids such as linolenic and linoleic acids (Fasina et al. 2006).

3.5.2 Waxes from Plant and Animal Sources

Waxes are composed of long-chain lipids of high molecular weight. Due to their hydrophobic character, waxes are highly efficient in moisture permeability reduction, help to decrease the water vapor permeability of films such as chitosan, and also contribute to retard the shrinkage and food spoilage (Gutiérrez-Pacheco et al. 2020). Waxes include many types of components being hydrocarbons, fatty alcohols, fatty acids, and long-chain esters the most commonly used to coated perishable foods (Fig. 3.6) (Avramescu et al. 2020).

Waxes reduce the surface abrasion during fruit handling and control soft scald formation (peel browning) in fruits through mechanical integrity and internal gas composition control (Lin and Zhao 2007). Waxes are insoluble in water but soluble in organic solvents because of their high hydrophobicity (Toxicology 1984; Lin and Zhao 2007; Huber and Embuscado 2009). The most popular method to elaborate wax microemulsions is to add water to the liquefied wax in fatty acid, and then, a base is added to invert the emulsion to wax in water (water-to-wax method) (Hagenmaier and Baker 1994). Fruits and vegetables acquire gloss by these formulations, but some disadvantages are poor mechanical properties, oily appearance in some products, and gloss loss during storage of coated food products (Suput et al. 2015; Spotti et al. 2016).

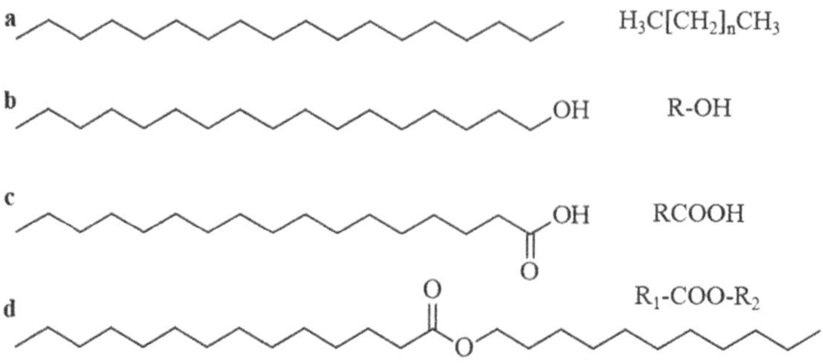

Fig. 3.6 Main components of waxes. (**a**) *n*-Alkanes ($n = 22–36$ C atoms), (**b**) fatty alcohols (R length is 12–34 C atoms), (**c**) fatty acids (R length is 12–34 C atoms), and (**d**) long-chain esters (R_1 and R_2 = 10–20 C atoms in length) (Avramescu et al. 2020)

The waxes commonly used are carnauba, bee, and candelilla (Huber and Embuscado 2009; Saucedo-Pompa et al. 2009). Fresh fruit and vegetables coated with waxes are sweet potatoes, apples, avocados, passion fruit, bell peppers, peaches, cantaloupes, pineapples, cucumbers, rutabagas, eggplants, yucca, grapefruits, turnips, lemons, tomatoes, limes, melons, squash, oranges, pumpkins, and parsnips (Alleyne and Hagenmaier 2000; Saucedo-Pompa et al. 2009; Vakkalanka et al. 2012; De León-Zapata et al. 2015; Aguirre-Joya et al. 2017; Ahmed et al. 2020). In this sense, coatings of carnauba wax added with oregano essential oil, applied by brushing onto cucumbers, decreased the weight loss and the microbial loads (yeasts, molds, and mesophilic bacteria) after 15 days of storage at 10 °C in comparison to uncoated fruit (Gutiérrez-Pacheco et al. 2020). Besides, bee and carnauba waxes are commonly used to preserve meat cuts, but few investigations have studied their effects on this food product (Bhagath and Manjula 2019). Even when it has been demonstrated that beeswax prevents hardening, mold development, and made the peel out easy, it also improves the oxidative stability and sensorial characteristics of long-ripened Italian salami (Trevisani et al. 2017).

3.5.3 Shellac

Shellac resin is the secretion from the insect *Laccifer lacca* Kerr; it is soluble in alcohol and basic solutions. Shellac is compatible with waxes; its incorporation into wax formulations increased the gloss of the coated food products (Huber and Embuscado 2009). Delicious apples commercialized in the USA are coated with shellac or shellac mixed with carnauba wax (Alleyne and Hagenmaier 2000). However, shellac has problems related to low gas permeability; consequently, ripening in some fruits is delayed and anaerobic conditions can cause undesirable flavor. Moreover, in the apple industry, shellac produces blushing or whitening where water is condensed on the coated fruit surface when apples are taken out from cold storage. However, shellac is recognized as a great coating to improve the appearance of apples (Chander et al. 2018). For this reason, it has been assessed in other products such as Valencia orange; in this fruit, it reduced the weight and firmness loss, without reducing flavor quality. In addition, the coatings dried quickly, forming a not sticky and odorless coating, and gave fruit gloss. Besides, did not show visible cracks after 60 days of storage at 5 °C (Khorram et al. 2017).

3.6 Composite Edible Films and Coatings for Food Packaging

Currently, researchers are focusing on the study of composite coatings and films to reduce the disadvantages of monocomponent packaging and merge the advantages of each component polymer. Composites coatings can be created by successive layers (layer by layer) or by conglomerates (emulsions or dispersions) to form one homogeneous layer. The layer-by-layer composite coatings comprise two or more layers that combine polysaccharide/polysaccharide, polysaccharide/protein, protein/protein, lipid/lipid, lipid/polysaccharides, and others (Abdelhedi et al. 2018). The main disadvantage of layer-by-layer coatings and films is the long process for application in foods: two dipping and two drying stages, limiting the application in the food industry. Furthermore, during the storage time, films and coatings are susceptible to develop cracks and exhibit nonuniform structures (Abdelhedi et al. 2018; Blachechen and Petri 2020).

The mixture of a hydrophilic structural matrix and a hydrophobic compound results in better functionality reflected in the enhancement of moisture and gas barrier properties in comparison with a packaging developed with one hydrocolloid. Otherwise, the mixture of hydrocolloids improves mainly the mechanical properties but shows weak moisture barriers because of the hydrophilic nature of proteins and polysaccharides compared with waxes or shellac (Maciel et al. 2020). For example, composite films of cellulose nanocrystals from pea hull (5%, w/w, CNC) and carboxymethyl cellulose (2%, w/w, CMC) showed an improved UV barrier, increased 50.8% tensile strength, and decreased 53.4% water vapor barrier compared to pure CMC film. The mixture also improved the thermal stability by increasing the melting temperature values of the composite films, which may be attributed to the generation of H-bonds between the CMC matrix and CNC as well as an increase in the crystallinity of composite films due to the crystallization of CNC (Li et al. 2020). Otherwise, the mechanical properties of starch films have been enhanced by the addition of other polysaccharides such as chitosan (Bonilla et al. 2013). In line with this, lignin nanoparticles into wheat gluten-based films increased the tensile strength by 141.8%, young modulus by 206.4%, and glass transition temperature by 18.4%, and decreased the water uptake by 37.7% (Yang et al. 2015). Food packaging made with hydrocolloids exhibits great barriers to O_2 due to their compact matrix and ordered the network structure H bonded (Yang and Paulson 2000).

The functional properties of the emulsified components, which will develop packaging, are closely related to the preparation method, type of components, the relation of hydrocolloid: lipid, the compatibility among components, and the method used to coat the food surface (Fabra et al. 2011). The addition of hydrocolloids to oils can improve the coating properties; for example, the incorporation of methylcellulose into palm oil makes an edible coating capable of maintaining the quality of sapota fruit for more than 7 days at 24 °C (Vishwasrao and Ananthanarayan 2017). Nevertheless, the addition of lipids to improve the barrier properties can affect other features of biopolymer films such as transparency (Jiménez et al. 2010). The physical and/or chemical interaction of the components in composite films and coatings results in an enhanced material with the better properties of each component (Han 2014; Yang et al. 2015; González et al. 2019).

The use of nanocomposite materials developed from biomaterials for edible packaging is a novel alternative to enhance mechanical and barrier properties in addition to what could be reached by macroscopic reinforcement of the components (Azeredo et al. 2009). Nanocomposite coatings are formed with at least two immiscible phases separated by an interface region. The nanometer scale of the material allows the complete dispersion into the major component (Nguyen-Tri et al. 2018). The nanoparticles dispersion provokes the formation of the matrix interfacial area that modifies the nanostructure, and consequently mechanical and thermal properties of the food packaging (Dalmas et al. 2007). In this way, the addition of zinc oxide nanoparticles into chitosan packaging improved the antimicrobial activity as well as antioxidant properties of the biopolymer extending the shelf life of fresh poultry (Gomes Lauriano Souza et al. 2020). Following this line, zinc oxide nanoparticles in sodium alginate coatings showed antioxidant and antimicrobial properties, decreasing the superoxide dismutase and peroxidase activity, and improving the maintenance of sensorial properties of the fruit by up to 20 days (Emamifar and Bavaisi 2020).

3.7 Functionalization of Edible Coatings for Perishable Foodstuff

The use of polysaccharides and proteins as main materials to develop films and coatings presents some disadvantages such as their high permeability to water vapor and low mechanical properties; this limits the use in foods that must be stored in high relative humidity conditions (Aider 2010). Then, some strategies such as modifications in the deacetylation degree, solvent type, pH, the addition of plasticizers, surfactants, cross-linking agents, and other main materials such as proteins, polysaccharides, or lipids as well have been used to improve their properties (Elsabee and Abdou 2013; Kumar et al. 2020). Besides, these food packagings are capable to carry biocontrol agents and bioactive compounds, such as antimicrobials, antibrowning agents, antioxidants, volatile precursors, vitamins, flavoring, and coloring compounds (Aloui et al. 2014; González-Estrada et al. 2017; Murrieta-Martínez et al. 2019). The incorporation of different additives into the films and coatings allows to controlling the release of these compounds from the packaging. The addition of these compounds promotes the functional performance of edible food packaging and enhances food stability, quality (reduces biochemical deterioration, enzymatic browning, and development of off-flavors), and the safety of the coated food (Valdés et al. 2015; Marín et al. 2017). However, active compounds could change the sensorial profile or the functional properties of the coated food. Active ingredients such as essential oils at high concentrations may produce toxicity in vegetal cells, and also the loss of their functionality should be considered as a result of interaction with outer factors or food constituents (Acevedo-Fani et al. 2017).

A cross-linker agent enhances the barrier and mechanical properties and decreases the solubility of edible films and coatings based on hydrocolloids by enhancing the polymer reticulation. The increase in cross-linking can be achieved through enzymatic, physical, or chemical treatments (Wang et al. 2017; Isleroglu et al. 2019; Rai and Poonia 2019). Some common cross-linking agents are di- and trivalent cations their activity varies between the different hydrocolloids; for example, the affinity of sodium alginate to cations increases in the order $Mg^{2+} < Ca^{2+-} < Sr^{2+} < Ba^{2+}$ (Kohn 1975). Other commonly used agents are glutaraldehyde, citric acid, lactic acid, tannin acid, genipin, UV radiation, and transglutaminase (Avramescu et al. 2020). In this sense, the addition of transglutaminase (20 units/g gluten) into wheat gluten films enhanced the tensile strength (~1 MPa), and the surface hydrophobicity increased from 88° to 113.08° (Cui et al. 2017).

Plasticizers are nonvolatile compounds with high boiling point liquids and low molecular weight (300–600 Da) and are formed by linear or cyclic carbon chains (14–40 C atoms) (Sejidov et al. 2005; Vieira et al. 2011). They are required for hydrocolloid-based edible food packaging in a proportion of 10–65% concerning dry basis depending on polymer rigidity (Sothornvit and Krochta 2005). Plasticizers increase flexibility and plasticity because the free volume or molecular mobility increases by intercalating between polymer chains, internal H bonding reduces among polymer chains, and intermolecular spacing increases. The addition of plasticizers affects the water attraction and increases film permeability to oxygen and water vapor as a result of decreased cohesion (Sothornvit 2005; Sánchez-Ortega et al. 2014; Avramescu et al. 2020). Plasticizers are classified into two categories: internal and external. Internal plasticizers react with the polymers, providing more space between the polymer chains; softening effect results from a decrement in the glass transition temperature and, thus, reducing the elastic modulus. Otherwise, external plasticizers do not chemically react with the polymers but interact producing swelling. Common plasticizers for edible coatings are monosaccharides, disaccharides, oligosaccharides, polyols, lipids, and their derivates. Besides, water is also an important plasticizer for food packaging; its content depends on the biopolymer and the selected external plasticizer (Sothornvit and Krochta 2005). For example, water solubility and wettability starch-based coatings and films are enhanced by the addition of hydrophilic plasticizers such as glycerol and sorbitol (Chiumarelli and Hubinger 2014; Parreidt et al. 2018). Besides, as their concentration increases, so does the percentage of elongation, water solubility, and water vapor permeability, while total color difference and puncture strength decreased (Ballesteros-Mártinez et al. 2020).

On the other hand, the principal concern in the design of food packaging is the insertion of substances with antimicrobial activity. Antifungal compounds incorporated into food packaging can control fungal decay, one of the principal causes of losses of perishable foods between their obtention and consumption. Besides protein foods such as meat, chicken, fish, and seafood are the main cause of foodborne diseases because possible contamination can occur during the slaughter process (Yemiş and Candoğan 2017; Sapper and Chiralt 2018; Bharti et al. 2020). Compounds such as organic acids, synthetic preservatives, chelating agents, metals, essential oils, plant extracts, and bacteriocins are aggregated in edible coatings and films to control the microbial loads (Bharti et al. 2020). The antimicrobial substances must be GRAS since they have to be consumed along with the food packaging (Khaneghah et al. 2018). In this sense, chitosan edible coatings added with bacteriocins (5.72 μg/mL) produced by *Pediococcus pentosaceus* 147 reduced 6.66 log CFU/g of the population of *Listeria monocytogenes* in fresh cheese in comparison to the uncoated control (Jutinico-Shubach et al. 2020).

Lipid addition into edible coatings and films can enhance cohesiveness, flexibility, and hydrophobicity, improving moisture barriers performance. This can extend the aroma, freshness, tenderness, color, and microbiological quality of perishable foodstuff (Sánchez-Ortega et al. 2014; Rocca-Smith et al. 2016). Several variables such as nature and concentration of the lipids, the size, and distribution of lipids into the film, and the method used to obtain the film or coating should be considered to obtain satisfactory results (Morillon et al. 2002). Moreover, the addition of oleic acid to egg coatings increases its tensile strength and elongation at break, and reduces the water vapor permeability (Taqi et al. 2011). However, this is dependent on the lipid added; for example, the incorporation of beeswax to pectin and gelatin films improved the oxygen barrier property and enhanced the antioxidant activity, without altering their mechanical properties (Bermúdez-Oria et al. 2019).

Essential oils (EOs) are secondary metabolites of low molecular weight rich in hydrophobic and volatile compounds obtained from natural sources. Antimicrobial and antioxidant activity of EOs are attributed to phenols, terpenoids, terpenes, phenylpropanoids, and other aromatic compounds found in them (Vigan 2010; Chaudhary et al. 2020). EOs commonly used in edible food packaging come from basil, oregano, marjoram, rosemary, coriander, thyme, basilica, clove, balm, and ginger due to the great antimicrobial activity and contribution to maintain or improve the sensorial properties and prolong the shelf life of perishable food products (Table 3.2) (Dávila-Rodríguez et al. 2020). Waxes and shellac added with cinnamon essential oil are used in commercial formulations to control citrus green and blue molds (Champa et al. 2020). Pectin coatings added with oregano essential oil and resveratrol extended the shelf life of pork by reducing the color change and pH, retarding protein and lipid oxidation, maintaining meat tenderness, and inhibiting microbial development after 15 days of storage at 4 °C (Xiong et al. 2020). Nonetheless, the added concentration should be optimized to avoid alterations in the sensorial properties, particularly in odors and flavors of coated foods or in the mechanical properties of the coatings and films (Chitranshi et al. 2020). Nevertheless, their strong organoleptic properties, low water solubility, low stability, high susceptibility to environmental conditions, and high volatility limit their use. Hence, the application of EOs using an encapsulation system in a suitable delivery system compatible with the food product offers a viable alternative for their use. Emulsions, liposomes, and solid lipid nanoparticles are alternatives to encapsulate Eos; emulsions are the most used techniques on fresh and minimally processed foods (Li et al. 2015; Prakash et al. 2018).

Another important characteristic of EOs are their antioxidant properties. Oxygen is responsible for common degradation processes in meats such as lipid oxidation, development of aerobic microorganisms, vitamin loss, and enzymatic browning (Ayranci and Tunc 2003). Antioxidant and antibrowning compounds delay the rancidity and discoloration of food products (Avramescu et al. 2020). In meat foods, the oxidation of fat produces off-flavor, off-color, as well as nutrient loss (Hong and Krochta 2006). Common antioxidants added into edibles packaging are green tea extract (Nisa et al. 2015), stearic, ascorbic, and citric acid (Ayranci and Tunc 2003), metal nanoparticles (Gomes Lauriano Souza et al. 2020), tea tree oil (Yue et al. 2020), and propolis extract (Moreno et al. 2020). Besides, recently the incorporation of metal nanoparticles has been proposed as a great alternative to improve their antioxidant properties (Emamifar and Bavaisi 2020; Gomes Lauriano Souza et al. 2020). The addition of canola oil and *Byrsonima crassifolia* (L.) extract in chitosan coatings applied on bell pepper increased the content of carotenoids in the fruit and the antioxidant capacity of coatings (González-Saucedo et al. 2019). Otherwise, the addition of butylated hydroxytoluene and green tea extract into starch coatings applied to beef decreased the metmyoglobin formation and lipid oxidation (Nisa et al. 2015).

3.8 Future Trends and Challenges in Edible Food Packaging for Preservation of Perishable Foodstuff

Nowadays, the application of natural coatings to preserve perishable foods shows advances in different aspects such as the use of new biopolymers and methods to apply them onto the food surface, and the incorporation of natural substances to improve their properties. The incorporation of herbs in food packaging is widely studied because they can have antimicrobial and antioxidant properties. Aloe vera gel, cinnamon, rosemary, tulsi, grapefruit, lemon, and thyme are the most common herbs used to formulate food packaging due to their content of vitamins, antioxidants, and essential minerals, also they perform as a nutraceutical. Aloe vera gel is an aqueous liquid obtained from *Aloe vera* leaves. This gel generates high interest as EC material because of its antifungal and antibacterial properties; besides, it reduces the loss of moisture and water, prolonging the shelf life of perishable foods such as vegetables (Mehedi et al. 2020), fruits (Pinzon et al. 2020), meat, chicken, and seafood (Ribeiro et al. 2020). Additionally, *aloe vera* contains 20 of the 22 amino acids required in the human diet and 7 of the 8 essential amino acids. Hence, it is a convenient source of vitamins (Farina et al. 2020).

The addition of herbs in food packaging is through the use of essential oils or extracts. Herb extracts are mainly composed of phenolic compounds responsible for their antimicrobial activities (Champa et al. 2020; Deshmukh et al. 2020). Besides, the employment of culinary and medicinal herbs in food packaging has been recently evaluated because of their antispasmodic, antimicrobial, anti-inflammatory, carminative, and mucolytic agent; as well as hormonal regulator properties. For example, cumin is used as a stimulant, carminative, and astringent against indigestion, flatulence, and diarrhea. Its addition (1700 mg/kg) in whey protein concentrate films inhibited the production of the aflatoxins B_1, B_2, G_1, and G_2 (Tavakolipour et al. 2020). However, extracts and essential oils are highly susceptible to environmental factors such as oxygen, pH, temperature, and lightenvironmental factors such as oxygen, pH, temperature, and light; hence, to maintain and enhance their properties the utilization of a polymeric matrix to emulsify or encapsulate and protect is needful (Aguilar-Veloz et al. 2020). An option to maintain and enhance the characteristics of essential oils and herb extracts is the use of nanoemulsions. Nanoencapsulation of bioactive compounds is used to control their release under specific conditions (Rao and McClements 2012; Hassan et al. 2018) and provide physical stability to the encapsulated essential oils, increasing their bioactivity attributed to a better diffusion and minimizing their effect on the organoleptic characteristics of the foods (Donsì et al. 2011). Nanoemulsions enable to disperse EOs in water to obtain homogenous coating-forming solutions and improve the wettability on the food surface (Donsì and Ferrari 2016; Acevedo-Fani et al. 2017). Besides, the addition of phenolic compounds to edible food packaging can provide benefits to human health because these compounds have been associated with antiinflammatory and anticancer properties, cardiovascular protection, prevention of diabetes, etc. (Bermúdez-Oria et al. 2019).

Otherwise, the addition of peptides into films and coatings can provide them antimicrobial and antioxidant properties. Peptides are short chains of 2–100 amino acids with amphiphilic properties. These polymers have been extracted from diverse natural sources included microorganisms, insects, plants, amphibians, birds, fish, and mammals (Jenssen et al. 2006). Bacteriocins are metabolites produced by lactic acid bacteria widely used for the enhancement of quality and safety in food products. Nisin is the most common bacteriocin used in food products; its addition to films provides antimicrobial properties against *Bacillus subtilis* but reduces the tensile strength of the films. The lower tensile strength could be attributed to the interaction between nisin and polysaccharide reducing the cross-linking of polysaccharide and softening the network structure (Guo et al. 2020). Otherwise, the addition of cyclolipopeptides of *Bacillus subtilis* to sodium alginate films increased the water solubility to 28.92%, and water vapor permeability was lowered to 398.10 g/m^2/day, making them a suitable coating for vulnerable berries, even when the cyclolipopeptides affected the mechanical properties (Xu et al. 2020). From another perspective, Calderón-Chiu et al. (2021) obtained protein hydrolysates from Jackfruit leaf protein concentrate using pancreatin. The hydrolysates were highly soluble (60–98%), and showed great emulsifying, foaming, and antioxidant properties. Hence, in the future, these peptides can be used to develop or upgrade the antioxidant activity of food packaging for perishables foods; besides, their use contributes to the use of waste bio-products. In line with this, the functionalization of food packaging by the addition of by-products such as galactomannan from *Gleditsia triacanthos* (Fabaceae) decreased the total soluble matter, moisture content, and swelling in water due to the hydrophobicity reduction, and also improved the mechanical properties of soy protein films (González et al. 2019). These alternatives contribute to reducing waste and the impact of synthetic packing on the environment.

On the other hand, some studies have demonstrated the functionalization of coatings and films by incorporation of biocontrol agents such as bacteria and yeast to prolong the shelf life of perishable food products (Marín et al. 2017). *Meyerozyma caribbica* has been used in sodium alginate to coat avocado fruit reducing up to 100% of the anthracnose caused by *Colletotrichum gloeosporioides* (Iñiguez-Moreno et al. 2020b). Besides, its microencapsulation with trehalose and application by spraying reduced 66.7% of the same disease in Ataulfo mangoes (Aguirre-Güitrón et al. 2018). Probiotics are bacteria and yeasts that provide multiple benefits to the consumers (FAO/WHO 2006). Probiotics are usually incorporated into dairy foods; this fact can limit their consumption by vegans, lactose intolerant, or allergic to milk protein population. Hence, their incorporation into edible packaging is an alternative to offer its consumption to susceptible population and contribute to human health. For example, edible coatings prepared with fructans, linseed mucilage, or alginate, added with *Lactobacillus casei* LC-01, were applied to extend the shelf life of fresh-cut yacon. The coating reduced the weight loss and darkening and preserved the physicochemical parameters of the vegetables after 15 days of storage. Besides, under simulated gastrointestinal conditions, its population was only reduced by 2.96 log CFU/g. Hence, edible packaging from natural sources can be considered as a viable matrix to carry probiotic bacteria (Rodrigues et al. 2018).

The process of coat food is based on deposition, adhesion, coalescence, and stabilization of the continuous layer by coacervation through drying, cooling, heating, or coagulation of the coating on the food surface (Huber and Embuscado 2009). Based on these principles and depending on the properties of the food, the main methods to coat perishable foods are dipping, spraying, and brushing. Nowadays, new methods such as electrospinning are used to develop coatings. Electrospinning is based on the formation of nanoscale fibers elaborated by high voltage ranging from 1 to 30 kV to generate an electrostatic field to induce the fibers formation from a viscoelastic polymer solution. Finally, the nanofibers are deposited on the surface of the foods by freezing of the mixture or evaporation of the solvent, providing gentle protection (Falco and Mallavia 2020). This technique has been used to coat foods highly susceptible to suffer mechanical damage such as strawberry, cherry tomatoes, and kumquat, allowing them to retain their freshness during the handling, storage, and transportation (Yue et al. 2018; Liu et al. 2020). Besides, other advantages of polymeric nanofibers are that they can be made with a wide range of raw materials and can modify their kinetics to release the bioactive compounds (Falco and Mallavia 2020). Moreover, electrospun nanofibers are superhydrophobic, have nanosize, great surface area, high porosity, a large aspect ratio, and superior mechanical and barrier properties. Also, provides better protection to thermolabile compounds because this process is carried out at ambient temperature (Zhao et al. 2020a). However, the morphology of the nanofiber and the properties of the food packaging are highly influenced by several processing parameters, such as the properties of polymer solutions (polymer type, viscosity, concentration, conductivity, surface tension, and solvent polarity), the processing conditions (flow rate, voltage, and distance to the collector), and the environmental conditions such as temperature and humidity. Coatings obtained by electrospinning can be formed by single-strand fiber spun with a single-axis needle. However, as a common method used to coat foods, this technique allows the combination of both hydrophilic and hydrophobic substances to obtain coatings with enhanced properties (Zhang et al. 2020); therefore, the multiaxis needles process can be employed to spin multicomponent fibers and coat foods. The control of the whole parameters and the high cost of the equipment represent the main disadvantages of this technique (Liu et al. 2020; Prabu and Dhurai 2020). However, the studies indicate that electrospinning is a great alternative to preserve perishable foods. For example, the mix of gelatin nanofibers with peppermint essential oil and chamomile essential oil could improve the hydrophobicity of the film surface. The addition of peppermint essential oil provides superior antimicrobial properties against *Staphylococcus aureus* and *Escherichia coli*, whereas the chamomile essential oil enhances the resistance to fiber oxidation (Tang et al. 2019). On foods such as strawberries, nanofibers of carboxymethyl chitosan/polyoxyethylene oxide (20:1) can prevent weight loss and postharvest diseases, improving the appearance of fruit at room conditions for 9 days while remaining non-toxic and harmless (Yue et al. 2018).

3.9 Conclusions

Several studies have demonstrated that food packaging based on materials obtained from animal or plant sources has great advantages, and represents a good alternative to reduce food waste and decreases the pollution by synthetic packaging obtained from non-renewable sources such as petroleum. Besides, these biopolymers and lipids obtained from renewable resources, are low cost, and non-toxic. Besides, the use of these materials is an alternative to use the agroindustrial waste and reduce its negative impact on the environment. However, they also have disadvantages that can be enhanced by the mixture of hydrocolloids or by the addition of antioxidant, antimicrobial, and antibrowning compounds of natural or synthetic origin. Then, each formulation must be optimized and assessed complying with the properties of the foods and should focus on improving the mechanical, barrier, thermal, and antimicrobial properties of food packaging. Because each food has a unique composition and requires specific conditions during their storage, any formulation aimed to develop films and coatings must consider the optimization of its mechanical, barrier, thermal, and antimicrobial properties prolong the shelf life while maintaining nutritional and organoleptic properties. In addition, it is important to consider the method used to develop the film or coating because it also affects their properties. Continuous research in the use of biopolymers to develop food packaging would contribute to the successful replacement of synthetic coatings to prolong the shelf life of fresh food products providing a green alternative to satisfy the global demand for fresh and minimally processed foods.

References

Abdelhedi O, Nasri R, Jridi M et al (2018) Composite bioactive films based on smooth-hound viscera proteins and gelatin: physicochemical characterization and antioxidant properties. Food Hydrocoll 74:176–186. https://doi.org/10.1016/j.foodhyd.2017.08.006

Acevedo-Fani A, Soliva-Fortuny R, Martín-Belloso O (2017) Nanoemulsions as edible coatings. Curr Opin Food Sci 15:43–49. https://doi.org/10.1016/j.cofs.2017.06.002

Aguilar-Veloz LM, Calderón-Santoyo M, Vázquez González Y, Ragazzo-Sánchez JA (2020) Application of essential oils and polyphenols as natural antimicrobial agents in postharvest treatments: advances and challenges. Food Sci Nutr 8:2555–2568. https://doi.org/10.1002/fsn3.1437

Aguirre-Güitrón L, Calderón-Santoyo M, Ortiz-Basurto RI et al (2018) Optimisation of the spray drying process of formulating the post-harvest biocontrol agent *Meyerozyma caribbica*. Biocontrol Sci Tech 28:574–590. https://doi.org/10.1080/09583157.2018.1468997

Aguirre-Joya JA, Ventura-Sobrevilla J, Martínez-Vazquez G et al (2017) Effects of a natural bioactive coating on the quality and shelf life prolongation at different storage conditions of avocado (*Persea americana* mill.) cv. Hass. Food Package Shelf Life 14:102–107. https://doi.org/10.1016/j.fpsl.2017.09.003

Ahmed A, Ali SW, Imran A et al (2020) Formulation of date pit oil-based edible wax coating for extending the storage stability of guava fruit. J Food Process Preserv 00:e14336. https://doi.org/10.1111/jfpp.14336

Aider M (2010) Chitosan application for active bio-based films production and potential in the food industry: review. LWT – Food Sci Technol 43:837–842. https://doi.org/10.1016/j.lwt.2010.01.021

Alleyne V, Hagenmaier RD (2000) Candelilla-shellac: an alternative formulation for coating apples. HortScience 35:691–693. https://doi.org/10.21273/hortsci.35.4.691

Aloui H, Khwaldia K, Sánchez-González L et al (2014) Alginate coatings containing grapefruit essential oil or grapefruit seed extract for grapes preservation. Int J Food Sci Technol 49:952–959. https://doi.org/10.1111/ijfs.12387

Alparslan Y, Baygar T (2017) Effect of chitosan film coating combined with orange peel essential oil on the shelf life of Deepwater pink shrimp. Food Bioprocess Technol 10:842–853. https://doi.org/10.1007/s11947-017-1862-y

Amirou S, Pizzi A (2020) Biosourced heat resistant coatings by cross-linking of proteins with triethyl phosphate. Prog Org Coatings 138:105403. https://doi.org/10.1016/j.porgcoat.2019.105403

Anjum MA, Akram H, Zaidi M, Ali S (2020) Effect of gum arabic and *Aloe vera* gel based edible coatings in combination with plant extracts on postharvest quality and storability of 'Gola' guava fruits. Sci Hortic 271:109506. https://doi.org/10.1016/j.scienta.2020.109506

Avramescu SM, Butean C, Popa CV et al (2020) Edible and functionalized films/coatings-performances and perspectives. Coatings 10:687. https://doi.org/10.3390/coatings10070687

Ayranci E, Tunc S (2003) A method for the measurement of the oxygen permeability and the development of edible films to reduce the rate of oxidative reactions in fresh foods. Food Chem 80:423–431. https://doi.org/10.1016/S0308-8146(02)00485-5

Azeredo HMC, Mattoso LHC, Wood D et al (2009) Nanocomposite edible films from mango puree reinforced with cellulose nanofibers. J Food Sci 74:31–35. https://doi.org/10.1111/j.1750-3841.2009.01186.x

Ballesteros-Mártinez L, Pérez-Cervera C, Andrade-Pizarro R (2020) Effect of glycerol and sorbitol concentrations on mechanical, optical, and barrier properties of sweet potato starch film. NFS J 20:1–9. https://doi.org/10.1016/j.nfs.2020.06.002

Barone JR, Schmidt WF (2006) Products of chemistry nonfood applications of proteinaceous renewable materials. J Chem Educ 83:1003–1009

Bastarrachea LJ, Denis-Rohr A, Goddard JM (2015) Antimicrobial food equipment coatings: applications and challenges. Annu Rev Food Sci Technol 6:97–118. https://doi.org/10.1146/annurev-food-022814-015453

Bermúdez-Oria A, Rodríguez-Gutiérrez G, Rubio-Senent F et al (2019) Effect of edible pectin-fish gelatin films containing the olive antioxidants hydroxytyrosol and 3,4-dihydroxyphenylglycol on beef meat during refrigerated storage. Meat Sci 148:213–218. https://doi.org/10.1016/j.meatsci.2018.07.003

Bhagath YB, Manjula K (2019) Influence of composite edible coating systems on preservation of fresh meat cuts and products: a brief review on their trends and applications. Int Food Res J 26: 377–392

Bharathi SKV, Maria ML, Moses JA, Anandharamakrishnan C (2020) Zein-based anti-browning cling wraps for fresh-cut apple slices. Int J Food Sci Technol 55:1238–1245. https://doi.org/10.1111/ijfs.14401

Bharti SK, Pathak V, Alam T et al (2020) Materiality of edible film packaging in muscle foods: a worthwhile conception. J Packag Technol Res 4:117–132. https://doi.org/10.1007/s41783-020-00087-9

Blachechen TS, Petri DFS (2020) Physicochemical and antimicrobial properties of in situ crosslinked alginate/hydroxypropyl methylcellulose/ε-polylysine films. J Appl Polym Sci 137: 48832. https://doi.org/10.1002/app.48832

Bockisch M (1998) Composition, structure, physical data, and chemical reactions of fats and oils, their derivatives, and their associates. In: Bockisch M (ed) Fats and oils handbook. AOCS Press, Hamburg, Germany, pp 53–120

Böhmer-Maas BW, Fonseca LM, Otero DM et al (2020) Photocatalytic zein-TiO2 nanofibers as ethylene absorbers for storage of cherry tomatoes. Food Packag Shelf Life 24:100508. https://doi.org/10.1016/j.fpsl.2020.100508

Bonilla J, Atarés L, Vargas M, Chiralt A (2012) Edible films and coatings to prevent the detrimental effect of oxygen on food quality: possibilities and limitations. J Food Eng 110:208–213. https://doi.org/10.1016/j.jfoodeng.2011.05.034

Bonilla J, Atarés L, Vargas M, Chiralt A (2013) Properties of wheat starch film-forming dispersions and films as affected by chitosan addition. J Food Eng 114:303–312. https://doi.org/10.1016/j.jfoodeng.2012.08.005

Bourtoom T (2008) Edible films and coatings: characteristics and properties. Int Food Res J 15: 237–248

Bradley DG, Min DB (1992) Singlet oxygen oxidation of foods. Crit Rev Food Sci Nutr 31:211–236. https://doi.org/10.1080/10408399209527570

Bravin B, Peressini D, Sensidoni A (2004) Influence of emulsifier type and content on functional properties of polysaccharicle lipid-based edible films. J Agric Food Chem 52:6448–6455. https://doi.org/10.1021/jf040065b

Buléon A, Colonna P, Planchot V, Ball S (1998) Starch granules: structure and biosynthesis. Int J Biol Macromol 23:85–112. https://doi.org/10.1016/S0141-8130(98)00040-3

Calderón-Chiu C, Calderón-Santoyo M, Herman-Lara E, Ragazzo-Sánchez JA (2021) Jackfruit (*Artocarpus heterophyllus* lam) leaf as a new source to obtain protein hydrolysates: physicochemical characterization, techno-functional properties and antioxidant capacity. Food Hydrocoll 112:106319. https://doi.org/10.1016/j.foodhyd.2020.106319

Casariego A, Souza BWS, Vicente AA et al (2008) Chitosan coating surface properties as affected by plasticizer, surfactant and polymer concentrations in relation to the surface properties of tomato and carrot. Food Hydrocoll 22:1452–1459. https://doi.org/10.1016/j.foodhyd.2007.09.010

Cazón P, Velazquez G, Ramírez JA, Vázquez M (2017) Polysaccharide-based films and coatings for food packaging: a review. Food Hydrocoll 68:136–148. https://doi.org/10.1016/j.foodhyd.2016.09.009

Chalak A, Abou-Daher C, Chaaban J, Abiad MG (2016) The global economic and regulatory determinants of household food waste generation: a cross-country analysis. Waste Manag 48: 418–422. https://doi.org/10.1016/j.wasman.2015.11.040

Champa WAH, Gunasekera N, Wilson WS et al (2020) Postharvest treatment of cinnamon (*Cinnamomum zeylanicum*) bark oil and hexanal incorporated bio-wax maintains quality and extends marketable life of lime (*Citrus aurantifolia* Swingle). Int J Fruit Sci 20:76–88. https://doi.org/10.1080/15538362.2019.1597823

Chander BV, Tandon R, Kapoor S, Sidhu MK (2018) Natural coatings for shelf-life enhancement and quality maintenance of fresh fruits and vegetables-a review. J Postharvest Technol 6:12–26

Chaudhary S, Kumar S, Kumar V, Sharma R (2020) Chitosan nanoemulsions as advanced edible coatings for fruits and vegetables: composition, fabrication and developments in last decade. Int J Biol Macromol 152:154–170. https://doi.org/10.1016/j.ijbiomac.2020.02.276

Chen M, Smith PM (2017) The U.S. cellulosic biofuels industry: expert views on commercialization drivers and barriers. Biomass Bioenergy 102:52–61. https://doi.org/10.1016/j.biombioe.2017.05.002

Chi K, Wang H, Catchmark JM (2020) Sustainable starch-based barrier coatings for packaging applications. Food Hydrocoll 103:105696. https://doi.org/10.1016/j.foodhyd.2020.105696

Chitranshi S, Dubey N, Sajjad M (2020) Sustainable botanical products for safe post-harvest management of perishable produce: a review. J Hortic Postharvest Res 3:125–140. https://doi.org/10.22077/jhpr.2019.2703.1083

Chiumarelli M, Hubinger MD (2014) Evaluation of edible films and coatings formulated with cassava starch, glycerol, carnauba wax and stearic acid. Food Hydrocoll 38:20–27. https://doi.org/10.1016/j.foodhyd.2013.11.013

Cordeiro de Azeredo HM (2012) Edible coatings. In: Rodrigues S, Fernandes FAN (eds) Advances in fruit processing technologies. CRC Press Inc., Boca Raton, pp 345–361

Criado P, Fraschini C, Salmieri S, Lacroix M (2020) Cellulose nanocrystals (CNCs) loaded alginate films against lipid oxidation of chicken breast. Food Res Int 132:109110. https://doi.org/10.1016/j.foodres.2020.109110

Cui L, Yuan J, Wang P et al (2017) Facilitation of α-polylysine in TGase-mediated crosslinking modification for gluten and its effect on properties of gluten films. J Cereal Sci 73:108–115. https://doi.org/10.1016/j.jcs.2016.12.006

Dalgleish DG (2011) On the structural models of bovine casein micelles—review and possible improvements. Soft Matter 7:2265–2272. https://doi.org/10.1039/c0sm00806k

Dalmas F, Cavaillé J-Y, Gauthier C et al (2007) Viscoelastic behavior and electrical properties of flexible nanofiber filled polymer nanocomposites. Influence of processing conditions. Compos Sci Technol 67:829–839. https://doi.org/10.1016/j.compscitech.2006.01.030

Dávila-Rodríguez M, López-Malo A, Palou E et al (2020) Essential oils microemulsions prepared with high-frequency ultrasound: physical properties and antimicrobial activity. J Food Sci Technol. https://doi.org/10.1007/s13197-020-04449-8

Day L, Augustin MA, Batey IL, Wrigley CW (2006) Wheat-gluten uses and industry needs. Trends Food Sci Technol 17:82–90. https://doi.org/10.1016/j.tifs.2005.10.003

De Cindio B, Gabriele D, Lupi FR (2015) Pectin: properties determination and uses. In: Encyclopedia of food and health, 1st edn. Elsevier Ltd., pp 294–300

de la Cruz A, Ortega D, García-Triana A et al (2020) A brief review of edible coating materials for the microencapsulation of probiotics. Coatings 10:197. https://doi.org/10.3390/coatings10030197

De León-Zapata MA, Sáenz-Galindo A, Rojas-Molina R et al (2015) Edible candelilla wax coating with fermented extract of tarbush improves the shelf life and quality of apples. Food Packag Shelf Life 3:70–75. https://doi.org/10.1016/j.fpsl.2015.01.001

Debeaufort F, Quezada-Gallo J-A, Voilley A (1998) Edible films and coatings: tomorrow's packagings: a review. Crit Rev Food Sci Nutr 38:299–313. https://doi.org/10.1080/10408699891274219
Defilippi BG, Ejsmentewicz T, Covarrubias MP et al (2018) Changes in cell wall pectins and their relation to postharvest mesocarp softening of "Hass" avocados (*Persea americana* mill.). Plant Physiol Biochem 128:142–151. https://doi.org/10.1016/j.plaphy.2018.05.018
Dehghani S, Hosseini SV, Regenstein JM (2018) Edible films and coatings in seafood preservation: a review. Food Chem 240:505–513. https://doi.org/10.1016/j.foodchem.2017.07.034
Delmer P (1983) Biosynthesis of cellulose. Adv Carbohydr Chem Biochem 41:105–153
Delmer DP (1987) Cellulose biosynthesis. Annu Rev Plant Physiol 38:259–290
Delmer DP, Amor Y (1995) Cellulose biosynthesis. Plant Cell 7:987–1000
Demers-Mathieu V, Gauthier SF, Britten M et al (2013) Antibacterial activity of peptides extracted from tryptic hydrolyzate of whey protein by nanofiltration. Int Dairy J 28:94–101. https://doi.org/10.1016/j.idairyj.2012.09.003
Deng Z, Jung J, Simonsen J, Zhao Y (2017) Cellulose nanomaterials emulsion coatings for controlling physiological activity, modifying surface morphology, and enhancing storability of postharvest bananas (*Musa acuminate*). Food Chem 232:359–368. https://doi.org/10.1016/j.foodchem.2017.04.028
Deshmukh SD, Patil SR, Rajvaidya RR (2020) Storage behavior of Nagpur mandarin fruits as affected by post-harvest application of plant leaf extracts under cold storage condition. Int J Chem Stud 8:877–880. https://doi.org/10.22271/chemi.2020.v8.i2m.8876
Doelker E (1993) Cellulose derivatives. Adv Polym Sci 107:199–265
Donsì F, Ferrari G (2016) Essential oil nanoemulsions as antimicrobial agents in food. J Biotechnol 233:106–120. https://doi.org/10.1016/j.jbiotec.2016.07.005
Donsì F, Sessa M, Ferrar G (2011) Nanoencapsulation of essential oils to enhance their antimicro bial activity in foods. LWT-Food Sci Technol 44:1908–1914. https://doi.org/10.1016/j.jbiotec.2010.08.173
Efenberger-Szmechtyk M, Nowak A, Czyzowska A (2020) Plant extracts rich in polyphenols: antibacterial agents and natural preservatives for meat and meat products. Crit Rev Food Sci Nutr:1–30. https://doi.org/10.1080/10408398.2020.1722060
Elik A, Yanik DK, Istanbullu Y, Guzelsoy NA (2019) Strategies to reduce post-harvest losses for fruits and vegetables. Int J Sci Technol Res 5:29–39. https://doi.org/10.7176/JSTR/5-3-04
Elsabee MZ, Abdou ES (2013) Chitosan based edible films and coatings: a review. Mater Sci Eng C 33:1819–1841. https://doi.org/10.1016/j.msec.2013.01.010
Emamifar A, Bavaisi S (2020) Nanocomposite coating based on sodium alginate and nano-ZnO for extending the storage life of fresh strawberries (*Fragaria* × *ananassa Duch.*). J Food Meas Charact 14:1012–1024. https://doi.org/10.1007/s11694-019-00350-x
Fabra MJ, Pérez-Masiá R, Talens P, Chiralt A (2011) Influence of the homogenization conditions and lipid self-association on properties of sodium caseinate based films containing oleic and stearic acids. Food Hydrocoll 25:1112–1121. https://doi.org/10.1016/j.foodhyd.2010.10.008
Falco A, Mallavia R (2020) Electrospun nanomaterials: applications in food, environmental remediation, and bioengineering. Nano 10:1714. https://doi.org/10.3390/nano10091714
Falguera V, Quintero JP, Jiménez A et al (2011) Edible films and coatings: structures, active functions and trends in their use. Food Sci Technol 22:292–303. https://doi.org/10.1016/j.tifs.2011.02.004
Fang Y, Al-Assaf S, Phillips GO et al (2007) Multiple steps and critical behaviors of the binding of calcium to alginate. J Phys Chem B 111:2456–2462
FAO/WHO (2006) Probiotics in food: Health and nutritional properties and guidelines for evaluation
Farina V, Passafiume R, Tinebra I et al (2020) Postharvest application of aloe vera gel-based edible coating to improve the quality and storage stability of fresh-cut papaya. J Food Qual:8303140. https://doi.org/10.1155/2020/8303140

Fasina OO, Hallman H, Craig-Schmidt M, Clements C (2006) Predicting temperature-dependence viscosity of vegetable oils from fatty acid composition. J Am Oil Chem Soc 83:899–903. https://doi.org/10.1007/s11746-006-5044-8
Feng X, Ng VK, Mikš-Krajnik M, Yang H (2017) Effects of fish gelatin and tea polyphenol coating on the spoilage and degradation of myofibril in fish fillet during cold storage. Food Bioprocess Technol 10:89–102. https://doi.org/10.1007/s11947-016-1798-7
Food and Agriculture Organization of the United Nations (FAO) (2009) How to feed the world in 2050
Food and Agriculture Organization of the United Nations (FAO) (2014) Food wastage footprint: fool cost-accounting. Rome, Italy
Food and Agriculture Organization of the United Nations (FAOSTAT) (2020) Cultivos. http://www.fao.org/faostat/es/#data/QC. Accessed 22 Feb 2020
Francisco CB, Pellá MG, Silva OA et al (2020) Shelf-life of guavas coated with biodegradable starch and cellulose-based films. Int J Biol Macromol 152:272–279. https://doi.org/10.1016/j.ijbiomac.2020.02.249
Garcia E, Barrett DM (2002) Preservative treatments for fresh-cut fruits and vegetables. In: Lamikanra O (ed) Fresh-cut fruits and vegetables. CRC Press Inc., Boca Raton, pp 267–304
Gennadios A, Brandenburg AH, Weller CL, Testin RF (1993) Effect of pH on properties of wheat gluten and soy protein isolate films. J Agric Food Chem 41:1835–1839. https://doi.org/10.1021/jf00035a006
Gennadios A, Weller CL, Gooding CH (1994) Measurement errors in water vapor permeability of highly permeable, hydrophilic edible films. J Food Eng 21:395–409
Gennadios A, Weller CL, Hanna MA, Froning GW (1996) Mechanical and barrier properties of egg albumen films. J Food Sci 61:585–589. https://doi.org/10.1037/h0056598
Gennadios A, Hanna MA, Kurth LB (1997) Application of edible coatings on meats, poultry and seafoods: a review. LWT – Food Sci Technol 30:337–350. https://doi.org/10.1006/fstl.1996.0202
Gomes Lauriano Souza V, Rodrigues C, Valente S et al (2020) Eco-friendly ZnO/chitosan bionanocomposites films for packaging of fresh poultry meat. Coatings 10:110
González A, Strumia MC, Alvarez CII (2011) Cross-linked soy protein as material for biodegradable films: synthesis, characterization and biodegradation. J Food Eng 106:331–338. https://doi.org/10.1016/j.jfoodeng.2011.05.030
González A, Barrera GN, Galimberti PI et al (2019) Development of edible films prepared by soy protein and the galactomannan fraction extracted from *Gleditsia triacanthos* (Fabaceae) seed. Food Hydrocoll 97:105227. https://doi.org/10.1016/j.foodhyd.2019.105227
González-Estrada RR, Carvajal-Millán E, Ragazzo-Sánchez JA et al (2017) Control of blue mold decay on *Persian lime*: application of covalently cross-linked arabinoxylans bioactive coatings with antagonistic yeast entrapped. LWT – Food Sci Technol 85:187–196. https://doi.org/10.1016/j.lwt.2017.07.019
González-Saucedo A, Barrera-Necha LL, Ventura-Aguilar RI et al (2019) Extension of the postharvest quality of bell pepper by applying nanostructured coatings of chitosan with *Byrsonima crassifolia* extract (L.) Kunth. Postharvest Biol Technol 149:74–82. https://doi.org/10.1016/j.postharvbio.2018.11.019
Grant GT, Morris ER, Rees DA et al (1973) Biological interactions between polysaccharides and divalent cations: the egg-box model. Rev Med Virol 16:195–198. https://doi.org/10.1002/rmv.519
Guérin-Dubiard C, Audic J-L (2007) Egg-protein-based films and coatings. In: Huopalahti R, López-Fandiño R, Anton M, Schade R (eds) Bioactive egg compounds. Springer-Verlag, Berlin, Heidelberg, pp 265–273
Guevara-Avendaño E, Carrillo JD, Ndinga-Muniania C et al (2018) Antifungal activity of avocado rhizobacteria against *fusarium euwallaceae* and *Graphium* spp., associated with *Euwallacea* spp. nr. *Fornicatus*, and *Phytophthora cinnamomi*. Antonie Van Leeuwenhoek 111:563–572. https://doi.org/10.1007/s10482-017-0977-5

Guo N, Zhu G, Chen D et al (2020) Preparation and characterization of gellan gum–guar gum blend films incorporated with nisin. J Food Sci 85:1799–1804. https://doi.org/10.1111/1750-3841.15143

Gustavsson J, Cederberg C, Sonesson U, et al (2011) Global food losses and food waste – extent, causes and prevention. Rome, Italy

Gutiérrez-Pacheco MM, Ortega-Ramírez LA, Silva-Espinoza BA et al (2020) Individual and combined coatings of chitosan and carnauba wax with oregano essential oil to avoid water loss and microbial decay of fresh cucumber. Coatings 10:614

Hagenmaier RD, Baker RA (1994) Wax microemulsions and emulsions as citrus coatings. J Agric Food Chem 42:899–902. https://doi.org/10.1021/jf00040a012

Hagenmaier RD, Shaw PE (1992) Gas permeability of fruit coating waxes. J Am Soc Hortic Sci 117:105–109. https://doi.org/10.21273/jashs.117.1.105

Han JH (2014) Edible films and coatings: a review. In: Innovations in food packaging. Elsevier Ltd., pp 211–254

Harholt J, Suttangkakul A, Scheller HV (2010) Biosynthesis of pectin. Plant Physiol 153:384–395. https://doi.org/10.1104/pp.110.156588

Hassan B, Chatha SAS, Hussain AI et al (2018) Recent advances on polysaccharides, lipids and protein based edible films and coatings: a review. Int J Biol Macromol 109:1095–1107. https://doi.org/10.1016/j.ijbiomac.2017.11.097

Hay ID, Rehman ZU, Moradali MF et al (2013) Microbial alginate production, modification and its applications. Microb Biotechnol 6:637–650. https://doi.org/10.1111/1751-7915.12076

Hemmatkhah F, Zeynali F, Almasi H (2020) Encapsulated cumin seed essential oil-loaded active papers: characterization and evaluation of the effect on quality attributes of beef hamburger. Food Bioprocess Technol 13:533–547. https://doi.org/10.1007/s11947-020-02418-9

Heralp TJ, Gnanasambandam R, Mcguire BH, Hachmeister KA (1995) Degradable wheat gluten films: preparation, properties and applications. J Food Sci 60:1147–1150. https://doi.org/10.1111/j.1365-2621.1995.tb06311.x

Hong SI, Krochta JM (2006) Oxygen barrier performance of whey-protein-coated plastic films as affected by temperature, relative humidity, base film and protein type. J Food Eng 77:739–745. https://doi.org/10.1016/j.jfoodeng.2005.07.034

Horne DS (2006) Casein micelle structure: models and muddles. Curr Opin Colloid Interface Sci 11:148–153. https://doi.org/10.1016/j.cocis.2005.11.004

Hossain I, Subahan A, Tuha M (2020) Biodegradable plastic production from daily household waste materials and comparison the decomposing time with synthetic polyethylene plastic. Int J Adv Life Sci Res 3:16–19

Huber KC, Embuscado M (2009) Edible films and coatings for food applications. Springer, New York

Iñiguez-Moreno M, González-Gutiérrez KN, Ragazzo-Sánchez JA et al (2020a) Morphological and molecular identification of the causal agents of post-harvest diseases in avocado fruit, and potential biocontrol with *Meyerozyma caribbica*. Arch Phytopathol Plant Prot:1834806. https://doi.org/10.1080/03235408.2020.1834806

Iñiguez-Moreno M, Ragazzo-Sánchez JA, Barros-Castillo JC et al (2020b) Sodium alginate coatings added with Meyerozyma caribbica: postharvest biocontrol of *Colletotrichum gloeosporioides* in avocado (*Persea americana* mill. cv. Hass). Postharvest Biol Technol 163:111123. https://doi.org/10.1016/j.postharvbio.2020.111123

Isleroglu H, Turker I, Koc B, Tokatli M (2019) Optimization of microencapsulation conditions of transglutaminase by freeze drying. J Food Sci Technol 56:4925–4937. https://doi.org/10.1007/s13197-019-03962-9

Izydorczyk MS, Cui SW, Wang Q (2005) Polysaccharide gums: structures, functional properties, and applications. In: Cui SW (ed) Food carbohydrates chemistry, physical properties, and applications. CRC Press Inc., Boca Raton

Jackson DS, Ratnayake WA (2006) Gelatinization and solubility of corn starch during heating in excess water: new insights. J Agric Food Chem 54:3712–3716

James MG, Denyer K, Myers AM (2003) Starch synthesis in the cereal endosperm. Curr Opin Plant Biol 6:215–222. https://doi.org/10.1016/S1369-5266(03)00042-6
Jenssen H, Hamill P, Hancock REW (2006) Peptide antimicrobial agents. Clin Microbiol Rev 19: 491–511. https://doi.org/10.1128/CMR.00056-05
Jiang M, Liu S, Du X, Wang Y (2010) Physical properties and internal microstructures of films made from catfish skin gelatin and triacetin mixtures. Food Hydrocoll 24:105–110. https://doi.org/10.1016/j.foodhyd.2009.08.011
Jiménez A, Fabra MJ, Talens P, Chiralt A (2010) Effect of lipid self-association on the microstructure and physical properties of hydroxypropyl-methylcellulose edible films containing fatty acids. Carbohydr Polym 82:585–593. https://doi.org/10.1016/j.carbpol.2010.05.014
Jiménez A, Fabra MJ, Talens P (2012) Edible and biodegradable starch films: a review. Food Bioprocess Technol 5:2058–2076. https://doi.org/10.1007/s11947-012-0835-4
Jones S, Thornton JM (1996) Principles of protein-protein interactions. Proc Natl Acad Sci 93:13–20
Jooyandeh H (2011) Whey protein films and coatings: a review. Pak J Nutr 10:296–301. https://doi.org/10.3923/pjn.2011.296.301
Jovanović S, Barać M, Maćejv O (2005) Whey proteins-properties and possibility of application. Mljekarstvo 55:215–233
Jung S, Cui Y, Barnes M et al (2020) Multifunctional bio-nanocomposite coatings for perishable fruits. Adv Mater 32:1908291. https://doi.org/10.1002/adma.201908291
Jutinico-Shubach A, Gutiérrez-Cortés C, Suarez H (2020) Antilisterial activity of chitosan-based edible coating incorporating cell-free supernatant from *Pediococcus pentosaceus* 147 on the preservation of fresh cheese. J Food Process Preserv 44:e14715. https://doi.org/10.1111/jfpp.14715
Karlton-Senaye BD, Ibrahim SA (2013) Impact of gums on the growth of probiotics. Agro Food Ind Hi Tech 24:10–14
Karupaiah T, Sundram K (2007) Effects of stereospecific positioning of fatty acids in triacylglycerol structures in native and randomized fats: a review of their nutritional implications. Nutr Metab 4: 16. https://doi.org/10.1186/1743-7075-4-16
Katakojwala R, Mohan SV (2020) Microcrystalline cellulose production from sugarcane bagasse: sustainable process development and life cycle assessment. J Clean Prod 249:119342. https://doi.org/10.1016/j.jclepro.2019.119342
Kaur S, Dhillon GS (2013) The versatile biopolymer chitosan: potential sources, evaluation of extraction methods and applications. Crit Rev Microbiol 40:155–175. https://doi.org/10.3109/1040841X.2013.770385
Kester JJ, Fennema OR (1986) Edible films and coatings: a review. Food Technol 40:47–59
Khalil AA, Deraz SF, Elrahman SA, El-Fawal G (2015) Enhancement of mechanical properties, microstructure, and antimicrobial activities of zein films cross-linked using succinic anhydride, eugenol, and citric acid. Prep Biochem Biotechnol 45:551–567. https://doi.org/10.1080/10826068.2014.940967
Khaneghah AM, Hashemi SMB, Es I et al (2018) Efficacy of antimicrobial agents for food contact applications: biological activity, incorporation into packaging, and assessment methods: a review. J Food Prot 81:1142–1156. https://doi.org/10.4315/0362-028X.JFP-17-509
Khorram F, Ramezanian A, Hosseini SMH (2017) Shellac, gelatin and Persian gum as alternative coating for orange fruit. Sci Hortic 225:22–28. https://doi.org/10.1016/j.scienta.2017.06.045
Khotimchenko YS, Kovalev VV, Savchenko OV, Ziganshina OA (2001) Physical-chemical properties, physiological activity, and usage of alginates, the polysaccharides of brown algae. Russ J Mar Biol 27:S53–S64. https://doi.org/10.1023/A:1013851022276
Kinsella JE, Morr CV (1984) Milk proteins: physicochemical and functional properties. C R C Crit Rev Food Sci Nutr 21:197–262. https://doi.org/10.1080/10408398409527401
Kohn R (1975) Ion binding on polyuronates — alginate and pectin. Pure Appl Chem 42:371–397. https://doi.org/10.1351/pac197542030371

Kolster P, Vereijken JM, de Graaf LA (1998) Protein modification and technical applications. In: Guéguen J, Popineau Y (eds) Plant proteins from European crops. Springer, Berlin, Heidelberg, pp 305–312

Kostik V, Memeti S, Bauer B (2013) Fatty acid composition of edible oils and fats. J Hyg Eng Des 4:112–116

Kumar MNVR (2000) A review of chitin and chitosan applications. React Funct Polym 46:1–27

Kumar R, Choudhary V, Mishra S et al (2002) Adhesives and plastics based on soy protein products. Ind Crop Prod 16:155–172. https://doi.org/10.1016/S0926-6690(02)00007-9

Kumar S, Mukherjee A, Dutta J (2020) Chitosan based nanocomposite films and coatings: emerging antimicrobial food packaging alternatives. Trends Food Sci Technol 97:196–209. https://doi.org/10.1016/j.tifs.2020.01.002

Kunte LA, Gennadios A, Cuppett SL et al (1997) Cast films from soy protein isolates and fractions. Cereal Chem 74:115–118

Lagrain B, Goderis B, Brijs K, Delcour JA (2010) Molecular basis of processing wheat gluten toward biobased materials. Biomacromolecules 11:533–541. https://doi.org/10.1021/bm100008p

Le Corre D, Bras J, Dufresne A (2010) Starch nanoparticles: a review. Biomacromolecules 11: 1139–1153

Lens JP, de Graaf LA, Stevels WM et al (2003) Influence of processing and storage conditions on the mechanical and barrier properties of films cast from aqueous wheat gluten dispersions. Ind Crop Prod 17:119–130. https://doi.org/10.1016/S0926-6690(02)00092-4

Li Y-D, Zheng J-B, Wang X-L et al (2008) Structure and properties of soy protein/poly(butylene succinate) blends with improved compatibility. Biomacromolecules 9:3157–3164. https://doi.org/10.1021/bm800745p

Li W, Chen H, He Z et al (2015) Influence of surfactant and oil composition on the stability and antibacterial activity of eugenol nanoemulsions. LWT – Food Sci Technol 62:39–47. https://doi.org/10.1016/j.lwt.2015.01.012

Li H, Shi H, He Y et al (2020) Preparation and characterization of carboxymethyl cellulose-based composite fi lms reinforced by cellulose nanocrystals derived from pea hull waste for food packaging applications. Int J Biol Macromol 164:4104–4112. https://doi.org/10.1016/j.ijbiomac.2020.09.010

Liakos I, Rizzello L, Scurr DJ et al (2014) All-natural composite wound dressing films of essential oils encapsulated in sodium alginate with antimicrobial properties. Int J Pharm 463:137–145. https://doi.org/10.1016/j.ijpharm.2013.10.046

Lin D, Zhao Y (2007) Innovations in the development and application of edible coatings for fresh and minimally processed fruits and vegetables. Compr Rev Food Sci Food Saf 6:60–75

Lin JF, Wei PJ, Pan JC, Ai CF (2004) Effect of nitrogen content at coating film and film thickness on nanohardness and Young's modulus of hydrogenated carbon films. Diam Relat Mater 13:42–53. https://doi.org/10.1016/j.diamond.2003.08.019

Liu H, Gough CR, Deng Q et al (2020) Recent advances in electrospun sustainable composites for biomedical, environmental, energy, and packaging applications. Int J Mol Sci 21:4019. https://doi.org/10.3390/ijms21114019

Löfgren C, Guillotin S, Evenbratt H et al (2005) Effects of calcium, pH, and blockiness on kinetic rheological behavior and microstructure of HM pectin gels. Biomacromolecules 6:646–652. https://doi.org/10.1021/bm049619+

Maciel VBV, Contini LRF, Yoshida CMP, Venturini AC (2020) Application of edible biopolymer coatings on meats, poultry, and seafood. Biopolym Membr Film:515–533. https://doi.org/10.1016/b978-0-12-818134-8.00021-3

Marín A, Atarés L, Chiralt A (2017) Improving function of biocontrol agents incorporated in antifungal fruit coatings: a review. Biocontrol Sci Technol:1390068. https://doi.org/10.1080/09583157.2017.1390068

Martin C, Smith AM (1995) Starch biosynthesis. Plant Cell 7:971–985. https://doi.org/10.1105/tpc.7.7.971

Matak KE, Tahergorabi R, Jaczynski J (2015) A review: protein isolates recovered by isoelectric solubilization/precipitation processing from muscle food by-products as a component of nutraceutical foods. Food Res Int 77:697–703. https://doi.org/10.1016/j.foodres.2015.05.048

Mehedi MNH, Halder A, Hasan MF et al (2020) Influence of edible starch and sodium bi-carbonate on postharvest quality of minimally processed carrot and potato. Asian J Agric Hortic Res 6:1–12. https://doi.org/10.9734/ajahr/2020/v6i230067

Milani JM, Tirgarian B (2020) An overview of edible protein-based packaging: main sources, advantages, drawbacks, recent progressions and food applications. J Packag Technol Res 4: 103–115. https://doi.org/10.1007/s41783-020-00086-w

Mohnen D (2008) Pectin structure and biosynthesis. Curr Opin Plant Biol 11:266–277. https://doi.org/10.1016/j.pbi.2008.03.006

Moreno MA, Vallejo AM, Ballester AR et al (2020) Antifungal edible coatings containing Argentinian propolis extract and their application in raspberries. Food Hydrocoll 107:105973. https://doi.org/10.1016/j.foodhyd.2020.105973

Morillon V, Debeaufort F, Blond G et al (2002) Factors affecting the moisture permeability of lipid-based edible films: a review. Crit Rev Food Sci Nutr 42:67–89. https://doi.org/10.1007/BF00668465

Morr CV, Ha EYW (1993) Whey protein concentrates and isolates: processing and functional properties. Crit Rev Food Sci Nutr 33:431–476

Munarin F, Tanzi MC, Petrini P (2012) Advances in biomedical applications of pectin gels. Int J Biol Macromol 51:681–689. https://doi.org/10.1016/j.ijbiomac.2012.07.002

Murrieta-Martínez C, Soto-Valdez H, Pacheco-Aguilar R et al (2019) Effect of different polyalcohols as plasticizers on the functional properties of squid protein film (*Dosidicus Gigas*). Coatings 9:77. https://doi.org/10.3390/coatings9020077

Netravali AN, Huang X, Mizuta K (2007) Advanced "green" composites. Adv Compos Mater 16: 269–282. https://doi.org/10.1163/156855107782325230

Nguyen-Tri P, Nguyen TA, Carriere P, Ngo Xuan C (2018) Nanocomposite coatings: preparation, characterization, properties, and applications. Int J Corros:1–19. https://doi.org/10.1155/2018/4749501

Nisa I, Ashwar BA, Shah A et al (2015) Development of potato starch based active packaging films loaded with antioxidants and its effect on shelf life of beef. J Food Sci Technol 52:7245–7253. https://doi.org/10.1007/s13197-015-1859-3

No HK, Meyers SP, Prinyawiwatkul W, Xu Z (2007) Applications of chitosan for improvement of quality and shelf life of foods: a review. J Food Sci 72:87–100. https://doi.org/10.1111/j.1750-3841.2007.00383.x

Obianom C, Romanazzi G, Sivakumar D (2019) Effects of chitosan treatment on avocado postharvest diseases and expression of phenylalanine ammonia-lyase, chitinase and lipoxygenase genes. Postharvest Biol Technol 147:214–221. https://doi.org/10.1016/j.postharvbio.2018.10.004

Orgaz B, Lobete MM, Puga CH, Jose CS (2011) Effectiveness of chitosan against mature biofilms formed by food related bacteria. Int J Mol Sci 12:817–828

Palacios H, Hernández JA, Esquivel M et al (2019) Isolation and characterization of Nanofibrillar cellulose from *Agave tequilana* weber bagasse. Adv Mater Sci Eng 1342547. https://doi.org/10.1155/2019/1342547

Palou L, Valencia-Chamorro SA, Pérez-Gago MB (2015) Antifungal edible coatings for fresh citrus fruit: a review. Coatings 5:962–986. https://doi.org/10.3390/coatings5040962

Parreidt TS, Schott M, Schmid M, Müller K (2018) Effect of presence and concentration of plasticizers, vegetable oils, and surfactants on the properties of sodium-alginate-based edible coatings. Int J Mol Sci 19:742. https://doi.org/10.3390/ijms19030742

Peressini D, Bravin B, Lapasin R et al (2003) Starch-methylcellulose based edible films: rheological properties of film-forming dispersions. J Food Eng 59:25–32. https://doi.org/10.1016/S0260-8774(02)00426-0

Pérez-Gago MB, Nadaud P, Krochta JM (1999) Water vapor permeability, solubility, and tensile properties of heat-denature versus native whey protein films. J Food Sci 64:1034–1037

Perez-Gago MB, Rojas C, Del Rio MA (2002) Effect of lipid type and amount of edible hydroxypropyl methylcellulose-lipid composite coatings used to protect postharvest quality of mandarins cv. Fortune J Food Sci 67:2903–2910. https://doi.org/10.1111/j.1365-2621.2002.tb08836.x

Petruccelli S, Añón MC (1994) Relationship between the method of obtention and the structural and functional properties of soy protein isolates. 2. Surface properties. J Agric Food Chem 42:2170–2176. https://doi.org/10.1021/jf00046a018

Phan The D, Debeaufort F, Luu D, Voilley A (2008) Moisture barrier, wetting and mechanical properties of shellac/agar or shellac/cassava starch bilayer bio-membrane for food applications. J Memb Sci 325:277–283. https://doi.org/10.1016/j.memsci.2008.07.052

Pilon L, Tetelboim MC, Gallo CR et al (2016) Physicochemical and microbiological changes in freshcut pineapple coated with wheat gluten and alginate. Acta Hortic 1111:221–226. https://doi.org/10.17660/ActaHortic.2016.1111.32

Pinzon MI, Sanchez LT, Garcia OR et al (2020) Increasing shelf life of strawberries (*Fragaria* ssp) by using a banana starch-chitosan-*Aloe vera* gel composite edible coating. Int J Food Sci Technol 55:92–98. https://doi.org/10.1111/ijfs.14254

Poppe J (1992) Gelatin. In: Imeson A (ed) Thickening and gelling agents for food. Springer, Boston, MA, pp 98–123

Prabu GTV, Dhurai B (2020) A novel profiled multi-pin electrospinning system for nanofiber production and encapsulation of nanoparticles into nanofibers. Sci Rep 10:4302. https://doi.org/10.1038/s41598-020-60752-6

Prakash A, Baskaran R, Paramasivam N, Vadivel V (2018) Essential oil based nanoemulsions to improve the microbial quality of minimally processed fruits and vegetables: a review. Food Res Int 111:509–523. https://doi.org/10.1016/j.foodres.2018.05.066

Rai S, Poonia A (2019) Formulation and characterization of edible films from pea starch and casein. J Pharmacogn Phytochem 8:317–321

Rao J, McClements DJ (2012) Food-grade microemulsions and nanoemulsions: role of oil phase composition on formation and stability. Food Hydrocoll 29:326–334. https://doi.org/10.1016/j.foodhyd.2012.04.008

Renkema JMS, Vliet TV (2002) Heat-induced gel formation by soy proteins at neutral pH. J Agric Food Chem 50:1569–1573. https://doi.org/10.1021/jf010763l

Rhim JW (2004) Physical and mechanical properties of water resistant sodium alginate films. LWT – Food Sci Technol 37:323–330. https://doi.org/10.1016/j.lwt.2003.09.008

Ribeiro AM, Estevinho BN, Rocha F (2020) Preparation and incorporation of functional ingredients in edible films and coatings. Food Bioprocess Technol. https://doi.org/10.1007/s11947-020-02528-4

Rocca-Smith JR, Marcuzzo E, Karbowiak T et al (2016) Effect of lipid incorporation on functional properties of wheat gluten based edible films. J Cereal Sci 69:275–282. https://doi.org/10.1016/j.jcs.2016.04.001

Rodrigues FJ, Cedran MF, Garcia S (2018) Influence of linseed mucilage incorporated into an alginate-base edible coating containing probiotic bacteria on shelf-life of fresh-cut yacon (*Smallanthus sonchifolius*). Food Bioprocess Technol 11:1605–1614. https://doi.org/10.1007/s11947-018-2128-z

Rozenberga L, Skute M, Belkova L et al (2016) Characterisation of films and nanopaper obtained from cellulose synthesised by acetic acid bacteria. Carbohydr Polym 144:33–40. https://doi.org/10.1016/j.carbpol.2016.02.025

Saito H, Sakakibara Y, Sakata A et al (2019) Antibacterial activity of lysozyme-chitosan oligosaccharide conjugates (LYZOX) against *Pseudomonas aeruginosa*, *Acinetobacter baumannii* and methicillin-resistant *Staphylococcus aureus*. PLoS One 14:e0217504. https://doi.org/10.1371/journal.pone.0217504

Sánchez-Ortega I, García-Almendárez BE, Santos-López EM et al (2014) Antimicrobial edible films and coatings for meat and meat products preservation. Sci World J 2014:1–18. https://doi.org/10.1155/2014/248935

Sapper M, Chiralt A (2018) Starch-based coatings for preservation of fruits and vegetables. Coatings 5:152. https://doi.org/10.3390/coatings8050152

Saucedo-Pompa S, Rojas-Molina R, Aguilera-Carbó AF et al (2009) Edible film based on candelilla wax to improve the shelf life and quality of avocado. Food Res Int 42:511–515. https://doi.org/10.1016/j.foodres.2009.02.017

Scartazzini L, Tosati JV, Cortez DHC et al (2019) Gelatin edible coatings with mint essential oil (*Mentha arvensis*): film characterization and antifungal properties. J Food Sci Technol 56:4045–4056. https://doi.org/10.1007/s13197-019-03873-9

Sejidov FT, Mansoori Y, Goodarzi N (2005) Esterification reaction using solid heterogeneous acid catalysts under solvent-less condition. J Mol Catal A Chem 240:186–190. https://doi.org/10.1016/j.molcata.2005.06.048

Semo E, Kesselman E, Danino D, Livney YD (2007) Casein micelle as a natural nano-capsular vehicle for nutraceuticals. Food Hydrocoll 21:936–942. https://doi.org/10.1016/j.foodhyd.2006.09.006

Shahbazi Y (2018) Application of carboxymethyl cellulose and chitosan coatings containing *Mentha spicata* essential oil in fresh strawberries. Int J Biol Macromol 112:264–272. https://doi.org/10.1016/j.ijbiomac.2018.01.186

Shewry PR, Halford NG (2002) Cereal seed storage proteins: structures, properties and role in grain utilization. J Exp Bot 53:947–958. https://doi.org/10.1093/jexbot/53.370.947

Shukla R, Cheryan M (2001) Zein: The industrial protein from corn. Ind Crop Prod 13:171–192. https://doi.org/10.1016/S0926-6690(00)00064-9

Sikorski P, Mo F, Skjåk-Bræk G, Stokke BT (2007) Evidence for egg-box-compatible interactions in calcium-alginate gels from fiber x-ray diffraction. Biomacromolecules 8:2098–2103. https://doi.org/10.1021/bm0701503

Song F, Tang D-L, Wang X-L, Wang Y-Z (2011) Biodegradable soy protein isolate-based materials: a review. Biomacromolecules 12:3369–3380. https://doi.org/10.1021/bm200904x

Sothornvit R (2005) Edible film formation and properties from different protein sources and orange coating application. Acta Hortic 682:1731–1738. https://doi.org/10.17660/ActaHortic.2005.682.231

Sothornvit R, Krochta JM (2005) Plasticizers in edible films and coatings. In: Han J (ed) Innovations in food packaging. Academic Press, London, UK, pp 403–433

Spotti ML, Cecchini JP, Spotti MJ, Carrara CR (2016) Brea gum (from *Cercidium praecox*) as a structural support for emulsion-based edible films. LWT – Food Sci Technol 68:127–134. https://doi.org/10.1016/j.lwt.2015.12.018

Suput D, Lazic V, Popovic S, Hromis N (2015) Edible films and coatings: sources, properties and application. Food Feed Res 42:11–22. https://doi.org/10.5937/ffr1501011s

Tang Y, Zhou Y, Lan X et al (2019) Electrospun gelatin nanofibers encapsulated with peppermint and chamomile essential oils as potential edible packaging. J Agric Food Chem 67:2227–2234. https://doi.org/10.1021/acs.jafc.8b06226

Taqi A, Askar KA, Nagy K et al (2011) Effect of different concentrations of olive oil and oleic acid on the mechanical properties of albumen (egg white) edible films. Afr J Biotechnol 10:12963–12972. https://doi.org/10.5897/ajb11.1971

Tavakolipour H, Kalbasi-Ashtari A, Mokhtarian M (2020) Effects of coating pistachio kernels with mixtures of whey protein and selected herbal plant extracts on growth inhibition of aspergillus flavus and prevention of aflatoxin during storage. J Food Saf 40:12711. https://doi.org/10.1111/jfs.12711

Tharanathan RN (2005) Starch - value addition by modification. Crit Rev Food Sci Nutr 45:371–384. https://doi.org/10.1080/10408390590967702

Tillekeratne M, Easteal AJ (2000) Modification of zein films by incorporation of poly(ethylene glycol)s. Polym Int 49:127–134. https://doi.org/10.1002/(sici)1097-0126(200001)49:1<127::aid-pi320>3.3.co;2-2

Travalini AP, Prestes E, Pinheiro LA, Demiate IM (2018) Extraction and characterization of nanocrystalline cellulose from cassava bagasse. J Polym Environ 26:789–797. https://doi.org/10.1007/s10924-017-0983-8

Trevisani M, Cecchini M, Siconolfi D et al (2017) Effects of beeswax coating on the oxidative stability of long-ripened Italian salami. J Food Qual:8089135. https://doi.org/10.1155/2017/8089135

Tripathi P, Dubey NK (2004) Exploitation of natural products as an alternative strategy to control postharvest fungal rotting of fruit and vegetables. Postharvest Biol Technol 32:235–245. https://doi.org/10.1016/j.postharvbio.2003.11.005

Vakkalanka MS, D'Souza T, Ray S et al (2012) Emerging packaging technologies for fresh produce. In: Yam KL, Lee DS (eds) . Woodhead Publishing Limited, Emerging food packaging technologies, pp 109–133

Valdés A, Burgos N, Jiménez A, Garrigós MC (2015) Natural pectin polysaccharides as edible coatings. Coatings 5:865–886. https://doi.org/10.3390/coatings5040865

Valencia-Chamorro SA, Pérez-Gago MB, Del Río MA, Palou L (2010) Effect of antifungal hydroxypropyl methylcellulose-lipid edible composite coatings on Penicillium decay development and postharvest quality of cold-stored "Ortanique" mandarins. J Food Sci 75:418–426. https://doi.org/10.1111/j.1750-3841.2010.01801.x

Vargas M, Pastor C, Chiralt A (2008) Recent advances in edible coatings for fresh and minimally. Crit Rev Food Sci Nutr 48:496–511. https://doi.org/10.1080/10408390701537344

Vieira MGA, da Silva MA, dos Santos LO, Beppu MM (2011) Natural-based plasticizers and biopolymer films: a review. Eur Polym J 47:254–263. https://doi.org/10.1016/j.eurpolymj.2010.12.011

Vigan M (2010) Essential oils: renewal of interest and toxicity. Eur J Dermatol 20:685–692. https://doi.org/10.1684/ejd.2010.1066

Villalobos R, Chanona J, Hernández P et al (2005) Gloss and transparency of hydroxypropyl methylcellulose films containing surfactants as affected by their microstructure. Food Hydrocoll 19:53–61. https://doi.org/10.1016/j.foodhyd.2004.04.014

Vishwasrao C, Ananthanarayan L (2017) Delayed post-harvest ripening-associated changes in *Manilkara zapota* L. var. Kalipatti with composite edible coating. J Sci Food Agric 97:536–542. https://doi.org/10.1002/jsfa.7758

Wang Y, Ma J, Xu Q, Zhang J (2017) Fabrication of antibacterial casein-based ZnO nanocomposite for flexible coatings. Mater Des 113:240–245. https://doi.org/10.1016/j.matdes.2016.09.082

Wang Q, Zhang L, Ding W (2020) Eugenol nanocapsules embedded with gelatin-chitosan for chilled pork preservation. Int J Biol Macromol 158:837–844. https://doi.org/10.1016/j.ijbiomac.2020.04.182

Wax C (1984) Toxicology AC of 1984. Final report on the safety assessment of candelilla wax, carnauba wax, Japan wax, and beeswax. J Am Coll Toxicol 3:1–41

Wei Q, Becherer T, Angioletti-Uberti S et al (2014) Protein interactions with polymer coatings and biomaterials. Angew Chem Int Ed 53:8004–8032. https://doi.org/10.1002/anie.201400546

Xiong Y, Li S, Warner RD, Fang Z (2020) Effect of oregano essential oil and resveratrol nanoemulsion loaded pectin edible coating on the preservation of pork loin in modified atmosphere packaging. Food Control 114:107226. https://doi.org/10.1016/j.foodcont.2020.107226

Xu L, Zhang B, Qin Y et al (2020) Preparation and characterization of antifungal coating films composed of sodium alginate and cyclolipopeptides produced by *Bacillus subtilis*. Int J Biol Macromol 143:602–609. https://doi.org/10.1016/j.ijbiomac.2019.12.051

Xue Y, Zhou S, Fan C et al (2019) Enhanced antifungal activities of eugenol-entrapped casein nanoparticles against anthracnose in postharvest fruits. Nano 9:1777. https://doi.org/10.3390/nano9121777

Yahaya SM, Mardiyya AY (2019) Review of post-harvest losses of fruits and vegetables. J Sci Tech Res 13:10192–10200. https://doi.org/10.26717/BJSTR.2019.13.002448

Yan C, Zhang W (2014) Coacervation processes. In: Microencapsulation in the food industry. Elsevier Inc., Amsterdam, The Netherlands, pp 125–137

Yang L, Paulson AT (2000) Mechanical and water vapour barrier properties of edible gellan films. Food Res Int 33:563–570. https://doi.org/10.1016/S0963-9969(00)00092-2

Yang W, Kenny JM, Puglia D (2015) Structure and properties of biodegradable wheat gluten bionanocomposites containing lignin nanoparticles. Ind Crop Prod 74:348–356. https://doi.org/10.1016/j.indcrop.2015.05.032

Yemiş GP, Candoğan K (2017) Antibacterial activity of soy edible coatings incorporated with thyme and oregano essential oils on beef against pathogenic bacteria. Food Sci Biotechnol 26: 1113–1121. https://doi.org/10.1007/s10068-017-0136-9

Yousuf B, Srivastava AK, Ahmad S (2020) Application of natural fruit extract and hydrocolloid-based coating to retain quality of fresh-cut melon. J Food Sci Technol:1–12. https://doi.org/10.1007/s13197-020-04397-3

Yue TT, Li X, Wang X-X et al (2018) Electrospinning of carboxymethyl chitosan/polyoxyethylene oxide nanofibers for fruit fresh-keeping. Nanoscale Res Lett 239. https://doi.org/10.1186/s11671-018-2642-y

Yue Q, Shao X, Wei Y et al (2020) Optimized preparation of tea tree oil complexation and their antifungal activity against *Botrytis cinerea*. Postharvest Biol Technol 162:111114. https://doi.org/10.1016/j.postharvbio.2019.111114

Zhang C, Li Y, Wang P, Zhang H (2020) Electrospinning of nanofibers: potentials and perspectives for active food packaging. Compr Rev Food Sci Food Saf 19:479–502. https://doi.org/10.1111/1541-4337.12536

Zhao L, Duan G, Zhang G et al (2020a) Electrospun functional materials toward food packaging applications: a review. Nano 10(1):150. https://doi.org/10.3390/nano10010150

Zhao M, Huang X, Zhang H et al (2020b) Probiotic encapsulation in water-in-water emulsion via heteroprotein complex coacervation of type-a gelatin/sodium caseinate. Food Hydrocoll 105: 105790. https://doi.org/10.1016/j.foodhyd.2020.105790

Scope, Functions, and Novelty of Packaging Edibles

4

Nitya Sharma, Aastha Bhardwaj, Prashant Said, Sukirti Joshi, Amrita Poonia, and J. K. Sahu

Abstract

The idea of edible packaging has been around for a while, but now is the right time to ripe the idea to take hold in the food industry. Due to the prevailing adverse conditions of environmental pollution caused by plastic wastes that end up in soil and freshwater, it has become imperative to find a sustainable packaging solution to replace single-use, lightweight polyethylene polymer plastics for retail marketing. Edible films have been in focus for this purpose because of their biodegradability and additional advantages like partial permeability to moisture and oxygen, along with its role as a carrier of functional ingredients (antimicrobials and antioxidants). Natural biopolymers like starch-based biodegradable edible films are widely accepted because of its competence and abundance. In addition to this, they are easily extractable with high yields, do not affect sensory properties of the food, and can be consumed without any health concerns. They are also found to be significantly cost-effective because of its availability from a wide range of agricultural sources such as cereals or legumes and their by-products, tubers, unripe fruits, and other plant storage organs. The

N. Sharma (✉) · S. Joshi · J. K. Sahu
Centre for Rural Development and Technology, IIT Delhi, Hauz Khas, New Delhi, India

A. Bhardwaj
Department of Food Technology, School of Interdisciplinary Sciences and Technology, Jamia Hamdard (Deemed to be University), Hamdard Nagar, New Delhi, India

P. Said
College of Agricultural Engineering and Post Harvest Technology, CAU Imphal Gangtok, Sikkim, India

A. Poonia
Department of Dairy Science and Food Technology, Institute of Agricultural Sciences, Banaras Hindu University, Varanasi, Uttar Pradesh, India

A. Poonia, T. Dhewa (eds.), *Edible Food Packaging*,
https://doi.org/10.1007/978-981-16-2383-7_4

literature suggests that starch-based edible films are based on five main raw materials: corn, maize, wheat, potato, sweet potato, and cassava. And in the recent years, corn starch has been widely used as a raw material for biodegradable polymer production. This chapter discusses the present status of the various sources used to produce starch-based edible films, novelty in starch-based edible packaging, and their effect on the shelf life of certain category of foods.

Keywords

Edible packaging · Coatings and films · Starch · Environment friendly

4.1 Introduction

An incredible amount of energy and engineering is involved in the manufacturing of petroleum-based materials and they take forever to degrade in nature. Modern day consumers and modern science are paving a way forward in sustainable packaging in the form of edible packaging, which have been driven by factors like sustainability, environment and ecological consciousness, food quality and safety, convenience, and product cost (Janjarasskul and Krochta 2010). Nature is the best packaging material manufacturer, for example, it does create skin of fruits to protect from microbes and other environmental factors and shielding them from physical, chemical, and biological degradation (Patel 2020). Natural sources have been utilized to develop edible proxies for the substitution of these multilayered plastic materials. Edible packaging films and coatings are developed from various food-grade, film-forming biopolymers such as proteins, polysaccharides, lipids, and/or resins or a combination of these. This book chapter gives a brief account on this packaging innovation that can be eaten by the consumer as a part of the food product, types, trends, advantages and disadvantages, and challenges related to their production and commercialization.

4.2 Background

The term edible packaging has taken rebirth in last 50 years. The most common definition among the many available definitions of edible packaging is "Any type of material used for enrobing (i.e., coating or wrapping) various foods to extend shelf life of the product that may be eaten together with food with or without further removal is considered an edible film or coating" (Pavlath and Orts 2009). Films may either be fabricated distinctly and then are coated onto the food material or the coatings are fabricated and applied directly onto the respective food material. In either of the cases, these edible films and coatings act as an obstruction for the easy movement of moisture, oxygen, and solute of the food, across the packaging material and also functions as packaging material which can be consumed along with inside content. Edible coatings are thin films that act as laminates and have intact association with food surface until product is consumed. Application of wax on citrus fruits, that is, edible packaging, is one of the first records available from Southern China in early twelfth century where citrus fruits were preserved and transported to North for the British Emperor's table by coating them with wax and packing in boxes (Hardenburg 1967). This process, then known as larding in Europe, was used to prevent water loss, reduce respiration/gas exchange, and elongate the keeping quality of seasonal fruits. Another example is of soy milk skin (known as *yuba*) that has been traditionally used in Asian countries since the fifteenth century (Park et al. 2002) as a packaging material. Since then, edible packaging has improved significantly with use of different strategies and technologies to improve the functional properties of packaging materials. Edible films and coatings are purposeful materials, chiefly formed from edible biopolymers and food-grade additives (GRAS). For example, proteins, polysaccharides, and lipids have been used to formulate and fabricate edible films and coatings as packaging materials, with various functionalities. Starch- and gum-based (polysaccharide) films and gelatin, collagen, gluten-based (protein) films provide hydrophilicity, while fats and wax-based (lipids) films provide exceptional water vapor barrier properties.

4.3 Functions and Scope

Edible coatings have been applied to food applications, intended for improvement in appearance by imparting gloss to the product, enhancing overall sensory attributes, retention of food texture, and preservation of food product against chemical and microbial spoilages (Lee et al. 2002; Janjarasskul and Krochta 2010). Edible films can also act as a carrier for various bioactive food components. List of functions and specific properties of such packaging edibles is given below.

4.3.1 Edible and Biodegradable Nature

The edibility and inherent biodegradability are two most beneficial characteristics of edible packaging (films and coatings). To maintain GRAS status and edibility of films and coatings, all packaging materials like biopolymers, plasticizers, and other additives should be food-grade along with acceptable food-processing facilities and equipment.

4.3.2 Containment and Protection

An ideal packaging is one which has better mechanical properties and serves function to shield food/content from mechanical injure, from harvesting to processing, during transit and storage, until consumption. The mechanical properties of edible films and coatings are dependent on nature and composition of film-forming materials, and their structural cohesion (Janjarasskul and Krochta 2010).

Simulated standard mechanical examinations are carried out to examine sustainability or strength of edible packaging material to physical impact, pressure, vibrations, or similar mechanical forces. The mass transfer between environment and food or vice versa depends on permeability properties which decide suitability of packaging material. The biopolymers are significantly influenced by the moisture content of food, humidity, and temperature. The physical strength of these biopolymers dramatically varies with temperature and humidity levels which is one of the limitations of edible packaging material being an ideal packaging material. Moisture or oil migration; oxygen, carbon dioxide, and other gases permeation; flavor and odor migration in and out of food; or leaching of packaging material components into food may retard the quality of food (Krochta 2002). Edible films have wide range of barrier properties, although these properties differ with the source of manufacturing like edible packaging have excellent oxygen barrier properties (except lipid-based materials).

4.3.3 Convenience and Preservation

Coating of fruits and vegetables can prevent brushing, cutting, and similar mechanical injuries during handling which results in improved handling convenience. Moreover, the protection from physical damage and prevention from chemical changes causes enhancement of quality of stored/coated foods, ultimately resulting in increased shelf life. The edible coating creates a barrier film between foreign contaminants and the food, thereby contamination and cross contamination are minimized.

4.3.4 Appearance and Printability

Esthetic appearance of a food package is a prime contributor to consumer purchase intention and acceptance. Edible films and coatings facilitate glossy appearance and a smoother texture for printing and labeling.

Edible packaging offers great hope to take control over use of plastic waste which is growing concern. Although having vast scope for edible packaging, consumers have psychological limit to eat wrappers along with wrappers or coatings resulting in hygienic concern. Besides all, the global market shows considerable interest to replace petroleum-based packaging films with edible and renewable biopolymers. Data Bridge Market Research predicted that edible packaging market will attain record hike of about 703.5 million by 2027, growing with the CAGR of 5.00% during 2020–2027. Furthermore, the global bioplastic market is estimated to grow at a CAGR of 25.8%, accounting to approximately US$ 215 billion, during 2020–2027. In Europe, edible packaging materials are included in the Regulations EC 1331/2008 and EU 234/2011 for food additives, enzymes, and flavorings. The raw materials must be part of this list and used according to legislations/permissions. To obtain GRAS status, manufacturer has to apply to FDA. There are three types of

Table 4.1 Commercially available edible packaging materials

Trade name of edible packaging material	Manufacturer
Semperfresh	AgriCoat NatureSeal Ltd.
Pürbloom	FruitSymbiose Inc.
FreshSeal	BASF
Bio-Fresh	De Leye Agro
BioCheeseCoat	Improveat
FibreCoat	Caragum International
Durkex 500	Loders Croklaan
Crystalac	Mantrose-Haeuser
WikiPearl	WikiFoods

Source: Cerqueira et al. (2016)

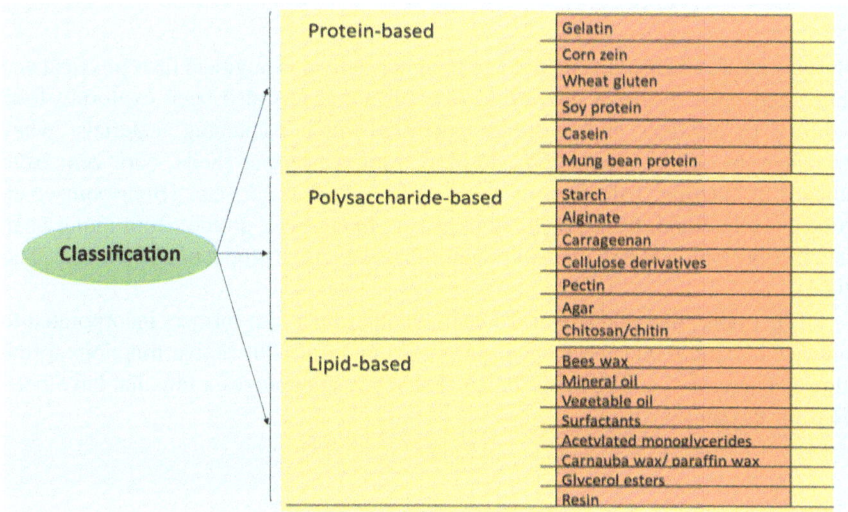

Fig. 4.1 Classification of edible packaging (Saklani et al. 2019)

GRAS status, that is, self-affirmed, FDA approved, and no comment response from FDA. However, GRAS does not guarantee complete food safety to allergenic or sensitive consumers. Cerqueira et al. (2016) has reported some commercially available packaging edibles along with their manufacturers (Table 4.1).

4.4 Types/Classification

Figure 4.1 summarizes the classification of edible films based on their source of origin. Hydrocolloids (e.g., proteins, polysaccharides, and alginate), lipids (e.g., fatty acids, acylglycerol, waxes), and miscellaneous composites are the major sources for formulation and fabrication of edible coatings and films. Additionally,

Table 4.2 Commonly used additives and their examples

Additive	Examples
Plasticizers	Mono-, di-, or oligosaccharides (e.g., glucose, fructose-glucose syrups, mannose, sucrose), polyols (e.g., glycerol, sorbitol, mannitol, xylitol, glyceryl derivatives, ethylene glycol, diethylene glycol, triethylene glycol, tetraethylene glycol, and polyethylene glycol), and lipids and derivatives (e.g., phospholipids, fatty acids, vegetable oils, etc.)
Emulsifiers	Acetylated monoglyceride, lecithin, glycerol monopalmitate, glycerol monostearate, polysorbate (60, 65, 80), sodium lauryl sulfate, sodium stearoyl lactylate, sorbitan monooleate, and sorbitan monostearate
Antimicrobial agents	Organic acids and their salts, chitosan, plant-based essential oil extracts, enzymes, and bacteriocins
Antioxidants	Phenolic antioxidants (butylated hydroxyanisole, butylated hydroxytoluene, propyl gallate, and tertiary butylhydroquinone), tocopherols, citric acid, phosphoric acid

plasticizers are often added to increase their elasticity and augment their physical and functional properties. An interesting research trend has also been exploring food industry by-products and waste as potential edible packaging materials: whey protein from cheese production, chitosan from crustacean shells, corn zein from ethanol production, fish proteins from surimi wash water, etc. (Bourtoom et al. 2006), potato starch from potato chips waste, mung bean protein from mung bean starch (Bourtoom 2008), and fruit pomace from beverage production (Park and Zhao 2006).

Furthermore, film additives are materials other than film formers incorporated to enhance structural and mechanical properties or to provide active functions to the films (Janjarasskul and Krochta 2010). Table 4.2 summarizes a few that have been used for the same.

4.5 Edible Packaging Materials and Their Composition

Edible films are mostly soluble formulations that can be applied to food surfaces to prevent microbial and mechanical spoilage/damage and maintain food quality. They are generally produced by either wet casting or dry extrusion processes; advanced methods such as spraying, fluidized bed coating, and panning are also employed for deposition of edible coatings on the surface of the food material (Suhag et al. 2020). Most preferred method for small-scale fabrication is solvent casting and dipping, while spraying and extrusion are commonly used methods for industrial scale. Henceforth, the various hydrocolloid-based films and coatings and their respective food applications are discussed as follows.

4.6 Protein-Based Films

Globular proteins, such as wheat gluten, soy protein, corn zein, whey protein, and mung bean protein, have been assessed for their film properties (Wittaya 2012). Proteins can be denatured with heat and organic solvents to bring about a change in their physical nature and subsequently formed into a film through solution casting method. The denatured polypeptide chains reassociate through new intermolecular interactions, ensuing modified film properties. Generally, protein-based films have good mechanical and optical properties and are optimal barriers of oxygen, carbon dioxide, and aroma. However, these films are susceptible to moisture, owing to their natural hydrophilic nature (Janjarasskul and Krochta 2010). Thus, incorporation of hydrophobic components by copolymerizing with a hydrophobic polymer or sandwiching between hydrophobic polymer layers to limit the ability of water absorption, seek to improve the hydrophilicity of the protein edible films and coatings. Additional methods include lamination, composite formation, nanoparticle addition, aging, annealing/heat curing, irradiation, cross-linking, ultrasound, microfluidization, etc., that have been attempted and have successfully demonstrated enhanced physicochemical functionality of protein films (Beikzadeh et al. 2020; Vachon et al. 2000).

4.6.1 Wheat Gluten Films and Coatings

They show versatile elastic and cohesive nature, thus providing a sturdy, heat sealable, and optimally transparent biodegradable material. Cross-linking (via enzymatic and chemical treatments) and incorporation of the nonpolar hydrophobic substance (such as mineral oil) have been sought after as potential ways to enhance the barrier properties of the films (Chen et al. 2019). These enhanced functionalities of wheat gluten-based films could be useful for active packaging, drug delivery systems, or modified atmosphere packaging (Guilbert et al. 2002). For instance, gluten-based edible film developed through thermoplastic processing with incorporation of thyme essential oil imparted the film in vitro antioxidant and antimicrobial properties (Ansorena et al. 2016).

4.6.2 Soy Protein Isolate (SPI)-Based Films

These are generally developed through solution casting, spinning, and extrusion methods, and structured via intermolecular disulfide and hydrophobic interactions. These have shown multiple functional attributes, such as adhesiveness, cohesiveness, water and fat absorption capability, fiber formation, and texturizing capacity (Chen et al. 2019). Additionally, they have documented various applications in edible packaging such as coatings on precooked meat and beef products to retard lipid oxidation and prevent meat surface shrinkage (Guerrero and O'Sullivan 2015), coatings on fresh horticultural produce to delay moisture evaporation (Shon and

Choi 2011), and coating on cheese (Al-Sahlany 2017). Composites of gluten protein and SPI have also been applied to peanuts in the form of a coating to prevent lipid oxidation. Addition of additives such as sodium dodecyl sulfate (SDS) and carboxymethyl cellulose (CMC) have also demonstrated improved extensibility and moisture barrier properties of SPI-based films. SPI films and coatings are also potential carriers of antimicrobials (such as essential oils) and flavoring agents for active packaging. A recent trend involves reinforcement of starch nanocrystals in SPI films with exceptional physical and mechanical properties that may also be employed in active food packaging applications (González and Igarzabal 2015).

4.6.3 Casein Protein-Based Coatings and Films

These films have the ability to form a continuous 3D network to form a cohesive film with flexibility, optimal mechanical properties, steady control of mass transfer, and appreciable sensory attributes. However, their hydrophilic nature limits their use in moisture barrier applications. Many techniques such as irradiation and cross-linking have aided in effectively improving their moisture barrier properties (Khwaldia et al. 2004) and have been applied as water-soluble food pouches for fresh produce, dried fruits, and frozen food products (Shendurse et al. 2018). These also have capacity to hold nutrients and bioactive ingredients that could be used to enhance organoleptic and shelf-life characteristics of foods. Due to their sorption ability, they facilitate controlled release of flavor and aroma components, thus enhancing the flavor profile of the food. For instance, irradiated calcium caseinate-based films have been used for microencapsulating flavors in coating of fruits, vegetables, and cheese (Shendurse et al. 2018).

4.6.4 Polysaccharide-Based Films

The monomeric units of polysaccharides are attached together by glycosidic bonds and the disruption of these bonds during the coacervation process to structure new intermolecular hydrogen bonds upon evaporation of the solvent results in fabrication of polysaccharide-based films and coatings. These have good film-forming properties and the presence of large number of hydrophilic moieties renders them permeable to water, while they evidence excellent mechanical as well as gas barrier properties. Thus, polysaccharides can easily be modified to improve their physiochemical properties by addition of salt and solvent, pH change, gelatinization, chemical modification of hydroxyl groups, cross-linking, hydrolysis, and employing nanotechnology (De Moura et al. 2009). They have an ability to form thermally induced gelatinous coatings. Methyl cellulose (MC) and hydroxypropylmethyl cellulose (HPMC) have been used as batter ingredients to minimize oil uptake and moisture loss during deep fat frying (Garcıa et al. 2002).

4.7 Starch-Based Coatings

Being water receptive demonstrates low water vapor barrier ability and has been applied as coatings of fruits and vegetables to augment their shelf life (Sapper and Chiralt 2018). Coatings reduce moisture migration, gaseous exchange, respiration and oxidative reaction rates, suppress physiological disorders, and retard textural changes (Versino et al. 2016). Emulsifiers and plasticizers are added to the film solution to improve the flexibility and extensibility of the final film structure. Owing to their biodegradability, edibility, and low cost, native and modified starch-based films have also received tremendous attention and applause for food packaging applications (Dai et al. 2019). Starch films have been demonstrated as a potential polymer matrix for controlled release of bioactive agents such as antioxidants and antimicrobial agents for active food packaging applications.

4.8 Chitosan-Based Films

These films have good mechanical properties and are selectively permeable to carbon dioxide and oxygen, however, are poor barriers to moisture and water vapor. Numerous strategies, such as cross-linking, irradiation, ultrasonication, and addition of neutral lipids and fatty acids waxes (Cieśla et al. 2006; Morillon et al. 2002), have been used to improve the functionality of such biopolymer-based films. Chitosan-based coatings are also associated with antifungal and antimicrobial activities that can be applied to antimicrobial food packaging applications. Direct addition of essential oils to chitosan films also show supplementary enhanced effects and applications in food packaging. For example, chitosan coating incorporated with essential oils to retard enzymatic browning in fresh produce (Yuan et al. 2016) and to extend the shelf life of meat products (Soares et al. 2015) have been reported.

4.9 Pectin-Based Edible Films and Coatings

Apple pomace, citrus albedo, sugar beet pulp, etc., are anticipated as potential by-products from food and beverage industries to extract pectin for film/coating formation. Pectin coatings are characterized by their good oxygen and carbon dioxide barrier, ability to retard lipid migration, and prevent moisture loss while retaining the sensory properties of the food product. The most recent trends in the field of pectin coating applications include the shelf-life extension of fresh cut, highly perishable, horticultural produce; the application of pectin coatings as prefrying treatment to reduce the oil consumption in deep fat fried products; and the use as predried treatments to improve the retention of nutrients and quality characteristics of dehydrated and lyophilized food (Valdés et al. 2015).

4.10 Alginate-Based Edible Coatings and Films

These films seek much curiosity and attention for packaging applications such as for improving product quality and extending the shelf life of fresh produce, meat, poultry, and seafoods, as well as cheese by reducing dehydration (as sacrificial moisture agent), regulating respiration, improving product appearance, and enhancing mechanical properties (Senturk Parreidt et al. 2018). A great variety of antimicrobial agents (e.g., essential oils) and antioxidants have been incorporated into alginate-based edible films and coatings that find application in food packaging to protect food against surface discoloration and scalding, mechanical degradation, and oxidative rancidity.

4.11 Lipid-Based Edible Coatings and Films

Lipid-based films are intended to prevent moisture migration. Unlike other films, lipid-based coatings exhibit water vapor barrier properties, owing to their hydrophobicity. Application of natural waxes on fresh horticultural produce prevents their desiccation during processing and handling, in addition to providing gloss to the product. Lipid compounds commonly used for the preparation of lipid-based edible films and coatings include neutral lipids, fatty acids (most fatty.

acids derived from vegetable oils), waxes (carnauba wax, candelilla wax, rice bran wax, beeswax, etc.), and resins (shellac, wood rosin, and coumarone indene). These are not cohesive and self-supporting structures, thus necessitating the addition of additives to film formulations such as plasticizers, emulsifiers, lubricants, binders, and defoaming agents. Acetylated glycerol monostearate-derived coatings have optimal oxygen barrier capability than other counterparts. A more common form of lipid-based film is a composite involving both lipid and hydrocolloid components; lipid imparts water vapor barrier property, while the hydrocolloid component provides selective barrier to oxygen and carbon dioxide, respectively. Lipids have been used as edible films and coatings for meat, poultry, seafood, fruits, vegetables, grains, candies, and fresh, cured, frozen, and processed foods (Singh et al. 2016; Galus and Kadzinska 2015). Composite edible coating systems based on herbs (such as plant exudates, gums, resins, essential oils) act as potential antimicrobial and antioxidant agents for meat- and chicken-based products (Matiacevich et al. 2015). Such films impart a waxy flavor and texture, slippery and greasy surface, and are associated with potential rancidity as a major disadvantage.

4.12 Technology Transfer/Patent

The number of edible packaging technologies developed in the last decade shows a growing interest of the scientists in this area of research. A list of patents that shows the production of edible films and coatings containing bioactives using different biopolymers and active compounds for different purposes is shown in Table 4.3.

Table 4.3 Technologies or patents in the area of packaging edibles

Patent publication number	Publication date	Inventor	Title
CN106317477B	November 5, 2019	Cui Haiying, Wu Juan, Lin Lin	Preparing method and usage of edible sodium alginate antibacterial coating
WO2017091095A1	June 1, 2017	Rui Miguel Nabeiro	Coffee-derived edible component, edible product system comprising said edible component, and use of said system
WO2017091094A1	June 1, 2017	Rui Miguel Nabeiro	Edible coating, edible product system provided with said edible coating, and use of said system
CN105199401B	December 26, 2017	Zhang Le, Wang Lihui, Yin Haisong, Pan Zhiheng, Tang Weihua, Chen Shan, Huang Yanling, Lv Chunhui, Dragon Tail, Liu Xin, Long Cheng Xiuwei, Chen Xi, Qi Fei, Yang Jingyuan, Peng Gao Lubao	Edible antibacterial calcium supplementing packaging film and preparation method thereof
CN104211975B	February 1, 2017	Yin Shouwei Shi, Weijian Tang Chuanhe, Yang Xiaoquan	Preparation method of water-blocking oxygen-blocking edible film
WO2016130376A1	August 18, 2016	Hector Gregorio Lara	Edible emulsion coating for extended shelf life
WO2016111659A1	July 14, 2016	Aykut Onder Barazi	Edible antimicrobial film made of pistachio resin
CN105461973A	April 6, 2016	Chen Yizhong	Preparation method of anti-oxidative edible orange peel fresh-keeping film
CN105061819A	November 18, 2015	Meng Lingwei Zhang, Dongjie Wang Xia, Gaofei Sun Tingting	Edible packaging film, preparation method and applications thereof
CN105086000A	November 25, 2015	Zou Xiaobo, Zhai Xiaodong, Shi Jiyong, Wang Sheng, Zhou Xucheng, Huang Xiaowei, Zhu Yaodi, Li Zhihua	Preparation method for edible packaging film
WO2015031663A1	March 5, 2015	Beverly A. Schad	Edible coating compositions, edible coatings, and methods for making and using the same
CN103589170A	February 19, 2014	Song Linxia	Edible chocolate film

(continued)

Table 4.3 (continued)

Patent publication number	Publication date	Inventor	Title
CN103159970A	June 19, 2013	Wang Xinwei, Zhao Renyong, Tian Shuangqi	Preparation method of edible film with antibacterial and antioxidant functions
WO2013089397A1	June 20, 2013	Gang Mo Sung, Won Jin Kim, Seok Hoon Chang, Jeong Jun Yu	Method and apparatus for manufacturing edible film product
WO2009045022A3	July 22, 2010	Seok Hoon Chang, Kyoung Tae Jung	Edible film
EP1692064B1	March 25, 2009	Jean-Pierre Giraud	Moisture-tight edible film dispenser and method thereof
US20070231441A1	October 4, 2007	Andrew Verrall, Stephen Goodrich, Solomon Brown	Edible film having improved sealing properties
WO2006103698A1	October 5, 2006	Gurudutt Prapulla Siddalingaiya, Mysore Nagarajaro Ramesh	Edible films and coatings based on fructooligosaccharides with probiotic properties
US6165521A	December 26, 2000	Walter G. Mayfield	Food products utilizing edible films and method of making and packaging same
US005620757A	April 15, 1997	Hirofumi Ninomiya, Shoji Suzuki, Kazuhiro Ishii	Edible film and method of making same
US5089307A	February 18, 1992	Hirofumi Ninomiya, Shoji Suzuki, Kazuhiro Ishii	Edible film and method of making same

4.13 Limitations and Challenges

The limiting factors like poor inferior physical characteristics and mechanical strength create hurdle in the growth of edible film market but provides a scope to scientific community for overcoming these limitations. As previously discussed, a number of techniques have been employed to modify/alter the native properties of films. Another major limitation is the compatibility of edible packaging materials with the consumers suffering from respective food allergies. Gluten allergy is a common prevalence among population and packaging coatings derived from wheat gluten will have an adverse effect on the health of the consumer; likewise for milk and lactose intolerant people who would face issues with coatings/films containing dairy ingredients in any form. Another limitation of usage of such films deals with consumer acceptance of edible packaging, who are accustomed to see plastic packages over the shelves, particularly for products that cannot be sold without a package.

4.14 Conclusion

Edible and biodegradable coatings and films are the most upcoming intervention in food packaging field that envisions creating sustainable approach for reducing packaging wastes, and additionally, improving stability, quality, safety, and variety for consumers. Ample researches have pointed out their importance and addressed respective issues related to the functional properties, particularly, the effect of miscellaneous factors (such as types of ingredients used, concentration of plasticizers, temperature, and pH values) on their properties and how these can be applied to food packaging applications. Regarding consumer acceptance, as more consumers are becoming mindful of their carbon footprint and waste contribution, there could be a shift to using edible food packaging as an alternative to harmful plastics. Intelligent technology and innovative marketing strategy, together, are quintessential for the success of edible food packaging.

References

Al-Sahlany ST (2017) Production of biodegradable film from soy protein and essential oil of lemon peel and use it as cheese preservative. Basrah JAgric Sci 30(2):27–35

Ansorena MR, Zubeldía F, Marcovich NE (2016) Active wheat gluten films obtained by thermoplastic processing. LWT-Food Sci Technol 69:47–54

Beikzadeh S, Ghorbani M, Shahbazi N, Izadi F, Pilevar Z, Mortazavian AM (2020) The effects of novel thermal and nonthermal technologies on the properties of edible food packaging. Food Eng Rev 12:333–345

Bourtoom T (2008) Factors affecting the properties of edible film prepared from mung bean proteins. Int Food Res J 15(2):167–180

Bourtoom T, Chinnan MS, Jantawat P, Sanguandeekul R (2006) Effect of select parameters on the properties of edible film from water-soluble fish proteins in surimi wash-water. Lebensm Wiss Technol 39(4):406–419

Cerqueira MÁPR, Teixeira J, Vicente A (2016) Edible packaging today, vol 36. CRC Press, Boca Raton, FL, p 1

Chen H, Wang J, Cheng Y, Wang C, Liu H, Bian H, Pan Y, Sun J, Han W (2019) Application of protein-based films and coatings for food packaging: a review. Polymers 11(12):2039

Cieśla K, Salmieri S, Lacroix M (2006) Modification of the properties of milk protein films by gamma radiation and polysaccharide addition. J Sci Food Agric 86(6):908–914

Dai L, Zhang J, Cheng F (2019) Effects of starches from different botanical sources and modification methods on physicochemical properties of starch-based edible films. Int J Biol Macromol 132:897–905

De Moura MR, Aouada FA, Avena-Bustillos RJ, McHugh TH, Krochta JM, Mattoso LHC (2009) Improved barrier and mechanical properties of novel hydroxypropyl methylcellulose edible films with chitosan/tripolyphosphate nanoparticles. J Food Eng 92(4):448–453

Galus S, Kadzinska J (2015) Food applications of emulsion-based edible films and coatings. Trends Food Sci Technol 45(2):273–283

Garcıa MA, Ferrero C, Bertola N, Martino M, Zaritzky N (2002) Edible coatings from cellulose derivatives to reduce oil uptake in fried products. Innov Food Sci Emerg Technol 3(4):391–397

González A, Igarzabal CIA (2015) Nanocrystal-reinforced soy protein films and their application as active packaging. Food Hydrocoll 43:777–784

Guerrero P, O'Sullivan MG, Kerry JP, de la Caba K (2015) Application of soy protein coatings and their effect on the quality and shelf-life stability of beef patties. RSC Adv 5(11):8182–8189

Guilbert S, Gontard N, Morel MH, Chalier P, Micard V, Redl A (2002) Formation and properties of wheat gluten films and coatings. Protein-Based Films Coat 3:69–122

Hardenburg RE (1967) Wax related coatings for horticultural products. A bibliography. Agricultural Research Service Bulletin 51–15, United States Department of Agriculture, Washington, DC

Janjarasskul T, Krochta JM (2010) Edible packaging materials. Ann Rev Food Sci Technol 1:415–448

Khwaldia K, Perez C, Banon S, Desobry S, Hardy J (2004) Milk proteins for edible films and coatings. Crit Rev Food Sci Nutr 44(4):239–251

Krochta JM (2002) Proteins as raw materials for films and coatings: definitions, current status, and opportunities. In: Gennadios A (ed) Protein-based films and coatings. CRC Press, Boca Raton, FL, pp 1–41

Lee SY, Dangaran KL, Krochta JM (2002) Gloss stability of whey-protein/plasticizer coating formulations on chocolate surface. J Food Sci 67(3):1121–1125

Matiacevich S, Acevedo N, Lopez D (2015) Characterization of edible active coating based on alginate-thyme oil-propionic acid for the preservation of fresh chicken breast fillets. J Food Process Preserv 39(6):2792–2801

Morillon V, Debeaufort F, Blond G, Capelle M, Voilley A (2002) Factors affecting the moisture permeability of lipid-based edible films: a review. Crit Rev Food Sci Nut 42(1):67–89

Park S, Zhao Y (2006) Development and characterization of edible films from cranberry pomace extracts. J Food Sci 71(2):E95–E101

Park SK, Hettiarachchy NS, Ju ZY, Gennadios A (2002) Formation and properties of soy protein films and coatings. In: Protein-based films and coatings. CRC, Boca Raton, FL, pp 123–137

Patel P (2020) Edible packaging. ACS Cent Sci. https://doi.org/10.1021/acscentsci.9b01251

Pavlath AE, Orts W (2009) Edible films and coatings: why, what, and how. In: Embuscado ME, Huber KC (eds) Edible films and coatings for food applications. Springer, London, New York. https://doi.org/10.1007/978-0-387-92824-1

Saklani P, Siddhnath Das SK, Singh SM (2019) A review of edible packaging for foods. Int J Curr Microbiol Appl Sci 8(7):2885–2895. https://doi.org/10.20546/ijcmas.2019.807.359

Sapper M, Chiralt A (2018) Starch-based coatings for preservation of fruits and vegetables. Coatings 8(5):152

Senturk Parreidt T, Müller K, Schmid M (2018) Alginate-based edible films and coatings for food packaging applications. Foods 7(10):170

Shendurse AM, Gopikrishna G, Patel AC, Pandya AJ (2018) Milk protein based edible films and coatings–preparation, properties and food applications. J Nut Health and Food Eng 8(2): 219–226

Shon JH, Choi YH (2011) Effect of edible coatings containing soy protein isolate (SPI) on the browning and moisture content of cut fruit and vegetables. J Appl Biol Chem 54(3):190–196

Singh S, Khemariya P, Rai A, Rai AC, Koley TK, Singh B (2016) Carnauba wax-based edible coating enhances shelf-life and retains quality of eggplant (*Solanum melongena*) fruits. LWT-Food Sci Technol 74:420–426

Soares NMF, Oliveira MS, Vicente AA (2015) Effects of glazing and chitosan-based coating application on frozen salmon preservation during six-month storage in industrial freezing chambers. LWT-Food Sci Technol 61(2):524–531

Suhag R, Kumar N, Petkoska AT, Upadhyay A (2020) Film formation and deposition methods of edible coating on food products: a review. Food Res Int 136:109582

Vachon C, Yu HL, Yefsah R, Alain R, St-Gelais D, Lacroix M (2000) Mechanical and structural properties of milk protein edible films cross-linked by heating and γ-irradiation. J Agric Food Chem 48(8):3202–3209

Valdés A, Burgos N, Jiménez A, Garrigós MC (2015) Natural pectin polysaccharides as edible coatings. Coatings 5(4):865–886
Versino F, Lopez OV, Garcia MA, Zaritzky NE (2016) Starch based films and food coatings: an overview. Starch-Stärke 68(11–12):1026–1037
Wittaya T (2012) Protein-based edible films: characteristics and improvement of properties. Struct Funct Food Eng:43–70
Yuan G, Chen X, Li D (2016) Chitosan films and coatings containing essential oils: the antioxidant and antimicrobial activity, and application in food systems. Food Res Int 89:117–128

Novel Microbial Sources of Packaging Edibles

5

Vandana, Para Dholakia, and Prabhjot Kaur Sabharwal

Abstract

In this contemporary epoch, consumers' quest for high-quality, convenient foods with long shelf life compelled to find the new possible solutions or techniques that do not change the food's natural properties dramatically. It has also been observed that in recent times, synthetic food packaging contributed a copious amount of wastes which poses a serious environmental threat. Nowadays, edible films and coatings are helpful to cater the current needs of consumers and also play a vital role in reducing the plastic footprints. These coatings are inexpensive, harmless and biodegradable and consumed together with the product. This chapter focuses on the novel sources of film materials, its origin, properties, effectiveness and uses.

Keywords

Polysaccharides · Microbial sources · Edibles · Gums · Packaging

5.1 Introduction

For many years, the world is grappling with the accumulation of plastic wastes which causes a severe ecological imbalance as they are resistant to degradation. Plastic bags stay in our oceans for centuries disrupting the wild life. Owing to this issue, the researchers are bound to search for an alternative choice that can beat the plastic as a packaging material. The biodegradable materials have been identified as a new way to address this problem by preparing renewable and low-cost materials.

Vandana (✉) · P. Dholakia · P. K. Sabharwal
Department of Food Technology, Shaheed Rajguru College of Applied Sciences for Women, University of Delhi, New Delhi, New Delhi, India

A. Poonia, T. Dhewa (eds.), *Edible Food Packaging*,
https://doi.org/10.1007/978-981-16-2383-7_5

Major emphasis had been put to produce an environment friendly packaging material that can reduce or eliminate the plastic footprints. Different sources for producing biodegradable material have been established. Gums are non-starch polysaccharides that impart viscous solution at low concentrations. Microbial polysaccharide gums are generally recognized as safe, approved by the Food and Drug Administration. On account of their rheological properties, it can be used as film formers, binders, emulsifiers and thickening and gelling agents. These carbohydrate polymers that make up a significant proportion of the cellular polymers are mainly placed in the outer layer of the cell wall (Pazur 1998). These polysaccharide gums have unique physical properties and are produced by a large array of microorganisms. These gums are additionally ordered into three groups including intercellular, exocellular and cell wall. The exocellular gums continually disseminated into the cell culture medium and can be effortlessly confined. The key elements of cell wall are intercellular polysaccharide gums and cell wall, which cannot be comfortably disconnected from cell biomass. The cell excretes the polysaccharides which are also known by the name of exopolysaccharides, which are of huge significance with respect to trade. Furthermore, the microbial polysaccharides can be acidic (like xanthan and gellan) and neutral (such as dextran and scleroglucan) in nature.

It can be ionic and non-ionic, containing linear polysaccharides to which side chains of different lengths or intricacy are attached. These are microbiologically stable and hence used as edible films or coating. Additionally, in association with bioactive compounds, it helps in extending the shelf life of highly perishable items.

5.2 Production of Microbial Polysaccharides

The amalgamation of polysaccharides favourably takes place in the limited nitrogen and in ample supply of carbon substrate in the growth medium. For optimal polysaccharide synthesis, a carbon/nitrogen ratio of around 10:1 is desired. The production (fermentation) process is generally executed by batch culture. By manipulating the nutrient supply, differential synthesis of polysaccharides can be accomplished. By regulating the supply of nitrogen and metal ions in the medium, primarily polysaccharides with no charges are generated and acidic polysaccharides are mainly integrated respectively. For the development and polysaccharide synthesis, around 90% saturation of molecular oxygen supply is adequate. The polysaccharide production is marked by the increase in consistency of the culture medium; it is recovered from appropriate techniques after precipitating with salts, acid or organic solvents. Table 5.1 shows the polysaccharides obtained from microorganisms.

Table 5.1 Polysaccharides obtained from microorganisms

S. no.	Name of the polysaccharide	Name of organism	Main chemical compounds	Application
1	Curdlan gum	*Agrobacterium* spp.	Glucose	Food additive, thickener, gelling agent
2	Cholic acid	*Escherichia coli*	Fucose, glucose, glucuronate and galactose	Viscosity enhancer
3	Gellan gum	*Sphingomonas* spp.	Glucose, rhamnose and glucuronate	Emulsion stabilizer, ophthalmic hydrogel
4	Xanthan gum	*Xanthomonas* spp.	D-glucose, D-mannose and glucuronic acid	Gelling agent in puddings, spreads etc.
5	K30 antigen	*Escherichia coli*	Mannose, galactose and glucuronate	Viscosity enhancer/ controlled drug release
6	Pullulan	*Aureobasidium pullulans*	Maltotriose (glucose)	Film forming, glazing agent, thickener
7	Levan	*Alcaligenes viscosus*	Fructose	Food additive with hypocholesterolemic effects and prebiotic
8	Alginates	*Pseudomonas* and *Azotobacter*	Mannuronate and guluronate	Stabilizer, gelling agent, thickener in soups, sauces and dairy products
9	Dextran	*Leuconostoc mesenteroides*	Glucose	Used in frozen foods
10	FucoPol	*Enterobacter* A47	Fucose, galactose, glucose, glucuronic acid, acetate, succinate and pyruvate	Inner layer in multilayer packaging

5.3 Applications of Microbial Gums in Edible Packaging

Microbial polysaccharides have extensive commercial significance as it could possibly be the alternative solution to synthetic and non-biodegradable packaging material. Its commercial value lies in its ability to change the flow of solutions or rheological properties. Polysaccharides can increase the viscosity of the solution even at low concentration and, are therefore, useful as thickening and gelling agents. These gums own coat-forming properties and can produce biodegradable materials, perhaps appurtenant in food packaging. Regardless of the sources that provide the focal points in utilizing polysaccharides, few disadvantages like low barrier to water, mechanical strength and cost limit its potential use. Nevertheless, with expanding thoughts and examination in this area, it has been conceivable to follow a few systems to overcome the issues. The motive to use biopolymers from biodegradable and inexhaustible resources is a welcoming step in packaging towards sustainable development.

5.4 Microbial Sources

Microorganisms (yeast, fungus or bacteria) produce polysaccharides which have film-forming ability, such as xanthan gum, FucoPol, pullulan, gellan gum, dextran etc.

5.4.1 Pullulan

Pullulan is created by yeast-like fungus *Aureobasidium pullulans*, taking sugars as source. It is neutral, linear, water-soluble exopolysaccharide (EPS) composed of maltotriose units connected by (α-1-4) glycosidic bonds albeit different other units connected with each other by (α-1-6) glycosidic bonds with a chemical formula $C_6H_{10}O_5$ (Singh et al. 2008). The polysaccharide molecular weight is approximately 100–200 kDa. It is an intermediate between the amylase and dextran structure (Farris et al. 2011). Many factors influenced the weight such as pH, temperature, carbon and nitrogen sources. It is white to off-white, tasteless or odourless powder which forms non-hygroscopic solution at 5–10% concentration. It was first discovered by Bernier in 1958 (Bernier 1958). It was commercially produced in the year 1976 by the Hayashibara Company (Okayama, Japan) and got commercialized in the year 1982. It has the ability to form film and it can also be harnessed as flocking agent, food additive or as substitute for blood plasma. Pullulan membranes have numerous properties, like homogeneous, clear, edible, printable, thermostable, not rigid and provide appreciable barriers to oxygen (Farris et al. 2011). Owing to these properties, pullulan membranes efficiently stops the growth of microbes, making a suitable choice in food sector. However, it does not provide barrier against water and is not mechanically strong.

Regardless of every favourable circumstances of pullulan, its significant expense has restricted the utilization of its films in some applications. Examinations are performed by cross linking and mixing polymer with different additives and biopolymers to make films with improved physicochemical attributes and mechanical properties. Many literatures reported that combining pullulan with other materials like alginate, chitosan, cellulose and starch showed significant improvements in thermal and mechanical properties, provides effective barrier against low water absorption and vapour permeability. Consecutively, it also showed improved performance when combined with lipids and proteins. By combining gelatine with pullulan layers exhibit greater elasticity and diminished oxygen porosity and value, whereas the use of rice wax has substantially improved moisture fume obstruction properties.

5.4.2 Gellan Gum

It was first discovered in 1970, is a food additive, an anionic exopolysaccharide, generated by the process of fermentation of saccharides by *Sphingomonas elodea*, previously recognized as *Pseudomonas elodea* or *Auromonas elodea*. This heteropolysaccharide is a compound with molecular weight around 5×10^5 Da. The chemical structure has a backbone consisting of one residue of β-D-glucuronic acid, one residue of α-L-rhamnose and two residues of β-D-glucose, and it is high priced (Chen et al. 2019). The approximate composition is glucuronic acid (20%), rhamnose (20%) and glucose (60%). Kelco (Atlanta, Georgia, USA) pointed out the commercial value during an extensive screening programme of soil and water bacteria. It has two acyl substituent (acetate and glycerate), that is, low acyl gellan gum is acquired by the elimination of acyl groups and tends to form inflexible, firm, brittle and thermostable gels. On the other hand, gellan with more acyl groups forms tender, extensible, non-fragile, heat-reversible gels (Cui 2005).

The gellan gum is comprehensively used as gelling, stabilizing and thickening agent, however, its applications can be stretched out to films and coatings for food industry, for example, batters for coating and adhesion in chicken, fish, cheese, vegetables and potatoes, adhesion systems. It is heat steady and impervious to acid medium and its clarity makes it appropriate for coating material. These membranes and coatings uncommonly offer benefits like it turn down the oil retention by giving an effective obstruction. In batters, for example, item freshness is kept up long after frying or baking.

5.4.3 Xanthan Gum

It is a non-toxic, negatively charged, water-soluble, naturally occurring EPS created by Gram-negative bacteria *Xanthomonas campestris* using simple sugars like sucrose and glucose as exclusive source of carbon. Researchers in Northern Regional Research Laboratories (Peoria, IL, USA), 1963, discovered it, and it was next to curdlan that get commercialized in 1970 (Wustenberg 2014). These days, it is the most comprehensively thought of and by and large recognized modern microbial biopolymer because of its biocompatible nature which makes it valuable in food area. It is more effective than other gums like locust gum and carrageenan (Wallingford and Labuza 1983). This heteropolysaccharide consists of basic rehashing pentasaccharide units made out of mannose, glucose and glucuronic acid (2:2:1 ratio) and pyruvate and acetyl substituent groups. The number of acetyl and pyruvate group varies as it depends on the strain of bacteria used. It gives greater resistance to flow with a strong shear-thinning behaviour even in minute concentrations in aqueous media. Inferable from its flow properties, its solutions are consistent in a wide assortment of ionic strength, pH and temperature values, bioreactor used and way of operation (Borges et al. 2008). It has a wide scope of industrial applications, such as textile, pharmaceutical, petroleum production, food, cosmetics or even slurry explosives.

It is principally used as an additive (suspending and thickening agent) in food industries. It has been approved by FDA without any specifying quantity (Kennedy and Bradshaw 1984). Many literatures reported that this gum is viable to be blended with wide range of materials; hence, it can be used as novel and natural material for making films which can act as a substitute to non-biodegradable food packaging material. Because of its significant expense of xanthan production, very little data are available on xanthan films for food packaging.

5.4.4 FucoPol

FucoPol is an exopolysaccharide with the molecular weight 5×10^6 Da (Torres et al. 2012), water-soluble, anionic and biodegradable (Freitas et al. 2011). It is acquired from natural sources, delivered by *Enterobacter* A47 utilizing glycerol, acquired as by-product from biodiesel industry. It is a characteristic and unadulterated source of monosaccharides; by virtue of this property, it can be used in the synthesis of various drugs, functional food and nutraceutical supplements like HMO (human milk oligosaccharides). This is composed of neutral sugars which are depicted in Table 5.2 (Torres et al. 2012).

At laboratory scale, FucoPol has shown positive growth and its outcome was reminiscent compared to commercial, microbial or bacterial polysaccharides present in the market, for example, gellan and xanthan. This polysaccharide is not available in the market, nonetheless, the production is still in its primary stage. Numerous new examinations exhibited emulsifying and flocculating capacity, equivalent to other polymers accessible in the market (Freitas et al. 2011). On account of its great thickening property in different aqueous formulations that comes with a wide array of ionic and pH strength, makes this polymer a decent choice in several food applications, cosmetics, pharmaceutical, paper, petroleum and textile industries. It is capable to form membrane and it is said to be transparent, with ductile behaviour, profoundly impervious to gases (in specific CO_2 and O_2) that has low water vapour barrier properties.

Taking into consideration, these films have smart potential to be included in a multilayer packaging material as an internal layer (Ferreira et al. 2016). As per the exploration, the layers comprised FucoPol, and chitosan have intensified properties in contrast to FucoPol-independent membranes. The film has several merits such as improved gas hindrance properties and mechanical strength, and lower solvency in liquid water. This membrane could be reasonably used in food packaging which contains low moisture content products.

Table 5.2 Composition of FucoPol

Fucose	36–38% mol
Galactose	22–24% mol
Glucose	27–33% mol
Glucuronic acid	9–10% mol
Acyl groups (acetate, succinate and pyruvate)	12–18 wt%

5.4.5 Dextran

Dextran is formed by the polymerization of α-D-glucopyranosyl moiety of sucrose in a reaction catalyzed by the enzyme dextran sucrase (Seymour et al. 1979). It is neutral in nature and produces solutions of low viscosity. For market needs, it is derived from the bacterial fermentation with *Leuconostoc mesenteroides* which is extricate from a Mexican beverage named 'pulque' (fermented cactus juice), though several microorganisms can produce dextran. It is refined by precipitation with methanol and further dispel in purified water. This system was done threefold to lessen carbohydrates other than dextrans.

According to the reported studies, the characteristic of dextrans in comparison to other biopolymers (potato starch, xanthan gum, carboxy methyl cellulose (CMC), sodium alginate and tragacanthin) unveiled better properties as gas obstruction. However, the limitation of dextran coating is that it does not provide significant water retention on fruits at room temperature rather stored at low temperature (4 °C). If water retentiveness is not the fundamental reason, dextran coating is a decent decision to settle on. A blend of biopolymers (multi-layer coating) as they are soluble in water and organic solvents or their application in an emulsified way could overcome the stated issue.

5.4.6 Curdlan

Curdlan is neutral, water-insoluble and high-molecular-weight polymer, comprises glucose rehashing units connected by β-(1,3) linkage. It has been authorized by the FDA (Food and Drug Administration) as a food additive. For the most part, it is obtained from *Alcaligenes faecalis* var. *myxogenes* and non-pathogenic *Agrobacterium* species. Curdlan is useful as a gelling agent, stabilizer, thickener and fat replacer (FR) in numerous items. Curdlan film (curdlan composite film) can show greater tensile strength, elongation, improved viscosity and good resistance against humidity after the alterations. This film can be used to keep fruits fresh.

5.4.7 GalactoPol

It is a negatively charged contemporary EPS formed by *Pseudomonas oleovorans*, possess different and useful properties which are of great interest in the food sector. It is made up of neutral monosugars and acyl group substituent. Additionally, it is derived from the bounty economical carbon source (glycerol), which is a by-product of the biodiesel industry. It has the potential to form thin films of desired and adequate mechanical characteristics, is appealing in nature and renewable.

5.4.8 Bacterial Cellulose

Cellulose is a basic structural material of the plants, but it has been observed that it is also synthesized by some of the Gram-negative bacteria principally of the genera *Agrobacterium*, *Acetobacter* and *Sarcina ventriculi*. The cellulose derived from the microbial sources have different properties unlike from plant sources and characterized by the ductility, purity, enhanced water holding capacity and have increased strength. Efforts have been made to synthesize bacterial cellulose in laboratory and can be tailor-made to cater the specific need. The films made from bacterial cellulose are flexible, strong, easy to print and impervious to water, but one major factor limiting its use is its high manufacturing cost.

5.4.9 Bacterial Alginate

Alginates are principally generated by *Pseudomonas* and *Azotobacter* and it contains acetyl group unlike those formed from algae. The alginates from algae have wider applications than the bacterial one. Attempts have been made to synthesize alginates commercially from fermentation as it possess desired physicochemical properties like mechanical and tensile strength equivalent to those produced from seaweeds. Alginate films are biodegradable; they protect the food from oxidation by providing shield against oxygen inclusion.

5.4.10 Levan

It is a fructan, which is naturally occurring, synthesized in bacteria, fungi and limited plant sources. For levan synthesis by bacteria, sucrose is used as primary carbon source. It has low sub-atomic weight and generated from monosugar fructose linked by 2–6 beta glycosidic linkage. In recent times, varieties of bacteria have been identified like LAB (lactic acid bacteria), *Bacillus* and *Pseudomonas* and also *Leuconostoc mesenteroides* for synthesis of the same. It is preferred over levan from other sources because of its higher yield and solubility in aqueous medium.

5.5 Regulatory and Safety Concerns

It has been established that these films are an indispensable part of the food and should comply all necessary regulations laid by the authorities. It should confirm the GRAS status and used within any specific limitation given by the US FDA. If any film does not have GRAS status but manufacturer demonstrated that it is safe to consume, can either file GRAS affirmation petition or proceed without FDA concurrence to the market. It is also important to label if it contains any known allergen even if it is used in minute quantities. These coatings should also cater the sensory acceptance of the consumers and does not pose any toxicity.

5.6 Conclusions and Future Perspectives

Exhaustive study and industry research is being administered to seek out novel and improved polymers, origin sources, production ways and characteristics to get biopolymers (in particular, polysaccharides) which will supplant non-biodegradable and traditional artificial ones as packaging materials as it has created havoc on ecological system. From futuristic approach, major emphasis is given on to produce which is competitive in terms of cost and efficiency. An in-depth analysis of life cycle, taking into consideration every facet from production costs to waste disposal and environmental threat, is crucial to adjudge the productivity of polysaccharide membranes than other non-biodegradable counterparts. Advancements in this accessible polysaccharide membrane sought to be done regarding mechanical properties, protection from permeability to water vapour and liquid water is important. The methodical approach can encompass the usability of various additives (including lipids), use of nanoparticles, design of multilayered membranes, mixing with different polymers and chemical changes. Subsequently, it becomes obligatory to search for sustainable method in the manufacturing of packaging for food products.

References

Bernier B (1958) The production of polysaccharides by fungi active in the decomposition of wood forest litter. Canadian J Microbio 4:195–204

Borges CD, Moreirada AS, Vendruscolo CT, Ayub MAZ (2008) Influence of agitation and aeration in xanthan production by *Xanthomonas campestris* pv pruni strain 101. Rev Argent Microbiol 40:81–85

Chen H, Wang J, Cheng Y, Wang C, Liu H, Bian H, Pan Y, Sun J, Han W (2019) Application of protein-based films and coatings for food packaging: a review. Polymers 11(12):2039. https://doi.org/10.3390/polym11122039

Cui SW (2005) Food carbohydrates. Taylor and Francis, Abingdon

Farris S, Introzzi L, Biagioni P, Holz SA, Piergiovanni L (2011) Wetting of biopolymer coatings: contact angle kinetics and image analysis investigation. Langmuir 27:7563–7574. https://doi.org/10.1021/la2017006

Ferreira ARV, Alves VD, Coelhoso IM (2016) Polysaccharide-based membranes in food packaging applications. Membranes 2016(6):22. https://doi.org/10.3390/membranes6020022

Freitas F, Alves VD, Torres CA, Cruz M, Sousa I, Melo MJ, Ramos AM, Reis MA (2011) Fucose-containing exopolysaccharide produced by the newly isolated *Enterobacter* strain A47 DSM 23139. Carbohydr Polym 83:159–165. https://doi.org/10.1016/j.carbpol.2010.07.034

Kennedy JF, Bradshaw IJ (1984) Production, properties and applications of xanthan. Prog Ind Microbiol 19:319–371

Pazur JH (1998) Anti-carbohydrate antibodies with specificity for monosaccharide and oligosaccharide unit of antigens. Adv Carbohydr Chem Biochem 53:201–261

Seymour FR, Knapp RD, Chen ECM, Bishop SH, Jeanes A (1979) Carbohydr Res 74:41–62

Singh R, Saini G, Kennedy J (2008) Pullulan: microbial sources, production and applications. Carbohydr Polym 73:515–531. https://doi.org/10.1016/j.carbpol.2008.01.003

Torres CA, Antunes S, Ricardo AR, Grandfils C, Alves VD, Freitas F, Reis MA (2012) Study of the interactive effect of temperature and pH on exopolysaccharide production by *Enterobacter* A47 using multivariate statistical analysis. Bioresour Technol 119:148–156

Wallingford L, Labuza TP (1983) Evaluation of the water-binding properties of food hydrocolloids by physical/chemical methods and in a low fat meat emulsion. J Food Sci 48:1–5

Wustenberg T (2014) General overview of food hydrocolloids. In: Cellulose and cellulose derivatives in the food derivatives in the food industry: fundamentals and applications. Wiley-VCH, Weinheim, Germany, pp 1–68

Part II

Sustainable Alternatives for Packaging Edibles, Sources & Applications

Fruit and Vegetable Waste: A Taste of Future Foods

6

Nisha Chaudhary and Priya Dangi

Abstract

Fruits and vegetables are the unexploited horticulture products that hold the largest share in the food waste produced globally. Regardless of being consumed in different forms such as raw, minimally processed and well-processed food and an abundant source of promising invaluable bioactive compounds, a considerable proportion in the form of peels, seeds, skins, rinds and pomace is wasted. This waste can be utilized for the extraction of functional compounds such as dietary fibre, carotenoids, polyphenols, essential oils, vitamins, minerals and certain enzymes. Based on the chemical nature of the compounds and the residue, diverse range of processing strategies, like enzyme-assisted extraction, supercritical fluid extraction, pulse electric field, microwave-assisted extraction and ultrasound-assisted extraction, have been employed for their extraction and purification. Such bioactive compounds exert a productive influence to improve human health, owing to their antioxidant, anticancer, anti-inflammatory and anti-allergic properties. With the technological advancement, these compounds pave the way for the production of enriched or fortified foods and food additives, the segment, which is expanding tremendously due to excellent consumer demand for naturally occurring, healthy and safe products. The chapter aims at providing a comprehensive knowledge on the upgradation of large volumes of fruit and vegetable waste to provide potential bioactive compounds and adopting a zero waste approach.

N. Chaudhary (✉)
Department of Food Science and Technology, College of Agriculture, Nagaur, Agriculture University, Jodhpur, Nagaur, India

P. Dangi
Department of Food Technology and Foods and Nutrition, Institute of Home Economics, University of Delhi, New Delhi, India

A. Poonia, T. Dhewa (eds.), *Edible Food Packaging*,
https://doi.org/10.1007/978-981-16-2383-7_6

Keywords

Fruits and vegetables waste · Bioactive compound · Zero waste approach · Naturally derived food additives · Nutraceuticals

6.1 Introduction

With the advent of increasing population size, changeable diet patterns and surging consumer demand for convenient and wholesome foods, the food-processing industry is expanding exponentially to deliver more and more variants in pre-processed and packaged foods. The demand for fruits and vegetables has amplified substantially, by the presence of natural biologically active compounds, which provides physiological benefits to human health. The technological advancement has brought an upsurge in the production of these seasonal products, yet, improper handling, poor infrastructure facilities and inadequate utilization strike way for extensive losses of these important commodities. The process of converting raw fruits and vegetables into value-added processed products requires the removal of unwanted portions such as stems, stalks, peels, rinds, stones, seeds etc. which additionally account for the production of significant wastes in the form of by-products and residues (Sagar et al. 2018). The waste so produced is highly complex due to the availability of a broad range of fruits and vegetables which undergo a series of processing operations from production to retail stage (Kodagoda and Marapana 2017). These losses and wastes indirectly squander critical reserves (water, land, energy, chemicals, fertilizers and labour), provoking a matter of global concern in terms of its disposal, economy, environmental impact and potential health implications (Vilariño et al. 2017).

According to the Food and Agriculture Organization of the United Nations, out of the total production of fruits and vegetables worldwide, 45% is wasted every year, which is the largest amongst all the food sectors. Regardless of this, India secured second position globally in the production of fruits and vegetables, beholds approximately 4.58–15.88% wastage of such products annually by cause of insufficiency of modern harvesting practices and cold storage rooms. Kummu et al. (2012) reported the quantitative segmentation of waste generated during various operations such as at the time of production (24–30%), post-harvest stage (20%) and consumption (30–35%). The quantity and quality of waste produced vary considerably from commodity to commodity, some of them may produce as high as 25–30% of waste containing potentially active biomolecules. The production of minimally processed fruits and vegetables requires operations such as peeling, slicing, cutting, dicing etc. which generate significant amounts of wastes, as exemplified in the case of mandarins, where peels account for 16% of the whole product. Dicing of papaya generates waste in the form of peel (8.5%), seeds (6.5%) and unusable pulp (32%) (Joshi et al. 2012). Many of the fruits and vegetables are converted into convenient processed products such as juices spawning roughly 5.5 million metric tonnes (MMT) of waste. Wine-producing industries utilize grapes as the key ingredient and produce around 5–9 MMT annually (Schieber et al. 2001). Furthermore, the shelf-life of horticulture produce is limited and can be extended to months and years by using preservation techniques like canning and freezing, which together generate about 6 MMT of solid waste annually, comprising leaves, stalks and stems.

As this enormous amount of waste generated poses a serious threat to the environment and country's economy, there is need to devise sustainable alternatives to further exploit such commodities. These wastes can act as a valuable source of phytochemicals, particularly phenolic compounds, phenolic acids, enzymes, organic acids, proteins and flavouring and colouring agents, which can be utilized as dietary supplements, nutraceutical compounds or functional foods and as source of food additives. The conversion of these wastes produced into a valuable product favours the horticulture-based industries to reduce their cost of treatment, generate additional profits and boost their competitiveness (Gowe 2015).

6.2 Potential Bioactive Components in Fruit and Vegetables Waste

The promising biologically active compounds extracted from fruits and vegetables waste are composed of an exceptional pool of molecules, comprising dietary fibre, phenolic compounds, antioxidants, pigments, enzymes and antimicrobial compounds, which are subsidized in the trimmings, seeds, peels, stems, shells, pomace after juice extraction and oil cakes. Table 6.1 demonstrates the wide range of bioactive factors from fruits and vegetables waste. Such waste is a vital source of various nutrients likewise carbohydrates, proteins, fats, minerals, fibres, etc. The various benign bioactive compounds present in waste are:

6.2.1 Phenolic Compounds

The waste obtained in the course of minimal processing of the fruits is enriched in total phenolics and flavonoids when compared to the final products, as more evident in mango seeds and peels than pulp. Phenolics are considered as the collateral metabolites of fruits and vegetables obligated for their sensory and nutritional profile. Phenolic compounds, constituting one or more aromatic rings and incorporating one or more hydroxyl groups in basic structure, are known to possess some sort of antioxidant activity by virtue of free radical inhibition activity (Balasundram et al. 2006). They serve fundamental duties in the reproduction and development of the plants, participate in defence mechanisms against parasites, pathogens and ultraviolet (UV) irradiation, and further provide colour to plants. Moreover, dietary phenolic compounds may bring additional health benefits related to the diminished risk of generating chronic diseases (Song et al. 2010).

Previous studies have also illustrated that peel and seeds parts are having an abundance of phytochemical compounds in relation to edible tissue. Czech et al. (2020) observed that lemon, orange, and grapefruit peels contain 15% greater total

Table 6.1 Bioactive compounds in fruits and vegetables waste

Fruit/ vegetable	Type of waste	Bioactive components	References
Apple	Pomace	Hydroxycinnamates, phloretin glycosides, quercetin, glycosides, catechins, procyanidins	Teleszko and Wojdyło (2015)
Banana	Bract	Cyanidin, anthocyanidins (delphinidin, pelargonidin, peonidin, petunidin, malvidin)	Alexandra Pazmiño-Durán et al. (2001)
Banana	Peel	Carotenoids (palmitate or caprate, xanthophylls, laurate)	Subagio et al. (1996)
Citrus fruits	Peel and solid residues	Eriocitrin, hesperidin, naringenin	Coll et al. (1998)
Cranberry	Leaves	Catechin, procyanidin B1, (−) epicatechin, myricetin-3-xylopyranoside, quercetin-3-O-galactoside, dimethoxymyricetin-hexoside, methoxyquercetin-pentoside	Teleszko and Wojdyło (2015)
Grapes	Seed	Procyanidins	Saito et al. (1998)
Grapes	Pomace	Catechins, anthocyanins, stilbenes, flavonol glycosides	Schieber et al. (2001)
Grapes	Skin	Catechin, epicatechin, epigallocatechin, epicatechin gallate	Souquet et al. (1996)
Mango	Seed Kernel	Gallates, gallotannins, gallic acid, ellagic acid	Schieber et al. (2000)
Mango	Peel	Flavonol glycosides	Schieber et al. (2000)
Olive	Peel and oil waste water	Myricetin, ferulic, sinapic, caffeic, gallic, and ellagic acids, oleuropein and hydroxytyrosol derivatives	Moo-Huchin et al. (2015)
Carrot	Pomace	Carotene (α and β)	Schieber et al. (2001)
Garlic	Husk	Di-ferulic acid, hydroxybenzoic acid, p-coumaric acid, caffeic acid-O-glucoside, coumaric acid-O-glucoside, N-caffeoylputrescine	Kallel et al. (2014)
Onion	Skin	Quercetin 3,40-O-diglucoside and quercetin 40-O-monoglucoside	Price and Rhodes (1997)
Potato	Peel	Chlorogenic, gallic, protocatechuic and caffeic acids, chlorogenic acid isomer II	Choi et al. (2016)
Red beet	Peel	L-tryptophan, *p*-coumaric and ferulic acids, cyclodopa glucoside derivatives	Kujala et al. (2001)
Tomato	Peel	Lycopene	Sharma and Le Maguer (1996)

phenolic compounds when compared with pulp of the same fruits. Likewise, peels of peaches, apples and pears along with yellow and white flesh nectarines yield double the quantity of total phenolic compounds as found in their fruit pulp. The phenolic compounds of bananas in the edible pulp part comprise of 232 mg/100 g of dry weight, which accounts approximately to 25% of the peel portion (Someya et al. 2002). Similarly, 249.4 mg/g of phenolic compounds is present in pomegranate peels, whilst 24.4 mg/g in the pomegranate pulp (Wolfe and Liu 2003). Interestingly, total phenolic compounds of seeds of various fruits as mangoes, longans, jackfruit and avocados were noticeably exceeding the edible portion (Soong and Barlow 2004). The seeds and peels of tomatoes are enriched with phenolic compounds than tomato pulp. Distinctly, 12 genotypes of tomato were under investigation for phenolic compound assessment, which resulted in curtailing levels in the flesh, 9.2–27.0 mg/100 g, with respect to 10.4–40.0 mg/100 g in the peel portion (Del Valle et al. 2006). Besides phenolic compounds, the peel also contains reasonable amounts of ascorbic acid, flavonoids and lycopene pigment when compared with pulp and seeds in several tomato cultivars (Toor and Savage 2005). Surprisingly, phenolic content of the waste of fruits and vegetables processing is leading tenfold than pulp. Such by-products could be treasured as the origin of phytochemicals for human health.

Zeyada et al. (2008) categorized the fruits and vegetables waste as per the richness in phenolic content in the following booming order: potato peel > watermelon peel > cucumber peel > tomato peel > olive leaves. Date seeds are also an exemplary source of antioxidants and phenolic compounds (Al-Farsi and Lee 2008). Date seed oil possesses a higher amount of phenolic compounds than nearly all edible oils, except the olives (Besbes et al. 2005). The seed waste extracts of cucumber, squash, bitter melon, bottle gourd and Indian round gourd have been observed to be extremely effective against certain microbes, such as *Escherichia coli*, *Fusarium oxysporium*, *Serratia marcescens*, *Streptococcus thermophilus* and *Trichoderma reesei* (Sonia et al. 2016), perhaps because of their high phenolic content. A huge waste produced by the citrus industry accounts for its peel and seed residues that is approximately 50% of the total fruit (Ignat et al. 2011). This waste encompasses phenolic compounds in supreme quantity when compared with the edible portion (Balasundram et al. 2006). Enriched concentrations of phenolics have also been found in peels of other fruits (apples, peaches and pears) in comparison to their edible parts (Gorinstein et al. 2001). It was reported that phenolic compounds found in banana pulp are merely the 25% of that contained in the peel (Someya et al. 2002). In conjunction with phenolic compounds, lofty portion of catecholamines, dopamine and levodopa were also present in banana peels (Gonzalez-Centeno et al. 2010). The peels of varied varieties of clingstone peaches were illustrated to obtain a larger concentration of phenolic compounds (more than 2.0–2.5 times) than the flesh (Chang et al. 2000).

6.2.2 Flavonoids

Polyphenolic compounds are classified into flavonoids, tannins, phenolic acids, stilbenes and lignans. Amongst diverse types of fruits, vegetables and other plant-based foods, flavonoids constitute one of the biggest bunches of phenolic compounds. More than 4000 flavonoids have been described up to now in the literature. The common generic structure of flavonoids is constituted of two aromatic rings (A and B rings) associated by three carbons that are generally present in an oxygenated heterocyclic ring. Distinction amongst generic structures of the heterocyclic ring categorizes them as flavonols, flavones, flavanols (catechins), flavanones, anthocyanidins and isoflavonoids. Flavonols (galangin, kaempferol, myricetin and quercetin), flavones (apigenin, luteolin and chrysin), flavanols (catechin, epicatechin, epigallocatechin (EGC) and epicatechin gallate (ECG)), flavanones (eriodictyol, naringenin and hesperidin), anthocyanidins (cyanidin, delphinidin, malvidin, peonidin, petunidin and pelargonidin) and isoflavonoids (genistein, daidzein, glycitein and formononetin) are commonly known flavonoids, which are present in our diet (Liu 2013). Naturally, flavonoids are present in the form of conjugates in glycosylated or esterified forms in normal situation, though they can also arise as aglycones, particularly due to the effects of food processing. Flavonoids have been recognized as bound to more than 80 different sugars (Hollman and Arts 2000). Anthocyanins are responsible for the red and blue colours of several fruits, vegetables and whole grains. Whereas, most of the occurrence of flavonoids is observed in oranges and orange juices and considered as the vital sources of flavonoid compounds such as hesperedin and naringenin. The dominant portions of flavonoids in apples are cyanides, quercetin and epicatechin. The most plentiful flavonoids in raisins are quercetin glycoside, quercetin, catechin, epicatechin, kaempferol glycoside, kaempferol and rutin (Parker et al. 2007; Zhao and Hall 3rd 2008). Apple pomace as fruit waste possesses an exceeding amount of flavonoids (2153–3734 mg/kg) in the form of isorhamnetin, kaempferol, quercetin, rhamnetin, glycoconjugates, procyanidin and epicatechin. Peel and pulp portion of citrus fruits contain flavones and flavanones in a very high amount, constituting major compounds such as apigenin-glucoside, diosmetin-glucoside, eriocitrin and hesperidin and narirutin. The flavonoid contents of mango kernel seed are profoundly better than pulp, as the former contains 7200–13,000 mg/kg of quercetin, isoquercetin and fisetin. Similarly, banana peel secures the vital content of flavonols (1019.6 mg/kg) in terms of rutin, quercetin, kaempferol, myricetin and laricitrin. Other than these, various berries' (bilberries, blueberries, cranberries and lingonberries) waste as press residue contains an immense quantity of anthocyanins (Ben-Othman et al. 2020). Grape juice and white wine industries produce by-products that include seeds and skins, and are vital sources of plentiful flavonoids, specifically monosaccharides, oligosaccharides and polymeric proanthocyanidins (Shrikhande 2000). In vegetables waste, beetroot pomace, stalks, stems and florets of broccoli and cauliflower withhold a fair quantity of flavonoids. Flavonoid composition of beetroot pomace is composed of catechin, epicatechin and rutin. Whilst flavonoid constituents of broccoli and cauliflower waste are kaempferol and quercetin (Thomas et al. 2018). Agarwal et al. (2012) considered the cucumber peel as an economical source of flavonoids for industrial purposes.

6.2.3 Phenolic Acids

Phenolic acids, which are one of the prime sources of dietary phenolics, can be subdivided into two main groups, as hydroxybenzoic acid and hydroxycinnamic acid derivatives (Liu 2013). Derivatives of hydroxybenzoic acid include p-hydroxybenzoic, gallic, vanillic, protocatechuic and syringic acids. The presence of phenolic acids is usually noticed in food in the bound form and particularly, these are constituents of cell wall structural components such as lignins and hydrolysable tannins, cellulose and proteins through ester bonds. These are commonly attached to fibre, protein, sugar, sugar derivatives and organic acids in different plant foods. The p-coumaric, ferulic, caffeic and sinapic acids are counted under the hydroxycinnamic acid derivatives (Liu 2004). In general, the primary abundance of ferulic acids is found in the seeds and leaves parts of plants, majorly conjugated via covalent bonds with mono- and disaccharides, glycoproteins, plant cell wall polysaccharides, polyamines, insoluble carbohydrate biopolymers, lignin and fibres (Liu 2004). Application of food processing with respect to thermal processing, pasteurization, freezing and fermentation assists in the release of the free and soluble forms of conjugated ferulic acids from the bound form of phenolic acids (Dewanto et al. 2002). In almost all the plants, caffeic, ferulic, p-coumaric, protocatechuic and vanillic acids are present in more or less quantity. Chlorogenic acids and curcumin are primary derivatives of hydroxycinnamic acids existing in plants. Chlorogenic acids, being the esters of caffeic acids, behave as substrates during enzymatic oxidation for inducing browning, especially in apples and potatoes. The most abundant phenolic acids of raisins are caftaric acid, coumaric acid, chlorogenic acid and gallic acid (Zhao and Hall 3rd 2008). The seed, rind and peel of fruits and vegetables acquire an exceeding quantity of phenolic compounds. Further, potato peel was exhibited to behold the 50% of phenolic compounds out of the total bioactive components' quantity (Friedman 1997). Choi et al. (2016) examined the "Superior" variety of the Korean potato and described the larger amount of phenols with respect to chlorogenic acid, chlorogenic acid isomer II and caffeic acid. All remaining fruits and vegetables waste is usually composed of a tough cell wall structure as stalks, stems, peel, seeds, kernel etc., where phenolic acids exist in various forms and quantities.

6.2.4 Organic Acids

A number of organic acids viz. citric acid, succinic acid, malic acid, acetic acid and tartaric acid are general elements of fruits and their successive by-products. These acids have been utilized conventionally as preservatives in the food industry, owing to their antimicrobial efficacy by changing the pH levels of food products. Commonly, bacteria best grow at pH around 6.5–7.5, yet tolerant to the pH range of 4–9, whereas yeasts and molds can grow conveniently at low pH values. Thus, the increment in acidity (by formulating organic acids) of individual food is effective for limiting microbial growth (Raybaudi-Massilia et al. 2009). Microorganisms' lysis occurs by H^+ attack on cell walls, membranes, protein synthesis systems, metabolic enzymes and DNA (Tripathi and Dubey 2004). Organic acids like citric and lactic acids have found applications in food, cosmetic and chemical industries. The production of citric acid can be accomplished efficiently by the fermentation of fruits and vegetables waste using distinct molds, yeasts and bacteria (Swain et al. 2011). Coffee husk and cassava bagasse are extraordinary substrates for *Aspergillus niger* to recover citric acid in good quantity (Vandenberghe et al. 2000). By-product of apple juice or wine industry, apple pomace has been a popular substrate material for *Aspergillus niger* to achieve up to 80% of citric acid as well (Dhillon et al. 2011). Pineapple, mandarins and mixed fruits waste yielded 51.4%, 50% and 46.5% of citric acid, respectively, with the same mold substrate (Prabha and Rangaiah 2014). Imandi et al. (2008) extracted utmost quantity of citric acid by processing pineapple waste via employment of *Yarrowia lipolytica* yeast. Lactic acid holds the prime importance in the carboxylic acid group by virtue of its distinguished usefulness in both kind of industries of food as well as of the non-food. Lactic acid works as an acidulant and preservative in food industry. The production of lactic acid is intricate with regard to the cost of raw material. John et al. (2006) determined that the *Lactobacillus delbrueckii* bacteria can convert total sugars of cassava bagasse into 99% of lactic acid under optimized conditions. Hence, the by-products of fruits and vegetables can easily be utilized for producing lactic acid by employing various microorganisms. Bacteria as *Lactobacillus delbrueckii*, *Lactobacillus casei* and *Lactobacillus plantarum* have been prominently employed to yield lactic acid from potato peel, mango, orange, green peas, sweet corn and cassava residue as substrates (Panda et al. 2016).

6.3 Isolation of Bioactive Elements From Fruits and Vegetables Waste

A classic approach to reduce the burden of high amounts of fruits and vegetables waste and to transform it into valuable products requires extraction of the bioactive compounds. Diverse ranges of techniques are available which can be utilized for isolation and purification of these compounds which can later be exploited as flavouring agents, colouring agents, additives, nutraceutical compounds, functional polymers, etc. The extraction method employed is analogous to the type of compound to be extracted and the quantities of such compounds in the waste material. These methods can be categorized into conventional/traditional and novel techniques.

6.3.1 Conventional Extraction Techniques

Conventional extraction techniques are the ones which are in operation for a long period of time and are usually dependent on the solvent, heat employed or a combination of both. Some of the commonly employed techniques are Soxhlet extraction, hydro-distillation and maceration.

6.3.1.1 Solvent Extraction Method

The solvent extraction method involves the use of various polar (ethanol, methanol, water, etc.) and non-polar solvents (hexane, acetone, etc.), which behave as transporters for the desirable compound amidst distinct phases. The disparity in the extent of solubility of different compounds in various solvents can be utilized as a way for separating the compounds of interest. Polyphenolic compounds solubilize promptly in polar solvents, whilst lipids have the ability to dissolve in non-polar solvents. Comprehensively, ethanol is the most favoured solvent in view of its low price and "GRAS" (generally recognized as safe) status (Galanakis 2013).

The extraction of target compounds using solvent can be accomplished in different ways: Soxhlet method or reflux extraction method. Soxhlet method is the oldest method which was initially devised for the extraction of lipids from a mixture. This method requires the repeated reflux of solvent through the bed of sample until complete extraction takes place. The process performance is dependent on operational parameters, for instance, chemical structure of the sample, composition of the solvent and temperature employed (Lafka et al. 2011). Reflux extraction method involves mixing the sample with a compatible solvent in an agitated vessel, pursued by centrifugation and filtration process. Process efficiency is governed by numerous factors such as the solvent-to-sample ratio, extraction time and pH.

Koubala et al. (2008) successfully extracted phenolic antioxidants from winery wastes using this technique. However, Chen et al. (2001) reported certain modifications in the process for the extraction of pectin and hemicelluloses where ethanol-induced precipitated compounds were further treated with an alkali or acid.

Advantages: Low processing cost and ease of operation of the solvent extraction technique make it beneficial over other techniques.

Limitations: This technique requires prolonged periods of processing which subject heat-labile bioactive compounds to the action of harsh solvents and heat, causing their thermal degradation; thereby reducing the quality and the quantity of the target products. To increase the yield, large volumes of solvent are required, which upscales the economy of process and poses a threat to the environment with regard to their disposal (Sagar et al. 2018; Singh et al. 2017).

6.3.1.2 Hydro-Distillation Technique

Unlike the solvent extraction method, which uses organic solvents for the extraction of bioactive compounds, hydro-distillation is a technique that employs water or steam to extract a wide variety of flavonoids and other potential bioactive compounds. The process exposes the sample to hot water/steam which frees the essential oils located in the oil glands of the plant tissue. Being highly volatile in nature, these essential oils get vaporized during the distillation process, and later condensed on cooling, resulting in an immiscible mixture of an oil and aqueous phase. The product obtained (a mixture of mainly odoriferous, coloured and other phytochemicals) is moved to a separator where the element of interest and fat get separated from water. The physicochemical processes involved are hydro-diffusion, hydrolysis and thermal disintegration.

Advantages: This process is quite easy to operate.

Limitations: The process is suitable for extracting heat-stable compounds, as heat-labile compounds may undergo degradation at high temperatures used in the extraction process. Furthermore, those compounds that exert an insignificant vapour pressure at 100 °C may co-distil with the water and lead to significant losses of target compounds. Overall, the process is highly energy- and time-driven (Vankar 2004).

6.3.1.3 Maceration

The beginning of the process marks with the grinding of the sample containing the product of interest into tiny particles in order to expand its surface area. Subsequently, menstruum (appropriate quantity of the solvent) is poured onto the sample with continuous agitation so as to accelerate the diffusion process and remove the concentrated solution from the surface. Large volumes of prepared solution are obtained after pressing the solid residue followed by the filtration process to remove impurities from the extract.

Advantages: The technique is of low cost.

Limitations: The technique is mostly suitable for carrying out low extractions at small-scale level only (Sagar et al. 2018).

6.3.2 Novel Extraction Techniques

With the upgradation in mechanization, novel techniques have been witnessed to overcome the limitations of conventional extraction techniques. The use of green technologies has brought about an upsurge in the overall yield, quality and purity of the product of interest, along with reduced process time and waste volumes. These major novel techniques are described below:

6.3.2.1 Supercritical Fluid Extraction (SFE)

SFE exposes the analyte between different phases (separation and stationary phase); and the extraction is governed by the fundamental thermodynamic properties of solvents at their supercritical point (Giannuzzo et al. 2003). This point is termed as "the specific temperature (Tc) or pressure (Pc), above which gas and liquid behaves as one phase", whereby the solvent exhibits the characteristics of liquids (density and salvation power) and gas (viscosity, diffusion and surface tension) simultaneously. This behaviour expedites greater extraction of bioactive compounds within short interval of time (Ameer et al. 2017). The selection of supercritical solvent is critical for efficacious working of this process. In comparison to ethane, butane, water and pentane, carbon dioxide is the preferred solvent as it is safe and the desirable conditions (30.9 °C and 73.8 bars) can easily be met, yet its low polarity limits the use as an individual solvent. However, this problem is possibly corrected by employing polar solvents as modifying agents, which certainly, by fixing the solvating capacity of carbon dioxide, amplify its extraction performance. These modifying agents can be methanol, ethanol, dichloromethane, acetone, etc. (Sihvonen et al. 1999). Supercritical carbon dioxide (SC-CO_2) when used in association with modifiers significantly enhances the product yield, as exemplified in the case of naringenin, a type of flavonoid from citrus waste, which was extracted in higher amounts at 9.5 MPa and 58.6 °C in the presence of ethanol than pure SC-CO_2, as reported by Giannuzzo et al. (2003). Similarly, procyanidins and polyphenols were extracted from seeds and peel of grapes by employing the methanol-modified CO_2 (Ashraf-Khorassani and Taylor 2004). Furthermore, this technique is commercially exploited for the recovery of hydroxytyrosol from olive mill waste (Lafka et al. 2011) and extraction of antioxidants from Brazilian cherry seeds (Santos et al. 2011).

Advantages: Carbon dioxide, by virtue of its non-explosive and non-toxic nature, is considered as an inexpensive alternate to organic solvents (Wang and Weller 2006).

Limitations: The scalability of the SFE technique is of major concern due to limited diffusivity of the solvent within the matrix, prolonged extraction period, immense-pressure specifications, intricacy amongst operational parameters, expensive infrastructure and divergence in the product characteristics (such as consistency). Moreover, separation of the solvent from the target product by the end of the SFE process is required that is achieved by several downstream processing steps. This difficulty is surmounted by reconciliating the SFE process with pre-processing steps involving fractionation and chemical/enzymatic conversion of wastes for adequate extraction and purification of bioactive compounds (Ameer et al. 2017).

6.3.2.2 Microwave-Assisted Extraction (MAE)

The extraction process is accomplished by exposing the sample to the electromagnetic field of microwaves ranging between 300 MHz and 300 GHz, although the widely employed frequency is 2450 MHz. The underlying principle of the technique is that as the microwaves pass through the solvent carrying the sample, energy is absorbed and this energy is later converted to thermal energy by the reason of dielectric properties of the solvent, effectuating the disruption of the hydrogen bonds. This action generates dipole moment amongst the molecules leading to the transfer of ions which diffuses solvent into the sample matrix, provoking the dissolution of the target components in the solvent (Datta et al. 2005; Zhang et al. 2011). The proficiency of the process is affected by many factors like the composition of solvent, microwave range, temperature during extraction, extraction time and sample matrix. The choice of solvent is dependent on its dielectric properties (dielectric constant and dielectric loss), and how it interacts with the sample. Dielectric constant relates with the capacity of the solvent to absorb microwave energy and dielectric loss refers to its ability to convert this absorbed energy into heat (Chen et al. 2001).

Inoue et al. (2010) has used the skin of *Citrus unshiu* fruit to extract significant amounts of hesperidin by using the microwaves. Likewise, Chandrasekar et al. (2015) extracted phenolic compounds from apple pomace and described the role of dissipation factor, dielectric constant, solubility and type of the solvent employed on the turnout of these compounds.

Advantages: Higher extraction in much shorter time, diminished solvent demand and low-priced, this technique is superior over conventional methods of extraction (Delazar et al. 2012). This makes the technique most suitable for extracting thermo-labile bioactive constituents by adopting a composite solvent that possesses a lower dielectric loss factor (Koubala et al. 2008).

Limitations: Presently, this process is not exploited to a great extent in the food industry, certainly, due to rigorous food quality and safety regulations. The solvent used during extraction is compelled to be removed which involves additional purification steps that hike up the process economy (Singh et al. 2017).

6.3.2.3 Ultrasound-Assisted Extraction (UAE)

This technique is based upon the exposure of fruits and vegetables waste to sound waves (frequency range 20–2000 kHz). As the sound waves travel through the sample matrix, expansion and compression cycles are induced which tear the molecules apart and unite them, respectively, creating bubbles that grow and collapse gradually. This cavitation process is known to disrupt the cell walls and boost the transfer of target compounds from cellular matrix into the extracting solvents. Such a technique is applied for processing liquid–liquid or liquid–solid samples and the competency of the process is influenced by operational parameters like pressure, temperature, frequency and sonication time (Kentish and Feng 2014). Shen et al. (2017) concluded frequency as the critical factor that influences the yield and the properties of the compound.

Anthocyanins and phenolic compounds from grape peel (Ghafoor et al. 2011) and other valuable elements from winter melon seeds (Bimakr et al. 2013) were successfully extracted by this technique.

Advantages: UAE is considered as an elementary and effective technique in comparison to traditional methods for the extraction of bioactive compounds from fruits and vegetables waste. The technique offers low cost in terms of the structure and operation scheme when compared to other novel techniques like microwave-assisted extraction and supercritical fluid extraction. Since the process works at low temperature conditions, thermal degradation of phytonutrients is prevented and extraction time is greatly reduced (Virot et al. 2010).

Limitations: The use of this technique is limited by its reduced capacity and lower yield compared to other methods.

6.3.2.4 Enzyme-Assisted Extraction (EAE)

The phytonutrients present in fruits and vegetables are majorly located in the cytoplasm of the cell where cell walls composed of polysaccharides (pectins, hemicellulose and cellulose) hinder the release of these targeted intracellular compounds. The treatment of these cells with enzymes (xylanase, β-glucosidase, cellulase, pectinase and β-gluconase) under the influence of mild conditions and aqueous solutions degrades the cell wall structures and aids in the release of target compounds (Gardossi et al. 2010; Moore et al. 2006). The performance of this technique is a function of enzyme concentration, sample composition, water/solid ratio, molecular size of target compounds and time required for hydrolysis.

EAE is quite useful for the separation of numerous compounds like carotenoids from pumpkin (Ghosh and Biswas 2015), anthocyanins from *Crocus sativus* (Lotfi et al. 2015) and grape skin (Muñoz et al. 2004), and lycopene from tomato peel (Zuorro et al. 2011).

Advantages: This technique uses water as an alternative to organic solvent, which distinguishes it from other techniques and referred to it as an environment-friendly technology for the separation of bioactive elements from the waste material.

Limitations: The use of this technique is restrained by the cost of the enzyme which poses difficulty in scaling up the process. However, this limitation is overcome by using an immobilization approach for enzymes which promotes enzyme reuse without compromising its specificity and activity (Puri et al. 2012).

6.3.2.5 Pulsed Electric Field (PEF)

It is a non-thermal processing technique that employs direct current to the sample rather than giving thermal treatment. When high-voltage current is allowed to pass through the sample matrix for fraction of seconds (usually in the range of microseconds to milliseconds), the molecules of cell membranes align themselves according to their respective charges. Over a short span of time, the critical value of transmembrane potential reaches 1 V, thereby increasing the repulsion amongst the charged molecules and widening the pores in the membrane resulting in increase in its permeability (Bryant and Wolfe 1987). The process is operational in either batch or continuous method and the product yield is a function of strength of the field applied, specific energy input, pulse frequency, material characteristics and treatment temperature (Heinz et al. 2003). An electric field ranging between 500 and 1000 V/cm impedes undue rise in temperature, thereby minimizing the degradation of heat-sensitive compounds (Ade-Omowaye et al. 2001).

The PEF technique is found to be most suitable for the extraction of anthocyanin monoglucosides from grape by-products (Corrales et al. 2008), anthocyanins and polyphenols from "Merlot" grapeskin (Delsart et al. 2012), phenolic and flavonoid compounds from orange peel (Luengo et al. 2013).

Advantages: The technique is eco-friendly and can be utilized for the extraction of high-valued bioactive compounds (anthocyanins, betanines, carotenoids, etc.) in substantial amounts within a short span of time. Moreover, as the process is non-thermal in nature, the quality of the final product is not compromised (Liu et al. 2018).

6.4 Utilization of Bioactive Compounds in Food

Non-edible portions of the fruits and vegetables lie between 25 and 30%; however in case of exotic fruits, this proportion further increases resulting in the generation of higher masses of by-products and waste when compared to the corresponding valuable edible portions, thereby altering the economics of the industry. This obstacle can be rectified to some extent by transforming the waste to a valued product that possesses reasonable monetary worth (Sun-Waterhouse et al. 2009).

Keeping this in consideration, Gowe (2015) concluded that fruits and vegetables waste can be presented as an innovative, natural and monetary source of nutraceutical compounds (viz protein, dietary fibre, etc.) and food additives (flavouring agents, colouring compound, antimicrobial compounds, etc.).

6.4.1 Nutraceutical and Functional Compounds

Nutraceutical and functional foods are emerging trends in the food industry due to the ever-increasing demand of consumers for "healthy" food. Such compounds besides providing nutrition play determinant task in boosting health and immunity and latterly halting and conjointly treating specific diseases. Fruits and vegetables are rich and unexploited sources of these functional ingredients. Mango peel constitutes dietary fibre, phenolic compounds, vitamin C, carotenoids, which are well identified for their effects in jeopardizing the cancer, cataract, Parkinson's and Alzheimer's disease (Ayala-Zavala et al. 2010). Núñez Selles et al. (2016) proclaimed that mangiferin (1,3,6,7-tetrahydroxyxanthone-C2-β-D-glucoside) which is a native bioactive xanthonoid of the mango tree *(Mangifera indica)*, either singly or in combination, is associated with positive effects in treating leukaemia and cervix, brain, breast, lung and prostate cancers. Grape seed oil is reported to be a satisfactory source of unsaturated fatty acids, specifically oleic and linoleic acids (8–15% w/w), which corresponds to over and above 89% of the aggregate oil composition (Davidov-Pardo and McClements 2015). Ismail et al. (2015) enlisted the role of grape seed oil as neuroprotective, hepatoprotective and effective in reducing cholesterol levels in liver. In consideration of this, grape seed oil is used in the meat industry as a functional ingredient in order to modify and formulate healthier food products (Choi et al. 2010). Papaya seeds accounting for 22% of the waste from papaya puree plants are known to cure sickle cell diseases and poisoning-related renal disorder (Imaga et al. 2009). Non-digestible oligosaccharides are another group of functional components, which are present in peels of many fruits and vegetables. These compounds are generally considered as prebiotics as they reach the colon undigested, where they are fermented by good microflora such as *Bifidobacteria* and *Lactic acid bacteria*, thus producing a positive effect on health (Kadirvelu et al. 2001). Dietary fibres (soluble and insoluble) are the major components of fruits and vegetables by-products and are known to produce health benefits in humans. Soluble dietary fibre is associated with blood cholesterol and restricts its intestinal absorption (Palafox-Carlos et al. 2011), whilst insoluble dietary fibre is associated with water absorption and intestinal regulation. Mango peel contains a high level of insoluble dietary fibres (45–78%) (Ajila et al. 2010). The peel of yellow passion fruit (*Passiflora edulis*) contains dietary fibre with high activity to protect individuals from diverticular diseases (Yapo and Koffi 2008).

6.4.2 Food Additives

Utilization of fruits and vegetables waste as a budding source of natural food additives has emerged out as an adequate substitute for combating environmental hazards. Besides enhancing the nutritional profile and serving potential health benefits, these compounds provide better acceptability, longevity, stability and safety to the product (Abuajah et al. 2015). Some of the commonly derived food additives are discussed below in detail.

6.4.3 Flavouring Agents

Flavour, being a key attribute of the food, is conferred by an amalgam of volatile compounds present in the matrix. These compounds are predominantly esters, aldehydes, alcohols, terpenes or their derivatives. The typical flavour of food is either due to a single compound called "impact compound" or it is the cumulative compound wherein several compounds interplay with the receptors from the nasal mucosa, thereby producing signals that are interpreted by the brain to conceive a sensory impression (Bicas et al. 2011). The waste parts of fruits and vegetables serve as wellspring of flavour and aroma compounds that meet consumer's expectations for natural, safe and healthy products. Vanilla, which is the prominent and commonly used flavour in food products, is composed of vanillin (4-hydroxy-3-methoxybenzaldehyde) and can be naturally derived from fermented pods of vanilla orchids. Nevertheless, vanillin can also be obtained from pineapple peel that contains the precursor for vanillic acid that is, ferulic acid. Use of the microbial biotransformation technique aids in the conversion of ferulic acid to vanillic acid, which can further be converted to vanillin (Priefert et al. 2001). Likewise, "furaneol" (2,5-dimethyl-4-hydroxy-3(2H)-furanone) responsible for the strawberry flavour can be derived from rhamnose which is obtained through chemical hydrolysis of citrus fruit peel (Haleva-Toledo et al. 1999). Lanza et al. (1976) manifested the use of *Ceratocystis* species to produce pineapple flavour component "ethyl butyrate" from apple pomace and coconut flavour component "δ-decalactone" from olive press cake. Oil extracted from lemon peel is commercially utilized as flavour enhancer for soft and alcoholic beverages (Lota et al. 2002).

6.4.4 Food Colorants

Colouring agents bestow colour to food products, making their physiognomy visually appealing/or assist in restoring colour losses on exposure to natural elements such as light, air, temperature, etc. or as a result of processing, storage, packaging and distribution. As majority of the fruits and vegetables are brightly coloured, their by-products turned out to be an excellent source of naturally occurring pigments such as chlorophylls, carotenoids, anthocyanins and betanins that can be produced at a relatively cheap rate and provide high stability and purity. Anthocyanins are responsible for a range of colours such as red, blue, violet, purple and magenta which can be obtained from the extracts of grape pomace, peels of red cabbage, radishes, purple sweet potatoes, red potatoes, black carrots, coffee husks, etc. (Rodriguez-Amaya 2017). Banana peels and bracts, rich in carotenoids (xanthophylls, α- and β-carotenoid) and anthocyanins (cyanidin and delphinidin) when incorporated in biscuits, culminated into a product that offers high dietary fibre content and less calories beyond affecting the colour, aroma and taste of biscuits. In this way, banana peels could be used as an effective colouring agent in combination with potent antioxidant and antimicrobial activities (Joshi 2007). Likewise, mango seed kernel and peel residue contain the carotenoids in abundance (Kodagoda and Marapana 2017). Kaur et al. (2011) utilized lycopene from tomato peels in dairy products, imparting colour to butter and ice cream. The European Union approved and regulated the commercial use of some of the colorants extracted from crude sources such as curcumin (E100), chlorophyll (E140), chlorophyllin (E141), carotenoids (β-carotene- E160a and lycopene- E160d), lutein (E161b), canthaxanthin (E161g), betanin (E162) and anthocyanins (E163) (Commission 2011).

6.4.5 Texture Modifiers

"Texture modifiers" is an umbrella term that includes all those types of agents that are involved in modifying the texture and mouthfeel of food products. Such agents encompass stabilizers, bulking agents, thickeners and emulsifiers (Saha and Bhattacharya 2010). Citrus fruits' by-products (seed powder and peel) are abundant source of pectin and dietary fibre (Sundar Raj et al. 2012), on such account these products are exclusively utilized as thickening, texturizing, gelling and stabilizing agents (M'hiri et al. 2018). Conjointly, plenty of dietary fibres in these peels fabricate them as carbohydrate-based fat replacers (Radi et al. 2009). Crizel et al. (2013) manifested the production of low-fat yogurts by incorporating fibre exclusively from orange by-products into yogurts. In a similar manner, tomato peels and seeds were incorporated into tomato sauce to improve its texture (Ortega et al. 2017). The husk of cocoa (*Theobroma cacao*) pod is another promising alternative to get an attractive supply of dietary fibre and pectin, which can undergo drying and grinding process before using them as texturizing agents. Supplementary, these pods can be used to extract juice that can be used in the preparation of hydrocolloids (Campos-Vega et al. 2018). These fibres are also extensively used in dairy products especially in ice creams, where they improve the texture and handling properties, administering smooth body and resistance to melting particularly by hindering crystal growth (Elleuch et al. 2011).

6.4.6 Antioxidants

Oxidation of food is one of the detrimental reactions that are capable of altering the flavour, colour, nutritional value and texture of food products. Besides this, the reaction may lead to the production of toxic compounds in the food which are fatal to human health. Subsequently, the need of antioxidant compounds arises, which prevents the occurrence of oxidatively induced degradation of foods and extends their shelf-life. Fruits and vegetables by-products are an abundant source of antioxidant compounds; however, their activity may vary particularly in terms of their behaviour as reducing and/or chelating agents, acidulants and enzyme inhibitors. Chemically, these compounds belong to two major groups: phenolic acid and flavonoids (Palafox-Carlos et al. 2011). Phenolic compounds being peroxide decomposers, free radical inhibitors, oxygen scavengers and metal chelating agents have shown abundant antioxidant properties. Apart from this, such compounds hold strong antitumoural, antibacterial, antiviral, anticancer, cardioprotective and antimutagenic activities (Joshi et al. 2012). Onion peels and stems (Roldán et al. 2008), mango peel and kernel (Ajila et al. 2007) have been studied extensively as hidden sources of antioxidant and antibrowning agents. Citrus fruits' peel and seeds are splendid on account of rare flavanones and many polymethoxylated flavones. Grape skin and pomace is enriched with catechin, epicatechin, epicatechin gallate and epigallocatechin, whilst grape seed extract containing flavonoids acts as a metal chelating agent, reducing agent for hydroperoxide, free radical scavengers such as superoxide, hydroxyl and 1,1-diphenyl2-picrylhydrazyl (Jacob et al. 2008). Grape pomace produced a new class of phenolic compounds with strong antioxidant properties, namely aminoethylthio-flavan-3-ol conjugates, by thiolysis of polymeric proanthocyanidins. Larrosa et al. (2002) reported on the addition of artichoke by-product extract to tomato juice and how it brings about longer shelf-life of the product, assuming high antioxidant activity of the former. Interestingly, eggplant peels rich in anthocyanins are a source of effective antioxidants (Sadilova et al. 2006). Numerous scientists practiced incorporation of these naturally derived antioxidants in a wide variety of products viz. meat, dairy and bakery products that confer additional functionality to the food products (Caleja et al. 2017; Gassara et al. 2016). Smith and Hong-Shum (2011) concluded that antioxidants, namely ascorbic acid (E300), tocopherol (E306), and β-carotene (E160a) derived from fruits and vegetables waste using adequate solvents, have been assigned GRAS (generally recognized as safe) status. Conclusively, the consumption of food with naturally occurring antioxidants, as well as antioxidant-enriched foods helps in the prevention of diseases caused by oxidative stress and such usefulness encourages the likelihood of their commercialization in favour of enhanced shelf-life of food (Pastrana-Bonilla et al. 2003).

6.4.7 Antimicrobial Compounds

Food products are highly susceptible to attack by a wide range of microorganisms and lead to spoilage, for this reason, foods are generally loaded with preservatives rendering the food safe for consumption purpose. An array of antimicrobial compounds (fungicides, salts, enzymes, essential oils and bacteriocins) can be effectively used as preservatives. Some of these compounds are obtained in significant amounts in fruits and vegetables waste such as leaves, roots, bark, pods, stalks, stems, hull, fruits, flowers, seeds, latex and rinds.

Olive leaves containing phenolic compounds exhibit some antimicrobial activity against *Bacillus cereus*, *Escherichia coli*, *Staphylococcus aureus*, *Candida albicans* and *Cryptococcus neoformans* (Talhaoui et al. 2015). Peel extracts of pomegranate (Al-Zoreky 2009) and avocado (Calderón-Oliver et al. 2016) show their activity against foodborne pathogens such as *E. coli*, *Listeria monocytogenes* and *Yersinia enterocolytica*. Papaya extract also shows good antimicrobial activity by virtue of sulphydroxyl protease that inhibits viral or microbial infection (Rajashekhara et al. 1990). Spices, being a source of phenolic compounds particularly capsaicin, gingerone and zingerone, are capable of restricting the growth and multiplication of bacterial spores (Burt 2004). The polyphenols present in green tea (*Camellia sinensis*) fight against different microorganisms such as *Shigella*, *Vibrio cholerae* and *Streptococcus mutans* (Si et al. 2006). The functionality of bioactive compounds as an antimicrobial preservative in foods is gaining interest amongst food processors in order to replace chemical preservatives, make the food nutritious and safe and supports reducing environmental impact. However, the major hitch in its vast utilization in food industry is limited due to lack of their approval for edible purposes.

6.5 Bioactive Compounds for Human Health

Bioactive compounds are food components primarily that influence physiological, metabolic or cellular activities of the humans or animals. Flavonoids, anthocyanins, betalains, tannins, carotenoids, plant sterols and glucosinolates are primary bioactive components. They possess antioxidant, anti-inflammatory and anticarcinogenic properties and can work as protective shield against numerous diseases and metabolic disorders. Bioactive compounds mainly find origin in fruits and vegetables. Such compounds have peculiar effects on various chronic diseases like cancer, cardiovascular disease, certain neurological conditions, diabetes mellitus, etc. A brief account of bioactive compounds, including their chemistry, sources and health benefits, is given in Table 6.2. Pivotal physiological and cellular activities of such bioactive compounds in the prevention of diseases are deliberated here in particular.

6.5.1 Effects of Bioactive Compounds on Chronic Diseases

Cancer is one of the dominant causes of death worldwide. The global cancer patients have multiplied to 18.1 million as new cases and 9.6 million cancer deaths took place in 2018, approximately (World Health Organization 2018). In addition, the increment in deaths because of cancer cases has predicted, which will continue over the next decade too. For a fact, the consumption of enough fruits and vegetables containing superb biological activity is one of the easiest courses to prevent and curb the risk of cancer. A number of investigations have demonstrated the positive

Table 6.2 Health benefits of various bioactive compounds in fruits and vegetables (Walia et al. 2019)

S. no. 4	Bioactive compounds	Chemistry	Waste sources	Health benefits
1.	Flavonoids	Two benzene rings linked to a three-carbon chain, which forms a closed pyran ring	Apple, berries, carrots, cabbage, leeks, onion, ginger, broccoli, grapefruit, kale, tomato, parsley, lemon	• Act as influential antioxidants • Neutralize free radicals and restrict damage to cells and other body tissues • Exert anti-inflammatory and anti-ageing characteristics • Studies depicted relationship amidst certain polyphenols and preventive effects on diseases as cancers, cardiovascular disease and neurodegenerative diseases • Potential to improvise the quality of blood vessel walls • Supportive effect on the nervous system • Help in regulating brain blood flow, result in better cognitive function
2.	Anthocyanins	A phenolic molecule containing 15 carbon atoms and perceived as two benzene rings joined by a chain with three carbons. Accumulation of flavylium nucleus is responsible for its hyperreactivity	Acai, blueberry, blackcurrant, bilberry, cherry, red grape and purple corn	• Conclusive effects on the cardiovascular state of health, anticancer and anti-inflammatory activities
3.	Tannins	Complex mixtures of polymeric polyphenols with gallic acid in terms of a base unit	Pomegranates, persimmons, berries such as cranberries, strawberries, blueberries, grapes, red wine, and spices such as cinnamon, vanilla, cloves, thyme	• Contain antioxidant properties that prevent the tissues from the free radicals by virtue of cellular ageing and other physiological processes • Bring about good effects on health by accelerating blood clotting, reducing blood pressure, decreasing serum lipid levels and immune response modulation • The quantity and quality of tannins are critical to these effects

(continued)

Table 6.2 (continued)

S. no. 4	Bioactive compounds	Chemistry	Waste sources	Health benefits
4.	Betalains	These are indole-derived pigments that are divided into the red-violet betacyanins and the yellow betaxanthins. These are water-soluble pigments	Red and yellow beetroot, leafy or green amaranth, coloured swiss chard, red pitahaya, prickly pear	• Progressive health effects by virtue of antioxidant, anticancer, antimicrobial and antilipidaemic activities • Potential functional foods to reinforce therapies in diseases related to inflammation, oxidative stress and dyslipidaemia like arterial stenosis, atherosclerosis, hypertension and cancer
5.	Carotenoids	Consist of eight isoprenoid units linked in an order that the placement of isoprene units is turned reverse in the molecule centre Cause the elements to be deeply coloured yellow, orange or red	Carrots, plums, mangoes, apricots, cantaloupes, sweet potatoes, kale, coriander, spinach, collard greens, turnip greens, fresh thyme and winter squash	• Exert protective effects for various types of cancers like lung cancer, breast cancer, prostate cancer, and head and neck cancers • Benefit vision and skin. • Being a precursor of vitamin, the carotenoids improve immune system, skin and mucous membranes, and eye health • As indicated by studies, amongst the risk-related biological conditions, the pro-oxidant influences of β-carotene are suspected in the onset of lung cancer • Carotenoid-rich food has been demonstrated to decline the signs of illness related to eye strain (dry eye, headaches and blurred vision) and boost night vision • Commercial supplements of β-carotene are considered like oral sunscreens due to their antioxidant action, exerting protective influence on skin from the sunlight and noxious ultraviolet (UV) radiation

6.	Glucosinolates	Contain a central carbon atom, bounded via a sulphur and a nitrogen atom to the thioglucose group and to a sulphate group, respectively, to form a sulphated aldoxime. Also, the central carbon is obligated to a specifically distinct side group. Glucosinolates with different side groups are responsible for the varied biological activities of these plant compounds	Cruciferous vegetables, such as wasabi, cabbage, broccoli, kale, watercress and garden cress	• Glucosinolates and their biologically active metabolites, specifically as isothiocyanates, work as protective agents against cancer and dementia • These peculiar biochemical substances, which are rarely found in other vegetables, have been designated as efficient in the destruction of various cancer cells without harming healthy cells • Risk of dementia is lowered and the cognitive decline rate slowed down in the elderly

relationship of carotenoid-rich fruits and vegetables consumption with decreased risk of cancer, specifically lung cancer, prostate cancer, breast cancer and head and neck cancers (Leoncini et al. 2015). In the background of lung cancer, several studies reported a decline in morbidity by β-carotene supplementation in non-smoking adults as subjects (Vieira et al. 2016). Further, the recent studies on the diet of non-smoker lung cancer patients revealed an inverse trend between the carotene-rich diet including α-carotene, β-carotene, lutein, lycopene and β-cryptoxanthin, and lung cancer risk (Albanes et al. 1996). Beta-carotene and retinol efficacy trial indicated a diverse tendency between the men and women who were at the stage of high-risk lung cancer development (smokers and asbestos workers), and highly alcoholic subjects (Goodman et al. 2004). All the while, a recent detailed analysis turned out the unforeseen "cancerogenic" (pro-oxidant) effects of supplementation of carotene in individuals with unhealthy lifestyle. Many epidemiological studies support the concept of bolstering in minimizing the risk of prostate cancer by several carotenes, and also their rich sources (Soares Nda et al. 2015). Amongst various carotenoids, the investigations came up with various manners of lycopene action with respect to enrichment of the oxidation stress defence system. Evidently, as per the recent meta-analysis of the observational studies related to the role-play of tomato-based products and lycopene, the roles played by them to inhibit prostate cancer were apparent. Recent human intervention and clinical trials supported the basis of this research (Stahl and Sies 2005). Several researches have shown the decisive role of β-carotene in preventing head, breast, mouth, pharynx and larynx cancers (Leoncini et al. 2015). Such studies further suggested the reduction in the risk of head and neck cancers nearly by 50% by consuming diet with high composition of fruits and vegetables. Some other research-based facts have also exhibited the correlations between the fruits and vegetables consumption and certainty of prevention from oesophageal, colon and other gut cancers (Stahl and Sies 2005). Previous experimental studies have presented increasing evidence for the health benefits of flavonoids on multiple biological pathways related to cancer including carcinogen bioactivation, cell signalling, cell cycle regulation, angiogenesis, oxidative stress and inflammation. Whereas the epidemiologic data of cancer in relation to flavonoids are still scantier and contradictory, a few protective correlations have been implied to flavonoid-rich foods such as soy-based products and pre-menopausal breast cancer; green tea extracts and stomach cancer; onion-based products and extracts and lung cancer (Le Marchand 2002). Lechner and Stoner (2019) has also demonstrated the positive effect of betanin pigments of red beetroot as an effective cancer chemopreventor in an animal study and suggested the requirement of a detailed human study in this regard.

6.5.2 Effects of Bioactive Compounds on Diabetes Mellitus

Bioactive compounds of fruits and vegetables are capable of evolving a potential defence system to counter several diseases and metabolic disorders due to their antioxidant, anti-inflammatory and immunoprotective effects. Such properties designate them as potential protective and preventive means for Type 1 and Type 2 diabetes mellitus also (Oh and Jun 2014). Flavonoids and isoflavonoids are common polyphenolic compounds of fruits and vegetables. Being efficient antioxidants with anti-inflammatory and immunoprotective properties, they can aid in the fight against certain metabolic diseases. Numerous collaborated investigations and randomized trials considered the flavonoids as a risk-lowing factor for type-2 diabetes and cardiovascular disease. Moreover, randomized trials under meta-analyses produce the evidences for beneficial role of flavonoids in curtailing LDL-cholesterol, insulin sensitivity and endothelial function (Grassi et al. 2015; Vandenberghe et al. 2000). Genistein is one of the supreme isoflavones with regard to diabetes, which is abundantly found in plant sources including fava beans, lupine, soybeans and their products. The promising demeanour of genistein for complementary approach for preventing or treating diabetes based on different cell-culture and animal-based investigations has established its productive influences on β-cell function. A human clinical trial on post-menopausal women after giving a dose of genistein (54 mg/day) illustrated the effect of reducing fasting glucose and enhancing the glucose tolerance and the insulin sensitivity (Gilbert and Liu 2013). Lycopene, being one of the strongest antioxidant amongst carotenoids, is acclaimed in promoting health by preventing chronic diseases as cancer and cardiovascular diseases. An animal study, covering exogenous administration of particular bit-by-bit doses of lycopene to hyperglycaemic rats, brought about a drop in glucose level that was dose-dependent, subsequent rise in the concentration of insulin and improvisation in serum lipid profile. In addition, total antioxidant status was multiplied with improved antioxidant enzyme activities. This investigation culminated in the fact that the antidiabetic activity of lycopene proceeds mainly by lowering the free radical activity (Ali and Agha 2009). Various fruits, vegetables and flowers in human diet gain their characteristic colour extensively via anthocyanins and anthocyanidins. Several in vitro studies have revealed the act of anthocyanins and anthocyanidins in stimulating insulin secretion and subsequent protective effects on β-cells of pancreas (Sun et al. 2012). A mass analysis of a total of 3,645,585 women and men was conducted by Wedick et al. (2012), who were not having diseases like diabetes, cardiovascular disease and cancer during outset of the study. In subsequent pursuit of the study, in 12,611 cases out of the masses type-2 diabetes was noted. An anthocyanin-based diet in this study, particularly containing blueberries, apples and pears, was found to be associated with a lower risk of type-2 diabetes in respective masses. The other investigation claimed that the consumption of pomegranate juice (384 mg/dL anthocyanins) by diabetic patients showed antioxidative effects accounting for a considerable decline in serum lipid peroxides (56%) and the oxidative state of their monocytes/macrophages (28%) (Rosenblat et al. 2006).

6.5.3 Effects of Bioactive Compounds on Cardiovascular Diseases

Oxidative stress and inflammation are the basic causes behind the prevalence of atherosclerosis and cardiovascular diseases. The protective roles of bioactive compounds of fruits and vegetables waste were proposed to be related to inhibit oxidative, inflammatory and metabolic stress. Whole berries are rich in components such as polyphenols, particularly anthocyanins, micronutrients and fibre. Several studies claimed the association of berries in promoting heart health. A few investigations using chokeberries, blueberries, cranberries and strawberries in various forms such as fresh form, juice, freeze-dried extracts or powders and purified anthocyanin extracts were carried out. Resulted berries-based products exhibited noticeable improvements in LDL (low-density lipoproteins) oxidation, lipid peroxidation, dyslipidaemia, glucose metabolism and total plasma antioxidant capacity. Whereas, these data depict the partial support in the recommendation of berries being an essential fruit group in relation to a healthy-heart diet (Basu et al. 2010). Studies in many disciplines and randomized trials have implied the part of flavonoids in minimizing the risk of cardiovascular disease, specifically by exerting their healthful influence on LDL-cholesterol, endothelial function and insulin sensitivity (van Dam et al. 2013). A methodical review concentrating on the meta-analyses of several prospectives has also suggested a significant decrease in the risk of cardiovascular disease by intake of six varied classes of flavonoids, viz. anthocyanidins, proanthocyanidins, flavonols, flavones, flavanones and flavan-3-ols (Wang et al. 2014).

6.6 Conclusions and Future Prospects

The multi-dimensional investigations concerning bioactive compounds derived from specific food residues as of fruits and vegetables and the accessibility of highly responsive measurement tools pave the way for great opportunities for appraising respective metabolites in varied food wastes. Diverse bioactive compounds have been evaluated and construed for their vigilant human health effects. They have peculiar antioxidant, anti-inflammatory and anticarcinogenic properties, besides linked with physiological and cellular effects that are involved in preventing varied chronic diseases and metabolic disorders likewise diabetes, cardiovascular disease, cancer, etc. Being the vital components of fruits and vegetables waste and their consumption as whole fruit and vegetable with opportune health effects compositely make them favourable for developing novel potential functional food. However, further intricate research is required to discern the explicit mechanisms of bioactive compounds' biological actions. At the same time, previous investigations deliberate great variation in realistic effects, as heterogeneity occurs in the randomized controlled trials undertaken beneath varied conditions. This postulates the scope of larger, longitudinal research further in order to ascertain rational health benefits and literal mechanism of bioactive compounds to improve human health.

Any food waste or its by-products can be utilized after extraction based on the feasibility of the process and raw material by incorporating any of the approaches aforementioned. Evidently, the supercritical fluid extraction technique was reported to be the most appropriate. Though, the feasibility of the extraction method profoundly has relied on the type of food waste and the outcome of optimization process. The utilization of bioactive compounds of fruits and vegetables waste by incorporating modern and adept bioprocessing techniques will be augmented in the value of food waste and ensued as cost reduction in formulated products. As consumers' demand for new, healthy, natural, innovative products is increasing day by day, such products can be utilized as colouring agents, flavouring agents, health-promoting supplements, nutraceuticals or as preservatives by replacing synthetic chemicals in foods. With the increasing production of fruits and vegetables, post-harvest losses and the industries based on them are also multiplying, which is generating colossal waste. This aforesaid waste turning into a raw material for producing bioactive compounds will be proved as an asset by generating revenue to the country, farmers and lucid utilization of natural resources. Fruits and vegetables waste usage will reduce the hardship in waste management of food-processing industries, since hazardous organic waste generation leads to crippling the environment. Extraction and utilization of bioactive compounds should be commercialized after conceptualizing the economic and technical background of it.

References

Abuajah CI, Ogbonna AC, Osuji CM (2015) Functional components and medicinal properties of food: a review. J Food Sci Technol 52(5):2522–2529. https://doi.org/10.1007/s13197-014-1396-5

Ade-Omowaye BIO, Angersbach A, Taiwo KA, Knorr D (2001) Use of pulsed electric field pre-treatment to improve dehydration characteristics of plant based foods. Trends Food Sci Technol 12:285–295

Agarwal M, Kumar A, Gupta R, Upadhyaya S (2012) Extraction of polyphenol, flavonoid from Emblica officinalis, Citrus limon, Cucumis sativus and evaluation of their antioxidant activity. Orient J Chem 28(2):993

Ajila CM, Naidu KA, Bhat SG, Rao UJSP (2007) Bioactive compounds and antioxidant potential of mango peel extract. Food Chem 105(3):982–988. https://doi.org/10.1016/j.foodchem.2007.04.052

Ajila C, Aalami M, Krishnarau L, Prasada Rao U (2010) Mango peel powder: a potential source of antioxidant and dietary fiber in macaroni preparations. Innov Food Sci Emerg Technol 11:219–224. https://doi.org/10.1016/j.ifset.2009.10.004

Albanes D, Heinonen OP, Taylor PR, Virtamo J, Edwards BK, Rautalahti M, Hartman AM, Palmgren J, Freedman LS, Haapakoski J, Barrett MJ, Pietinen P, Malila N, Tala E, Liippo K, Salomaa ER, Tangrea JA, Teppo L, Askin FB, Taskinen E, Erozan Y, Greenwald P, Huttunen JK (1996) Alpha-tocopherol and beta-carotene supplements and lung cancer incidence in the alpha-tocopherol, beta-carotene cancer prevention study: effects of base-line characteristics and study compliance. J Natl Cancer Inst 88(21):1560–1570. https://doi.org/10.1093/jnci/88.21.1560

Alexandra Pazmiño-Durán E, Giusti MM, Wrolstad RE, Glória MBA (2001) Anthocyanins from banana bracts (Musa X paradisiaca) as potential food colorants. Food Chem 73(3):327–332. https://doi.org/10.1016/S0308-8146(00)00305-8

Al-Farsi MA, Lee CY (2008) Optimization of phenolics and dietary fibre extraction from date seeds. Food Chem 108:977–985

Ali MM, Agha FG (2009) Amelioration of streptozotocin-induced diabetes mellitus, oxidative stress and dyslipidemia in rats by tomato extract lycopene. Scand J Clin Lab Invest 69(3):371–379. https://doi.org/10.1080/00365510802658473

Al-Zoreky NS (2009) Antimicrobial activity of pomegranate (Punica granatum *L.*) fruit peels. Int J Food Microbiol 134(3):244–248. https://doi.org/10.1016/j.ijfoodmicro.2009.07.002

Ameer K, Shahbaz HM, Kwon J-H (2017) Green extraction methods for polyphenols from plant matrices and their byproducts: a review. Comp Rev Food Sci Food Safety 16(2):295–315. https://doi.org/10.1111/1541-4337.12253

Ashraf-Khorassani M, Taylor LT (2004) Sequential fractionation of grape seeds into oils, polyphenols, and procyanidins via a single system employing CO_2-based fluids. J Agric Food Chem 52(9):2440–2444. https://doi.org/10.1021/jf030510n

Ayala-Zavala JF, Rosas-Domínguez C, Vega-Vega V, González-Aguilar GA (2010) Antioxidant enrichment and antimicrobial protection of fresh-cut fruits using their own byproducts: looking for integral exploitation. J Food Sci 75(8):R175–R181. https://doi.org/10.1111/j.1750-3841.2010.01792.x

Balasundram N, Sundram K, Samman S (2006) Phenolic compounds in plants and agri-industrial by-products: antioxidant activity, occurrence, and potential uses. Food Chem 99:191–203

Basu A, Rhone M, Lyons TJ (2010) Berries: emerging impact on cardiovascular health. Nutr Rev 68(3):168–177. https://doi.org/10.1111/j.1753-4887.2010.00273.x

Ben-Othman S, Jõudu I, Bhat R (2020) Bioactives from agri-food wastes: present insights and future challenges. Molecules (Basel, Switzerland) 25(3):510. https://doi.org/10.3390/molecules25030510

Besbes S, Blecker C, Deroanne C, Lognay G, Drira NE, Attia H (2005) Heating effects on some quality characteristics of date seed oil. Food Chem 91:469–476

Bicas JL, Molina G, Dionísio AP, Barros FFC, Wagner R, Maróstica MR, Pastore GM (2011) Volatile constituents of exotic fruits from Brazil. Food Res Int 44(7):1843–1855

Bimakr M, Rahman RA, Taip F, Adzahan NM, Sarker ZI, Ganjloo A (2013) Ultrasound-assisted extraction of valuable compounds from winter melon (*Benincasa hispida*) seeds. Int Food Res J 20:331–338

Bryant G, Wolfe J (1987) Electromechanical stresses produced in the plasma membranes of suspended cells by applied electric fields. J Membr Biol 96:129–139

Burt S (2004) Essential oils: their antibacterial properties and potential applications in foods—a review. Food Microbiol 94:223–254

Calderón-Oliver M, Escalona-Buendía HB, Medina-Campos ON, Pedraza-Chaverri J, Pedroza-Islas R, Ponce-Alquicira E (2016) Optimization of the antioxidant and antimicrobial response of the combined effect of nisin and avocado byproducts. LWT—Food Sci Technol 65:46–52. https://doi.org/10.1016/j.lwt.2015.07.048

Caleja C, Barros L, Antonio AL, Oliveira MBPP, Ferreira ICFR (2017) A comparative study between natural and synthetic antioxidants: evaluation of their performance after incorporation into biscuits. Food Chem 216:342–346. https://doi.org/10.1016/j.foodchem.2016.08.075

Campos-Vega R, Nieto-Figueroa KH, Oomah BD (2018) Cocoa (Theobroma cacao *L.*) pod husk: renewable source of bioactive compounds. Trends Food Sci Technol 81:172–184. https://doi.org/10.1016/j.tifs.2018.09.022

Chandrasekar V, Martín-González MFS, Hirst P, Ballard TS (2015) Optimizing microwave-assisted extraction of phenolic antioxidants from red delicious and Jonathan apple pomace. J Food Process Eng 38(6):571–582. https://doi.org/10.1111/jfpe.12187

Chang S, Tan C, Franke EN, Barrett DM (2000) Low-density lipoprotein antioxidant activity of phenolic compounds and polyphenol oxidase activity in selected clingstone peach cultivar. J Agric Food Chem 48:47–51

Chen T, Xuegang S, Xiao W, Liu X, Zhang W, Ma K, Zhu Y (2001) Optimization of microwave-assisted extraction of solanesol from potato leaves and stems. Med Chem Res 19:732–742. https://doi.org/10.1007/s00044-009-9226-4

Choi YS, Choi JH, Han DJ, Kim HY, Lee MA, Jeong JY, Chung HJ, Kim CJ (2010) Effects of replacing pork back fat with vegetable oils and rice bran fiber on the quality of reduced-fat frankfurters. Meat Sci 84(3):557–563. https://doi.org/10.1016/j.meatsci.2009.10.012

Choi S-H, Kozukue N, Kim H, Friedman M (2016) Analysis of protein amino acids, non-protein amino acids and metabolites, dietary protein, glucose, fructose, sucrose, phenolic, and flavonoid content and antioxidative properties of potato tubers, peels, and cortexes (pulps). J Food Comp Anal 50:77–87

Coll MD, Coll L, Laencina J, Tomás-Barberán FA (1998) Recovery of flavanones from wastes of industrially processed lemons. Zeitschrift für Lebensmitteluntersuchung und -Forschung A 206(6):404–407. https://doi.org/10.1007/s002170050282

Commission E (2011) Commission Regulation (EU) No 1129/2011 of 11 November 2011 amending Annex II to Regulation (EC) No 1333/2008 of the European Parliament and of the Council by Establishing a Union List of Food Additives Text with EEA Relevance. European Commission, Brussels, Belgium

Corrales M, Toepfl S, Butz P, Knorr D, Tauscher B (2008) Extraction of anthocyanins from grape by-product assisted by ultrasonics, high hydrostatic pressure or pulsed electric fields: a comparison. Innov Food Sci Emerg Technol 9:85–91. https://doi.org/10.1016/j.ifset.2007.06.002

Crizel TM, Jablonski A, Rios AO, Rech R, Flôres SH (2013) Dietary fiber from orange byproducts as a potential fat replacer. LWT—Food Sci Technol 53(1):9–14. https://doi.org/10.1016/j.lwt.2013.02.002

Czech A, Zarycka E, Yanovych D, Zasadna Z, Grzegorczyk I, Kłys S (2020) Mineral content of the pulp and peel of various citrus fruit cultivars. Biol Trace Elem Res 193(2):555–563. https://doi.org/10.1007/s12011-019-01727-1

Datta AK, Sumnu G, Raghavan GSV (2005) Dielectric properties of foods. In: Rao MA, Rizvi SSH, Datta AK (eds) Engineering properties of foods, 3rd edn. Taylor & Francis, Boca Raton

Davidov-Pardo G, McClements DJ (2015) Nutraceutical delivery systems: resveratrol encapsulation in grape seed oil nanoemulsions formed by spontaneous emulsification. Food Chem 167: 205–212. https://doi.org/10.1016/j.foodchem.2014.06.082

Del Valle M, Camara M, Torija ME (2006) Chemical characterization of tomato pomace. J Sci Food Agric 86:1232–1236

Delazar A, Nahar L, Hamedeyazdan S, Sarker SD (2012) Microwave-assisted extraction in natural products isolation. Methods Mol Biol 864:89–115. https://doi.org/10.1007/978-1-61779-624-1_5

Delsart C, Ghidossi R, Poupot C, Cholet C, Grimi N, Vorobiev E, Milisic V, Peuchot MM (2012) Enhanced extraction of valuable compounds from merlot grapes by pulsed electric field. Am J Enol Viticul 63:205–211

Dewanto V, Wu X, Liu RH (2002) Processed sweet corn has higher antioxidant activity. J Agric Food Chem 50(17):4959–4964. https://doi.org/10.1021/jf0255937

Dhillon GS, Brar SK, Verma M, Tyagi RD (2011) Enhanced solid-state citric acid bio-production using apple pomace waste through surface response methodology. J Appl Microbiol 110:1045–1055

Elleuch M, Bedigian D, Roiseux O, Besbes S, Blecker C, Attia H (2011) Dietary fibre and fibre-rich by-products of food processing: characterisation, technological functionality and commercial applications: a review. Food Chem 124(2):411–421

Friedman M (1997) Chemistry, biochemistry, and dietary role of potato polyphenols: a review. J Agric Food Chem 45:1523–1540

Galanakis CM (2013) Emerging technologies for the production of nutraceuticals from agricultural by-products: a viewpoint of opportunities and challenges. Food Bioprod Process 91(4):575–579

Gardossi L, Poulsen PB, Ballesteros A, Hult K, Svedas VK, Vasić-Racki D, Carrea G, Magnusson A, Schmid A, Wohlgemuth R, Halling PJ (2010) Guidelines for reporting of biocatalytic reactions. Trends Biotechnol 28(4):171–180. https://doi.org/10.1016/j.tibtech.2010.01.001

Gassara F, Kouassi AP, Brar SK, Belkacemi K (2016) Green alternatives to nitrates and nitrites in meat-based products—a review. Crit Rev Food Sci Nutr 56(13):2133–2148. https://doi.org/10.1080/10408398.2013.812610

Ghafoor K, Hui T, Choi YH (2011) Optimization of ultrasonic-assisted extraction of total anthocyanins from grape peel using response surface methodology. J Food Biochem 35(3):735–746. https://doi.org/10.1111/j.1745-4514.2010.00413.x

Ghosh D, Biswas PK (2015) Enzyme-aided extraction of carotenoids from pumpkin tissues. Indian Chem Eng 58:1–11

Giannuzzo AN, Boggetti HJ, Nazareno MA, Mishima HT (2003) Supercritical fluid extraction of naringin from the peel of Citrus paradisi. Phytochem Anal 14(4):221–223. https://doi.org/10.1002/pca.706

Gilbert ER, Liu D (2013) Anti-diabetic functions of soy isoflavone genistein: mechanisms underlying its effects on pancreatic β-cell function. Food Funct 4(2):200–212. https://doi.org/10.1039/c2fo30199g

Gonzalez-Centeno MR, Rossell C, Simal S, Garau MC, Lopez F, Femenia A (2010) Physico-chemical properties of cell wall materials obtained from ten grape varieties and their byproducts: grape pomaces and stems. LWT-Food Sci Technol 43:1580–1586

Goodman GE, Thornquist MD, Balmes J, Cullen MR, Meyskens FL Jr, Omenn GS, Valanis B, Williams JH Jr (2004) The beta-carotene and retinol efficacy trial: incidence of lung cancer and cardiovascular disease mortality during 6-year follow-up after stopping beta-carotene and retinol supplements. J Natl Cancer Inst 96(23):1743–1750. https://doi.org/10.1093/jnci/djh320

Gorinstein S, Martín-Belloso O, Park YS, Haruenkit R, Lojek A, Cız M, Caspi A, Libman I, Trakhtenberg S (2001) Comparison of some biochemical characteristics of different citrus fruits. Food Chem 74:309–315

Gowe C (2015) Review on potential use of fruit and vegetables by-products as a valuable source of natural food additives. Food Sci Qual Manage 45:47–61

Grassi D, Desideri G, Mai F, Martella L, De Feo M, Soddu D, Fellini E, Veneri M, Stamerra CA, Ferri C (2015) Cocoa, glucose tolerance, and insulin signaling: cardiometabolic Protection. J Agric Food Chem 63(45):9919–9926. https://doi.org/10.1021/acs.jafc.5b00913

Haleva-Toledo E, Naim M, Zehavi U, Rouseff RL (1999) Effects of L-cysteine and N-acetyl-L-cysteine on 4-hydroxy-2, 5-dimethyl-3 (2H)-furanone (furaneol), 5-(hydroxymethyl) furfural, and 5-methylfurfural formation and browning in buffer solutions containing either rhamnose or glucose and arginine. J Agric Food Chem 47:4140–4145

Heinz V, Toepfl S, Knorr D (2003) Impact of temperature on lethality and energy efficiency of apple juice pasteurization by pulsed electric fields treatment. Innov Food Sci Emerg Technol 4: 167–175. https://doi.org/10.1016/S1466-8564(03)00017-1

Hollman PCH, Arts ICW (2000) Flavonols, flavones and flavanols—nature, occurrence and dietary burden. J Sci Food Agric 80(7):1081–1093. https://doi.org/10.1002/(sici)1097-0010(20000515)80:7<1081::aid-jsfa566>3.0.co;2-g

Ignat I, Volf I, Popa VI (2011) A critical review of methods for characterisation of polyphenolic compounds in fruits and vegetables. Food Chem 126:1821–1835

Imaga N, Gbenle G, Okochi V, Akanbi S, Edeoghon S, Oigbochie V, Kehinde M, Bamiro S (2009) Antisickling property of Carica papaya leaf extract. Afr J Biochem Res 3:102–106

Imandi SB, Bandaru VVR, Somalanka SR, Bandaru SR, Garapati HR (2008) Application of statistical experimental designs for the optimization of medium constituents for the production of citric acid from pineapple waste. Bioresour Technol 99:4445–4450

Inoue T, Tsubaki S, Ogawa K, Onishi K, Azuma J (2010) Isolation of hesperidin from peels of thinned Citrus unshiu fruits by microwave-assisted extraction. Food Chem 123:542–547

Ismail AF, Moawed FS, Mohamed MA (2015) Protective mechanism of grape seed oil on carbon tetrachloride-induced brain damage in γ-irradiated rats. J Photochem Photobiol B Biol 153:317–323. https://doi.org/10.1016/j.jphotobiol.2015.10.005

Jacob JK, Hakimuddin F, Paliyath G, Fisher H (2008) Antioxidant and antiproliferative activity of polyphenols in novel high polyphenol grape lines. Food Res Int 41(4):419

John RP, Nampoothiri KM, Pandey A (2006) Solid-state fermentation for L-lactic acid production from agro wastes using *Lactobacillus delbrueckii*. Process Biochem 41:759–763

Joshi RVJ (2007) Low calorie biscuits from banana peel pulp. Solid Waste Technol Manage 33: 142–147
Joshi V, Kumar A, Kumar V (2012) Antimicrobial, antioxidant and phyto-chemicals from fruit and vegetable wastes: a review. Int J Food Ferment Technol 2:123–136
Kadirvelu K, Thamaraiselvi K, Namasivayam C (2001) Removal of heavy metals from industrial wastewaters by adsorption onto activated carbon prepared from an agricultural solid waste. Bioresour Technol 76(1):63–65. https://doi.org/10.1016/s0960-8524(00)00072-9
Kallel F, Driss D, Chaari F, Belghith L, Bouaziz F, Ghorbel R, Chaabouni SE (2014) Garlic (*Allium sativum* L.) husk waste as a potential source of phenolic compounds: influence of extracting solvents on its antimicrobial and antioxidant properties. Ind Crop Prod 62:34–41. https://doi.org/10.1016/j.indcrop.2014.07.047
Kaur D, Wani AA, Singh DP, Sogi DS (2011) Shelf life enhancement of butter, ice-cream, and mayonnaise by addition of lycopene. Int J Food Properties 14(6):1217–1231. https://doi.org/10.1080/10942911003637335
Kentish S, Feng H (2014) Applications of power ultrasound in food processing. Annu Rev Food Sci Technol 5:263–284
Kodagoda G, Marapana U (2017) Utilization of fruit processing by-products for industrial applications: a review. Int J Food Sci Nutr 24:2455–4898
Koubala BB, Mbome LI, Kansci G, Tchouanguep Mbiapo F, Crepeau MJ, Thibault JF, Ralet MC (2008) Physicochemical properties of pectins from ambarella peels (*Spondias cytherea*) obtained using different extraction conditions. Food Chem 106(3):1202–1207. https://doi.org/10.1016/j.foodchem.2007.07.065
Kujala T, Loponen J, Pihlaja K (2001) Betalains and phenolics in red beetroot (Beta vulgaris) peel extracts: extraction and characterisation. Zeitschrift für Naturforschung C 56(5–6):343. https://doi.org/10.1515/znc-2001-5-604
Kummu M, de Moel H, Porkka M, Siebert S, Varis O, Ward PJ (2012) Lost food, wasted resources: global food supply chain losses and their impacts on freshwater, cropland, and fertiliser use. Sci Total Environ 438:477–489
Lafka TL, Lazou AE, Sinanoglou VJ, Lazos ES (2011) Phenolic and antioxidant potential of olive oil mill wastes. Food Chem 125(1):92–98. https://doi.org/10.1016/j.foodchem.2010.08.041
Lanza E, Ko KH, Palmer JK (1976) Aroma production by cultures of *Ceratocystis moniliformis*. J Agric Food Chem 24:1247–1250
Larrosa M, Llorach R, Espín JC, Tomás-Barberán FA (2002) Increase of antioxidant activity of tomato juice upon functionalisation with vegetable by-product extracts. LWT—Food Sci Technol 35(6):532–542. https://doi.org/10.1006/fstl.2002.0907
Le Marchand L (2002) Cancer preventive effects of flavonoids-a review. Biomed Pharmacother 56(6):296–301. https://doi.org/10.1016/s0753-3322(02)00186-5
Lechner JF, Stoner GD (2019) Red beetroot and betalains as cancer chemopreventative agents. Molecules (Basel, Switzerland) 24(8):1602. https://doi.org/10.3390/molecules24081602
Leoncini E, Nedovic D, Panic N, Pastorino R, Edefonti V, Boccia S (2015) Carotenoid intake from natural sources and head and neck cancer: a systematic review and meta-analysis of epidemiological studies. Cancer Epidemiol Biomark Prev 24(7):1003. https://doi.org/10.1158/1055-9965.EPI-15-0053
Liu RH (2004) Potential synergy of phytochemicals in cancer prevention: mechanism of action. J Nutr 134(12 Suppl):3479S–3485S. https://doi.org/10.1093/jn/134.12.3479S
Liu RH (2013) Dietary bioactive compounds and their health implications. J Food Sci 78(1):18–25. https://doi.org/10.1111/1750-3841.12101
Liu Z-W, Zeng X-A, Ngadi M (2018) Enhanced extraction of phenolic compounds from onion by pulsed electric field (PEF). J Food Process Preserv 42(9):e13755. https://doi.org/10.1111/jfpp.13755
Lota M-L, de Rocca SD, Tomi F, Jacquemond C, Casanova J (2002) Volatile components of peel and leaf oils of lemon and lime species. J Agric Food Chem 50(4):796–805. https://doi.org/10.1021/jf010924l

Lotfi L, Kalbasi-Ashtari A, Hamedi M, Ghorbani F (2015) Effects of enzymatic extraction on anthocyanins yield of saffron tepals (*Crocos sativus*) along with its color properties and structural stability. J Food Drug Anal 23(2):210–218. https://doi.org/10.1016/j.jfda.2014.10.011

Luengo E, Álvarez I, Raso J (2013) Improving the pressing extraction of polyphenols of orange peel by pulsed electric fields. Innov Food Sci Emerg Technol 17:79–84. https://doi.org/10.1016/j.ifset.2012.10.005

M'hiri N, Ghali R, Ben Nasr I, Boudhrioua N (2018) Effect of different drying processes on functional properties of industrial lemon byproduct. Process Safety Environ Protect 116:450–460. https://doi.org/10.1016/j.psep.2018.03.004

Moo-Huchin VM, Moo-Huchin MI, Estrada-León RJ, Cuevas-Glory L, Estrada-Mota IA, Ortiz-Vázquez E, Betancur-Ancona D, Sauri-Duch E (2015) Antioxidant compounds, antioxidant activity and phenolic content in peel from three tropical fruits from Yucatan, Mexico. Food Chem 166:17–22. https://doi.org/10.1016/j.foodchem.2014.05.127

Moore J, Cheng Z, Su L, Yu LL (2006) Effects of solid-state enzymatic treatments on the antioxidant properties of wheat bran. J Agric Food Chem 54(24):9032–9045. https://doi.org/10.1021/jf0616715

Muñoz O, Sepúlveda M, Schwartz M (2004) Effects of enzymatic treatment on anthocyanic pigments from grapes skin from Chilean wine. Food Chem 87:487–490. https://doi.org/10.1016/j.foodchem.2003.12.024

Núñez Selles AJ, Daglia M, Rastrelli L (2016) The potential role of mangiferin in cancer treatment through its immunomodulatory, anti-angiogenic, apoptopic, and gene regulatory effects. Biofactors 42(5):475–491. https://doi.org/10.1002/biof.1299

Oh YS, Jun HS (2014) Role of bioactive food components in diabetes prevention: effects on Beta-cell function and preservation. Nutr Metab Insights 7:51–59. https://doi.org/10.4137/nmi.s13589

Ortega MGG, Ruiz-Roso MR, Bermejo AC. (2017). Obtaining enriched fried tomato sauce using tomato byproducts. In: Paper presented at the Proceedings of the XIV International Symposium on Processing Tomato, Leuven, Belgium

Palafox-Carlos H, Ayala-Zavala JF, González-Aguilar GA (2011) The role of dietary fiber in the bioaccessibility and bioavailability of fruit and vegetable antioxidants. J Food Sci 76(1):R6–R15. https://doi.org/10.1111/j.1750-3841.2010.01957.x

Panda SK, Mishra SS, Kayitesi E, Ray RC (2016) Microbial-processing of fruit and vegetable wastes for production of vital enzymes and organic acids: biotechnology and scopes. Environ Res 146:161–172

Parker TL, Wang XH, Pazmiño J, Engeseth NJ (2007) Antioxidant capacity and phenolic content of grapes, sun-dried raisins, and golden raisins and their effect on ex vivo serum antioxidant capacity. J Agric Food Chem 55(21):8472–8477. https://doi.org/10.1021/jf071468p

Pastrana-Bonilla E, Akoh CC, Sellappan S, Krewer G (2003) Phenolic content and antioxidant capacity of muscadine grapes. J Agric Food Chem 51:4497–4503

Prabha MS, Rangaiah GS (2014) Citric acid production using Ananas comosusn and its waste with the effect of alcohols. Int J Curr Microbiolo Appl 3:747–754

Price KR, Rhodes MJC (1997) Analysis of the major flavonol glycosides present in four varieties of onion (*Allium cepa*) and changes in composition resulting from autolysis. J Sci Food Agric 74(3):331–339. https://doi.org/10.1002/(sici)1097-0010(199707)74:3<331::aid-jsfa806>3.0.co;2-c

Priefert H, Rabenhorst J, Steinbuchel A (2001) Biotechnology production of vanillin. Appl Microbiol Biotechnol 56:296–314

Puri M, Sharma D, Barrow CJ (2012) Enzyme-assisted extraction of bioactives from plants. Trends Biotechnol 30(1):37–44. https://doi.org/10.1016/j.tibtech.2011.06.014

Radi M, Niakousari M, Amiri S (2009) Physicochemical, textural and sensory properties of low-fat yogurt produced by using modified wheat starch as a fat replacer. J Appl Sci 9:2194–2197

Rajashekhara E, Tippannavar CM, Sreenivasa MN, Sharma JS (1990) Inhibition activity of papain on facultative pathogens. Zentralblatt Fuer Microbiologie 145:455–456
Raybaudi-Massilia RM, Mosqueda-Melgar J, Soliva-Fortuny R, Martín-Belloso O (2009) Control of pathogenic and spoilage microorganisms in fresh-cut fruits and fruit juices by traditional and alternative natural antimicrobials. Comp Rev Food Sci Food Safety 8(3):157–180. https://doi.org/10.1111/j.1541-4337.2009.00076.x
Rodriguez-Amaya DB (ed) (2017) Natural food pigments and colorants. Springer International Publishing, Switzerland
Roldán E, Sánchez-Moreno C, de Ancos B, Cano MP (2008) Characterisation of onion (Allium cepa *L.*) by-products as food ingredients with antioxidant and antibrowning properties. Food Chem 108(3):907–916. https://doi.org/10.1016/j.foodchem.2007.11.058
Rosenblat M, Hayek T, Aviram M (2006) Anti-oxidative effects of pomegranate juice (PJ) consumption by diabetic patients on serum and on macrophages. Atherosclerosis 187(2):363–371. https://doi.org/10.1016/j.atherosclerosis.2005.09.006
Sadilova E, Stintzing FC, Carle R (2006) Anthocyanins, colour and antioxidant properties of eggplant (Solanum melongena *L.*) and violet pepper (Capsicum annuum *L.*) peel extracts. Zeitschrift für Naturforschung C 61:527–535
Sagar NA, Pareek S, Sharma S, Yahia EM, Lobo MG (2018) Fruit and vegetable waste: bioactive compounds, their extraction, and possible utilization. Comp Rev Food Sci Food Safety 17(3):512–531. https://doi.org/10.1111/1541-4337.12330
Saha D, Bhattacharya S (2010) Hydrocolloids as thickening and gelling agents in food: a critical review. J Food Sci Technol 47(6):587–597. https://doi.org/10.1007/s13197-010-0162-6
Saito M, Hosoyama H, Ariga T, Kataoka S, Yamaji N (1998) Antiulcer activity of grape seed extract and procyanidins. J Agric Food Chem 46(4):1460–1464. https://doi.org/10.1021/jf9709156
Santos DN, De Souza LL, Ferreira NJ, De Oliveira AL (2011) Process variables study on supercritical CO_2 extraction of Brazilian cherry seeds (Eugenia uniflora *L.*) rich in bioactive volatile. Food Process Eng Chang World:1669–1670
Schieber A, Ullrich W, Carle R (2000) Characterization of polyphenols in mango puree concentrate by HPLC with diode array and mass spectrometric detection. Innov Food Sci Emerg Technol 1(2):161–166. https://doi.org/10.1016/S1466-8564(00)00015-1
Schieber A, Stintzing FC, Carle R (2001) By-products of plant food processing as a source of functional compounds--recent developments. Trends Food Sci Technol 12(11):401–413. https://doi.org/10.1016/s0924-2244(02)00012-2
Sharma SK, Le Maguer M (1996) Lycopene in tomatoes and tomato pulp fractions. Italian J Food Sci 8(2):107–113
Shen X, Shao S, Guo M (2017) Ultrasound-induced changes in physical and functional properties of whey proteins. Int J Food Sci Technol 52(2):381–388. https://doi.org/10.1111/ijfs.13292
Shrikhande AJ (2000) Wine by-products with health benefits. Food Res Int 33:469–474
Si WD, Gong J, Tsao R, Kalab M, Yang R, Yin YL (2006) Bioassayguided purification and identification of antimicrobial components in Chinese green tea extract. J Chromatogr 1125: 204–210
Sihvonen M, Järvenpää E, Hietaniemi V, Huopalahti R (1999) Advances in supercritical carbon dioxide technologies. Trends Food Sci Technol 10(6):217–222
Singh A, Raghavan V, Md Salim NS (2017) Potential utilization of fruit and vegetable wastes for food through drying or extraction techniques. Novel Tech Nutr Food Sci 1:15–27. https://doi.org/10.31031/NTNF.2017.01.000506
Smith J, Hong-Shum L (2011) Food additives data book. John Wiley & Sons, New York, NY
Soares Nda C, Teodoro AJ, Lotsch PF, Granjeiro JM, Borojevic R (2015) Anticancer properties of carotenoids in prostate cancer. a review. Histol Histopathol 30(10):1143–1154. https://doi.org/10.14670/hh-11-635
Someya S, Yoshiki Y, Okubo K (2002) Antioxidant compounds from bananas (Musa cavendish). Food Chem 79:351–355

Song W, Derito CM, Liu MK, He X, Dong M, Liu RH (2010) Cellular antioxidant activity of common vegetables. J Agric Food Chem 58(11):6621–6629. https://doi.org/10.1021/jf9035832

Sonia NS, Mini C, Geethalekshmi PR (2016) Vegetable peels as natural antioxidants for processed foods: a review. Agric Rev 37:35–41

Soong YY, Barlow PJ (2004) Antioxidant activity and phenolic content of selected fruit seeds. Food Chem 88:411–417

Souquet J-M, Cheynier V, Brossaud F, Moutounet M (1996) Polymeric proanthocyanidins from grape skins. Phytochemistry 43(2):509–512. https://doi.org/10.1016/0031-9422(96)00301-9

Stahl W, Sies H (2005) Bioactivity and protective effects of natural carotenoids. Biochim Biophys Acta 1740(2):101–107. https://doi.org/10.1016/j.bbadis.2004.12.006

Subagio A, Morita N, Sawada S (1996) Carotenoids and their fatty-acid esters in banana peel. J Nutr Sci Vitaminol 42(6):553–566. https://doi.org/10.3177/jnsv.42.553

Sun C-D, Zhang B, Zhang J-K, Xu C-J, Wu Y-L, Li X, Chen K-S (2012) Cyanidin-3-glucoside-rich extract from Chinese bayberry fruit protects pancreatic β cells and ameliorates hyperglycemia in streptozotocin-induced diabetic mice. J Med Food 15(3):288–298. https://doi.org/10.1089/jmf.2011.1806

Sundar Raj A, Rubila S, Jayabalan R, Ranganathan T (2012) A review on pectin: chemistry due to general properties of pectin and its pharmaceutical uses. Sci Rep 1:550–551

Sun-Waterhouse D, Wen I, Wibisono R, Melton LD, Wadhwa S (2009) Evaluation of the extraction efficiency for polyphenol extracts from by-products of green kiwifruit juicing. Int J Food Sci Technol 44(12):2644–2652. https://doi.org/10.1111/j.1365-2621.2009.02097.x

Swain MR, Ray RC, Patra JK (2011) Citric acid: microbial production and applications in food and pharmaceutical industries. Nova Science Publishers, Hauppauge, New York

Talhaoui N, Taamalli A, Gómez-Caravaca AM, Fernández-Gutiérrez A, Segura-Carretero A (2015) Phenolic compounds in olive leaves: Analytical determination, biotic and abiotic influence, and health benefits. Food Res Int 77:92–108. https://doi.org/10.1016/j.foodres.2015.09.011

Teleszko M, Wojdyło A (2015) Comparison of phenolic compounds and antioxidant potential between selected edible fruits and their leaves. J Funct Foods 14:736–746. https://doi.org/10.1016/j.jff.2015.02.041

Thomas M, Badr A, Desjardins Y, Gosselin A, Angers P (2018) Characterization of industrial broccoli discards (Brassica oleracea var. italica) for their glucosinolate, polyphenol and flavonoid contents using UPLC MS/MS and spectrophotometric methods. Food Chem 245:1204–1211. https://doi.org/10.1016/j.foodchem.2017.11.021

Toor RK, Savage GP (2005) Antioxidant activity in different fractions of tomatoes. Food Res Int 38:487–494

Tripathi P, Dubey NK (2004) Exploitation of natural products as an alternative strategy to control postharvest fungal rotting of fruit and vegetables. Postharvest Biol Technol 32(3):235–245. https://doi.org/10.1016/j.postharvbio.2003.11.005

van Dam RM, Naidoo N, Landberg R (2013) Dietary flavonoids and the development of type 2 diabetes and cardiovascular diseases: review of recent findings. Current Opin Lipidol 24(1):25–33. https://doi.org/10.1097/mol.0b013e32835bcdff

Vandenberghe LP, Soccol CR, Pandey A, Lebeault JM (2000) Solid-state fermentation for the synthesis of citric acid by *Aspergillus niger*. Bioresour Technol 74:175–178

Vankar PS (2004) Essential oils and fragrances from natural sources. Resonance 9(4):30–41. https://doi.org/10.1007/BF02834854

Vieira AR, Abar L, Vingeliene S, Chan DS, Aune D, Navarro-Rosenblatt D, Stevens C, Greenwood D, Norat T (2016) Fruits, vegetables and lung cancer risk: a systematic review and meta-analysis. Ann Oncol 27(1):81–96. https://doi.org/10.1093/annonc/mdv381

Vilariño MV, Franco C, Quarrington C (2017) Food loss and waste reduction as an integral part of a circular economy. Front Environ Sci 5(21)

Virot M, Tomao V, Le Bourvellec C, Renard CM, Chemat F (2010) Towards the industrial production of antioxidants from food processing by-products with ultrasound-assisted extraction. Ultrason Sonochem 17(6):1066–1074

Walia A, Gupta AK, Sharma V (2019) Role of bioactive compounds in human health. Acta Scientific Med Sci 3(9)
Wang L, Weller CL (2006) Recent advances in extraction of nutraceuticals from plants. Trends Food Sci Technol 17:300–312
Wang X, Ouyang YY, Liu J, Zhao G (2014) Flavonoid intake and risk of CVD: a systematic review and meta-analysis of prospective cohort studies. Br J Nutr 111(1):1–11. https://doi.org/10.1017/s000711451300278x
Wedick NM, Pan A, Cassidy A, Rimm EB, Sampson L, Rosner B, Willett W, Hu FB, Sun Q, van Dam RM (2012) Dietary flavonoid intakes and risk of type 2 diabetes in US men and women. Am J Clin Nutr 95(4):925–933. https://doi.org/10.3945/ajcn.111.028894
Wolfe KL, Liu RH (2003) Apple peels as a value-added food ingredient. J Agric Food Chem 51: 1676–1683
World Health Organization (2018) Latest global cancer data: Cancer burden rises to 18.1 million new cases and 9.6 million cancer deaths in 2018. World Health Organization, Geneva
Yapo BM, Koffi KL (2008) Dietary fiber components in yellow passion fruit rind- a potential fiber source. J Agric and Food Chem 56(14):5880–5883. https://doi.org/10.1021/jf073247p
Zeyada NN, Zeitoum MAM, Barbary OM (2008) Utilization of some vegetables and fruit waste as natural antioxidants. Alex J Food Sci Technol 5:1–11
Zhang W, Yao Y, Sullivan N, Chen Y (2011) Modeling the primary size effects of citrate-coated silver nanoparticles on their ion release kinetics. Environ Sci Technol 45(10):4422–4428. https://doi.org/10.1021/es104205a
Zhao B, Hall CA 3rd (2008) Composition and antioxidant activity of raisin extracts obtained from various solvents. Food Chem 108(2):511–518. https://doi.org/10.1016/j.foodchem.2007.11.003
Zuorro A, Fidaleo M, Lavecchia R (2011) Enzyme-assisted extraction of lycopene from tomato processing waste. Enzyme Microbial Technol 49(6–7):567–573. https://doi.org/10.1016/j.enzmictec.2011.04.020

Potato Waste-Based Packaging Edibles: A Sustainable Approach for Food Preservation

7

Pragya Mishra and Amrita Poonia

Abstract

At present, majority of food packaging solutions are fossil-based plastics, which are obsessively produced, and used in an unsustainable manner. Low recycling proportion, non-degradability, and toxic emissions are some of plastic packaging disadvantages which limits its usage and opens an era of opportunity to discover sustainable edible packaging material from non-toxic, biocompatible, biodegradable, natural agricultural, or food resources along with waste utilization as current trend. Potato (*Solanum tuberosum* L.) belongs to the Solanaceae family; regarded as major agricultural crops throughout the world. Potato peel and starch are the major byproduct wastes from potato-based processing industries. Various industrial applications are known, among which the recent one is its usage as a biodegradable antimicrobial, edible food packaging material. Potato waste is the best possible sustainable alternative to replace plasticized paper packaging. With secondary function, potato peel could be used further like an excellent fertilizer by returning through biological cycle; reducing the environmental impact as waste. Potato wastes are being utilized as matrix for the development of composite edible films, coatings, wrappers, capsules, and straws via many energy-efficient methods. Other ingredients are also used in combination for improvement in the potato waste properties by addition of plasticizers and cross-linkers and implanting the polymer with fillers like nanomaterials, fibers,

P. Mishra (✉)
Food Processing and Management, DDU Kaushal Kendra, RGSC, Banaras Hindu University, Varanasi, Uttar Pradesh, India
e-mail: pragyafpm@bhu.ac.in

A. Poonia
Department of Dairy Science and Food Technology, Institute of Agricultural Sciences, Banaras Hindu University, Varanasi, Uttar Pradesh, India

A. Poonia, T. Dhewa (eds.), *Edible Food Packaging*,
https://doi.org/10.1007/978-981-16-2383-7_7

capsules, and whiskers. Blending such natural and synthetic polymers with natural extracts and essential oils also helps in improving physicochemical properties. A number of growing innovators, scientists, researchers, and entrepreneurs are trying to make edible packaging from natural sources such as potato wastes despite the fact that easy convenience, mechanical properties, production, and cost of plastic food packaging are hard to beat. The present article piles up all necessary literature related to the most recent developments of potato waste in form of potential remedy for food packaging applications along with its role in preserving the food material with potential as active food packaging when enclosed in it.

Keywords

Potato · Waste · Starch · Peel · Edible · Packaging · Biodegradable

7.1 Introduction

Plastic packaging has more advantages over metals, papers, and other materials due to its unique physicochemical properties, such as moisture barrier property, light weight, cheapness, and mechanical durability. Therefore, these have wide utilization in nearly every industrial field all over the world for packaging purposes. The major disadvantage of plastic has been related to economic and environmental problems. These may include shrinking of the landfill capacity due to increased plastic waste amount which has damaging effects on land as well as marine ecosystem, non-degradability, toxic emissions during plastic incineration, etc. This may put pressure on researchers to find out new biodegradable materials of natural origin that may be alternatives of plastics. The ability of material disintegration or its natural breakdown to carbon dioxide and water through the action of microbial environment and humidity is called as biodegradability of that material (El Kadi 2010). Bio-based plastics manufactured from natural renewable resources like corn and potatoes starch, peel, other fruit and vegetable waste, or micro-organisms under certain standardized conditions are considered as green materials due to its biodegradability. The potato peel with antimicrobial and antioxidant activities opens up a prospect for its usage in the form of additives in different food products and also in packing material applications (Sanchez Maldonado et al. 2014). One of recent study had reported the bio plastics production process from potato peels which was found to be biodegradable within 28 days (Arıkan and Bilgen 2019).

Potato, belonging to Solanaceae family, referred as fourth leading crop after rice, wheat, and maize, constitutes major part in human diet all around the world. Prior to processing of potatoes, the initial preprocessing step is usually peeling of potatoes resulting in huge quantities of potato peel waste (PPW) every year, which mainly depends on the employed peeling method (Arapoglou et al. 2009). PPW as zero value by-product is a matter of serious concern round the world as such horticultural waste ranges from 15% to 40% (70 and 140 thousand tons) of initial product mass

(Chang 2011). Such big amounts of PPW obtained after industrial potato processing need to be given proper attention for its management and further utilization as sustainable value-added product (Sepelev and Galoburda 2015). Although PPW is not suitable for consumption among non-ruminants due to highly fibrous nature, PPW has economic value assistances and proved to be one of the most remarkable raw materials for its usage in edible packaging developments. Low-cost PPW by-product encompasses huge amount of phyto-constituents, such as cellulose, fermentable sugars, hemicellulose, starch, nonstarch polysaccharides (NSP), lignin, polyphenols, protein, and lipids which made it a valuable base material like natural antioxidants, antimicrobial, dietary fiber, biopolymers, etc. (Arapoglou et al. 2009; Al-Weshahy and Rao 2012; Wu et al. 2012) for development of biodegradable, edible product as an innovative step in its sustainable utilization. Consumable thin layers are referred as edible packaging film which is prepared by ingredients that are eatable; renewable that covers the food material and forms a barrier between food and its surrounding environment (Skurtys et al. 2010). Edibles films have advantages like biodegradability, increased shelf life of food in which it is incorporated/packaged, increased protection from moisture removal, and the browning effect occurs due to varying environmental changes (Robertson 2012).

7.2 Chemical Composition of Potato Peel

Arapoglou et al. (2009) reported the proximate composition of raw potato peel in g 100 g^{-1}. They have mentioned that moisture content ranged from 83.3 to 85.1 with an average value of 84.2, protein ranged from 1.2 to 2.3 with average value of 1.8 total lipids content ranged from 0.1 to 0.4 with average value of 0.3, the total carbohydrate in raw potato peel ranged from 8.7 to 12.4 with average value of 10.6, total ash content ranged from 0.9 to 1.6 with average value of 1.3 along with starch, and total dietary fiber 7.8 and 2.5 g 100 g^{-1}, respectively. Apart from this, PPW contains high quantities of polyphenols (up to 50%) in the form of phenolic acids and flavonoids which act as plant protectant against phytopathogens due to its antioxidant activity. Potato peel also contains secondary metabolites such as glycoalkaloids which are toxic to humans and animals and hence limits its usage as edibles. The glycoalkaloids present in potato peel are majorly α-solanine, α-chaconine, and solanidine in which concentration of solanine and chaconine ranged from 84 to 3526 mg kg^{-1} (α-solanine:α-chaconine; 2.1:2.4). The permissible acceptable limit for human consumption of glycoalkaloid was found to be 20 mg 100 g^{-1} (Papathanasiou et al. 1999). The potato peel contains very low fermentable reducing sugar which accounts for only 0.6 g 100 g^{-1} of dry weight.

7.3 Potential Benefits of Potato Peel Waste for Its Usage as Edibles

Priedniece et al. (2017) reported the potential phytochemicals present in potato and its peel starch and protein with possible usages as bioplastic having characteristics similar to low-density polyethylene (LDPE). Some other phytochemicals, such as glycoalkaloids, suberin, dietary fiber, and phenols also have promising usage as food supplements, pharmaceutics, and packaging applications along with health benefits (Table 7.1). The phytochemicals present in potato peel may be used as gluten-free raw material having immense potential for the production of gluten-free food supplements and packaging edibles with no harm to environment as they are biodegradable.

7.4 Technology Employed for Treatment of PPW

For treatment and increased bioavailability of antioxidants present in PPW, various technologies such as pressurized liquid extraction (PLE), subcritical fluid extraction, supercritical carbon dioxide, ultrasound-assisted extraction, accelerated solvent, and also microwave-assisted extraction are utilized. The ultrasonic treatment of PPW helps in improving its free radical scavenging activity and serves as an innovative, low-cost, potential, and simple method for depolymerization of pectin present in PPW which may be relocated to other polymers (Dai and Mumper 2010; Singh et al.

Table 7.1 Potato peel phyto-constituents with its health benefits and potential usage as packaging edibles

S. no.	Chemical constituent	Health benefits	Usage	References
1	Dietary fiber	Lowers plasma cholesterol Positive effect on blood Glucose profile	Packaging	Camire et al. (1993), Singh et al. (2005)
2	Phenols	Natural antioxidant to prevent lipid oxidation	Medicine, capsule packaging material	Singh and Rajini (2004), Mohdaly et al. (2010)
3	PPW extract	Bacteriostatic effect – inhibits multiplication of bacteria	Edible packaging	Sotillo et al. (1998).
4	Glycoalkaloids	Lower cholesterol levels, reduced allergic reaction, pains, and inflammation, lowering of blood sugar, and glycogen level in livers	Food supplements and packaging	Priedniece et al. (2017)
5	Suberin	Protection against damage, infections, and draught	Bioplastic	Priedniece et al. (2017)
6	Starch and protein	healing abilities	Bioplastic	Priedniece et al. (2017)

2011; Proestos and Komaitis 2008; Singh and Saldana 2011; Ogutu and Mu 2017). Another novel method reported was pressurized liquid extraction (PLE) or accelerated solvent extraction.

7.5 Processing Employed to Develop PPW-Based Packaging Edibles

Bio-based plastics production intended for packaging purpose in 2011 have a small market which account for approximately 1.2 million tons as compared to the petroleum-based plastic production worldwide which was 250 million tons. One of the studies (Tammineni et al. 2013) reported the production of edible film from PPW powder. The preparation method is represented in the form of schematic diagram as given in Fig. 7.1. The PPW powder was pre-homogenized and then the biopolymer homogenate was destructed by three ways, namely, high-pressure homogenization, gamma irradiation, and ultrasound method. After destruction the biopolymers are mixed with glycerol and soy lecithin which act as plasticizer and emulsifier, respectively. The developed edible film via utilizing high-pressure homogenization process is reported to have improved moisture barrier, tensile, and physicochemical properties although more research is needed to explore its application at commercial level for usage in food packaging.

One of the recent study (Othman et al. 2017) was also done to develop potato peel base edible film for identification of its properties such as water sorption and water permeability. The development process includes following steps as given below in flow sheet (Fig. 7.2).

Tammineni et al. (2013) developed PPW-based-oregano oil-incorporating film (PPW-OO film) with Oregano essential oil (OEO) matrix which was applied on cold-smoked salmon and it was reported that Listeria population decreased from 6.7 to 4.7 log CFU/g. A number of research studies have pointed toward the potential physicochemical and barrier properties of potato peel (Borah et al. 2017), and other fruit peels (Hanani et al. 2018) which may prove to be helpful in designing biodegradable edible film with value-added commercial usages.

7.6 Conclusion

Present scenario needs exploration of the more sustainable solution for management of potato peel waste that could further utilized with full potential to support in attainment of the socio-economic and environmental benefits. Moreover, potato peel waste utilization in developing potentially value-added products, for example, biodegradable edible films coating with prebiotic potential may be an alternative eco-friendly and sustainable approach which have open up innovative prospects for business purposes. Most of such attempts are still in its early stages stage mainly due to absences of technology advancement, scientific verdicts, and validation. Therefore, it's an urgent need to generate awareness among group of researcher and

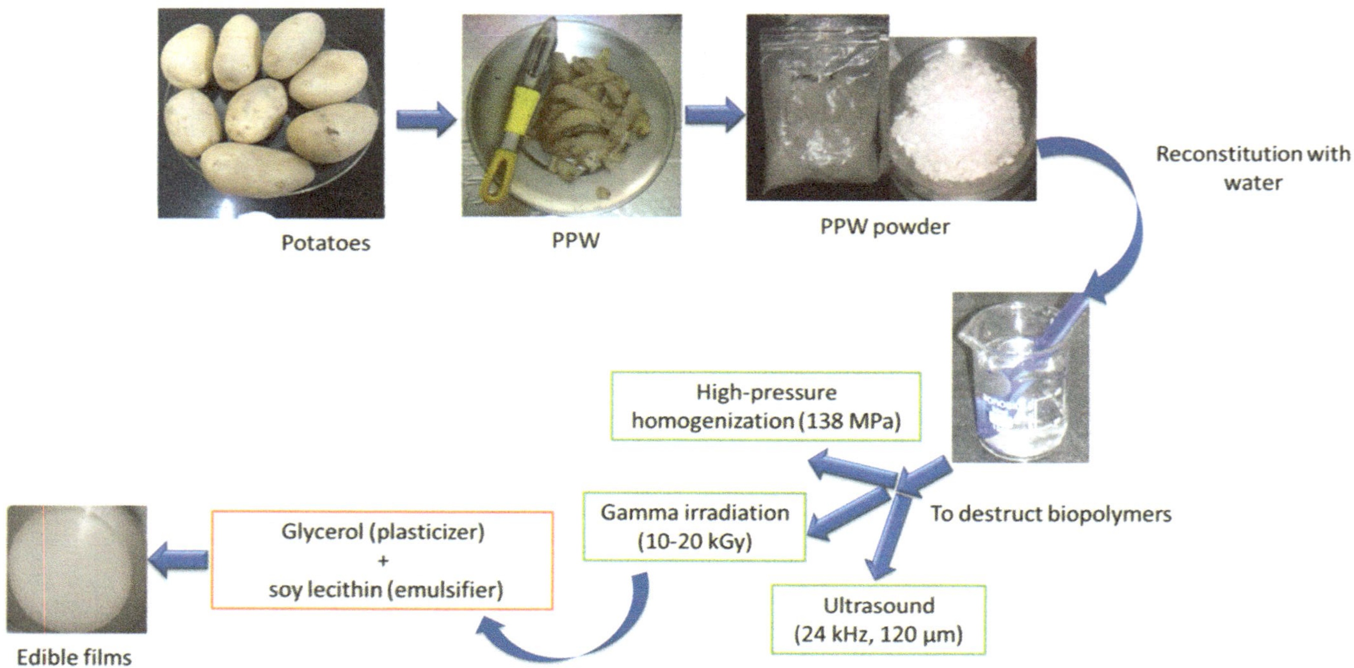

Fig. 7.1 Diagrammatic representation of the processing steps for the development of PPW edible film (Tammineni et al. 2013)

Potato peel

↓

Washed

↓

Oven dried (45 °C; 72 h)

↓

The dried potato peel ground into fine powder

↓

Sieved (64 μm mesh size sieve)

↓

5 g PPW powder + 100 ml distilled water + 2 ml plasticiser (glycerol) + 1 g anti-browning agent (ascorbic acid).

↓

Mixture transferred to a water bath shaker (90 °C; 60 rpm; 30 minutes)

↓

Cooled (room temperature)

↓

The cooled mixture solution was poured into an aluminium casting tray (484 cm^2)

↓

Covered with transparent film at the bottom of the casting tray

↓

Casting tray was placed into a natural convection oven (50 °C; 26 h) to the edible film peeled off from the transparent film (thickness 0.10 ± 0.01 mm).

Fig. 7.2 Processing steps flow chart for the development of PPW edible film

industrialist for continuous improvement in promotion and development of value-added bio-based edibles via utilizing potato peel waste.

References

Al-Weshahy A, Rao VA (2012) Potato peel as a source of important phytochemical antioxidant nutraceuticals and their role in human health – a review. Available at: http://www.intechopen.com/books/phytochemicals-as-nutraceuticals-global-approaches-to-their-role-in-nutrition-and-health/potato-peelas-a-source-of-important-phytochemical-antioxidant-nutraceuticals-and-their-role-in-huma, 23 January 2015

Arapoglou D, Vlyssides A, Haidemenaki K, Malli V, Marchant R, Israilides C, Road C (2009) Alternative ways for potato industries waste utilisation. In: Proceedings of the 11th International Conference on Environmental Science and Technology, Chania, Crete, Greece, 3–5 September 2009, pp. 3–5

Arıkan EB, Bilgen HD (2019) Production of bioplastic from potato peel waste and investigation of its biodegradability. Int Adv Res Eng J 3(2):093–097

Borah PP, Das P, Badwaik LS (2017) Ultrasound treated potato peel and sweet lime pomace based biopolymer film development. Ultrason Sonochem 36:11–19

Camire ME, Zhao J, Violette DA (1993) In vitro binding of bile acids by extruded potato peels. J Agri Food Chem 41:2391–2394

Chang KC (2011) Polyphenol antioxidants from potato peels: extraction, optimization and application to stabilizing lipid oxidation in foods. In: Proceedings of the National Conference on Undergraduate Research (NCUR), Ithaca College, New York, USA

Dai J, Mumper RJ (2010) Plant phenolics: extraction, analysis and their antioxidant and anticancer properties. Molecules 15(10):7313–7352

El Kadi S (2010) Bioplastic production form inexpensive sources bacterial biosynthesis, cultivation system, production and biodegradability. VDM Publishing House, USA

Hanani ZAN, Husna AA, Syahida SN, Khaizura MN, Jamilah B (2018) Effect of different fruit peels on the functional properties of gelatin/polyethylene bilayer films for active packaging. Food Packag Shelf Life 18:201–211

Mohdaly AAA, Sarhan MA, Smetanska I, Mahmoud A (2010) Antioxidant properties of various solvent extracts of potato peel, sugar beet pulp and sesame cake. J Sci Food Agric 90:218–226

Ogutu FO, Mu TH (2017) Ultrasonic degradation of sweet potato pectin and its antioxidant activity. Ultraso Sonochem 38:726–734

Othman SH, Edwal SAM, Risyon NP, Basha RK, Talib RA (2017) Water sorption and water permeability properties of edible film made from potato peel waste. Food Sci Technol Campinas 37(Suppl. 1):63–70

Papathanasiou F, Mitchell SH, Watson S, Harvey BMR (1999) Effect of environmental stress during tuber development on accumulation of glycoalkaloids in potato (*Solanum tuberosum L*). J Sci Food Agric 79:1183–1189

Priedniece V, Spalvins K, Ivanovs K, Pubule J, Blumberga D (2017) Bioproducts from potatoes a review. Environ Clim Techn 21:18–27

Proestos C, Komaitis M (2008) Application of microwave-assisted extraction to the fast extraction of plant phenolic compounds. LWT – Food Sci Technol 41(4):652–659

Robertson GL (2012) Food packaging principles and practice. CRC Press, Florida

Sanchez Maldonado AF, Mudge E, Ganzle MG, Schieber A (2014) Extraction and fractionation of phenolic acids and glycoalkaloids from potato peels using acidified water/ethanol-based solvents. Food Res Int 65:27–34

Sepelev I, Galoburda R (2015) Industrial potato peel waste application in food production: a review. Res rural deve 1:130–136

Singh A, Sabally K, Kubow S, Donnelly DJ, Gariepy Y, Orsat V, Raghavan GSV (2011) Microwave-assisted extraction of phenolic antioxidants from potato peels. Molecules 16(3):2218–2232

Singh N, Kamath V, Rajini PS (2005) Attenuation of hyperglycemia and associated biochemical parameters in STZ-induced diabetic rats by dietary supplementation of potato peel powder. Clin Chim Acta Int J Clin Chem 353:165–175

Singh N, Rajini P (2004) Free radical scavenging activity of an aqueous extract of potato peel. Food Chem 85:611–616

Singh PP, Saldana MDA (2011) Subcritical water extraction of phenolic compounds from potato peel. Food Res Int 44(8):2452–2458

Skurtys O, Acevedo C, Pedreschi F, Enrione J, Osorio F, Aguilera JM (2010) Food hydrocolloid edible films and coatings. Nova Science Publishers, New York

Sotillo DR, Hadley M, Wolf-Hall C (1998) Potato peel extract a nonmutagenic antioxidant with potential antimicrobial activity. J Food Sci 63:907–910

Tammineni N, Ünlü G, Min SC (2013) Development of antimicrobial potato peel waste-based edible films with oregano essential oil to inhibit *Listeria monocytogenes* on cold-smoked salmon. Int J Food Sci Tech 48:211–214

Wu ZG, Xu HY, Ma Q, Cao Y, Ma JN, Ma CM (2012) Isolation, identification and quantification of unsaturated fatty acids, amides, phenolic compounds and glycoalkaloids from potato peel. Food Chem 135:2425–2429

8 By-products Utilization of Fruits and Vegetables as Edible Packaging

Ekta Sehgal, Sangeeta C. Sindhu, H. Roopa, Anju Kumari, Rakesh Kumar, Partibha Ahlawat, and Ritu Sindhu

Abstract

A major problem with the disposal of fruits and vegetable industry is their utilization of by-products. Millions of tons of residue are underutilized each year. It is possible to explore fruits, such as bananas, apples, watermelon, and grapes and vegetables, such as okra, tomatoes, etc., for their by-products for packaging different edibles. For the production of edible packagings, the extraction of components such as, pectin, cellulosic material from skin residues, peels, kernels, low-cost pomace, abundance and renewable characteristics, can be carried out. Various processes in the food industry involving the production of fruit juices, the expression of oils, and the production of wine result in the production of residues that are a source of many bioactive compounds, namely, polyphenols, pigments, vitamins, and minerals, which can be used for feedstock, fruit powder, etc., which can be beneficial for a wide range of food industries. Because of environmental factors, the use of these edible nutritious residues as edible packages to contain foods can protect them from deterioration. This not only serves to improve the physico-mechanical properties of the finished product, but also the sensory and nutritional properties. The aim of the review is to identify the characteristics, application, and future perspectives of different residues of fruits and vegetables that have been extensively studied by different researchers, with a greater emphasis on the importance of the role that these residues can also

E. Sehgal · H. Roopa · A. Kumari · P. Ahlawat · R. Sindhu
Centre of Food Science and Technology, CCS HAU, Hisar, Haryana, India

S. C. Sindhu (✉)
Department of Food and Nutrition, CCS HAU, Hisar, Haryana, India
e-mail: chahalsangeeta@hau.ac.in

R. Kumar
Department of Microbiology, CCS HAU, Hisar, Haryana, India

A. Poonia, T. Dhewa (eds.), *Edible Food Packaging*,
https://doi.org/10.1007/978-981-16-2383-7_8

play as part of packaging edibles and understanding the properties of different packaging materials intended for food applications in the industry.

Keywords

Pectin · Polyphenols · Pigments · Cellulosic material · Physico-mechanical · Nutritional · Sensory · Edible packages

8.1 Introduction

Due to their nutritional nature and richness of antioxidants and fiber, the demand for fruit and vegetables is increasingly gaining market every day. The exponential growth in the fruit and vegetable sector may also lead to an increase in the by-products of waste. After processing of fruit and vegetables, such as pods, peels (outer-rind), seeds, skins (potato skin), purees, etc., the non-edible portion of these by-products accounted for approximately 10–60% of the total product weight (Sharma et al. 2016). Instead of dumping or disposal, the main global issue is the storage and disposal of waste or waste management, the use of by-products can be cost-effective, can reduce the bulk of waste, decrease environmental pollution, and will have good environmental impact (China et al. 2020). Residues can be used in many ways such as encapsulation of bioactive compounds which can further be involved in producing isolates, edible oil, edible films, edible coats, etc. (Majerska et al. 2019). Edible packaging is a film or coating formed on, placed on, or between food and food components that is a continuous layer of edible material. The package forms an integral part of the food that can be consumed as part of the entire food product. This is not a new technique of preservation, as it has been used for centuries in the food industry to preserve food, first used on fruits and vegetables. The role of by-products obtained from fruit and vegetables, their typical classification, and different properties of edible packaging materials were therefore studied in this work. Figure 8.1 talks about the fate of fruit and vegetable by-products. The challenges facing the food industry in relation to fruit and vegetable residues and the roles of different legal regulations are also discussed. In addition, possible applications, future trends, and important issues have also been addressed during commercialization (Fig. 8.1).

8.2 Fruit and Vegetable Residues

After processing (waste), the non-edible portion of fruits and vegetables accounts for about 10–60% of the total initial weight of the product (Sharma et al. 2016). This non-edible part is rich in nutraceutical properties of organic matter and photochemicals. From the processing of fruits and vegetables, a significant amount of residues in the form of pods, peels, pulp, stones, and seeds are generated. The problem of dumping or disposal of these residues is further worsened by legal

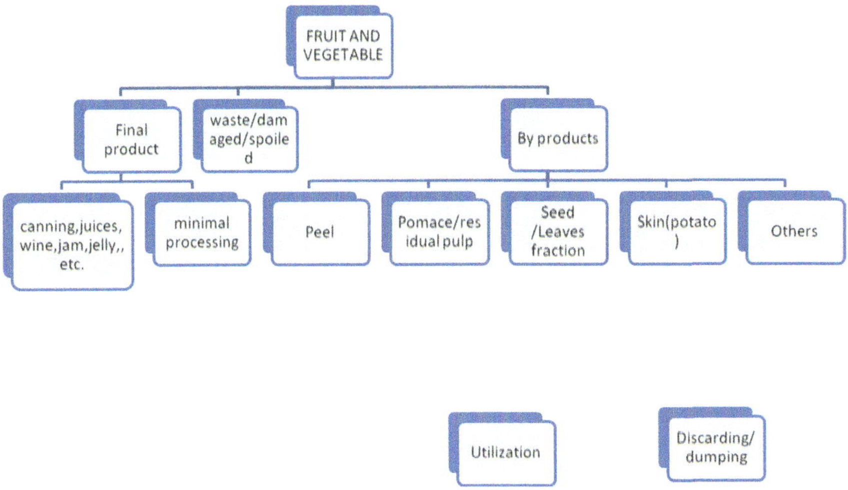

Fig. 8.1 Graphical illustration of the fate of fruit and vegetable waste occurrence

restrictions. An important step toward sustainable development is the use of waste for the production of different crucial bioactive components. Thus, utilising these residues involves many benefits such as inexpensiveness, efficiency/efficacy and no harm to environment and creating abundant job opportunities (Sharma et al. 2016). Increased production, inadequate methods of handling, and infrastructure have led to enormous losses and waste of these important food commodities, as well as their components, by-products, and residues, both in developing and developed countries. Purées of fresh produce have been widely used as hydrocolloid coating material components. However, they are still a subject of study (Galus et al. 2020). The physical properties of the meal and starch edible films of ripe banana flour were studied by Pelissari et al. (2014). A large volume of solid residues is produced by the food industry every year. They are superfood residues and contain bioactive compounds, which is why they are used as ingredients in the manufacture of functional food products. A few fruits have higher antioxidant concentrations than pulp, peel, pomace, stones, and seed fractions (Sagar et al. 2018). The use of fruits and vegetables as components of edible films and coatings has been a topic of numerous research papers in recent decades. The use of residues in the preparation of edible films and coatings, however, has recently been of great importance. Food by-product biopolymers have been targeted for their film-forming potential, which has been the focus of recent reports (Galus et al. 2020).

In their use as film-forming components, fruit and vegetable by-products, which are normally processed into flour, have shown great results. Otoni et al. (2017) researched edible films focusing on flour made from various residues, including courgetti, carrot, lettuce, passion fruit, spinach, orange, mint, taro, cucumber, rocket, and watermelon. The developed homogeneous, elastic films had a good unique packaging properties. The incorporation of residue flour from potato skins resulted

Table 8.1 Some fruits and vegetables, their residues, and uses

S. no	Fruits and vegetable	Residue	Used in different product	Reference
1	Apple, grape, pear, etc.	Fruit purees	Edible films	McHugh et al. (1996)
2	Guava	Starch, pectin, cellulose, and hemicellulose	Film-forming hydrocolloids	Azeredo et al. (2009)
3	Grapes, peach, pear, apple, and apricot	Fruit leathers	Films or coatings for nuts	Lin and Zhao (2007)
4	Pectin-rich fruit	Pectin	Fresh-cut melon	Ferrari et al. (2013)
5	Grape	Flour and extract	Hydrophobic films	Gutierrez et al. (2015)
6	Potato	Skins residue flour	Film	Brito et al. (2019)

in improved mechanical resistance. Arquelau et al. (2019) studied the production and characterization of films based on flour made from ripe banana peels and corn starch.

Grapefruit albedo showed strong functional properties of the film-forming substance as well as manufactured films when investigated. Brito et al. (2019) studied production of solid waste edible films from the isotonic drink manufacturing unit, using both fruits (passion fruit, sweet orange, and watermelon) and vegetables (arugula, lettuce, carrot, spinach, mint, yams, cucumber, and zucchini). The resulting films, with high water solubility, were homogeneous, yellowish, and malleable. Improved color, mechanical, and barrier properties and significantly reduced film hygroscopicity were provided by the small amounts of pectin present in them (Mendes et al. 2020). In general, plant residues are complex materials, which is why different residue fractions may have different uses, either as dietary fibers or as primary components for edible films and coatings, depending on their structure and particle size. For edible films and coatings, these barrier requirements depend on their use and the characteristics of the foods to be covered. Fresh fruit and vegetable films or coatings must have low permeability to water vapor (WVP) in order to minimize the rate of desiccation, while oxygen (O_2P) permeability should also be low enough to delay breathing, but not too low to produce anaerobic conditions (Sapper and Chiralt 2018). Some commonly available fruits and vegetables alonside their uses can be found in Table 8.1. It should have low O_2P and low WVP to reduce lipid oxidation rates, as well as to minimize water absorption and crunchy texture loss (Debeaufort and Voilley 2009). Similar to the cost of processing fruit or vegetables, the cost of processing waste, energy demand is not high, as the processing differences are based on established techniques and there is no need to advance technology (Andrade et al. 2016) (Table 8.1).

8.3 Classification of Packaging Edibles

In everyday life, we can see edible containers in foods, such as ice cream cones, waxing on fruit and vegetables, and cellulose coating in meat casings. Edible packaging is nothing but food wrapped in a food that is primary packaging made of edible components.

It is possible to divide edible packaging into two types, one coating and another film. On the basis of output and implementation methods, both of these are distinguished.

8.3.1 Edible Films

Edible films are a free-standing sheet placed between the food component layers. They are obtained from egg yolk, yolk fraction and protein fraction (purified form) with a thickness of 5.0–25 μm and further can be used to make pouches and bags or wrap the product. To form laminated sheets, several films can be combined; it is produced by various methods such as casting technique where film-forming solution is dried on a solid surface. Drying is done through conduction too (Díaz-Montes and Castro-Muñoz 2021).

8.3.2 Edible Coatings (EC)

Edible coatings are also a thin layer of edible materials, but they are typically applied to the outer surface of the product as a liquid of viscosity through spraying, dipping, dripping, foaming, fluidized bed coating, panning, etc. (Suhag et al. 2020).

Spraying:- This is the most popular method of coating whole fruits and vegetables, particularly when a thin, uniform coating is required. It finds utility on being applied as film to a specific side or applied through cross linkages as practiced with alginate coating.

Dipping:- Edible coatings may be applied by dipping products into coating solutions and then allowing the drying and solidification of excess solutions. It has been used extensively to coat fruits and vegetables.

Dripping:-The most cost-effective is this coating application technique. However, only when the product has sufficient tumbling action over many brushes that are saturated with the coatings, can good uniform coverage be achieved due to relatively large droplet sizes. For fruits and vegetables, dripping is used.

Foaming:- For some emulsion coatings, foam application is used. A foaming agent is added to the coating or the applicator tank is blown into compressed air.

Fluidized bed coating:- A technique that can be used to apply a very thin layer to very low-density or small-sized dry particles. In order to increase the effect of functional ingredients and additives, such as processing aids, preservatives, fortifiers, flavors, and other handling additives, improved aesthetics, taste, and color may be used. For bakery products, it is commonly used (Andrade et al. 2012).

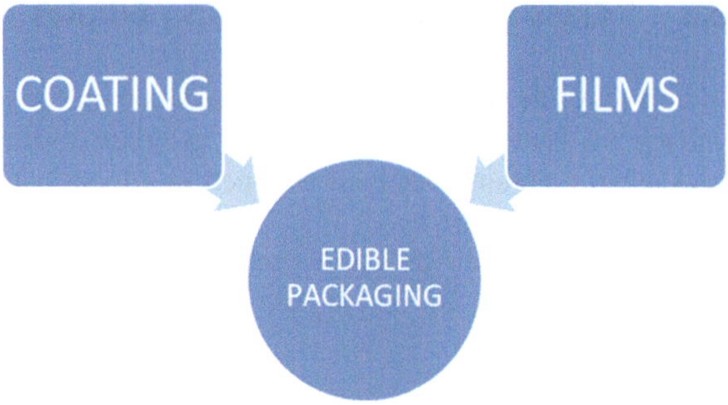

Fig. 8.2 Edible packaging

Panning:- Part of the technology is a stainless steel pan that is sealed and drilled along the side panels. The coating is delivered by a pump to spray guns installed in various sections of the pan. Panning is a slow process in which the pan's speeds vary depending on the size of the center.

Wet or dry methods are used for manufacturing edible packaging products. All films and coatings can be obtained for wet production by solvent evaporation (Galus et al. 2020). The main difference is that a casting method produces films that dry the film-forming solution on a solid surface (Galus et al. 2020). The most common techniques on a laboratory scale are this casting process, although heat conduction or convection and infrared heating can also perform some industrial scale application drying. Figure 8.2 identifies a relation between coating and films. By drying the film-forming solution over food products, edible coatings, on the other hand, are often produced. The coating process can be dipping (immersing), spraying, spreading, or brushing. Nevertheless, dry technology, such as thermo-pressing/thermoforming or extrusion, is only used for the production of edible films, especially for the purposes of edible packaging. In order to control the different functional properties of edible materials, which depend on several factors, these conventional methods may be modified. New techniques and techniques, however, are studied as a response to consumer preferences for healthier foods with maintained quality and environmentally friendly packaging (Galus et al. 2020). Edible films are self-standing structures in nature, while edible coatings adhere to the food's surface. Edible materials are distribution mechanisms for various bioactive and nutraceutical substances (Akbari-Alavijeh et al. 2020) (Fig. 8.2).

In edible coatings and films, distinct types of nano-emulsions are favored. This can act as a barrier to moisture, flavor, oxygen, nutrients, and also act as an antioxidant, antibrowning, antimicrobial agent, etc. (Benbettaïe et al. 2019). There are three main components of the typical material or edible packaging material: polysaccharides, proteins, fats (oils, free fatty acids, beeswax, etc.), polysaccharides (starch and modified starch, chitin and chitosan, pectin, kefiran, etc.), lipids and wax, agro-industrial residues, and proteins of animals and vegetables.

8.4 Fruit and Vegetable Residues as Edibles for Packaging

Fruit and vegetable solid waste mainly contains soluble sugars and other fiber and hydrolyzable materials. These residues are disposed of or left to rot in municipal bins due to the lack of adequate infrastructure available to handle such a huge amount of biomass and/or an established commercial use for such waste. The disposal of such waste can result in additional costs for processors, and direct disposal of soil or landfill can cause serious environmental problems (A sustainable option for reducing environmental pollution, reducing energy consumption, and reducing greenhouse emissions is the conversion of abundant fruit and vegetable processing residue into value-added products.).

Sensory properties such as taste and texture may be affected by edible packaging. Edible packages can be crispy or brittle, crunchy or soft, flavored, can have sweet taste, and bitter taste can also be involved. Many processes and researches carried out to reduce the effect of edible packaging on food products may constitute soundness in edible packaging material if citric fruit residue is used. More than 36 different fruits and vegetables have been used for the production of edible or biodegradable coatings and films for the production of 13 other polymers, such as starch, cellulose derivatives, gelatin, soy protein, chitosan and pectin, and alginate and carrageenan, which use plasticizers, such as glycerol, sorbitol, sucrose, inverted sugar, and syrup (Otoni et al. 2014).

The Food and Agriculture Organization of the United Nations (FAO) has estimated that fruit and vegetable losses and waste are the highest among all food types and can reach up to 60%. Fruit and vegetable processing operations produce substantial by-product waste, which makes up about 25–30% of the entire group of commodities. The waste consists primarily of seeds, skin, rind, and pomace, containing good sources of bioactive compounds which are potentially valuable, such as carotenoids, polyphenols, dietary fibers, vitamins, enzymes, and oils. The standard composition of edible films comprises four major kinds of materials: lipids, resins/gums, polysaccharides, and protein. Their barrier properties depend on their source characteristics, for example, protein-based films and polysaccharides are very effective barriers to oxygen and carbon dioxide, while their resistance to transmission of water vapor is limited. Multi-component films have also been made in an effort to combine the advantages of individual film-forming materials. The use of fruits and vegetables can also help mitigate the significant loss of fruits and vegetables due to defects or insufficient ripening phase after harvesting. Using a single compound type (starch film) or several components (proteins, carbohydrates, and lipids), film/coating solutions can be produced depending on the final use (Chen et al. 2000). In order to produce edible films from fruit and vegetables (components, extracts, juices, purees, and processing waste), an attempt to obtain materials with unique sensory and nutritional properties has also been made. Indeed, more than 35 species of plants have already been used to produce edible films. Pomace was produced from the manufacture of an isotonic drink consisting of orange, passion fruit, watermelon, lettuce, courgette, carrot, spinach, mint, taro, cucumber, and rocket, and was used for the manufacture of edible films (Hernalsteens 2020).

Table 8.2 Application and functions of edible package by fruit residues

S. no	Edible package by fruit residues	Application	Functions	Reference
1	Peach puree film	Nuts, confections, and baked goods	Oxygen barriers	McHugh et al. (1996)
2	Starch, pectin, and chitosan films	Cured meat products	Pathogenic micro-organism growth inhibition	Quintavalla and Vicini (2002)
3	Papaya puree films (cinnamaldehyde nanoemulsion)	Papaya purees	Water-barrier properties, plasticizer	Otoni et al. (2014)
4	Mango peel, antioxidant extract of mango seed kernel, and glycerol	Coating peaches	Less ethylene and CO_2 production and less O_2 consumption	Torres-León et al. (2018)
5	Ripe "Prata" banana (Musa spp.) peels		Low water vapor permeability and high flexibility	Arquelau et al. (2019)
6	Pomelo (Citrus grandis, or Chinese grapefruit) peels	Increased the soybean oil shelf life	Delaying oil oxidation during storage	Wu et al. (2019), Liu et al. (2018), Sârbu et al. (2012)
7	Arrowroot starch and blackberry pulp	Promote the stability of anthocyanins	Good appearance and flexibility in handling	Nogueira et al. (2019)

The application of coatings was emphasized immediately after harvesting as one means to alleviate water loss from fresh produce. Successful cases have been shown where the coating not only reduced water loss and delayed senescence, but also enhanced the antimicrobial properties of the coated product. Therefore, edible coatings based on plant extracts from horticultural products can be applied to provide extensive consumer relief, but there are still some gaps that need to be addressed and properties that are not well assessed, such as product surface adhesion (both capacity and time required), particularly for products intended for distant markets or long-term storage (Lin and Zhao 2007). Table 8.2 talks about application of fruit residues and Table 8.3 talks that of vegetable residues. There is also a need to make a proper comparison with commercially used coatings and diversify applications in order to focus on benefits such as inhibiting physiological disorders while relying on the quality preservation and antimicrobial properties already obtained (Hernalsteens 2020) (Table 8.2).

Although edible and biodegradable films have weaker mechanical and water-barrier properties than those based on polymers derived from petroleum, they require the addition of another component to act as filler for reinforcement, such as ethylene-vinyl acetate, ethylene-vinyl alcohol (EVOH), high-density polyethylene (HDPE), low-density polyethylene (LDPE), polyamide (nylon), polybutylene adipate-co-terephthalate, polyethylene, polyethylene terephthalate, polylactic acid,

Table 8.3 Application and functions of edible package by vegetables residues

S. no	Edible package	Application	Functions	References
1	Carrot processing waste	Food packaging	Higher tensile strength	Wang et al. (2015)
2	Potato peels	Food packaging	Film packaging	Tammineni et al. (2013)
3	Pumpkin	Residue extract	Antimicrobial	Andrade et al. (2016)

polypropylene, polystyrene (PS), polyvinylene, polybutylene adipate-co-terephthalate, polyethylene, polyethylene terephthalate, polylactic acid, and polypropylene (Bastarrachea et al. 2011).

8.5 Properties of Packaging Edibles Utilized in Food Sector

Different packaging edibles derived from fruit and vegetable residues can be used on the basis of their barrier properties (e.g., vapor, gas, and aromatic components), optical properties (e.g., opacity, clarity, and translucency), and even mechanical properties. Therefore, for accurate film performance, packaging film specifications, a major temperature and relative humidity (RH) influence is required for final packaging edibles preparation.

8.5.1 Barrier Properties

This includes certain features of packaging edibles to protect foods from various environmental influences, such as water, vapor, and gas, fragrance compounds, oil, that is essential for many foods but has been less studied. Permeability is the rate of vapor transmission across the unit area of material per unit thickness due to the difference in unit pressure between the two sides (Yang and Paulson 2000). Active diffusion is the main mechanism involved in gas (or vapor) transmission, where the penetrant dissolves on the high concentration side of the matrix in liquid form, which then diffuses through the film due to a concentration gradient within the liquid phase, and finally evaporates or is released at the other surface.

8.5.2 Water Vapor Permeability

Water activity (WA) is a consideration based on the moisture content and association of water molecules with other food molecules that not only influences the sensory consistency, but also the shelf-life of a food object. Different chemical and enzymatic reactions occur at rates controlled by the predominant rate (such as lipid oxidation, Maillard reaction, and enzymatic browning) and pathogen development.

Solute polymer associations are also allowed by this property in edible films and mass transfer processes are clarified. The hydrophilic nature of edible films creates good moisture-barrier properties, thereby contributing to the weight loss of food products. The RH material also significantly affects the moisture permeability of various edible films used (Ramos et al. 2012). The moisture sorption isotherm is used to convey an edible film's water sorption capability, which indicates the amount of water that has been transmitted inside the product. The isotherm not only recognizes various changes in stability and quality occurring through packaging and storage, but also promotes water interaction with film components (Srinivasa et al. 2007). To decrease water vapor permeability, hydrophobic compounds need to be used in the composition of edible packaging film. Due to high surface energy between the two layers, diverse composite films tend to delaminate. The water vapor permeability of peach puree edible films could be enhanced by calcium addition.

8.5.3 Gas Permeability

Gas permeability is one of the most commonly studied properties of edible films where gas is involved during degradation reactions in foods, for example, lipid oxidation, microbial growth, enzymatic browning, and vitamin loss. That's why various packaging residues exclude oxygen from the vicinity of the food. This property is affected by RH, temperature, and thickness .The Banana flour films basically, designed for dried products, possessed excellent mechanical and oxygen-barrier properties (Sothornvit and Pitak 2007). The mango purée has outstanding oxygen-barrier properties, so mango films can be used to increase the shelf life of freshly cut mangoes (Sothornvit and Rodsamran 2010). Also, gas transfer rate in peaches could also be improved (Torres-Le6n et al. 2018).

8.5.4 Aroma Compound Permeability

When aromatic compounds and their perception are involved, this property is of considerable interest:

1. To control degradation and inhibit molecular adsorption
2. To preserve the sensory content of foodstuffs and foodstuffs
3. To move molecules in a regulated manner

This property requires to be paid further attention from the science community (Chalier et al. 2006; Ramos et al. 2012).

8.5.5 Mechanical Properties

In addition to resilience when used to separate layers of homogeneous food (Sonti 2000) during manufacturing, handling, and storage, packaging edible films should have an accurate mechanical strength and extension capacity to attack various external factors (Yang and Paulson 2000). These properties also depend on conditions of film-forming, including type of process and solvent, rate of cooling or evaporation, and technique of coating, such as spraying, foaming, etc. (Guilbert et al. 1996). These properties may be capable of film-forming conditions, including type of process and solvent, rate of cooling or evaporation, and technique of coating, such as spraying, foaming, etc. (Guilbert et al. 1996). Such features may vary with the thickness of the film and the speed of the test used as close control over working conditions is required during testing.

Different fruit purees like apricot, apple, banana, mango, and pear can easily form edible films. Cellulose and pectin are the foundations of their preparation. Films made from fruit purees usually offer little versatility. Plasticizers may help to improve their texture and processing. Fruit purees obtained from composite films with fiber or particle attachment may be used to improve their properties. Water, pectin, cellulose, and protein are included, which serve as an effective component of edible films and packets. Because these films made up of carrot pulp can be brittle, with certain cracks and holes, rigid, the inclusion of plasticizers could help to improve the flexibility of films made of carrot pulp. These show excellent oxygen tolerance and can be added to oxidation-susceptible foods (primarily cooked foods and confectionery products) (Wang et al. 2015).

8.6 Implementation of Packaging Edibles Used in Food Industry

Fruit and vegetable purees have been extensively involved in numerous studies as an important component in packaging edibles. Every year in the food industry, a huge yield of solid residues containing tons of nutrients and bioactive components is generated. Biopolymers such as polysaccharides or dietary fibers are often included, including fractions of peel, pomace, and fruit seeds with higher antioxidants than the fractions of pulp.

Because plant residues are an excellent source of nutrients and are readily available, they are used to produce food products that are usable. Biopolymers containing by-products from various food industries are recognized for their film-forming ability. Different fruit and vegetable residues can be turned into flour, showing their ability as film-forming agents. However, the use of contaminants from the cultivation of fruit or vegetables is gaining significance when preparing edible films and coatings. The various physical properties of plantain bananas-based edible films of flour and starch were studied by Pelissari et al. (2013). In 2018, Andrade et al. examined new edible flour-based films made from various residues, including orange, passion fruit, watermelon, etc. The researchers obtained homogeneous, versatile films without incorporating plasticizers with promising properties. Moreover, through the addition of potato skins, residue flour resulted in improved mechanical resistance. Tibolla et al. (2014) have developed and characterized films based on flour made from ripe banana peels and corn starch. Sub-products from citrus fruits such as grapefruit albedo were studied as a film-forming substance with strong functional properties of processed films. For films obtained from pomelo peel flour, similar findings were produced. Brito et al. (2019) collected edible films based on solid residues that included all fruits (sweet orange, passion fruit, and watermelon) and vegetables for isotonic drink preparation (zucchini, broccoli, cabbage, spinach, mint, yams, cucumber, and arugula). The films were homogeneous, yellowish, and malleable with high water solubility. The pectin-enriched formulation showed tremendous technical and functional potential for the production of biodegradable films. On the other hand, films containing small amounts of pectin offered enhanced color, mechanical, and barrier properties and slightly lower film hygroscopicity. Prata banana peels were used to promote higher mechanical resistance during packaging along with corn starch and glycerol (Arquelau et al. 2019). Nogueira et al. (2019) studied film-forming solutions created by the use of blackberry pulp to provide the final product with reddish color, thus adding attractiveness during final packaging. In the presence of spoilage micro-organisms in food, there have been a number of harmful improvements such as oxidative rancidity, which leads to changes in the organoleptic properties of the product, while pathogenic micro-organisms have resulted in food-borne disease. Mali and Grossmann (2003) employed yam starch films as wrapping to prolong the shelf life of strawberries. The authors have concluded that starch films minimize the decay and microbiological degradation of preserved samples.

8.7 Challenges Faced by Food Industry

From the beginning of the 90s, the main issues related to product use and disposal was raised for the research community (Kroyer 1995). Most horticultural products are renewable and the content of cellulose and lignin in high amounts is cellulosic in nature. Different researchers have reported the conversion of various fruit and vegetable residues such as apple pomace peel into biofuel (Parmar and Rupasinghe 2013; Liguori et al. 2015). The breakdown of cellulose and hemicellulose enzymatically to release xylose and glucose and the conversion of fermentative micro-organisms to ethanol is considered an important way of using by-products. Pyrolytic decomposition also yields hydrogen and methane under anaerobic conditions (Das et al. 2012; Azadi et al. 2013). There are many valuable chemicals in food waste that can be used by other industries, but face a challenge of proper control. Many fundamental processes of the food industry (production, till storage) generate waste in a large quantity that is organic in nature and suffers from disposal problems. Incorrect handling can also lead to microbial (bacterial) contamination due to the high level of biochemical oxygen demand (BOD) and chemical oxygen demand (COD), varying composition in chemical aspects and other changes such as pH. This adds to the cumbersome use of by-products and additional costs associated with waste management (Pfaltzgraff et al. 2013).

In addition, the preparation of packaging material based on fruit and vegetable residues needs to be monitored from preparation of formulations, drying, and storage until application to the final product. Du et al. (2009) studied the residues of fruits and vegetables, where studies were based on specific components of the fruit and vegetables involved and on different challenges facing the fruit and vegetable industry. The consideration of fruit and vegetable extract as part of a functional additive was included. The lack of knowledge or less knowledge of constituents (binding agents), such as pectin, cellulose, and hemicellulose and their properties to be studied for use as packaging material can sometimes have an impact on final packaging of products by ignoring the knowledge of the species and variety of crops used (Kadzińska et al. 2019). (Sistrunk 1985). On the basis of microbial analysis, low WA obtained by using fruit and vegetables packaging residue makes it potentially secure. Shelf-life research on the basis of various properties of edible packaging material (i.e., thermal, sorption) can be beneficial, but it is currently a challenge to achieve a product with great consumer acceptance successfully (Kadzińska et al. 2019).

8.8 Regulations on Food Safety

As part of the edible packaging material, various food safety policies with regard to packaging residues (peels, pomace from different fruits and vegetables) should be of great importance depending on the recognition of the total amount of vegetarian products from fillers to the outer packaging material. Not only does this contribute to consumer acceptance, but to popularity as well. Although in both developed and developing countries, this packaging of residue involving whole fruit and vegetables is still benign due to the risk of harboring microbiological contaminants (Han 2003). Both the Food and Drug Administration (FDA) and FSSAI have laws for different edible coatings that can be used for human use as "food additives with "generally accepted as safe" GRAS. Edible coatings are specified as coatings made from food items, chemicals, ingredients, packaging materials, or contact substances by the European Directive, CODEX and US regulations (FDA). Codex NFS downloaded by York University revised 2018 (PT) alimentarius standard 04.1.1 and 04.2.1 for fresh fruits and vegetables stated that fresh fruit/vegetables are generally free of additives. Codex standard for packaging (of pre-labeled food, revised 2018) marks the need of fresh produce labeling and best before for proper shelf life and consumer safety (under clause 4.7). The Food Safety and Standards Act 2006 is virtually non-existent, but there should be stringent enforcement of the FSSA and the Laws of State Governments. The health and food safety agencies involved should keep a close eye on the use of unsafe post-harvest therapies. Any civil actions for breach of the rules of the act should be given.

In the meantime, companies that voluntarily work on fruits and vegetables through the use of products must undergo compliance that can promote the vegetarian aspect of diets and be free from the presence of toxicants that can naturally occur as part of the food residue defense system. Food inspectors and food safety officers are trained under FSSA 2006 to inspect and examine food products according to the consumer. FSSAI has started training programs on food handling and healthy food processing for food handlers in organized and unorganized industries. Proper preparation should be given to fruit merchants and vendors, and knowledge of potential health risks with a sense of moral obligation should be given to society. Infrastructural facilities for the proper handling, storing, and marketing of fresh produce should be improved by governments (specifically in developed countries). For encouraging biodegradable, environmentally safe and non-hazardous cheap material to protect dignity from farm to fork, conducting the study on the basis of packaging material is respected. Further, considering food safety regulation as a beacon of promoting safe food for all alongside better food for all requires more stringent laws which are right now gullibly present.

8.9 Future Trends in Edibles for Packaging

Biodegradable in accordance with edible films from fruit and vegetable residues requires a sustainable solution to fresh, high-quality, and cleaner food packages in order to meet the warning and vegetarian nutritional requirements, which is a concern for researchers as it falls within the criteria of consumer non-compliant demands. Edible films and coatings obtained from edible ingredients are becoming more and more relevant in the food industry because of their properties and good performance in food packaging. These are derived from natural edible constituents that can be eaten by all mammals, without any health risk. By better understanding breathing, there have been many advanced performances that are effective in extending the shelf life of fresh fruits. Furthermore, these natural, edible polymers are classified by lipids of different carbohydrates (polysaccharides) and protein-dependent moiety (Mohamed et al. 2020). Edible coatings keep foods fresh with the same advantage as altered atmospheric packaging. In order to prolong the shelf life of fresh fruit, an innovative technique with a deeper understanding of the respiratory tract must be successfully investigated. Recent research in this field needs interdisciplinary action from the point of view of food technologists, food chemists, nutritionists, and toxicologists. On practical grounds such as food packaging materials, a few topics need to be resolved. Also, processing difficulties need to be considered. Undoubtedly, packaging edibles will be a complete rescue for those seeking a health-assisted complete vegetarian diet, but are still scarce to regulate them. In any event, proper standardization, based on principles of sound science, consumer health, economic viability, and health claims need to be thoroughly considered on a wider and wider scale. The future is hopefully promising in the coming years with such sunshine strategies to meet the demands of consumers.

8.10 Conclusion

As fruits and vegetables are of significant nutritional importance, their residues can serve as great boon as packaging edible for consumers. The vegetarian dietary standards are remarkably gaining attention due to shift in the capacity of health risk related to fatty foods. While packaging edibles not only protect the food before consumption but also adds to nutritional delight, flavor, and aroma of the final product, thus, retaining its characteristics which may be of importance when the shelf-life extension and prevention of various biochemical and enzymatic reactions are needed. Thus, packaging edibles from fruits and vegetables residue aid in goals related to sustainability in the food sector.

References

Akbari-Alavijeh S, Shaddel R, Jafari SM (2020) Encapsulation of food bioactives and nutraceuticals by various chitosan-based nanocarriers. Food Hydrocoll 05:105774
Andrade RD, Skurtys O, Osorio FA (2012) Atomizing spray systems for application of edible coatings. Compr Rev Food Sci Food Saf 11:323–337
Andrade RM, Ferreira MS, Gonçalves ÉC (2016) Development and characterization of edible films based on fruit and vegetable residues. J Food Sci 81:E412–E418
Azadi P, Inderwildi OR, Farnood R, King DA (2013) Liquid fuels, hydrogen and chemicals from lignin: a critical review. Renew Sustain Energy Rev 21:506–523
Azeredo HM, Mattoso LHC, Wood D, Williams TG, Avena-Bustillos RJ, McHugh TH (2009) Nanocomposite edible films from mango puree reinforced with cellulose nanofibers. J Food Sci 74(5):N31–N35
Bastarrachea L, Dhawan S, Sablani SS (2011) Engineering properties of polymeric-based antimicrobial films for food packaging: a review. Food Eng Rev 3:79–93
Benbettaïe N, Debeaufort F, Karbowiak T (2019) Bioactive edible films for food applications: Mechanisms of antimicrobial and antioxidant activity. Crit Rev Food Sci Nutr 59:3431–3455
Brito TB, Carrajola JF, Gonçalves ECBA, Martelli-Tosi M, Ferreira MSL (2019) Fruit and vegetable residues flours with different granulometry range as raw material for pectin-enriched biodegradable film preparation. Int Food Res J 121:412–421
Chalier P, Tunc S, Gastaldi E, Gontard N (2006) Control of aroma transfer by biopolymer based materials. Dev Food Sci 43:437–440
Chen PL, Ruan RR, Addis PB, Xu L, Yi L (2000) U.S. Patent No. 6,083,582. U.S. Patent and Trademark Office, Washington, DC
China CR, Maguta MM, Nyandoro SS, Hilonga A, Kanth SV, Njau KN (2020) Alternative tanning technologies and their suitability in curbing environmental pollution from the leather industry: a comprehensive review. Chemosphere 254:126804
Das SP, Ravindran R, Ahmed S, Das D, Goyal D, Fontes CM, Goyal A (2012) Bioethanol production involving recombinant C. thermocellum hydrolytic hemicellulase and fermentative microbes. Appl Biochem Biotechnol 167:1475–1488
de Faria Arquelau PB, Silva VDM, Garcia MAVT, de Araújo RLB, Fante CA (2019) Characterization of edible coatings based on ripe Prata banana peel flour. Food Hydrocol 89:570–578
Debeaufort F, Voilley A (2009) Lipid-based edible films and coatings. In: Edible films and coatings for food applications, pp. 135–168.
Díaz-Montes E, Castro-Muñoz R (2021) Edible films and coatings as food-quality preservers: an overview. Foods 10:249

Du WX, Olsen CW, Avena-Bustillos RJ, McHugh TH, Levin CE, Mandrell R, Friedman M (2009) Antibacterial effects of allspice, garlic, and oregano essential oils in tomato films determined by overlay and vapor-phase methods. J Food Sci 74:M390–M397

Ferrari CC, Sarantópoulos CI, Carmello-Guerreiro SM, Hubinger MD (2013) Effect of osmotic dehydration and pectin edible coatings on quality and shelf life of fresh-cut melon. Food Bioproc Tech 6:80–91

Galus S, Arik Kibar EA, Gniewosz M, Kraśniewska K (2020) Novel materials in the preparation of edible films and coatings: a review. Coatings 10:674

Guilbert S, Gontard N, Gorris LG (1996) Prolongation of the shelf-life of perishable food products using biodegradable films and coatings. LWT-Food Sci Technol 29(1–2):10–17

Gutierrez TJ, Tapia MS, Perez E, Fama L (2015) Structural and mechanical properties of edible films made from native and modified cush-cush yam and cassava starch. Food Hydrocoll 45: 211–217

Han JH (2003) Antimicrobial food packaging. Novel Food Packag Tech 8:50–70

Hernalsteens S (2020) Edible films and coatings made up of fruits and vegetables. Biopolym Membr Films:575–588

Kadzińska J, Janowicz M, Kalisz S, Bryś J, Lenart A (2019) An overview of fruit and vegetable edible packaging materials. Packag Technol Sci 32:483–495

Kroyer GT (1995) Impact of food processing on the environment—an overview. LWT-Food Sci Technol 28(6):547–552

Liguori R, Soccol CR, Porto de Souza Vandenberghe L, Woiciechowski AL, Faraco V (2015) Second generation ethanol production from brewers' spent grain. Energies 8:2575–2586

Lin D, Zhao Y (2007) Innovations in the development and application of edible coatings for fresh and minimally processed fruits and vegetables. Compr RevFood Sci F 6:60–75

Liu Y, Liu A, Ibrahim SA, Yang H, Huang W (2018) Isolation and characterization of microcrystalline cellulose from pomelo peel. Int J Biol Macromol 111:717–721

Majerska J, Michalska A, Figiel A (2019) A review of new directions in managing fruit and vegetable processing by-products. Trends Food Sci Technol 88:207–219

Mali S, Grossmann MVE (2003) Effects of yam starch films on storability and quality of fresh strawberries (*Fragaria ananassa*). J Agr Food Chem 51:7005–7011

McHugh TH, Huxsoll CC, Krochta JM (1996) Permeability properties of fruit puree edible films. J Food Sci 61:88–91

Mendes JF, Norcin LB, Manrich A, Pinheiro ACM, Oliveira JE, Mattoso LHC (2020) Characterization of pectin films integrated with cocoa butter by continuous casting: physical, thermal and barrier properties. J Polym Environ 28:2905–2917

Mohamed SA, El-Sakhawy M, El-Sakhawy MAM (2020) Polysaccharides, protein and lipid-based natural edible films in food packaging: a review. Carbohydr Polym 238:116178

Nogueira GF, Soares CT, Cavasini R, Fakhouri FM, de Oliveira RA (2019) Bioactive films of arrowroot starch and blackberry pulp: physical, mechanical and barrier properties and stability to pH and sterilization. Food Chem 275:417–425

Otoni CG, de Moura MR, Aouada FA, Camilloto GP, Cruz RS, Lorevice MV, Mattoso LH (2014) Antimicrobial and physical-mechanical properties of pectin/papaya puree/cinnamaldehyde nanoemulsion edible composite films. Food Hydrocoll 41:188–194

Otoni CG, Avena-Bustillos RJ, Azeredo HM, Lorevice MV, Moura MR, Mattoso LH, McHugh TH (2017) Recent advances on edible films based on fruits and vegetables—a review. Compr Rev Food Sci Food Saf 1:1151–1169

Parmar I, Rupasinghe HV (2013) Bio-conversion of apple pomace into ethanol and acetic acid: enzymatic hydrolysis and fermentation. Bioresour Technol 130:613–620

Pelissari FM, Andrade-Mahecha MM, Amaral Sobral PJ, Menegalli FC (2013) Comparative study on the properties of flour and starch films of plantain bananas (*Musa paradisiaca*). Food Hydrocoll 30:681–690

Pelissari FM, do Amaral Sobral PJ, Menegalli FC (2014) Isolation and characterization of cellulose nanofibers from banana peels. Cellulose 21(1):417–432

Pfaltzgraff LA, Cooper EC, Budarin V, Clark JH (2013) Food waste biomass: a resource for high-value chemicals. Green Chem 15(2):307–314
Quintavalla S, Vicini L (2002) Antimicrobial food packaging in the meat industry. Meat Sci 62: 373–380
Ramos OL, Fernandes JC, Silva SI, Pintado ME, Malcata FX (2012) Edible films and coatings from whey proteins: a review on formulation, and on mechanical and bioactive properties. Crit Rev Food Sci Nutr 52:533–552
Sagar NA, Pareek S, Sharma S, Yahia EM, Lobo MG (2018) Fruit and vegetable waste: bioactive compounds, their extraction, and possible utilization. Compr Rev Food Sci Food Saf 17:512–531
Sapper M, Chiralt A (2018) Starch-based coatings for preservation of fruits and vegetables. Coatings 8:152
Sârbu C, Naşcu-Briciu RD, Kot-Wasik A, Gorinstein S, Wasik A, Namieśnik J (2012) Classification and fingerprinting of kiwi and pomelo fruits by multivariate analysis of chromatographic and spectroscopic data. Food Chem 130(4):994–1002
Sharma R, Oberoi HS, Dhillon GS (2016) Fruit and vegetable processing waste: renewable feed stocks for enzyme production. In: Agro-industrial wastes as feedstock for enzyme production. Academic Press, pp 23–59
Sistrunk WA (1985) Peach quality assessment: fresh and processed. In: Evaluation of quality of fruits and vegetables. Springer, Boston, MA, pp 1–46
Sonti S (2000) Consumer perception and application of edible coatings on fresh cut fruits and vegetables, p. 143
Sothornvit R, Pitak N (2007) Oxygen permeability and mechanical properties of banana films. Food Res Int 40:365–370
Sothornvit R, Rodsamran P (2010) Mango film coated for fresh-cut mango in modified atmosphere packaging. Int J Food Sci Technol 45(8):1689–1695
Srinivasa PC, Ramesh MN, Tharanathan RN (2007) Effect of plasticizers and fatty acids on mechanical and permeability characteristics of chitosan films. Food Hydrocol 21:1113–1122
Suhag R, Kumar N, Petkoska AT, Upadhyay A (2020) Film formation and deposition methods of edible coating on food products: a review. Food Res Int 136:109582
Tammineni N, Ünlü G, Min SC (2013) Development of antimicrobial potato peel waste-based edible films with oregano essential oil to inhibit Listeria monocytogenes on cold-smoked salmon. Int J Food Sci 48:211–214
Tibolla H, Pelissari FM, Menegalli FC (2014) Cellulose nanofibers produced from banana peel by chemical and enzymatic treatment. LWT-Food Sci Technol 59:1311–1318
Torres-León C, Vicente AA, Flores-López ML, Rojas R, Serna-Cock L, Alvarez-Pérez OB, Aguilar CN (2018) Edible films and coatings based on mango (var. Ataulfo) by-products to improve gas transfer rate of peach. LWT Food Sci Technol 97:624–631
Wang X, Kong D, Ma Z, Zhao R (2015) Effect of carrot puree edible films on quality preservation of fresh-cut carrots. Irish J Agric Food Res 54:64–71
Wu H, Lei Y, Zhu R, Zhao M, Lu J, Xiao D, Li S (2019) Preparation and characterization of bioactive edible packaging films based on pomelo peel flours incorporating tea polyphenol. Food Hydrocol 90:41–49
Yang L, Paulson AT (2000) Effects of lipids on mechanical and moisture barrier properties of edible gellan film. Food Res Int 33:571–578

Fruit Purees, Extracts and Juices: Sustainable Source of Edible Packaging

9

Neha Thakur and Vandana Chaudhary

Abstract

Plastic waste from synthetic polymer-based food packaging industry has detrimental environmental impact due to continued and injudicious use during the past century. Agriculture sector generates agri-residues in large amounts which also find their way to pollute environment. This chapter explores the potential of novel plant-based materials (fruits and vegetables) and their residues which can be used as source of bio-based edible packaging. Hence, the waste from one industry serving as the raw material for the other can be projected as a viable solution to the plastic pollution and an important step towards sustainability. Recent developments made towards edible films derived from vegetables and fruits, their properties, potential applications and future scopes are discussed in this chapter.

Keywords

Edible packaging · Sustainable sources · Barrier properties · Mechanical properties

N. Thakur
Department of Livestock Product Technology, Lala Lajpat Rai University of Veterinary and Animal Sciences, Hisar, Haryana, India

V. Chaudhary (✉)
College of Dairy Science and Technology, Lala Lajpat Rai University of Veterinary and Animal Sciences, Hisar, Haryana, India

A. Poonia, T. Dhewa (eds.), *Edible Food Packaging*,
https://doi.org/10.1007/978-981-16-2383-7_9

9.1 Introduction

A food package serves important functions like containment, protection from hazards, attracting to and informing the consumer about the product, to name a few. The basic role of a food packet/package is to provide a barrier against the external environment, to reduce exposure to chemical and biological agents and to prevent losses of desirable compounds. Synthetic polymers are important to the packaging industry, but their use raises aesthetic and environmental concerns, particularly with regard to solid waste accumulation problems and the threat to wildlife (Jayasekara et al. 2005). As plastic pollution globally is gaining increased attention, the use of bio-based plastics, especially in the food packaging sector, is growing in popularity (Gerassimidou et al. 2020). There is a dire need to look for novel packaging materials that are environment friendly as well as well-accepted by the consumers.

The concept of edible packaging using novel materials stems from this fact itself. Edible films and coatings can be made up of several active ingredients which can be incorporated into the matrix and consumed with the food, improving safety or nutritional and sensory attributes; the tendencies are to use edible coatings as carriers of functional ingredients by incorporating antimicrobials, antibrownings and nutraceuticals to improve the quality of fruits and vegetables (López et al. 2010).

Bio-based packaging materials have been industrially feasible on a large scale for more than a decade (Babu et al. 2013). Proteins and carbohydrates can be used as matrices for edible films and coatings, which are supposed to be ingested with the food (Krochta 2002), along with agri-waste which can be utilized for the development of edible packaging.

Food industry produces a large amount of residues, roughly amounting to one third of the total meant for human consumption. This huge waste so generated amounting to 1.3 billion ton per year (Gustavsson et al. 2011) can serve as raw material for edible food packaging industry. The residues from fruits and vegetable processing industry include peels, seeds, pulp and stones. These residues are a rich source of biopolymers (such as polysaccharides and dietary fibres) and bioactive compounds (Babbar et al. 2011; Ajila and Prasada Rao 2013). These agro-industrial by-products are a great amalgamation of film-forming properties provided by biopolymers (e.g., polysaccharides, pectin and derivatives of cellulose) and of the rich colour contributed by plant pigments and flavour provided by the volatile compounds of plant origin (Azeredo et al. 2009; Kaya and Maskan 2003). Edible films formed from these materials can provide a viable alternative to the plastic films, along with nutritional supplementation depending on the nature and composition of the raw material used to develop the dibble film. Such films have an important role to play in ensuring and enhancing the sustainability of the packaging industry (Perez-Mateos et al. 2009; Hu et al. 2009). There is an increase in the production of residues from the agro-industries and the massive detrimental impact of the plastic packaging industry, and there is a huge potential for diverting the waste of the former industry as the raw material for the latter. Hence, there is a need to study and explore development of edible films based on vegetable and fruit sources.

Fruits and vegetables are a rich source of polyphenols and pigments having high antioxidant potential, which is beneficial for health of the end consumer (Deng and Zhao 2011). Hence, films derived from these would have higher acceptability among the consumers (Espitia et al. 2014). This chapter deals with story of films based on films developed from novel plant materials (vegetables and fruits) from the beginning to the products currently available in the market and the future prospects of the same.

9.2 History of Plants as Edible Films

A number of organically derived polymers which are biodegradable in nature have been used to produce edible coatings and films. Novel sources like agri-wastes and food processing industry by-products along with recent advances in the field of food processing have become a promising area of research due to their potential application in the food packaging industry (Galus et al. 2020). It is a challenging task for the post-harvest industry to develop novel edible packaging materials that are locally available, easy to process and having excellent performance as viable film-forming material (López et al. 2010). Whole grain flours, fruit and vegetable residues, root plants, plant gums and wild plants are a few sources of such novel edible packaging materials. Plants as source of bio-based films need to be explored because they are easily available, eco-friendly, healthy and familiar to consumer and can be modified easily in accordance with consumer preference (Galus et al. 2020).

Vegetables and fruits provide numerous health benefits; hence, films formed from these act as additional source of intake of nutritive substances (Espitia et al. 2014), due to the presence of polyphenols and natural pigments (Deng and Zhao 2011).

Numerous studies with fruits, vegetables and their respective residues as sources of raw materials for edible film manufacture have been conducted in the last decade.

Various fruits and vegetables like taro, courgette, passion fruit, rocket, orange, cucumber, watermelon, mint, lettuce and spinach and their residues in powder form have been used to develop edible films (Andrade et al. 2016). The flours prepared from these materials led to formation of films which were homogenous in nature. These films did not require addition of a plasticizer in the formulation, which is a highly desirable quality. By-products like potato skin residue can be used to improve the mechanical properties of the films. Other examples of by-product that improves the film properties are ripened banana peels and starches obtained from corn, potato, etc. (de Faria Arquelau et al. 2019). Citrus fruits with high vitamin content can also be used for the same. Mariniello et al. (2010) obtained good results by using albedo, a grapefruit for the manufacture of edible films. Additionally, materials like flour made from pomelo peel (Wu et al. 2019), solid residues from vegetables (lettuce, spinach, mint, carrot, arugula, yams, cucumber and zucchini) and fruits (watermelon, sweet orange and passion fruit) have been used to develop edible films (Brito et al. 2019).

Vegetable and fruits or residues, thereof, are a tricky material to work with because the properties and behaviours of the films so obtained are highly variable in nature and yet to be characterized. The variation depends on the part used, composition, concentration, degree of homogeneity of the film obtained and processing technology.

9.3 Various Vegetables and Fruits as Sources of Edible Coatings/Films

The search for novel materials to develop edible films has led to emergence of vegetables and fruits as primary raw material due to their exceptional nutritional and sensory properties. Numerous plant species have been studied for their potential use in edible films (Table 9.1). The films can be developed using either a part or whole of the fruit/vegetable or residue left after processing. The residues or components used usually comprise of purees, extracts/juices or pomaces. World food sector faces an important challenge in terms of waste management, and need for valourization is now more than ever. For instance, the apple peels from apple processing industry amount to 9000 tonnes annually (Shin et al. 2014). The residues generated at such high rates still remain underutilized because of low market value. An optimal approach of reduction of agri-waste is via use of residues from agricultural sector and food processing industry as biopolymers for edible films along with conventional materials (Esparza et al. 2020). The conventional methods of disposal underutilize these raw materials as fertilizers or as raw material for animal feeds or simply discarded into environment or landfills. Open disposal of such waste leads to environmental pollution and also wastage of valuable by-products which are rich in nutrients (Park and Zhao 2006).

Vegetable and fruit pomace might not be appealing in terms of sensory evaluation, yet their rich nutrient profile makes them an excellent source of biodegradable material to be used in food packing systems. By-products from fruit juice processing industry are a rich source of dietary fibres, phytochemicals and pectin (Park and Zhao 2006). Vegetable matters derived from of mint, like taro, courgette, passion fruit, rocket, orange, cucumber, watermelon, lettuce and spinach have been successfully used to produce edible films (Andrade et al. 2016). Besides purees and pomaces, specific components like phenolic groups or pigments extracted from wastes produced by food processing industry can also be used. A major drawback of using extracts is that they do not ensure 100% valourization of food waste because post-extraction, a major portion of residue, rich in nutrients like dietary fibre, is still left which could have been utilized more judiciously in the manufacture of whole fruit/vegetable-based films. Use of whole plant material is a more sustainable approach and should be promoted while research or its application. But if the waste material is rich in compounds like lipids or hydrophobic elements, then extraction becomes a useful processing step in film manufacture and improves film-forming properties like moisture barrier (Deng and Zhao 2011).

Table 9.1 Commonly found fruits and vegetables for potential use as edible films in food industry (Otoni et al. 2017)

Common name	Scientific name
[A] Purees of commonly found fruits and vegetables in edible films/coatings	
Acai	Euterpe oleracea Mart.
Apple	Malus domestica Borkh., M. pumila Mill., Pyrus malus L.
Apricot	Prunus armeniaca L.
Banana	*Musa* × *paradisiaca* L., *M. cavendishii* Lamb.
Barbados cherry (acerola)	Malpighia emarginata DC., *M. glabra* L., *M. punicifolia* L.
Broccoli	Brassica oleracea
Carrot	Daucus carota L
Celery	*A. graveolens* L
Gooseberry	*Ribes uva-crispa* L
Guava	Psidium guajava L
Hibiscus	Hibiscus sabdariffa L
Mango	Mangifera indica L.
Papaya	Carica papaya L.
Passion fruit	Passiflora edulis Sims.
Peach	Prunus persica (L.) Batsch
Pear	Pyrus communis L.
Red mombin	Spondias purpurea L
Strawberry	*Fragaria* × *ananassa* Duch., *F. vesca* L
Tomato	Solanum lycopersicum L.
Watermelon	Citrullus lanatus (Thunb.)
[B] Pomaces of commonly found fruits and vegetables in edible films/coatings	
Apple	Malus domestica Borkh., M. pumila Mill., Pyrus malus L.
Banana	*Musa* × *paradisiaca* L., *M. cavendishii* Lamb.
Carrot	Daucus carota L
Courgette	Cucurbita pepo L
Cranberry	Vaccinium macrocarpon Aiton
Cucumber	Cucumis sativus L.
Fennel	Foeniculum vulgare Mill.
Grape	Vitis vinifera L
Lettuce	Lactuca sativa L
Mint	*Mentha* sp
Orange	Citrus sinensis (L.) Osbeck
Rocket	Eruca sativa Mill.
Spinach	*Spinacea oleracea* L.
Taro	Colocasia esculenta (L.) Schott
Watermelon	Citrullus lanatus (Thunb.)
[C] Extract of commonly found fruits and vegetables in edible films/coatings	
Broccoli	Brassica oleracea
Cashew apple	*A. occidentale* L
Corn	Zea mays L.
Orange	Citrus sinensis (L.) Osbeck
Papaya	Carica papaya L.

9.3.1 Purees in Edible Films

Preparation of edible sheets from the puree of fruits and vegetables could be considered an innovative idea. Studies have suggested that edible sheets can be manufactured from the squash puree (Torabi et al. 2020). Edible films from fruit puree were first invented by McHugh et al. (1996) who studied various barrier properties of the edible films. Rojas-Graü et al. (2007) developed a new method to extend the shelf life and improve the quality of fresh apple slices. Their sheets were made out of apple puree and contained various amounts of fatty acids, waxes, vegetable oils and pectin. Dehydrated papaya-based edible films with moringa as source of antioxidant have been well-accepted by the sensory panellists (Rodríguez et al. 2020). The addition of fruit purees to the production of biodegradable edible films allows obtaining edible films with sensory acceptance and nutritional value.

Various fruits and vegetable purees have been applied for the preparation of edible sheets: carrot (Wang et al. 2011), peach (McHugh et al. 1996), apple (Rojas-Graü et al. 2007), banana (Perez-gago and Krochta 2001) and tomato (Du et al. 2009) are some of these examples. Edible sheets prepared from the puree of fruits and vegetables are cheap and their utilization – especially in seasons when the respective fruit or vegetable is highly harvested – could be a method to prevent agricultural waste by processing surplus raw material into products with extended shelf life.

9.3.2 Pomaces in Edible Films

The addition of pomaces into edible packaging films provides an alternate source for valourization of the fruit/vegetable industry waste stream, as well as the added benefits of migration of antioxidant compounds into the food products can increase their nutritional and health benefits. Singh et al. (2020) developed edible films containing blueberry pomace powder (BPP) and found that the films containing higher BPP content had higher UV absorption, which could absorb UV light and protect packaged food content from UV deterioration. Similarly, films developed using apple pomace extract (APE) show excellent antioxidant and antibacterial, and they also improved mechanical properties and the pH sensitivity (Lan et al. 2020). Wine grape pomace (WGP) extract-based edible films with the addition of a small amount of commercial polysaccharides showed attractive colour and comparable mechanical and water barrier properties to other edible films. The films also demonstrated their potential antioxidant and antimicrobial functions (Deng and Zhao 2011). Hence, the use of pomaces in edible films is desirable for they can act as attractive wraps for food or pharmaceutical industry.

9.3.3 Extract/Juices in Edible Films

Like whole vegetables and fruits, the extracts from juice-based industry or food processing industry can help in creating edible films with attractive colours and desirable flavours. Extracts or juices of vegetables and fruits have the potential to replace synthetic additives like antioxidants or flavour enhancers. The extract/juice-based films present attractive colour due to natural pigments (Chambi et al. 2020). Biobased packaging fabricated from aronia juice had better antimicrobial performance than the aronia freeze powder based group (Lee et al. 2020). Extracts/juices can help achieving the desired functional properties of the films so formed and can be tailor-made to suit the needs of the food material which is to be wrapped in these (Park and Zhao 2006).

9.4 Components of Fruit and Vegetable-Based Film Forming Formulations (FFF)

Vegetable- and fruit-based edible films can be derived from a single plant or more (blends), or even multi-component/composite in nature having a part of conventional polymers or specific additives (binders, plasticizers, flavour enhancers, antioxidants, antimicrobials and essences). Such films can be single layered or multi-layered depending on the target or intended use. Constituents of fruit- and vegetable-based film formulations include various components (Table 9.2) which contribute to different properties of the films.

Table 9.2 Components of fruit- and vegetable-based FFF with examples

Binding agents	Plasticizers	Fillers	Functional additives	Other additives
Cassava starch Chitosan Carboxymethylcellulose Corn starch Gelatin High-methoxyl pectin Hydroxypropyl methylcellulose (HPMC) Low-methoxyl pectin Methylcellulose Phaseolin Polylactide Sodium alginate Soy protein	Glycerol Sucrose Inverted sugar Sorbitol	Cellulose nanocrystals (CNCs) Microcrystalline cellulose (MCC) Montmorillonite (MMT) nanoclays Chitosan nanoparticles	Essential oils Tannins Flavonoids Phenolic acid derivatives	Browning inhibitors Cross-linkers

9.5 Film-Forming Procedure

The film formulation using fruit and vegetable components can be summarized into following steps (Fig. 9.1):

1. Firstly, film-forming solution is developed by dispersing all the ingredients. All the ingredients should be so suspended so as to get homogenous solution.
2. This step is followed by removal of microbubbles of air, known as degassing or defoaming. It is an extremely important step in order to prevent structural defects due to entrapped air.
3. Films are then casted upon a smooth surface (bench or continuous casting). Casting is the most preferred method of production of vegetable- or fruit-based edible films because of thermosensitive nature of the components.

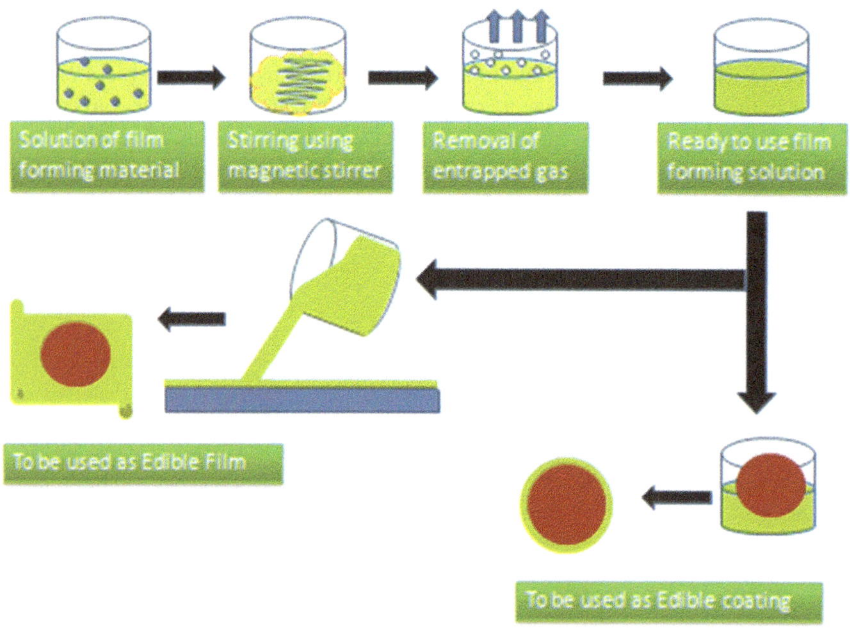

Fig. 9.1 Schematic representation of the edible coating and edible film manufacture (Otoni et al. 2017)

9.6 Characteristics of Fruit- and Vegetable-Based Edible Films

9.6.1 Barrier Properties of Vegetable- and Fruit-Based Edible Films

9.6.1.1 Barrier to Moisture

A good moisture barrier is an important pre-requisite for preservation of foods. Measurement of water vapour transmission rate (WVTR) or water vapour permeation (WVP) gives an idea about the water diffusion. Vegetables and fruits are made up of biomolecules that are highly polar, hydrophilic and having high water holding capacity (WHC) (Azeredo et al. 2012; Deng and Zhao 2011). This leads to poor barrier properties with respect to water/moisture (Shin et al. 2014). WVP is considered to be a more accurate measure of the moisture barrier property and is influenced by the constituents of the film. Fruit- and vegetable-based films have higher WVP in comparison to the synthetic polymers (McHugh et al. 1996).

With a view to improve the WVTR or WVP of the films, nanoemulsions or nanocomposites can be formed (Azeredo et al. 2012). Addition of cellulose nanofibre significantly improved the WVP of mango puree-based edible films (Azeredo et al. 2009), and acerola puree edible films showed better barrier properties upon the use of cellulose nanocrystals (CNC) or montmorillonite (MMT) nanoclay (Azeredo et al. 2012). Similar results were obtained for guava puree edible films (Lorevice et al. 2012) and banana puree edible films (Martelli et al. 2013) where lower WVP was observed after making use of nanoclays in the film formulation due to increased twists (tortuosity) in the pathway of vapour diffusion leading to reduced permeability to water molecules (Azeredo et al. 2009; Azeredo et al. 2012). Use of pectin to form hydrogen and covalent bonds with the polyphenols present in fruits like apple (Du et al. 2009) lowers the hydrophilicity, hence improving the WVP.

Other measures include use of hydrophobic additives like lipids (McHugh et al. 1996), vegetable oils, beeswax (McHugh and Senesi 2000) and essential oils (Rojas-Graü et al. 2007) in fruit-based film formulations.

9.6.1.2 Barrier to Oxygen

Oxidative rancidity in foods is one of the major causes of its spoilage. Hence, a good oxygen barrier is a desirable trait in edible films depending on the nature of food product in order to prevent loss of nutrients and preserve its sensory attributes (Sothornvit and Pitak 2007). Hydrocolloids present in vegetable- and fruit-based film formulations lead to polarity, resulting in good barrier properties to gases that are non-polar in nature like oxygen (Wang et al. 2011). In comparison to synthetic polymer based or conventional packaging materials, edible films based on vegetables and fruits have better permeability to oxygen (McHugh et al. 1996). Augmentation of the oxygen barrier of novel edible films based on fruits like apple has been observed upon the addition of essential oils (EOs) like carvacrol (Du et al. 2009), lemongrass EO and citral EO (Rojas-Graü et al. 2007). Such films display good oxygen barrier properties in spite the addition of active compounds due to presence of natural antioxidants in them (McHugh et al. 1996), which slow down the oxidation process by reducing the respiration rates of foods (McHugh and Senesi 2000).

9.6.2 Stability, Shelf Life and Antioxidant Potential

Vegetable- and fruit-based edible films are rich in nutrients, and hence, they are prone to oxidation and microbial contamination. But edible films have low water activity which retards the microbial growth and extends their shelf life. As a matter of fact, edible films in themselves are dehydrated food products; hence, they do not tend to spoil easily (McHugh et al. 1996). Low water activity enables them to be stored under dry conditions for long periods of time. Thermoplastic starch-based films containing acerola, red mombin and mango purees showed higher water activity (Dantas et al. 2015). Presence of natural sugars in fruits and its crystallization further affects the physical performance of the films so developed.

9.6.3 Mechanical Properties

The mechanical properties are a function of their composition. Films derived from vegetables tend to be stronger and less extensible than fruit films due to the presence of high content of dietary fibres in comparison to total sugar content (McHugh and Olsen 2004).

Elongation at break and tensile strength are the two main indicators of the mechanical performance of films. Tensile strength indicates the ability to defend against tensile deformation, whereas the capacity of film stretching is known as elongation at break. The mechanical properties of films are related to the intermolecular and intramolecular interactions of polymer molecules, but also the film thickness (Zhang and Li 2020) as well as the purity of the ingredients used. These properties can be modified by addition of cross-linkers, binding agents, fillers and/or plasticizers as mentioned in Table 9.2.

9.6.4 Thermal Characteristics

The physical performance of the final film, which may be brittle, hard, tough, rubbery plastic or viscous fluid in nature, is a function of the polymer chain mobility. Chain mobility is expected to increase with temperature (Lorevice et al. 2012). The glass transition temperature (T_g) is influenced by the presence of short-chain sugars present in the fruit purees (Azeredo et al. 2009; Souza et al. 2011), and pectin addition further lowers T_g values in fruit purees (Martelli et al. 2013). Although low T_g values are undesirable from the chemical stability point of view, they provide films which are flexible even under refrigeration temperatures (Azeredo et al. 2009). Edible films based on using agri-waste can be produced on a commercial scale with increased yield and reduced processing time and production cost. Note should be taken that lower temperature for degradation is used a top priority while industrial manufacture of films (Martelli et al. 2013). Addition of binding agents like HPMC presents a wider window for processing by increasing degradation temperature (190 °C) (Lorevice et al. 2012). In order to manufacture these edible films, the processability at higher temperatures needs to be explored by further research.

9.6.5 Nutritional Characteristics

Nutritional and health-promoting functional properties of edible films based on vegetables and fruits are of utmost importance. As mentioned earlier, fruits and vegetables are a rich source of pigments, flavours, antioxidants, vitamins and other nutrients. For instance, acerola incorporated starch films were found to be rich in vitamin C and β-carotene containing 3.8 times and 56 times the recommended daily intakes, respectively (Farias et al. 2012). Such edible films are rich source of polyphenols which retard the process of auto-oxidation of the food products (Deng and Zhao 2011). The release of active compounds is dependent on interactions between active compounds and film-forming bio-macromolecules (Reis et al. 2015). Films based on mango and acerola purees with higher vitamin C content (65.4 to approximately 600 μg/g) and good radical-scavenging activity (low peroxide value) (Souza et al. 2011) and films from grape pomace (Hayashi et al. 2006) showed same results. Nutritional properties of vegetable- and fruit-based films give them an added advantage because they are naturally loaded with beneficial nutrients, and this property needs to be exploited the most at a commercial scale.

9.6.6 Antimicrobial Potential

Fruit- and vegetable-based edible films are carriers of antimicrobials, which occur naturally in the raw material used (Rojas-Graü et al. 2007) or naturally occurring antimicrobials as additives (Ravishankar et al. 2012). Essential oils from naturally occurring antimicrobials enhance the bactericidal potential of these films (Zhu et al. 2014).

9.7 Potential Applications

From past three decades, a plethora of fruits and vegetables has been identified as potential sources of raw materials for edible films (Table 9.1). Fruit leather derived by drying fruit puree, from apple, peach, grapes and apricot, has been traditionally used to wrap nuts or used as snack. By studying the properties of various fruit puree blends and their respective film properties, we can arrive at industry friendly film formulations. The composite films have relatively excellent combination properties due to the stronger intermolecular interactions and more compact microstructure, showing potential application as a green alternative to bioactive packaging materials for oil products (Wu et al. 2019). The sensory attributes like texture, flavour and colour need to be explored for they play a major role in deciding the consumer preference and acceptability. These films so developed can find their role in nutraceutical industry due to an attractive nutrient profile (Tontul and Topuz 2017) and potential use in various drug delivery systems in pharmaceutical industry. Another area of potential application is in development of new food products be supported by extensive sensory study to arrive at formulations that are superior to the pre-existing ones: for example, mango leathers produced using the cast-tape drying process and conditioned at different RH values (da Silva et al. 2019; Torres et al. 2015).

Another example is addition of binding agents like maltodextrin to produce fruit leather with different textures and sensations when chewed and bitten (Valenzuela and Aguilera 2015). The research using vegetable/fruit extracts as edible films and coatings needs to focus on characterizing the mechanical properties of the films and their mass transfer barriers (McHugh et al. 1996).

The scientific fraternity should aim at using whole fruits, fruit wastes or fruit surpluses in development of composite films. Nanoemulsion technology in the development of blended films is another promising area for research and application in food industry (Otoni et al. 2014).

Film formulations also vary in accordance with the part of fruit/vegetable being used. For example, composition of mango seed kernel is different from the fruit itself, as it contains a high amount of carbohydrate and lipids and a tiny amount of protein. It leads to different film-forming characteristics than the fruit itself. The films obtained from mango seed kernel possess higher antioxidant activities (Adilah et al. 2018) and could serve as an exciting alternative to divert the waste of mango processing industry into raw material for food packaging industry.

9.8 Future Prospects and Conclusion

During the boom in development of polymer-based food packaging industry, the bio-based polymer industry was not prioritized. The polymer-based food packaging industry has highly evolved over the past century, but the ecosystem paid a high price for this development. The times and trends are shifting towards a more sustainable packaging industry, and it is high time that the bio-based polymer industry should become a priority. The biopolymer-based industry is catching up and growing tremendously over the past two decades. Advancements in the field of food science technology and biotechnology have made it possible to utilize novel plant-based materials for use in edible films as an alternative to the conventional packaging system that is taking a toll on the environment. Earlier technologies focused on food-based edible packaging. Newer technologies are now focused on turning agri-waste/residues into raw materials for the bio-based polymer industry.

Further research is required to study the properties of blends obtained from fruits and vegetables and needs to be improved upon. Use of nanoemulsions to stabilize the film matrix to improve film-forming abilities is the pipeline as well. Fruits/

vegetables that are easily or indigenously available should be targeted to form films that are product and, if possible, region specific. Use of locally available materials would promote self-reliance among the people and increase acceptability by the end consumer. Since synthetic polymer-based industry is well established, it is tough for the bioplastics to entirely replace the former. But, combinations of synthetic and novel biomaterials can be worked upon to bring about a significant reduction in the plastic load of the environment. Functional films derived from fruits and vegetables are and attractive arena to venture out for the budding agripreneurs who can propagate the well-established scientific findings by applying them in small ventures which might later be expanded as and when required. Use of fruits and vegetable films in ready-to-eat products might increase its use by the millennials. Additionally, cost economics of the edible films is yet to be worked out since these films tend to be expensive than the conventional packaging. But their health benefits and eco-friendly nature might attract consumers with higher purchasing power. To conclude, it can be said a holistic approach considering all the above-mentioned factors should be undertaken to ensure that fruit- and vegetable-based packaging makes it to the mainstream and becomes a household item.

References

Adilah ZM, Jamilah B, Hanani ZN (2018) Functional and antioxidant properties of protein-based films incorporated with mango kernel extract for active packaging. Food Hydrocoll 74:207–218

Ajila CM, Prasada Rao UJS (2013) Mango peel dietary fibre: composition and associated bound phenolics. J Funct Foods 5(1):444–450

Andrade RMS, Ferreira MSL, Gonçalves ECBA (2016) Development and characterization of edible films based on fruit and vegetable residues. J Food Sci 81(2):412–418

Azeredo HM, Mattoso LH, Wood D, Williams TG, Avena-Bustillos RJ, McHugh TH (2009) Nanocomposite edible films from mango puree reinforced with cellulose nanofibers. J Food Sci 74(5):31–35

Azeredo HMC, Miranda KWE, Ribeiro HL, Rosa MF, Nascimento DM (2012) Nanoreinforced alginate—acerola puree coatings on acerola fruits. J Food Eng 113(4):505–510

Babbar N, Oberoi HS, Uppal DS, Patil RT (2011) Total phenolic content and antioxidant capacity of extracts obtained from 6 important fruit residues. Food Res Int 44(1):391–396

Babu RP, O'connor K, Seeram R (2013) Current progress on bio-based polymers and their future trends. Prog Biomater 2(1):8

Brito TB, Carrajola JF, Gonçalves EC, Martelli-Tosi M, Ferreira MS (2019) Fruit and vegetable residues flours with different granulometry range as raw material for pectin-enriched biodegradable film preparation. Food Res Int 121(1):412–421

Chambi HN, da Costa BS, de Lima WC, Kassardjian DC, Schmidt FA (2020) Fruit juices in polysaccharides edible films. Afr J Food Sci 14(3):53–62

da Silva SR, de Moraes JO, de Souza PG, Carciofi BAM, Laurindo JB (2019) Production of mango leathers by cast-tape drying. Product characteristics and sensory evaluation. LWT 99:445–452

Dantas EA, Costa SS, Cruz LS, Bramont WB, Costa AS, Druzian FF, Machado BA (2015) Characterization and evaluation of the antioxidant properties of biodegradable films incorporated with tropical fruit pulps/Caracterizacao e avaliacao das propriedades antioxidantes de filmes biodegradaveis incorporados com polpas de frutas tropicais. Ciência Rural 45(1):142–149

de Faria Arquelau PB, Silva VDM, Garcia MAVT, de Araújo RLB, Fante CA (2019) Characterization of edible coatings based on ripe "Prata" banana peel flour. Food Hydrocoll 89:570–578
Deng Q, Zhao Y (2011) Physicochemical, nutritional, and antimicrobial properties of wine grape (cv. Merlot) pomace extract-based films. J Food Sci 76(3):309–317
Du WX, Olsen C, Avena-Bustillos R, McHugh T, Levin C, Friedman M (2009) Effects of allspice, cinnamon, and clove bud essential oils in edible apple films on physical properties and antimicrobial activities. J Food Sci 74:372–378
Esparza I, Jiménez-Moreno N, Bimbela F, Ancín-Azpilicueta C, Gandía LM (2020) Fruit and vegetable waste management: Conventional and emerging approaches. J Environ Manag 265: 110510
Espitia PJP, Avena-Bustillos RJ, Du W-X, Chiou B-S, Williams TG, Wood D, McHugh TH, Soares NFF (2014) Physical and antibacterial properties of açaí edible films formulated with thyme essential oil and apple skin polyphenols. J Food Sci 79(5):903–910
Farias MG, Fakhouri FM, de Carvalho CW, Ascheri JL (2012) Physicochemical characterization of edible starch films with barbados cherry (*Malphigia emarginata* DC). Quim Nova 35(3):546–552
Galus S, Arik Kibar EA, Gniewosz M, Kraśniewska K (2020) Novel materials in the preparation of edible films and coatings—a review. Coatings 10(7):674
Gerassimidou S, Martin OV, Chapman SP, Hahladakis JN, Iacovidou E (2020) Development of an integrated sustainability matrix to depict challenges and trade-offs of introducing bio-based plastics in the food packaging value chain. J Clean Prod:125378
Gustavsson J, Cederberg C, Sonesson U, Van Otterdijk RAM (2011) Global food losses and food waste: extent, causes, and prevention. Save Food! Food and Agriculture Organization of the United Nations (FAO), Rome, Italy Dusseldorf, Germany
Hayashi A, Veiga-Santos P, Ditchfield C, Tadini CC (2006) Investigation of antioxidant activity of cassava starch biobased materials. In: Proceedings of the 2nd CIGR International Symposium on Future of Food Engineering
Hu G, Chen J, Gao J (2009) Preparation and characteristics of oxidized potato starch films. Carbohydr Polym 76(2):291–298
Jayasekara R, Harding I, Bowater I, Lonergan G (2005) Biodegradability of a selected range of polymers and polymer blends and standard methods for assessment of biodegradation. J Polym Environ 13(3):231–251
Kaya S, Maskan A (2003) Water vapor permeability of "pestil" (a fruit leather) made from boiled grape juice with starch. J Food Eng 57:295–299
Krochta JM (2002) Proteins as raw materials for films and coatings: definitions, current status, and opportunities. Protein-Based Films Coat 1:1–40
Lan W, Wang S, Zhang Z, Liang X, Liu X, Zhang J (2020) Development of red apple pomace extract/chitosan-based films reinforced by TiO_2 nanoparticles as a multifunctional packaging material. Int J Biol Macromol
Lee KH, Chun Y, Jang YW, Lee SK, Kim HR, Lee JH, Kim SW, Park C, Yoo HY (2020) Fabrication of functional bioelastomer for food packaging from aronia (*Aronia melanocarpa*) juice processing by-products. Foods 9(11):1565
López O, García MA, Zaritzky NE (2010) Novel sources of edible films and coatings. Stewart Postharvest Rev 6(3):1–8
Lorevice MV, Moura MR, Aouada FA, Mattoso LH (2012) Development of novel guava puree films containing chitosan nanoparticles. J Nanosci Nanotechnol 12(3):2711–2717
Mariniello L, Giosafatto CV, Di Pierro P, Sorrentino A, Porta R (2010) Swelling, mechanical, and barrier properties of albedo-based films prepared in the presence of phaseolin cross-linked or not by transglutaminase. Biomacromolecules 11(9):2394–2398
Martelli MR, Barros TT, de Moura MR, Mattoso LH, Assis OB (2013) Effect of chitosan nanoparticles and pectin content on mechanical properties and water vapor permeability of banana puree films. J Food Sci 78(1):N98–N104

McHugh TH, Olsen CW (2004) Tensile properties of fruit and vegetable edible films. United States–Japan Cooper Progr Nat Resour 11:104–108
McHugh TH, Senesi E (2000) Apple wraps: a novel method to improve the quality and extend the shelf life of fresh-cut apples. J Food Sci 65(3):480–485
McHugh T, Huxsoll CC, Krochta JM (1996) Permeability properties of fruit puree edible films. J Food Sci 61:88–91
Otoni CG, de Moura MR, Aouada FA, Camilloto GP, Cruz RS, Lorevice MV, de FF Soares N, Mattoso LH (2014) Antimicrobial and physical-mechanical properties of pectin/papaya puree/ cinnamaldehyde nanoemulsion edible composite films. Food Hydrocoll 41:188–194
Otoni CG, Avena-Bustillos RJ, Azeredo HM, Lorevice MV, Moura MR, Mattoso LH, McHugh TH (2017) Recent advances on edible films based on fruits and vegetables—a review. Comp Rev Food Sci Food Safety 16(5):1151–1169
Park SI, Zhao Y (2006) Development and characterization of edible films from cranberry pomace extracts. J Food Sci 71(2):95–101
Perez-gago M, Krochta J (2001) Denaturation time and temperature effects on solubility, tensile properties, and oxygen permeability of whey protein edible films. J Food Sci 66:705–710
Perez-Mateos M, Montero P, Gómez-Guillén MC (2009) Formulation and stability of biodegradable films made from cod gelatin and sunflower oil blends. Food Hydrocoll 23(1):53–61
Ravishankar S, Jaroni D, Zhu L, Olsen C, McHugh T, Friedman M (2012) Inactivation of *Listeria monocytogenes* on ham and bologna using pectin-based apple, carrot, and hibiscus edible films containing carvacrol and cinnamaldehyde. J Food Sci 77(7):M377–M382
Reis LC, de Souza CO, da Silva JB, Martins AC, Nunes IL, Druzian JI (2015) Active biocomposites of cassava starch: the effect of yerba mate extract and mango pulp as antioxidant additives on the properties and the stability of a packaged product. Food Bioprod Process 94:382–391
Rodríguez GM, Sibaja JC, Espitia PJ, Otoni CG (2020) Antioxidant active packaging based on papaya edible films incorporated with *Moringa oleifera* and ascorbic acid for food preservation. Food Hydrocoll 103:105630
Rojas-Graü MA, Raybaudi-Massilia RM, Soliva-Fortuny RC, Avena-Bustillos RJ, McHugh TH, Martín-Belloso O (2007) Apple puree-alginate edible coating as carrier of antimicrobial agents to prolong shelf-life of fresh-cut apples. Postharvest Biol Technol 45:254–264
Shin S-H, Kim S-J, Lee S-H, Park K-M, Han J (2014) Apple peel and carboxymethylcellulose-based nanocomposite films containing different nanoclays. J Food Sci 79(3):342–353
Singh A, Gu Y, Castellarin SD, Kitts DD, Pratap-Singh A (2020) Development and characterization of the edible packaging films incorporated with blueberry pomace. Foods 9(11):1599
Sothornvit R, Pitak N (2007) Oxygen permeability and mechanical properties of banana films. Food Res Int 40(3):365–370
Souza CO, Silva LT, Silva JR, López JA, Veiga-Santos P, Druzian JI (2011) Mango and acerola pulps as antioxidant additives in cassava starch bio-based film. J Agric Food Chem 59(6):2248–2254
Tontul I, Topuz A (2017) Effects of different drying methods on the physicochemical properties of pomegranate leather (pestil). LWT 80:294–303
Torabi A, Mohebbi M, Tabatabaei-Yazdi F, Shahidi F, Khalilian-Movahhed M, Zahedi Y (2020) Application of different carbohydrates to produce squash puree based edible sheet. J Food Sci Technol 57(2):673–682
Torres CA, Romero LA, Diaz RI (2015) Quality and sensory attributes of apple and quince leathers made without preservatives and with enhanced antioxidant activity. LWT—Food Sci Technol 62(2):996–1003
Valenzuela C, Aguilera JM (2015) Effects of maltodextrin on hygroscopicity and crispness of apple leathers. J Food Eng 144:1–9

Wang X, Sun X, Liu H, Li M, Ma Z (2011) Barrier and mechanical properties of carrot puree films. Food Bioprod Process 89(2):149–156

Wu H, Lei Y, Zhu R, Zhao M, Lu J, Xiao D, Jiao C, Zhang Z, Shen G, Li S (2019) Preparation and characterization of bioactive edible packaging films based on pomelo peel flours incorporating tea polyphenol. Food Hydrocoll 90:41–49

Zhang Y, Li Y (2020) Comparison of physicochemical and mechanical properties of edible films made from navy bean and corn starches. J Sci Food Agric 101(4):1538–1545

Zhu L, Olsen C, McHugh T, Friedman M, Jaroni D, Ravishankar S (2014) Apple, carrot, and hibiscus edible films containing the plant antimicrobials carvacrol and cinnamaldehyde inactivate Salmonella Newport on organic leafy greens in sealed plastic bags. J Food Sci 79(1):61–66

Edible Films and Coatings for Fruits and Vegetables: Composition, Functions, and Regulatory Aspects

10

Neeraj Panchal, Kalyan Das, Pramod K. Prabhakar, and Neeraj Ghanghas

Abstract

Edible coatings and film are sustainable and environment friendly solutions and possess the potential to replace the convention packaging material. In the past two decades, a lot of advancement has been made in edible food packaging material. Many researchers are working on improving the physicochemical properties of edible packaging material. Various biochemical reactions and uncertain environment conditions are the major cause of the postharvest losses of horticulture commodities. Edible film (EF) and edible coating (EC) can provide a tool for preventing postharvest losses of fruits and vegetables. Horticulture commodities greatly vary in their composition and requirements for their postharvest preservation; therefore, product-specific EF or EC is required for preservation and preventing losses. These can be incorporated with bio-actives and other ingredients, for example, antimicrobials, flavoring compounds, pigments, nutrients, etc. The resulting film/coating not only acts as an effective barrier but also enhances its functions. Edible packaging materials are effective in

N. Panchal · K. Das
Department of Basic and Applied Sciences, National Institute of Food Technology Entrepreneurship & Management, Kundli, Sonipat, Haryana, India

P. K. Prabhakar (✉)
Department of Food Science and Technology, National Institute of Food Technology Entrepreneurship & Management, Kundli, Sonipat, Haryana, India
e-mail: pramodkp@niftem.ac.in

N. Ghanghas
Department of Food Science and Technology, National Institute of Food Technology Entrepreneurship & Management, Kundli, Sonipat, Haryana, India

Department of Food and Human Nutritional Sciences, University of Manitoba, Winnipeg, MB, Canada

A. Poonia, T. Dhewa (eds.), *Edible Food Packaging*,
https://doi.org/10.1007/978-981-16-2383-7_10

controlling growth of microbes, biochemical changes, preserving texture, and sensory attributes of the fruits and vegetables, which ultimately results in enhanced shelf-life and better acceptability by consumers. This review article discusses various types of edible films and coating for fruits & vegetables, their composition, functions, properties, and effects as well.

Keywords

Edible film · Edible coating · Fruit and vegetables · Packaging material · Shelf-life · Active ingredients

10.1 Introduction

India is one of the countries whose economy is dependent on agriculture. According to the report published by the National Horticulture Board (NHB 2017) of India, of the total horticulture produce, 90% is accounted for fruits and vegetables. In total world's fruits and vegetable production, India holds a second position and a leader in the production of several horticulture crops. The production of horticulture crops has been significantly increased from 1991–1992 to 2016–2017 and mainly fruits and vegetable production increased from 2874 to 6480 MT and 5593–10,290 MT, respectively. This enormous increase in the production also resulted in an increase in post-harvest losses of horticulture crops (NHB 2017). These are highly perishable commodities; all fruits and vegetables contain 65–95% water. The transpiration and respiration rate affect the quality of the commodity. The high rate of water loss and use up of stored food accelerates the decay process and the produce becomes inedible. The main cause of physiological deteriorations is respiration, transpiration, ripening, physical damage, diseases, and pests. Also, the external conditions such as temperature and relative humidity affect the freshness of fruits and vegetables (FAO 1985; Baldwin 1994; Oms-Oliu et al. 2008b). Due to these reasons, a lot of care is required after harvesting, transportation, and storage, otherwise produce becomes unfit for consumption. Methods are required for a reduction in wastage for economical purposes also. There are several methods available to minimize the loss in quality. To increase the safety and to minimize wastage of horticulture produce, various processing operations and techniques such as drying, canning, packaging, blanching, and other value addition methods are generally exercised. These processing methods more or less result in the loss of vitamin C and other heat-labile bioactives. The enrobing of polymeric material, for example, starch, proteins, oils/fats, and herbal extracts, on fruits and vegetables surface also enhances shelf-life by maintaining the organoleptic attributes. These films and coatings are treated as packaging material that can be consumed along and easily biodegradable if not consumed. It also provides preservation like an encapsulation of antioxidants, antimicrobials, flavoring compounds, etc. (Seydim and Sarikus 2006). If the edible films and coatings are applied in multilayer, this can constrain the metabolic reactions by providing a gas barrier, degradation of nutritional content such as vitamin A, C, etc. (Brasil et al. 2012). The edible packaging materials have many advantages, but it cannot be considered as an ideal packaging material due to some limitations such as fruits and vegetables coated with edible packaging material requires over coating of other packaging material for handling and hygienic purpose (Debeaufort et al. 1998). This paper is focused upon the need for edible packaging material, components, limitations, and regulatory issues related to edible films and coatings.

10.2 Edible Films and Coatings

Edible packaging material can be defined as "a film, sheet, thin layer or coating which is an integral part of a food and is eaten with it is termed as edible coating" (Debeaufort et al. 1998). In other words, the edible packaging material is that which is made of edible material formed into thin sheets, which is consumed, provides a barrier to transfer phenomena (Du et al. 2011; Bourtoom 2008). Edible films and coatings terms are often used synonymously but there exists a difference between them. Edible film is independent, performing thin structures which are then wrapped on the surface of fruits and vegetables may or may not be removable. On the other hand, edible coating is packaging material that is formed directly on the surface and consumed along (Debeaufort et al. 1998). Edible film is formed by spreading the film-forming solution which is followed by drying and removal of solvent by adopting the process of extrusion, co-extrusion, lamination which is only for multi-layer coatings whereas edible coating methods involve spray fluidization, falling, pan coating, spraying, dipping, brushing, etc. (Debeaufort et al. 1998; Kester and Fennema 1986; Cuq et al. 1995; Falguera et al. 2011). For the structural matrix of edible films and coatings, polymers like polysaccharides, proteins, lipids, and composites are preferred (Falguera et al. 2011; Bourtoom 2008; Cuq et al. 1995; Gennadios et al. 1993). Plasticizers are also added to improve the properties of film formation. Edible films and coatings can be obtained via two processes such as wet or dry methods (Guilbert et al. 1997). In a dry process, the polymeric material is being heated above their melting points which make material to start flow by the process of molten casting, extrusion, heat processing (Gennadios et al. 1993). It does not involve the use of solvent but requires the knowledge of thermoplastic properties of material for film formation. Wet film formation involves the dispersion of film-forming solution in solvent which is sprayed, spread, or dipped followed by drying for removal of solvent (Cuq et al. 1995; Kester and Fennema 1986). The solvent for film formation should be food grades like water, ethanol, biodegradables, etc. For homogenous film structure, various factors are responsible such as pH, temperature, relative humidity, drying conditions, solvent used, solubility of solvent concentration, matrix, thermo-plasticity of polymer such as phase transition temperature, glass transition, gelatinization temperature, hydrophilic properties such as surface tension, viscosity, solid surface energy (Cuq et al. 1995; Guilbert et al. 1997; Kester and Fennema 1986). The chemistry of film formation is not yet well explained. Film-forming solution gelatinizes to form a rigid structure as a result of solvent removal

Table 10.1 Functions of edible film and coatings

Selective properties	Active properties
• Barrier to gases (oxygen, carbon dioxide, and nitrogen) • Light (UV) • Organic vapors (aroma compounds and solvents) • Water vapor • Solutes (salts, lipids, additives, pigments)	• Encapsulation or carriage of flavorings, antimicrobials, antioxidants, pigments, and other additives • Enhancement of mechanical and optical properties • Individual protection of small pieces of food (Separation of food by individual portion, soluble package for pre-dosed food ingredients or additives)

Source: Debeaufort et al. (1998)

supported by interactions like covalent, electrostatic, and ionic form (Gennadios et al. 1993). There is a phase transition from the polymer in water or solvent to water in polymer (Kester and Fennema 1986). Edible films and coatings provide elective and active properties which are summarized in Table 10.1.

10.3 Development of Edible Film and Coatings

The terminology of edible films and coatings may be new but the concept is not new. These have been used for so long without having the knowledge of its chemistry. Even edible films were used to write instructions that if they get caught, can be easily destroyed by eating. In Japan, the first-ever commercial edible film named "Yuba" made with the skin of boiled soy milk was used since the fifteenth century (Gennadios et al. 1993). Edible films and coatings were used for fruits and vegetables since ages to protect them from the moisture loss. The first use of fruits and vegetables was seen in China in the twelfth century for citrus fruits (Guilbert et al. 1997). Citrus fruits were kept inside the boxes and wax was poured on that (Hardenburg 1967; Embuscado and Huber 2009). Later on, wax coatings were used to add shine and prevent moisture loss from the surface (Baldwin 1994). Even since the 1930s, wax emulsion and oil in water were also coated on the surfaces of fruits and vegetables to improve appearance and preservation (Debeaufort et al. 1998). In Europe, larding was used to preserve fruits and vegetables (Embuscado and Huber 2009).

Edible films and coatings were used for various foods other than fruits and vegetables since ages but an edible coating finds more of an application than edible films. To prevent the shrinkages in meat and cheese larding (animal fat), coatings were used (Kester and Fennema 1986; Debeaufort et al. 1998). Candies coated with zein (corn protein), nuts coated with sugar were most common (Krochta and Mulder-Johnston 1997). Cellulose coatings are used in the bakery, confectionary, and meat industry. Meat was coated with the edible film made with gelatin (Embuscado and Huber 2009). Other than the food industry, edible films and coatings find application in the pharmaceutical industry also such as sugar and gelatin coatings on pills and soft capsules, respectively (Gennadios et al. 1993; Krochta and Mulder-Johnston

1997). Sausage casings are made of collagen. Edible films and coatings are gaining importance day by day and find its excellent potential in food processing.

10.4 Need for Edible Packaging Material for Fruits and Vegetables

There are enormous numbers of reasons which trigger the use of edible packaging material. The primary reason is that packaging material that is being used to date are petroleum-, metal-, or paper-based manufactured using nonrenewable sources. These take thousands of years to decompose resulting in an increase of non-biodegradable waste. According to a report prepared on plastic industry by FICCI (2014), out of the total plastic generation, 43% is used /consumed for manufacturing of packaging material for food materials which makes a significant part of the total. In the food processing industry, the consumption of plastic as packaging material is in different forms. For instance, 33% is used in polyethylene form, 29% as polypropylene (FICCI 2014). This industry is expected to grow at higher rates due to increased income of consumers, change in their eating habits, and inclination toward the consumption of the packaged product. An alternative to this plastic waste generation is recycling, but it is also a high cost and energy demanding process which includes cost of transportation and disposal. According to a report by USEPA (2015), the use of edible packaging material can result in the following:

- Reduce cost of waste disposal.
- Save money of recycling.
- Reduce environmental impacts.
- Reduce health-related issues.

For fruits and vegetables, there are various reasons as they undergo several changes post-harvest. The primary reason is the physiological changes. After the harvesting of fruits and vegetables, physiological processes continue to occur but in a different manner. Once produce is harvested, fruits and vegetables have no source to obtain nutrients so draw nutrients from reserves. These results in the aging process and slowly fruits and vegetables decay, that is, become unacceptable for consumption (FAO 1985). The two major physiological processes are respiration and transpiration (water loss) (Baldwin 1994). These changes also cause a decrease in nutrition content of these commodities such as dry matter content as reserves are used continuously and vitamin C content also decreases with time.

During respiration, the plant takes up the oxygen and releases carbon dioxide. It is one of the necessary reactions in plants, in which oxygen breaks down the carbohydrates into carbon dioxide, water, and energy, that is, heat which is in a continuous process in the plant, but after harvesting continues on the account of the reserves. Once this reserve ends, fruits and vegetables die (Baldwin 1994; FAO 1985). Air movement while storing the product is necessary to remove heat from respiration. When the produce is kept in poor ventilation, carbon dioxide accumulation leads to quick deterioration in the quality (FAO 1985).

Before harvesting, the plant continues to replace water that is lost through transpiration to take water from the soil, but once produce is harvested, there is continuous loss of water. Fruits and vegetables shrink, and loss of weight occurs. The crucial factor that affects the rate of water loss is the ratio of surface area to the volume of produce. The type of produce is another important factor that affects the rate at which water is lost, for example, the presence of waxy skin with large number of pores in green leafy vegetables is the main cause of rapid loss of water. In contrast, the rate of water loss is gradual in potatoes because of thick corky skin with less pores (FAO 1985; Baldwin 1994).

Fruits and vegetables are basically of two kinds. One is climacteric which ripens after being detached from the parent plant and other are non-climacteric which ripens only when attached to the parent plant. Fruits and vegetables continue to ripen after harvest which is followed by senescence which degrades the quality. During ripening, ethylene gas is produced which is triggered by consumption of sugars. In climacteric fruits, the produce is harvested before ripening (green stage) and then transported to remote locations. But once fruits and vegetables become fully mature, ethylene production continues which deteriorates quality. Any injury during transportation, handling, or because of attack by parasites increases the production rate of ethylene (Baldwin 1994; Oms-Oliu et al. 2008a, b; FAO 1985). In earlier times, people used to grow fruits and vegetables for their own needs and can be consumed fresh. But in the last few decades, fruits and vegetables are being transported and stored. To cover this whole food supply chain, it takes a considerable amount of time to reach the final consumer. During this time period, commodities can lose their color, appearance which reduces its aesthetic value (Embuscado and Huber 2009). Edible packaging can be one of the solutions to the entire above-mentioned problem.

10.5 Components and Composition of Edible Film and Coating

Like other packaging materials, polymers are also the basic material for formation of edible films and coatings. The major polymeric contributing materials are polysaccharides, protein, lipids, and resins shown in Table 10.2 (Falguera et al. 2011; Bourtoom 2008; Cuq et al. 1995; Gennadios et al. 1993). It is not necessary that the edible film will be based upon polymer only. Waxes and resins with or without addition of polymers are used to develop bilayer or composite films and coatings (Baldwin 1994).

Plasticizers are added to film-forming solution as a component. Plasticizers can be defined as the agents that are low in molecular weight which when added to the film-forming solutions decrease its glass transition temperature (Gennadios et al. 1993). They are basically of two types one which interrupts the polymer-polymer bonding to keep them farther and itself interacts with polymer by forming hydrogen bonds. It interacts with water molecules and results in bigger hydrodynamic radius and high moisture content (Sothornvit and Krochta, 2001). There are various factors

Table 10.2 Components of edible film and coatings

Components	Material
Basic material	**Polysaccharides:** Starch (maize, cassava), modified starch, modified cellulose (CMC, MC, HPC, HPMC, ethyl cellulose), alginates (sodium alginate), pectin, carrageenan, pullulan, chitosan, galactomonnas, gums such as gellan, xanthan, guar, locust, etc. **Proteins:** Soy proteins, pea proteins, rice proteins, pea proteins, gluten, casein, whey proteins, collagen, etc. **Lipids (waxes and resins):** Rice bran wax, candelilla wax, shellac, paraffin wax, terpene, acetoglycerides, etc.
Plasticizers	Glycerin, propylene glycol, sorbitol, sucrose, polyethylene glycol, corn syrup, water
Active ingredient	Antioxidants, antimicrobials, nutrients, anti-browning agents, flavorings, pigments, bacteriocins (nicin or pediocin or netamycin)
Other additives	Emulsifying agents such as lecithin, Tweens, and Spans; lipid emulsions such as edible waxes and fatty acids; crosslinkers such as calcium chloride

that affect the plasticizer's action such as size, shape, number of oxygen atoms, and efficiency of plasticizer (Sothornvit and Krochta 2001). Plasticizers work by breaking the polymer-polymer interaction increases the distance between the polymers and rather forms bonds with the polymer. Thus, resulting in increase in the mobility of polymer, decrease in crystalline region to amorphous especially in the case of polysaccharide, protein films, decrease in gelatinization temperature, increase in mechanical properties and resistance to vapor or gases. Water also works as a plasticizer, but it dehydrates at low humidity, add plasticizer with hydrophilic properties. In charged molecules, repulsive forces whereas between the same charged molecules, polar and non-polar molecules also provide plasticization effect by increasing the distance (Gennadios et al. 1993).

Selection and use of polymers in producing edible packaging material is often decided by studying tensile properties. These properties are accomplished by drying the produced film or coating on flat surfaces or glass. In case of fruits and vegetables, water barrier efficiency and oxygen barrier are required. Prior is necessary to prevent the surface dehydration and latter to prevent the ripening of fresh fruits and vegetables. There are some safety and health issues associated with the selection of polymer material. These are generally regarded as safe (GRAS), shelf-life, light processing, respiration, transpiration, controlled atmospheric pressure (CAP) and modified atmospheric pressure (MAP), allergenicity (Attila et al. 2009). Materials which have been selected for the formation of edible films and coatings should have a GRAS status as material will be in direct contact with the food material. The selection of polymers should solely be not on the basis of GRAS as there are also other factors to consider. The polymers selected for formation of edible packaging material should have shelf-life compatible with fruits and vegetables. According to Attila, the edible material should have at least shelf-life of 2 weeks during which fruits and vegetables can go through the whole supply chain and still should be in the condition that can be consumed by the consumer. Edible material also should have some durability to light processing (Attila et al. 2009). Light processing includes a variety of processes that are used for fruits and vegetables to convert them in edible form without much change in their original quality (Shewfelt 1987). During the light processing, certain changes occur such as the tissues get disrupted which creates many membrane-related problems and eventually become perishable in nature. The polymer can be subjected to light processing without undergoing any change in itself and also in fruits and vegetables. As mentioned earlier, fruits and vegetables undergo various reactions after harvesting. These reactions result in loss of texture which is undesirable by the consumers. Thus, decreases the value in the market so polymer should be such that it can prevent these kinds of reactions after harvesting. Temperature is one of the major factors which affect the quality during transportation as fruits and vegetables are exposed to temperature abuse so, as a result, they are packaged in modified atmosphere packaging (MAP) or controlled atmospheric storage (CAS). But during the temperature fluctuations, the water condensates on the surface and causes spoilage. An edible film or coating should not impart any taste to fruits and vegetables should not be detectable upon consumption, that is, good sensory characteristic. It should provide good barrier properties like gases and water vapor, confers mechanical stability which may be able to withstand pressure and prevent mechanical injury during transporting and handling. Polymeric material should have enough biochemical, physicochemical, and microbial stability. It should not degrade with passage of time and produce substances that are harmful upon consumption for humans. Ideal edible films and coatings are that which fulfill all the functions performed by packaging material but cost is also an important criterion. The raw material and process required for producing such types of packaging material should be less. Moreover, it should require simple technology for production (Debeaufort et al. 1998).

10.5.1 Polysaccharides

Different types of polysaccharides such as starch, non-starch carbohydrates such as cellulose, pectin derivatives, chitosan, etc., gums, and fibers (Han 2014; Bourtoom 2008; Xiao et al. 2011) which are used for edible films and coatings. Major polysaccharides are neutral in nature while some of the gums are positively charged with the exception of negative charge (Han 2014). Neutral carbohydrates have hydroxyl or hydrophilic moieties on the surface in large numbers which is useful in film formation. The presence of hydrophilic moieties on polysaccharides structures confers poor water vapor barrier properties (Kester and Fennema 1986; Brasil et al. 2012; Vásconez et al. 2009). Alginate- and gellan-based coating when coated upon papaya showed poor water vapor barrier (Tapia et al. 2007). Even the addition of sunflower oil did not show significant difference in water vapor pressure. But due to this hydrophilic nature, polysaccharide-based coatings can be used to prevent the moisture loss from the surface of fruits and vegetables (Kester and Fennema 1986). This prevents the texture loss which is also desirable. In one of the studies used the polysaccharides such as alginate-, pectin-, and gellan-based edible film was formed to coat the fresh-cut melon (Oms-Oliu et al. 2008a, b). The film was investigated for the gas exchange, antioxidant properties, color, firmness, sensory quality, and microbial quality for 15 days at 4 °C. The film showed the increased water vapor resistance which prevents the dehydration of cut melon pieces. Polysaccharides such as alginate- and gellan-based edible coatings were formed to coat the fresh-cut pieces of fuji apples and effect on shelf-life was studied (Rojas-Graü et al. 2008). Then, alginate- and gellan-based film wrapped on cut fuji apples were evaluated for changes in headspace gas composition, color changes, firmness, and microbiological quality which was stored for 23 days at 4 °C. It was found that edible coating was able to effectively prolong the shelf-life of cut fuji apples by 2 weeks. In the same study, the significant reduction in the mesophilic and psychrophilic microorganisms was also observed.

Closed tightly packed and hydrogen bonding in polysaccharides confers good gas barrier properties to edible films (Yang and Paulson 2000). These are able to modify gas atmosphere in fruits that delay the ripening (Baldwin 1994). Gas, especially the oxygen barrier property is highly dependent on the relative humidity. More the relative humidity, the more the mobility, the more the mass transfer across the film due to plasticizing effect (Bonilla et al. 2012). As if the humidity is high, it causes the swelling in polysaccharide matrix which in result increases the permeability and diffusivity across the film (Kumins 1965). Polysaccharide-based edible films also provide other additional functions such as it minimizes the browning when incorporated with anti-browning agents and also the textural changes (Rojas-Graü et al. 2008). The different polysaccharide materials used for fruits and vegetables are summarized and shown in Table 10.3.

10.5.2 Proteins

Proteins are generally used for formation of edible films and coatings. Different types of proteins which are used for formation can either be plant proteins or animal proteins. Animal proteins include casein protein, whey protein, meat proteins such as gelatin, feather keratin whereas plant protein includes wheat, soy, corn proteins, mung beans proteins, and peanut proteins (Gennadios et al. 1993; Bourtoom 2008). A protein generally in their native state exists as fibrous and globular proteins. Out of both the types of fibrous proteins are insoluble in water whereas globular proteins are soluble in water (Scope 1994). During edible film formation, fibrous proteins due to their fully extended structure these proteins can associate them in parallel. This association is through bonds such as hydrogen, ionic, hydrophobic, and covalent bonding. On the other hand, globular proteins form spherical structures held by bonds such as hydrogen, ionic, hydrophobic, and covalent, that is, disulfide bonds which are undesirable for edible film formation (Scope 1994). Proteins have unique characteristics such as their conformational denaturation, electrostatic charges, and amphiphilic nature (Han 2014). So, proteins primary, secondary, and tertiary structure can be treated with acids, alkalis, metal ions, salts, chemical hydrolysis, enzymatic treatment, heat denaturation, pressure, irradiation, mechanical treatment, and chemical crosslinking to bring about the desirable changes which are required

Table 10.3 Polysaccharide-based films for fruits and vegetables

Polysaccharide	Fruit/vegetable	Properties/functions	Reference
Cashew gum (20 g/ 100 mL water)	Mango	• Decreases weight loss, Prevent the change in TSS and pH	Souza et al. (2010)
Chitosan coating (0.2 g/ mL)	Fresh-cut papayas	• Reduce deteriorative process, maintains quality • Enhances shelf-life	Gonzalez-Aguilar et al. (2009)
Calcium-alginate coating (sodium alginate 4%, $CaCl_2$ 5%)	Fresh-cut carrot	• Prevent microbial growth, preserve sensory qualities • Enhance shelf-life	Costa et al. (2012)
Pectin (1%, 1 g/33.33 mL water, 15 g/500 mLwater)	Mellon, peach, mango	Improves fruit sensory acceptance • Promote the reduction of product respiration rate • Maintenance color characteristics • Pectin edible film formulated with cinnamon leaf oil decreases bacterial growth, increases antioxidants status • Reduces rate of texture softening, increases shelf-life	Ferrari et al. (2013), Ayala-Zavala et al. (2013)
Carrageenan (0.2–0.8%, w/v)	Papaya	• Delays ripening, extends shelf-life	Hamzah et al. (2013)
Alginate (1–5%)	Button mushroom, mango, cherry, persimmon fresh-cut apples, fresh-cut melon, fresh-cut mango	• Good gas barrier property, poor water barrier property • Effective in maintaining firmness of mushroom • Effective in delaying browning, preserve color, and sensory qualities, control microbial growth, higher elasticity	Jiang et al. (2013), Robles-Sánchez et al. (2013), Díaz-Mula et al. (2012), Neves et al. (2012), Chiabrando and Giacalone (2014), Raybaudi-Massilia et al. (2008), Salinas-Roca et al. (2016)
Basil-seed gum (5%) Coatings	Fresh-cut apricots	• Enhances shelf-life, control microbial growth	Hashemi et al. (2017)
Sucrose-based polymer (0.1–1%)	Mandarin	• Increase concentration of soluble solids, Vit. C content, sugar content, and carotenoids	Tao et al. (2012)

(continued)

Table 10.3 (continued)

Polysaccharide	Fruit/vegetable	Properties/functions	Reference
		• Surges catalase, peroxidase, and superoxidase dismutase activities • Reduces decay rate, TA and polyphenol oxidase activity	
Gelatin (5%)	Persimmon	• Good gas barrier, good for preserving flavor, affects the appearance	Neves et al. (2012)
Aloe vera gel (5%, 15%, 50%, 100%)	Fresh-cut apple, kiwifruit, pomegranate arils	• Reduces deterioration and respiration rate • Reduces microbial load, good sensory attributes • Delays browning	Song et al. (2013), Benitez et al. (2013), Martinez-Romero et al. (2013)
Maize starch (20 g/L)	Brussels sprouts	• No change in the contents of ascorbic acid and total flavonoids • Increases the radical-scavenging activity • No reduction in nutritional quality	Vina et al. (2007)

for edible films and coatings formation (Han 2014). Globular protein structure can be subjected to any of the above-mentioned treatments to get more of the extended structures that are required for edible film formation. More the number of extended structures is more the number of interactions. As a result of interactions, the edible films and coatings forms are stronger. But at the same time, these are less flexible, less permeable to gases, water vapors, and liquids (Kester and Fennema 1986).

On the surface of some proteins, there are some groups that are bonded to polymers via hydrogen bonding or ionic bonds, the edible films or coating forms are susceptible to moisture but barrier to oxygen. So, the edible films or coatings based on proteins provide a barrier to oxygen at low humidity (Bourtoom 2008). Edible films and coatings based on protein show high water vapor permeability due to their hydrophilic nature which is not desirable in fruits and vegetables. Film formed with high concentration of gelatin showed high water vapor permeability (Fakhoury et al. 2012; Jongjareonrak et al. 2006). Similarly, edible film formed with cornstarch and gelatin plasticized with glycerol, with increase in gelatin amount had increased water vapor permeability but less opacity (Fakhouri et al. 2015). The different protein materials used for coating of fruits and vegetables are summarized and shown in Table 10.4.

Table 10.4 Protein-based edible films and coatings for fruits and vegetables

Protein	Composition	Method of application	Fruits/ vegetables	Properties/functions	References
Rice bran protein	Rice bran protein (4 g), gelatin (4 g), fructose (2 g), grape seed extract (1%), and water (100 mL)	Wrapping in film	Strawberry	• Increases shelf-life by 2 days • No significant change in color and sensory properties	Shin et al. (2012)
SPI (soy protein isolate)	SPI to Glycerol in 2:1 ratio, cysteine (0.5%)	Dipping in coating	Apple	• Retain firmness • Controlled enzymatic browning • Extend shelf-life	Ghidelli et al. (2010)
Caseinate	NaCas: Glycerol (1:0.3 ratio), NaCas to lipid ratio 1:0.5, and lipid fraction composed of oleic acid to beeswax in 70:30 mass ratio	By dipping	Dried pineapple	• Extend shelf-life	Talens et al. (2012)
Whey protein isolate nano Fibrils	WPNF (5%, w/v), glycerol (4%, w/v), and Trehalose (3%, w/v)	Dipping	Apples	• Retard change total phenolic content • Prevents browning, and product weight loss	Feng et al. (2018)
Whey protein	5% protein (calcium caseinate or whey protein powder), 2.5% glycerol, 0.25% CMC, and 0.125% CaCl2	Dipping	Apple and potato slices	• Delays browning • Whey protein shows better antioxidant property than calcium caseinate	Tien et al. (2001)
Protein isolate of white mouth croaker (CPI)	CPI (35 g), MMT (5 g), and glycerol (10.5 g)	Immersion	Fresh-cut papaya	• Reduces weight loss • Preserve firmness • Effective in controlling color and microbial growth	Cortez-Vega et al. (2014)
Soy protein	SPI (5% w/v), SPI:Glycerol ratio 2:1	Immersion	Fresh-cut eggplant	• Controls enzymatic browning • Extend shelf-life	Ghidelli et al. (2014)

10.5.3 Lipids and Resins

Lipids and resins used for producing edible packaging materials are waxes and paraffin, acetoglycerides, and shellac resins (Bourtoom 2008). Edible film made with waxes to coat fruits and vegetables has been used since long. Polysaccharides and proteins are polymers, but lipids and resins are not polymers. But lipids and resins are edible, biodegradable, and cohesive biomaterials which make lipid a suitable material for film formation (Han 2014). Lipids and resins have the property to be semisolid at room temperature which is undesirable for formation of edible films and coatings formation. Along with this property, lipids and resins also have the property of characteristic phase transition temperature. Due to this property of phase transition, lipids and resins can be modified into any shape (Han 2014). The transition in lipids and resins can be brought about by subjecting them to heat treatment which causes the change in transition between solid, semisolid, and liquid state. Lipids are hydrophobic in nature hence, the film or coatings made are high in water resistance and low in surface energy (Han 2014). Film or coating blocks the transport of water due to low polarity. Due to this moisture barrier property, these films are also able to reduce weight loss of fruits and vegetables due to desiccation (Krochta and Mulder-Johnston 1997). The migration of water across the film depends on various factors such as length of chain, degree of unsaturation, and number of acyl groups. Length of hydrocarbon chain is in direct relation with water migration whereas rest other factors are in inverse relation (Krochta and Mulder-Johnston 1997; Coma et al. 2001). One of the disadvantages associated with the edible films and coatings made with lipids and resins is that they are brittle and thick (Bourtoom 2008).

This disadvantage can be overcome by making them in association with other polymers such as association of lipid film with polysaccharides results in formation of film with increased mechanical strength (Bourtoom 2008). Similarly, this can be associated with other polymers which can result in desirable properties required for film formation. Other issues related to fatty acid-based edible films its vulnerability to oxidation and sensory issues (Brody et al. 2001). Lipid-based edible films are able to provide selective barriers to gases. More is the length of hydrocarbon more the oxygen permeability across the membrane. Unsaturation and permeability of oxygen show inverse relation (Krochta and Mulder-Johnston 1997). Lipids being the water-insoluble, the water-soluble pigments of fruits and vegetables will not migrate to the membrane and will retain with fruits and vegetables. The different lipids and resins used for coating of edible films and coatings are summarized and shown in Table 10.5.

10.5.4 Composites

Sometimes above discussed polymers, lipids, and resins are not effective in preserving the quality of fruits and vegetables by themselves. So, these are made effective by using them in combination. These combinations can be protein-carbohydrate,

Table 10.5 Lipid- and resin-based edible films and coatings for fruits and vegetables

Lipids/ resin film	Composition	Method of application	Fruits/ vegetables	Properties/ functions	References
Beeswax emulsion	–	By emitting	Sweet pepper, mango, avocado	• Improved peal resistance to penetration. • Improved sensory firmness. • Reduced rate of decay.	Bustan and Lahav et al. (Bustan and Lahav 2012)
Wax and various oils	Mustard oil (100%) or coconut oil (100%) or castor oil (100%) or sesamum oil (100%) or liquid paraffin wax (100%)	Dipping	Lime	• Coconut oil was most effective in preserving quality. • Acceptable sensory and optical properties. • No incidence of molds & their growth up to 18 days of storage.	Bisen et al. (2012)
Carnauba and mineral oil coatings	Commercial carnauba Stafresh 2505™ (SF 2505) and mineral oil Stafresh 151™ (SF 151)	Manually applied using ArtexMR brushes	Tomato	• Preserves freshness without any adverse effect on nutrients.	Dávila-Aviña et al. (Dávila-Aviña et al. 2014)

protein-lipid, or carbohydrate-lipid (Bourtoom 2008). As far as the structure of composite film is concerned, it can be bilayer or conglomerate. In bilayer, there is one layer above another. On the other hand, conglomerate is a mixture of several layers in one layer. Conglomerate can be further classified into components distinct and components intermixing. In component distinct conglomerate composites, all the components are distinctly visible whereas in components intermixing conglomerate composites not all the components are distinctly visible shown in Fig. 10.1.

Combination of different films for formation of edible films and coatings results in enhanced properties such as mechanical and permeability property of edible film. For instance, polysaccharide-based edible film alone cannot provide good water vapor properties (WVP) due to its hydrophilic nature but addition of lipid can produce a film with water vapor barrier properties and also high gas permeability specifically at high RH (Coma et al. 2001). An edible film based upon hydroxypropyl methylcellulose (HPMC), a polysaccharide was formulated with

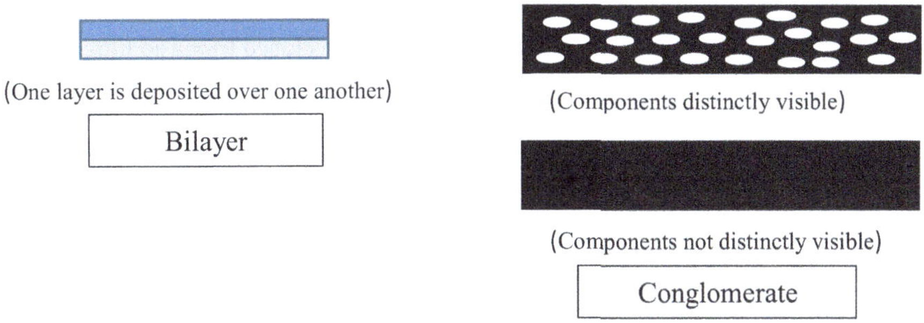

Fig. 10.1 Types of composite edible film and coatings

the incorporation of stearic acid known as fatty acid showed high water vapor barrier properties (Coma et al. 2001). One more α-nanoparticles ionic gelation was formulated (Medina et al. 2019). The prepared film was applied to blueberries and tomatoes as an internal coating to PET. Quinoa does not possess good film property whereas chitosan films do not have good mechanical property and have antibacterial property. Film formed with both results in the hydrophilic interaction as H- bonding also reported by Abugoch et al. (2011) as FTIR peak between quinoa and chitosan. Moreover, film showed high antibacterial against *Botrytis cinerea.* The interaction between these two gives synergistic effects. The anionic sulfide group of chitosan interacts ionically with protonated amino group of the chitosan which results in the production of film with completely new mechanical, barrier, elongation, and adhesive properties even without the addition of plasticizer (Abugoch et al. 2011). Composite edible film produced with rapeseed protein hydrolysate and chitosan results in the increase of α helix structure between 15.4% and 25% as addition of chitosan causes conformational changes in protein. This change affects the properties of film in a positive manner, that is, the tensile strength of film increased from 16.04 to 23.46 MPa. Moreover, the hydrolysis of protein results in the production of smaller proteins which increases the interaction of the protein with chitosan (Zhang et al. 2019). It is not necessary the edible film can only be produced with above-mentioned combinations. Piccirilli et al. (2019) formulated a film with the combination of whey protein concentrate (WPC) and liquid smoke (LS). WPC film showed high UV barrier properties in the range of 200–400 nm, but the addition of LS increases the range up to 400 nm which is highly desirable in the case of fruits and vegetables with light sensitive components. The increased range for UV light barrier is attributed to the presence of phenolic and carbonyl groups in LS (Piccirilli et al. 2019). Transparency is also one of the important characteristics of the edible film for consumer acceptance. The solvent used for preparation has an effect on the film formation (Tsai and Weng 2019). Whey protein and zein edible film when prepared with deionized water showed more transparency as compared to when prepared with ethanol. But sometimes this combination works less. For instance, carboxymethylcellulose film in combination with soy protein showed even poorer water barrier properties due to their hydrophilic nature (Baldwin et al. 1996). Hence, composite edible films and coatings can be better for fruits and vegetables because these can serve multipurpose. Dhumal et al. (2019) developed triphasic film, that is, composite film of chitosan, guar gum, and whey protein isolate with the addition of different antimicrobial essential oils such as eugenol, carvacrol, and citral. The addition of essential oils resulted in the reduction of the tensile strength due to their lipophilic nature which is not required for fruits and vegetables packaging. On the other hand, WPI composite film developed with the addition of rapeseed oil resulted in more oxygen and carbon dioxide permeability which is a desirable characteristic for fruits and vegetables as they respire after harvesting (Galus and Kadzińska 2019). The different composite materials used for coating of fruits and vegetables are summarized and shown in Table 10.6.

10.6 Regulatory Aspect of Edible Films and Coatings

Regarding the regulatory aspect of edible films and coatings, there is not much information available. As these are consumed along with the food product, the requirement should be that are being followed for the food products. Even the component in Table 1 used for forming film solution should be safe to consume and does not cause any health hazard to consumers. These should hold GRAS which is generally regarded as safe and have status as this will be in direct contact with the food material. Polymer used should be approved by FDA and if the polymer has not been approved by FDA, then the manufacturer of edible film and coating can file the petition for approval of material. Attila et al. (2009) stated that, "there are three types of GRAS designations. (1) Self – affirmed, where the manufacturer has carried out necessary work and is ready to defend GRAS status if challenged, (2) FDA pending, where results of research have been submitted to FDA for approval, (3) No comment – which is the response of FDA if after review, it has no challenges."

According to Debeaufort et al. (1998), edible films and coatings can be classified as foods or food ingredients. If this is applied on any food product, then there should be proper labelling containing the information about the ingredients used for functional ingredients also and any allergic compounds (Gennadios et al. 1997; Falguera et al. 2011). Since, there is some migration of chemicals from the packaging material to the food product. There is a need to set the overall migration limits (OMLs) and specific migration limits (SMLs) (Falguera et al. 2011). However, Food Safety and Standards Authority of India (FSSAI 2019) has recommended maximum permissible limit of food additives for coating the surface of fresh fruits and vegetables. There are various food additives (beeswax, candelilla wax, carnauba wax, glycerol ester of wood rosin, iron oxide, microcrystalline wax, ortho-phenylphenol, sodium ortho-phenylphenol, polyethylene glycol, polyvinylpyrrolidone, sulfites, shellac, bleached, sucroglycerides, beeswax, lauric arginate ethyl ester, phosphates, shellac) enlisted by FSSAI under Food Safety and Standard Regulation 2011 (Food Products Standards and Food Additives) (FSSAI 2019). The permissible limit of all food additives is well mentioned in Table 10.7.

Table 10.6 Composite edible films and coatings for fruits and vegetables

Composition of coating/films	Method of application	Fruits/vegetables	Properties/functions	References
Ca-Caseinate (2.5% w/v), whey protein (2.5% w/v), and glycerol (2.5% w/v)	By spraying	Strawberry	• Enhance shelf-life.	Vu et al. (2012), Robledo et al. (2018a , 2018b)
Gum tragacanth and aloe vera in 50:50 ratio	By immersion	Mushroom	• Extend shelf-life. • Preserve texture. • Retard ripening process.	Mohebbi et al. (2012)
Palm Oil and Beeswax in 1:1 ratio	By Hand-wipe technique	Guava	• Reduces the rate of weight loss and yellowness and. • Slow down the decline of glossiness, lightness, and greenness of guava.	Ruzaina et al. (2013)
80 g kg – 1 SC of Hydroxypropyl methylcellulose and beeswax: shellac in ratio of 1:3	Dip coated by immersion	Mandarin	• Effective in controlling weight loss. • Extend shelf-life. • Can result in off flavor development.	Contreras-Oliva et al. (2012)
Corn starch (waxy, modified, and native) and GEL in 1:4 ratio; GEL contains plasticizer (sorbitol or glycerol) (100 g/kg of the GEL)	By Immersion	Red crimson grapes	• Gelatin significantly increased mechanical strength, solubility in water, permeability to water vapor, and thickness of the biofilms, while also decreasing the opacity. • Sorbitol had significantly lower permeability to water vapor and higher tensile strength than the films plasticized with glycerol.	Fakhouri et al. (2015)
Whey protein concentrate and Glycerol in 3:1 ratio and bee wax (20% dry basis)	By immersion	Fresh-cut apple	• Facilitates incorporation of antioxidants in coating. • Antioxidant incorporated edible coating was better in reducing browning in comparison to that of antioxidants used alone. • Coating does not affect the weight loss.	Perez-Gago et al. (2006)

(continued)

Table 10.6 (continued)

Composition of coating/films	Method of application	Fruits/vegetables	Properties/functions	References
Methylcellulose (3 g) Polyethylene glycol (1 ml), Stearic acid (0.6 g)	By immersion	Mushroom and cauliflower; apricots and green peppers	• Lowers the water loss rate. • Coating incorporated with antioxidants was effective in slowing browning reactions and lowers Vitamin C loss.	Ayranci and Tunc (2003); Ayranci and Tunc (2004)
Cassava starch (3% w/w), Glycerol (1.5% w/w), Carnauba wax (0.2% w/w) and Stearic acid (0.8% w/w)	By immersion	Apples	• Better mechanical properties. • Effective barrier to moisture and gas exchange.	Chiumarelli and Hubinger (2014)
Alginate or pectin (2 g/100 ml water) and Gellan (0.5 g/100 ml water)	By immersion	Melon (Piel de Sapo)	• Increase the water vapor resistance. • Prevent desiccation and • Maintain fruit firmness throughout storage	Oms-Oliu et al. (2008b)
Alginate (1 g/100 g), trans-cinnamaldehyde (2 g/100 g), and Pectin (2 g/100 g)	Layer by layer technique	Water melon	• Extend shelf-life, • Prevent weight loss and preserve texture during storage. • Effective carrier of natural antimicrobial compounds.	Sipahi et al. (2013)
Sodium Alginate (1.6 g), Acerola puree (100 g), and water (50 g)	Film	Acerola fruits	• Better moisture barrier property. • Higher strength and modulus of film compared to other fruit puree films.	Azeredo et al. (2012)
Chitosan (2 g/100 g) and pectin (1 g/100 g)	Layer by layer deposition by immersion method	Fresh-cut papaya	• Reduces losses of Vitamin C and total carotenoids content. • Improves microbiological & physicochemical quality.	Brasil et al. (2012)
Alginate (1.5%), Chitosan (1.5%), and CaCl2 (5%)	Layer by layer	Fresh-cut melon	• Shown good adhesion with antimicrobial property, retains firmness.	Poverenov et al. (2014)
Agar agar based (1%), chitosan (0.2%), and acetic acid (0.2%)	By immersion	Garlic	• Reduces respiration and moisture loss.	Geraldine et al. (2008)

Cassava starch (10 g/L) and glycerol (10 g/L)	By immersion	Fresh-cut mango	• Promotes a higher weight loss of cut mangoes. • Affects fruit texture characteristics. • Increasing carotenogenesis, and. • Favor growth of microbes during storage,	Chiumarelli et al. (Chiumarelli et al. 2010, 2011), Chiumarelli and Hubinger (2014)
150 mL honey/L of water +50 g SPI/L of water	Dip coating	Fresh-cut kajari melon	• Better sensory results. • Effective in maintaining color and ascorbic acid. • Limits the microbial growth.	Yousuf and Srivastava (2017)

Table 10.7 Recommend limits of food additives for surface treatment of fresh fruit (As per FSSAI, 2019)

Food category name	Food additive	INS No	Max. recommended level
Surface-treated fresh fruit[a]	Beeswax	901	Good manufacturing practice
	Candelilla wax	902	Good manufacturing practice
	Carnauba wax	903	Good manufacturing practice
	Glycerol ester of wood rosin	445(iii)	110 mg/kg
	Iron oxide		1000 mg/kg
	Microcrystalline wax	905c(i)	50 mg/kg
	ortho-Phenylphenol	231	12 mg/kg
	Sodium ortho-phenylphenol	232	12 mg/kg
	Polyethylene glycol	1521	Good manufacturing practice
	Polyvinylpyrrolidone	1201	Good manufacturing practice
	Sulfites		30 mg/kg
	Shellac, bleached	904	Good manufacturing practice
	Sucroglycerides	474	Good manufacturing practice
Surface-treated fresh vegetables (including mushrooms and fungi, roots and tubers, fresh pulses and legumes, and aloe vera) sea weeds, nuts and seeds	Beeswax	901	Good manufacturing practice
	Candelilla wax	902	Good manufacturing practice
	Carnauba wax	903	Good manufacturing practice
	Glycerol ester of wood rosin	445(iii)	110 mg/kg
	Lauric arginate ethyl ester	243	200 mg/kg
	Microcrystalline wax	905c(i)	50 mg/kg
	Phosphates		1760 mg/kg
	Shellac, bleached	904	Good manufacturing practice

[a]As per FSSAI (2019), **surface-treated fresh fruit defined as** "the surfaces of certain fresh fruit are coated with glazes or waxes or are treated with other food additives that act as protective coatings and/or help to preserve the freshness and quality of the fruit. Examples include apples, oranges, dates, and longans, etc."

10.7 Concluding Remarks

Edible films and coatings can serve as a potential packaging material for the whole fruits and vegetables as well as for the cut. These can perform almost all the functions that a packaging material can perform. Along with the basic functions of packaging material, it can perform other functions too. These can be formulated with the incorporation of various active ingredients for example antimicrobials, antioxidants, etc. Furthermore, research is required in this field as these are solely not able to protect fruits and vegetables due to their mechanical stability issues. Moreover, edible packaging material based on proteins has potential to cause allergic reactions to the person's vulnerable to allergens. This should be studied before or must be labeled properly for consumer's sake. Edible packaging materials can only be considered as a primary packaging material. Edible films give poor heat sealability which requires more research in this concern. Another limitation associated with the use is their adhesion to surface. This can be overcome by either multilayer electrodeposition which involves layer by layer coating. Some issues are there in commercialization because of the requirement of whole new machinery, new processes which will add cost on the manufacturer's part. Consumer acceptance is also necessary as they are somewhat reluctant in accepting the new product. However, edible films and coatings have enormous potential to substitute the existing packaging material.

References

Abugoch LE, Tapia C, Villamán MC, Yazdani-Pedram M, Díaz-Dosque M (2011) Characterization of quinoa protein–chitosan blends edible films. Food Hydrocoll 25(5):879–886

Attila E, Orts W, Embuscado ME, Huber KC (2009) Edible films and coatings: why, what, and how? Edible films and coatings for food applications. Springer, NY, pp 1–23

Ayala-Zavala JF, Silva-Espinoza BA, Cruz-Valenzuela MR, Leyva JM, Ortega-Ramírez LA, Carrazco-Lugo DK, Pérez-Carlón JJ, Melgarejo-Flores BG, González-Aguilar GA, Miranda MRA (2013) Pectin-cinnamon leaf oil coatings add antioxidant and antibacterial properties to fresh-cut peach. Flavour Frag J 28(1):39–45

Ayranci E, Tunc S (2003) A method for the measurement of the oxygen permeability and the development of edible films to reduce the rate of oxidative reactions in fresh foods. Food Chem 80(3):423–431

Ayranci E, Tunc S (2004) The effect of edible coatings on water and vitamin C loss of apricots (*Armeniaca vulgaris* Lam.) and green peppers (*Capsicum annuum* L.). Food Chem 87(3):339–342

Azeredo HM, Miranda KW, Rosa MF, Nascimento DM, de Moura MR (2012) Edible films from alginate-acerola puree reinforced with cellulose whiskers. LWT-Food Sci Technol 46(1):294–297

Baldwin EA (1994) Edible coatings for fresh fruits and vegetables: past, present, and future. In: Edible coatings and films to improve food quality. CRC Press, pp 25–63

Baldwin EA, Nisperos MO, Chen XIUHUA, Hagenmaier RD (1996) Improving storage life of cut apple and potato with edible coating. Postharvest Biol Technol 9(2):151–163

Benítez S, Achaerandio I, Sepulcre F, Pujolà M (2013) Aloe vera based edible coatings improve the quality of minimally processed 'Hayward'kiwifruit. Postharvest Biol Technol 81:29–36

Bisen A, Pandey SK, Patel N (2012) Effect of skin coatings on prolonging shelf life of kagzi lime fruits (*Citrus aurantifolia* Swingle). J Food Sci Technol 49(6):753–759
Bonilla J, Atarés L, Vargas M, Chiralt A (2012) Edible films and coatings to prevent the detrimental effect of oxygen on food quality: possibilities and limitations. J Food Eng 110(2):208–213
Bourtoom T (2008) Edible films and coatings: characteristics and properties. Int Food Res J 15(3):237–248
Brasil IM, Gomes C, Puerta-Gomez A, Castell-Perez ME, Moreira RG (2012) Polysaccharide-based multilayered antimicrobial edible coating enhances quality of fresh-cut papaya. LWT-Food Sci and Technol 47(1):39–45
Brody AL, Strupinsky ER, Kline LR (2001) Active packaging for food applications. Technomic Publishing Co., Inc., Lancaster, p 218
Bustan A, Lahav J (2012) Influence of organic beeswax-based coating emulsion on the post storage quality of sweet pepper, mango, and avocado fruit. Acta Hortic 934:807–814
Chiabrando V, Giacalone G (2014) Effect of essential oils incorporated into an alginate-based edible coating on fresh-cut apple quality during storage. Qual Assur Safety Crops Foods 7(3):251–259
Chiumarelli M, Hubinger MD (2014) Evaluation of edible films and coatings formulated with cassava starch, glycerol, carnauba wax and stearic acid. Food Hydrocoll 38:20–27
Chiumarelli M, Pereira LM, Ferrari CC, Sarantópoulos CI, Hubinger MD (2010) Cassava starch coating and citric acid to preserve quality parameters of fresh-cut "Tommy Atkins" mango. J Food Sci 75(5):E297–E304
Chiumarelli M, Ferrari CC, Sarantópoulos CI, Hubinger MD (2011) Fresh-cut 'Tommy Atkins' mango pre-treated with citric acid and coated with cassava (*Manihot esculenta* Crantz) starch or sodium alginate. Innov Food Sci Emerg Technol 12(3):381–387
Coma V, Sebti I, Pardon P, Deschamps A, Pichavant FH (2001) Antimicrobial edible packaging based on cellulosic ethers, fatty acids, and nisin incorporation to inhibit *Listeria innocua* and *Staphylococcus aureus*. J Food Prot 64(4):470–475
Contreras-Oliva A, Rojas-Argudo C, Pérez-Gago MB (2012) Effect of solid content and composition of hydroxypropyl methylcellulose-lipid edible coatings on physico-chemical and nutritional quality of 'Oronules' mandarins. J Sci Food Agric 92(4):794–802
Cortez-Vega WR, Pizato S, de Souza JTA, Prentice C (2014) Using edible coatings from Whitemouth croaker (*Micropogonias furnieri*) protein isolate and organo-clay nanocomposite for improve the conservation properties of fresh-cut 'Formosa' papaya. Innov Food Sci Emerg Technol 22:197–202
Costa C, Conte A, Buonocore GG, Lavorgna M, Del Nobile MA (2012) Calcium-alginate coating loaded with silver-montmorillonite nanoparticles to prolong the shelf-life of fresh-cut carrots. Food Res Int 48(1):164–169
Cuq B, Gontard N, Guilbert S (1995) Edible films and coatings as active layers. In: Active food packaging. Springer, Boston, MA, pp 111–142
Dávila-Aviña JE, Villa-Rodríguez JA, Villegas-Ochoa MA, Tortoledo-Ortiz O, Olivas GI, Ayala-Zavala JF, González-Aguilar GA (2014) Effect of edible coatings on bioactive compounds and antioxidant capacity of tomatoes at different maturity stages. J Food Sci Technol 51(10):2706–2712
Debeaufort F, Quezada-Gallo JA, Voilley A (1998) Edible films and coatings: tomorrow's packagings: a review. Crit Rev Food Sci 38(4):299–313
Dhumal CV, Pal K, Sarkar P (2019) Characterization of tri-phasic edible films from chitosan, guar gum, and whey protein isolate loaded with plant-based antimicrobial compounds. Polym-Plast Technol Mater 58(3):255–269
Díaz-Mula HM, Serrano M, Valero D (2012) Alginate coatings preserve fruit quality and bioactive compounds during storage of sweet cherry fruit. Food Bioprocess Technol 5(8):2990–2997
Du WX, Avena-Bustillos RJ, Hua SST, McHugh TH (2011) Antimicrobial volatile essential oils in edible films for food safety. Sci Against Microbial Pathogens: Commun Curr Res Technol Adv 2:1124–1134

Embuscado ME, Huber KC (2009) Edible films and coatings for food applications, vol 9. Springer, New York, NY
Fakhouri FM, Martelli SM, Caon T, Velasco JI, Mei LHI (2015) Edible films and coatings based on starch/gelatin: film properties and effect of coatings on quality of refrigerated Red Crimson grapes. Postharvest Biol Technol 109:57–64
Fakhoury FM, Martelli SM, Bertan LC, Yamashita F, Mei LHI, Queiroz FPC (2012) Edible films made from blends of manioc starch and gelatin–influence of different types of plasticizer and different levels of macromolecules on their properties. LWT-Food Sci Technol 49(1):149–154
Falguera V, Quintero JP, Jiménez A, Muñoz JA, Ibarz A (2011) Edible films and coatings: structures, active functions and trends in their use. Trends Food Sci Technol 22(6):292–303
FAO (1985) The prevention of post-harvest food losses in fruits, vegetables and root crops, Food and Agriculture Organization, Rome
Feng Z, Wu G, Liu C, Li D, Jiang B, Zhang X (2018) Edible coating based on whey protein isolate nanofibrils for antioxidation and inhibition of product browning. Food Hydrocoll 79:179–188
Ferrari CC, Sarantópoulos CI, Carmello-Guerreiro SM, Hubinger MD (2013) Effect of osmotic dehydration and pectin edible coatings on quality and shelf life of fresh-cut melon. Food Bioprocess Technol 6(1):80–91
FICCI (2014) potential of plastic industry in northern India with special focus on plasticulture and food processing. The Federation of Indian Chambers of Commerce and Industry, India
FSSAI (2019) The food safety and standards (Food Products Standards and Food Additives) Regulations- 2011, Compendium of Food Additives and Regulations, (Version-IX (29.03.2019)). https://archive.fssai.gov.in/dam/jcr:f3124f1e-bc84-4835-8463-7347fcc71bfe/Compendium_Food_Additives_Regulations_29_03_2019.pdf
Galus S, Kadzińska J (2019) Gas barrier and wetting properties of whey protein isolate-based emulsion films. Polym Eng Sci 59(S1):E375–E383
Gennadios A, Brandenburg AH, Weller CL, Testin RF (1993) Effect of pH on properties of wheat gluten and soy protein isolate films. J Agric Food Chem 41:1835–1839
Geraldine RM, Soares NDFF, Botrel DA, de Almeida GL (2008) Characterization and effect of edible coatings on minimally processed garlic quality. Carbohydr Polym 72(3):403–409
Ghidelli C, Sanchís E, Rojas-Argudo C, del Río MA, Pérez-Gago MB, Mateos M (2010) Effect of edible coating and modified atmosphere packaging on enzymatic browning of fresh-cut apple. In: XXVIII International Horticultural Congress on Science and Horticulture for People (IHC2010): International Symposium, vol 34, pp 515–522
Ghidelli C, Mateos M, Rojas-Argudo C, Pérez-Gago MB (2014) Extending the shelf life of fresh-cut eggplant with a soy protein–cysteine based edible coating and modified atmosphere packaging. Postharvest Biol Technol 95:81–87
González-Aguilar GA, Valenzuela-Soto E, Lizardi-Mendoza J, Goycoolea F, Martínez-Téllez MA, Villegas-Ochoa MA, Ayala-Zavala JF (2009) Effect of chitosan coating in preventing deterioration and preserving the quality of fresh-cut papaya 'Maradol'. J Sci Food Agric 89(1):15–23
Guilbert S, Cuq B, Gontard N (1997) Recent innovations in edible and/or biodegradable packaging materials. Food Addit Contam 14:741–751
Hamzah HM, Osman A, Tan CP, Ghazali FM (2013) Carrageenan as an alternative coating for papaya (*Carica papaya* L. cv. Eksotika). Postharvest Biol Technol 75:142–146
Han JH (2014) Edible films and coatings: a review. In: Innovations in food packaging, 2nd ed., pp. 213–255
Hardenburg RE (1967) Wax and related coatings for horticultural products; a bibliography
Hashemi SMB, Khaneghah AM, Ghahfarrokhi MG, Eş I (2017) Basil-seed gum containing origanumvulgare subsp. Viride essential oil as edible coating for fresh-cut apricots. Postharvest Biol Technol 125:26–34
Jiang T (2013) Effect of alginate coating on physicochemical and sensory qualities of button mushrooms (*Agaricus bisporus*) under a high oxygen modified atmosphere. Postharvest Biol Technol 76:91–97

Jongjareonrak A, Benjakul S, Visessanguan W, Tanaka M (2006) Effects of plasticizers on the properties of edible films from skin gelatin of bigeye snapper and brownstripe red snapper. Eur Food Res Technol 222(3–4):229–235
Kester JJ, Fennema OR (1986) Edible films and coatings: a review. Food Technology (USA)
Krochta JM, Mulder-Johnston CD (1997) Edible and biodegradable polymer films: challenges and opportunities. Food Technol 51(2):61–74
Kumins CA (1965) Transport through polymer films. In Journal of Polymer Science Part C: Polymer Symposia, vol. 10(1), 1–9. New York: Wiley Subscription Services, Inc., A Wiley Company
Martínez-Romero D, Castillo S, Guillén F, Díaz-Mula HM, Zapata PJ, Valero D, Serrano M (2013) Aloe vera gel coating maintains quality and safety of ready-to-eat pomegranate arils. Postharvest Biol Technol 86:107–112
Medina E, Caro N, Abugoch L, Gamboa A, Díaz-Dosque M, Tapia C (2019) Chitosan thymol nanoparticles improve the antimicrobial effect and the water vapour barrier of chitosan-quinoa protein films. J Food Eng 240:191–198
Mohebbi M, Ansarifar E, Hasanpour N, Amiryousefi MR (2012) Suitability of aloe vera and gum tragacanth as edible coatings for extending the shelf life of button mushroom. Food Bioproc Technol 5(8):3193–3202
Neves ACV Jr, Coneglian RCC, Soares AG, Freitas DGC, Fonseca MJO, Barreira FR et al (2012) Physical and sensory characterization of edible coatings applied to minimally processed persimmon. Acta Hortic 934:537–542
NHB (2017) Horticultural statistics at a glance. National horticulture board, India
Oms-Oliu G, Soliva-Fortuny R, Martín-Belloso O (2008a) Edible coatings with antibrowning agents to maintain sensory quality and antioxidant properties of fresh-cut pears. Postharvest Biol Technol 50(1):87–94
Oms-Oliu G, Soliva-Fortuny R, Martín-Belloso O (2008b) Using polysaccharide-based edible coatings to enhance quality and antioxidant properties of fresh-cut melon. LWT-Food Sci Technol 41(10):1862–1870
Perez-Gago MB, Serra M, Del Rio MA (2006) Color change of fresh-cut apples coated with whey protein concentrate-based edible coatings. Postharvest Biol Technol 39(1):84–92
Piccirilli GN, Soazo M, Pérez LM, Delorenzi NJ, Verdini RA (2019) Effect of storage conditions on the physicochemical characteristics of edible films based on whey protein concentrate and liquid smoke. Food Hydrocoll 87:221–228
Poverenov E, Danino S, Horev B, Granit R, Vinokur Y, Rodov V (2014) Layer-by-layer electrostatic deposition of edible coating on fresh-cut melon model: anticipated and unexpected effects of alginate–chitosan combination. Food Bioprocess Technol 7(5):1424–1432
Raybaudi-Massilia RM, Mosqueda-Melgar J, Martín-Belloso O (2008) Edible alginate-based coating as carrier of antimicrobials to improve shelf-life and safety of fresh-cut melon. Int J Food Microbiol 121(3):313–327
Robledo N, López L, Bunger A, Tapia C, Abugoch L (2018a) Effects of antimicrobial edible coating of thymol-nano-emulsion/quinoa protein/chitosan on the safety, sensorial properties, and quality of refrigerated strawberries (Fragaria ananassa) under commercial storage environment. Food Bioprocess Technol 11(8):1566–1574
Robledo N, Vera P, López L, Yazdani-Pedram M, Tapia C, Abugoch L (2018b) Thymolnanoemulsions incorporated in quinoa protein/chitosan edible films; antifungal effect in cherry tomatoes. Food Chem 246:211–219
Robles-Sánchez RM, Rojas-Graü MA, Odriozola-Serrano I, González-Aguilar G, Martin-Belloso O (2013) Influence of alginate-based edible coating as carrier of antibrowning agents on bioactive compounds and antioxidant activity in fresh-cut Kent mangoes. LWT-Food Sci Technol 50(1):240–246
Rojas-Graü MA, Tapia MS, Martín-Belloso O (2008) Using polysaccharide-based edible coatings to maintain quality of fresh-cut Fuji apples. LWT-Food Sci Technol 41(1):139–147

Ruzaina I, Norizzah AR, Halimahton Zahrah MS, Cheow CS, Adi MS, Noorakmar AW, Mohd. Zahid, A (2013) Utilisation of palm-based and beeswax coating on the postharvest-life of guava (*Psidium guajava* L.) during ambient and chilled storage. J Int Food Res 20 (1):265–274

Salinas-Roca B, Soliva-Fortuny R, Welti-Chanes J, Martín-Belloso O (2016) Combined effect of pulsed light, edible coating and malic acid dipping to improve fresh-cut mango safety and quality. Food Control 66:190–197

Scope RK (1994) Separation by precipitation. In: Scope RK (ed) Protein purification; principles and practice. Springer-Verlag, New York, pp 71–101

Seydim AC, Sarikus G (2006) Antimicrobial activity of whey protein based edible films incorporated with oregano, rosemary and garlic essential oils. Food Res Int 39(5):639–644

Shewfelt RL (1987) Quality of minimally processed fruits and vegetables. J Food Qual 10(3):143–156

Shin YJ, Song HY, Song KB (2012) Effect of a combined treatment of rice bran protein film packaging with aqueous chlorine dioxide washing and ultraviolet-C irradiation on the postharvest quality of 'Goha' strawberries. J Food Eng 113(3):374–379

Sipahi RE, Castell-Perez ME, Moreira RG, Gomes C, Castillo A (2013) Improved multilayered antimicrobial alginate-based edible coating extends the shelf life of fresh-cut watermelon (Citrullus lanatus). LWT-Food Sci Technol 51(1):9–15

Song HY, Jo WS, Song NB, Min SC, Song KB (2013) Quality change of apple slices coated with Aloe vera gel during storage. J Food Sci 78(6):C817–C822

Sothornvit R, Krochta JM (2001) Plasticizer effect on mechanical properties of β-lactoglobulin films. J Food Eng 50(3):149–155

Souza M, Cerqueira M, Souza B, Teixeira J, Porto A, Vicente A, Carneiro-da-Cunha M (2010) Polysaccharide from *Anacardium occidentale* L. tree gum (Policaju) as a coating for Tommy Atkins mangoes. Chem Pap 64(4):475–481

Talens P, Pérez-Masía R, Fabra MJ, Vargas M, Chiralt A (2012) Application of edible coatings to partially dehydrated pineapple for use in fruit-cereal products. J Food Eng 112(12):86–93

Tao N-G, Ao T-T, Liu Y-J, Huang S-R (2012) Effect of sucrose-based polymers on quality of Satsuma mandarin fruit (Citrus unshiu Marc. cv. Miyagawa Wase). Int J Food Sci Technol 47(5):997–1003

Tapia MS, Rojas-Graü MA, Rodríguez FJ, Ramírez J, Carmona A, Martin-Belloso O (2007) Alginate-and gellan-based edible films for probiotic coatings on fresh-cut fruits. J Food Sci 72(4):E190–E196

Tien L, Vachon C, Mateescu MA, Lacroix M (2001) Milk protein coatings prevent oxidative browning of apples and potatoes. J Food Sci 66(2001):512–516

Tsai MJ, Weng YM (2019) Novel edible composite films fabricated with whey protein isolate and zein: Preparation and physicochemical property evaluation. LWT-Food Sci Technol 101: 567–574

USEPA (2015) Reducing wasted food & packaging: a guide for food services and restaurants. United States Environmental Protection Agency, 1200 Pennsylvania Avenue, N.W

Vásconez MB, Flores SK, Campos CA, Alvarado J, Gerschenson LN (2009) Antimicrobial activity and physical properties of chitosan–tapioca starch based edible films and coatings. Food Res Int 42(7):762–769

Viña SZ, Mugridge A, García MA, Ferreyra RM, Martino MN, Chaves AR, Zaritzky NE (2007) Effects of polyvinylchloride films and edible starch coatings on quality aspects of refrigerated Brussels sprouts. Food Chem 103(3):701–709

Vu KD, Hollingsworth RG, Salmier S, Takala PN, Lacroix M (2012) Development of bioactive coatings based on γ-irradiated proteins to preserve strawberries. Radiat Phys Chem 81(8):1211–1214

Xiao Z, Luo Y, Luo Y, Wang Q (2011) Combined effects of sodium chlorite dip treatment and chitosan coatings on the quality of fresh-cut d'Anjou pears. Postharvest Biol Technol 62(3):319–326

Yang L, Paulson AT (2000) Mechanical and water vapour barrier properties of edible gellan films. Food Res Int 33(7):563–570

Yousuf B, Srivastava AK (2017) A novel approach for quality maintenance and shelf-life extension of fresh-cut Kajari melon: Effect of treatments with honey and soy protein isolate. LWT-Food Sci Technol 79:568–578

Zhang C, Wang Z, Li Y, Yang Y, Ju X, He R (2019) The preparation and physiochemical characterization of rapeseed protein hydrolysate-chitosan composite films. Food Chem 272: 694–701

Milk Protein-Based Edible Coatings: Properties and Applications

11

Alok Mishra, Bimlesh Mann, Amrita Poonia, Dinesh Chandra Rai, and Ankita Hooda

Abstract

Dairy industry produces a lot of by-products which when not utilised suitably, get added to the waste. The largest component produced during the coagulation process of casein is the blue-greenish liquid containing an important milk protein 'whey proteins'. Similarly, 'casein proteins' are produced as a by-product during commercial production of dairy whey. However, the casein so obtained needs to be converted into edible form before using in food applications. Research on edible coatings has been in the focus for its ability to reduce moisture losses, impart good barrier properties (gases), moderate elongation, good tensile strength, high flexibility and having no effect on taste or flavour most often. Whey and casein proteins can be effectively used in the development of edible coatings with improved physicochemical and textural properties. Plasticisation is required during the production of whey and casein protein-based coatings and films as they often lack flexibility. Conjugates and additives are added to modify the binding characteristics of the coatings. Milk proteins are a field of interest for developing edible coatings as these have good tensile strength, good oxygen barrier properties, moderate elongation, reduce moisture loss and, hence, preserve the original flavour and sensory profile. These also do not impart their own taste or flavour. Applications of whey protein isolates/concentrates and edible casein/caseinates in the formation of edible coatings have yielded good results. Various

A. Mishra · A. Poonia (✉) · D. C. Rai · A. Hooda
Department of Dairy Science and Food Technology, Institute of Agricultural Sciences, Banaras Hindu University, Varanasi, Uttar Pradesh, India
e-mail: amrita12@bhu.ac.in

B. Mann
Department of Dairy Chemistry, National Dairy Research Institute, Karnal, Haryana, India

A. Poonia, T. Dhewa (eds.), *Edible Food Packaging*,
https://doi.org/10.1007/978-981-16-2383-7_11

developments are constantly being made to prepare value-added products using whey and edible casein with enhanced nutritional value and better shelf life.

Keywords

Milk protein · Whey proteins · By-products · Edible coating · Applications

11.1 Introduction

Dairy industry has its presence in almost every corner of the world having a wide range of products to offer such as milk and milk products, cheese, butter, yoghurt and so on. Incidentally, a lot of waste and by-products are also generated by the dairy industry (Jaganmai and Jinka 2017). Anything other than main product falls into the category of by-product. The common dairy by-products are buttermilk, whey, skim milk and ghee residue. In the era of industrialisation and commercialisation, every by-product has its worth. Thus, these by-products are often processed into some edible form which have high potential in terms of health and economic value. Further processing yields derived by-products such as whey protein concentrates, casein and caseinates, protein hydrolysates, co-precipitates, lactose and low lactose powders (Gupta 2008).

The demand of consumers for high-quality safe food with less/minimal processing using natural ingredients is ever increasing. Edible coatings (EC) offer an edge over traditional packaging materials as it improve the food quality, safety and stability (Shendurse et al. 2018). These coatings are entirely consumable; that is, product can be consumed along with coating. The raw materials used in their production are edible biopolymers such as proteins, lipids, polysaccharides and their combinations along with various food-grade additives (Han 2014; Suput et al. 2015). Most often these coatings are not visible to naked eyes. Thus, the sensory appeal and visual appearance are also very high. Except that, EC may extend the shelf life and improve overall food quality (Shendurse et al. 2018). EC can efficiently control the mass transfer within the food components and between the food and environment (Paviath and Orts 2009). Also, EC restricts migration of moisture and gases (CO_2, O_2 and N_2) and/or improves the rheological and mechanical characteristics of the food (Lacroix and Vu Khanh 2014; Ncama et al. 2018).

Use of proteins for edible coating development either alone or in combination with polysaccharides and lipids has been studied over time. Milk proteins (whey, casein and caseinates) have proven to be excellent components in edible coating preparation (Shendurse et al. 2018). Milk protein constitutes of 80% casein and 20% whey protein, which have further been characterised into different components. Milk proteins either in total or their components can be used for formation of edible food coatings. For forming a 3D network of cohesive films, a biopolymer based (protein–protein) interactions are crucial. Coatings based on milk protein can act as protective barrier for the food as well as between the food components. The functional properties like mechanical strength, sensory properties and regulating mass transfer are important. Mechanical properties are necessary to maintain integrity of package and mass transfer properties to prevent migration of ingredients either way or sensory properties to maintain the flavour profile of product. The complex intermolecular bindings of milk protein make them excellent in gas barrier properties also (Shendurse et al. 2018). Owing to the cross-linked structure of proteins, such ECs offer greater stability and longer durability as compared to the polysaccharide-based coatings (Barone and Schmidt 2006). The hydrophilicity of proteins along with the numerous bonds and interactions between protein chains results in densely packed network structures. It is responsible for promising barrier properties against gases (Schmid et al. 2015).

The chapter has put in efforts to discuss the relevant techniques, functionalities and properties of dairy by-products–based edible coatings considering the recent literature available.

11.2 Milk Proteins as Dairy By-Products

Milk is a rich source of protein, especially for the lacto-vegetarians. Of the total bovine milk proteins, casein is 80% and whey proteins are 20% (Shendurse et al. 2018). Although whey proteins are present in small quantity, they have high biological value (104), efficiency ratio (3.6) and net protein utilisation (95). However, whey proteins are the derived by-products from the largest by-product of dairy world which is 'whey'. Whey is a highly perishable dilute greenish-yellow liquid which is drained out during production of cheese, paneer, chakka, chhana and co-precipitates (Gupta 2008). Global market estimates of whey products are about \$6.5 billion in sales. Casein is a derived by-product of a dairy set-up majorly involved in cream manufacturing, yielding skim milk as a by-product. There is a huge market of casein in the food industry accounting for up to 2.5 lakh tonnes, with United Sates of America having the largest demand of 20,000 tonnes. As an estimate, 20% of casein demand is for applications in nutraceutical development (Gupta 2008).

11.2.1 Whey Proteins

Dairy whey accounts for the 85–95% of the total milk volume containing nearly 55% of the whole milk nutrients (Yadav et al. 2015). This protein fraction in dairy is heterogeneous mixture of various proteins such as α-lactalbumin, β-lactoglobulin, immunoglobulins, bovine serum albumin (BSA), lactoferrin, proteose peptone, lactoperoxidase and glycol macropeptide. The protein fractions are having potential functional and nutraceutical properties. A brief account of these whey protein fractions is given in Table 11.1.

As an estimate, 180 million tonnes/year of whey is produced worldwide and the majority amounts (~70%) to that from United Sates of America and European Union (Mollea et al. 2013; Yadav et al. 2015). Whey proteins are concentrated or isolated

Table 11.1 Different fractions of dairy whey protein (Di Pierro et al. 2018)

Protein fraction	Quantity (% TWP)	Mol. wt. (kDa)	I.P.	Form	Other features
β-Lactoglobulin	>50	18.3	5.4	Dimeric: 2 identical subunits; 162 amino acid each	Forms aggregate at around 80 °C
α-Lactalbumin	10	14	4.4	Globular: single polypeptide chain of 123 amino acid	More resistant to heat and denaturation is reversible
Bovine serum albumin	5	66.3	5.1	Protein of 582 amino acid	High capacity to reversibly bind several ligands
Immunoglobulins (Ig)	0.7	150–1000	5.5 to 8.3	Monomeric: IgG Polymeric: IgA and IgM	IgG is up to 80% (w/w) of total Ig
Lactoferrin	<2	76.5	9.5–10	Single polypeptide chain of 700 amino acids	Resistant to heat, acid, trypsin and chymotrypsin Easily hydrolysed by pepsin
Lactoperoxidase	0.5	78	9.5	Single polypeptide chain of 612 amino acids	Completely loses catalytic activity at temperature > 70 °C
Proteose peptone	25	4–22	–	Four components: three originate from β-casein, and another derives from fat globule membrane	Protein fraction remaining soluble to milk at 95 °C and pH 4.7
Glycomacropeptide	10–20	6.8	4.3–4.6	64 amino acid residues, mostly rich in branched chains	Originates from κ-casein by the action of the milk clotting chymosin during cheese processing

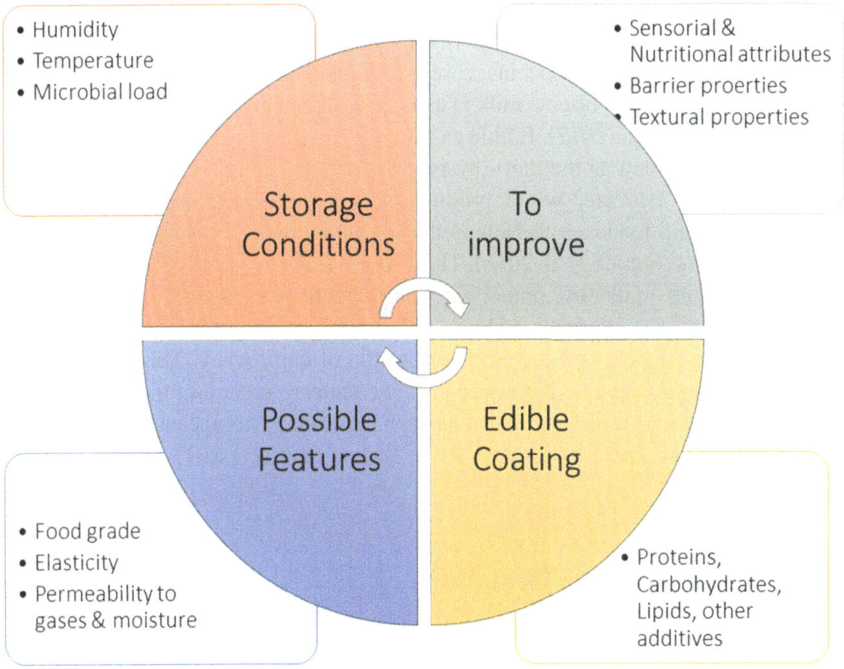

Fig. 11.1 Edible coatings: Functions and attributes of their application on food products

using different technologies to yield whey protein concentrates (WPC) and whey protein isolate (WPI), respectively, based on the purity; that is, the percentage of protein content. The purity of WPC ranges from 30% to 80% and that of WPI is more than 90% (United States Dairy Export Council 2006). Membrane filtration methods are the most preferred methods used in whey protein concentration. Ultrafiltration and nanofiltration are economical methods that yield high-quality products with proteins in their native state having good functional properties (Limsawat and Pruksasri 2010).

TWP total whey protein, *Mol. wt.* molecular weight, *kDa* kilo-dalton, *I.P.* isoelectric point

11.2.2 Casein and Caseinates

Casein is one of the most important protein found exclusively in milk. It amounts for approximately 80% of the total milk protein (Sarode et al. 2016). It is present in the form of micelles (20–300 nm in diameter) of which 93% is casein and rest are the salts of casein (caseinates) with calcium, sodium, magnesium, potassium, phosphate and other minerals (Sarode et al. 2016; Badem and Uçar 2017). Casein gets precipitated at pH 4.6 by acidification as it is the isoelectric point. Casein is constituted of four monomers, namely α_{S1} (~38%), α_{S2} (~10%), β (~34%) and κ-caseins (~15%), which are held together by non-covalent interactions such as hydrogen bonding, hydrophobic and electrostatic interactions (Atamer et al. 2017; Sarode et al. 2016). Except cysteine, casein is a rich source of essential amino acids. The amount of casein in bovine milk is usually in the range 2.4–2.9% (Atamer et al. 2017; Badem and Uçar 2017). Edible casein and caseinates can be isolated from the milk and are regarded as the dairy by-products finding many applications in food industry. As an estimate, world production of casein and caseinates could be 430–460 thousand tonnes with New Zealand, Netherlands and Germany being the largest producers (Sarode et al. 2016). The commercial production of edible casein is done from skim milk by rennet or acid coagulation methods followed by demineralisation and purification steps. The feasibility of edible casein production depends on the efficient and economic utilisation of dairy whey. The production of caseinates has not picked up in many dairying countries, including India, because of its high drying cost, low bulk density and high packaging, storage and transportation costs (Sarode et al. 2016; Southward 2016; Badem and Uçar 2017).

11.3 Edible Coating Formation

Coating formation using proteins majorly depends upon the chain alignment and structure of the proteins which itself depends on the initial arrangement of amino acids. This initial amino acid sequence determines the interactions between the protein chains and other film components (Belitz et al. 2008; Schmid and Müller 2019). The protein-based coatings are good barriers of gases (O_2 and CO_2) while permeable to moisture, making these ideal for hydrophilic surfaces such as meat and dairy products. However, in foods requiring enhanced moisture protection, hydrophobic materials like beeswax or oleic acid can be added (Wagh et al. 2014; Chen et al. 2019; Dinika et al. 2020). Protein-based coatings are usually stiff and brittle with low flexibility. Thus, plasticisers like polyethylene glycol or glycerol can be added to enhance their flexibility (Ramos 2011; Dinika et al. 2020). The plasticisers help to reduce the protein chain-to-chain interaction that intensifies the free volume and increases chain movements (Henriques et al. 2016). However, the addition of plasticiser adversely affects the barrier properties of the film against gases, aroma, moisture, antioxidants, oils, bacteriocins and other solutes (Bodnar et al. 2007).

The formation and properties of a coating are dependent on two forms of interactions: cohesion and adhesion. Cohesion forces act between the polymer (protein) molecules while the adhesion forces act between the coating and the substrate (Henriques et al. 2016). The efficacy of these forces is dependent on the properties of coating material (protein) such as molecular weight, chain structure and polarity (Sothornvit and Krochta 2005). Edible coating formation is mainly performed by two types of mechanisms: wet (solvent casting) and dry (extrusion and compression moulding) processing (Henriques et al. 2016).

11.3.1 Wet Coating

Wet processing of edible coatings, also called solvent casting, is relatively simple, quick and widely used method at laboratory scale. The solvent casting of protein-based coatings is cost-efficient and can be performed on various scales (Henriques et al. 2016). In case of solvent casting, coating suspension is prepared by the dissolution of proteins into the solvent. The solvent generally used is water, alcohol or a blend of other solvents. Further, additives like plasticisers, emulsifiers, antioxidants, etc. are added to the suspension followed by its homogenisation. The suspension is adjusted for pH, if required (Henriques et al. 2016; Schmid and Müller 2019). The proteins are denatured to initiate the cross linking necessary to enhance the coating formation by giving heat treatment before, after or during the coating formation (Schmid et al. 2013). The application of suspension to form coating is done by spraying, rolling, dipping, spinning, brushing, physical vapour deposition or enrobing the food in the coating suspension (Coltelli et al. 2016; Henriques et al. 2016). Further, removal of solvent is performed by hot air, microwave or infrared drying (Embuscado and Huber 2009). The functional properties of the coating depend on the concentration of protein in suspension, additives, solvent, pH and the method of drying (Schmid and Müller 2019).

11.3.2 Dry Coating

Dry coating formation is performed using the thermoplastic properties of the polymers. The coatings that are low in water content are produced by the extrusion and compression moulding. Such techniques are common in industrial settings and have been adapted to the production of protein-based films for mass production of packing materials such as water-soluble cups and pouches for foods and ingredients intended for individual servings (Verbeek and van den Berg 2010; Henriques et al. 2016).

Various researchers have suggested the thermoplastic behaviour of some proteins, but this property remains underexplored in the domain of edible coating formation (Hernandez et al. 2005, 2006). Most proteins possess thermoset behaviour especially globular proteins (whey proteins), which, under the influence of heat, tend to unfold and cross-link (Domininghaus et al. 2008). The proteins lacking natural thermoplastic behaviour may undergo modifications. Studies have suggested various additives which may aid in protein extrusion (Verbeek and van den Berg 2010; Schmid et al. 2014a, b, 2016). The plasticisers and chemical additives (reducing agents) are used to tweak formulation adjustments. These modifiers reduce intermolecular interactions, which results in reduced thermal decomposition temperature and enhances the flexibility of thermoplastic protein polymer. The plasticisers generally used are water, sucrose, sorbitol and glycerol (Schmid and Müller 2019).

11.4 Applications of Whey Protein-Based Edible Coating

Whey protein films possess excellent barrier properties against oxygen, aroma and oil at low-to-intermediate relative humidity. Besides, whey protein films are adequately durable due to their mechanical properties. Hence, they can be used to coat food products, form films in the form of pouches for food ingredients and form separating layers in case of multi-ingredients food products (Shendurse et al. 2018). Based on desired functions, dairy whey edible coatings can perform following applications:

- Act as barrier against moisture, oxygen and gases on foods.
- Control movement of moisture within the food components; thereby preventing unwanted chemical reactions, avoiding sogginess and restrain microbial growth.
- Effectively used in coating sweets, confectionary, fast foods, pet foods, etc.
- Formation of active coatings and films.
- Delivery of flavours, antioxidants, antimicrobials, functional compounds, etc.

Broadly, these applications can be classified into passive and active packaging. The passive packaging is traditional approach of providing mechanical and barrier protection, while active packing is a novel approach to extend shelf life of the food product and create certain enhancements in characteristics without affecting the overall quality (Henriques et al. 2016). Edible films and coatings using dairy whey protein can be prepared using either whey protein concentrates or isolates. Isolates are purer and have more protein concentrations as compared to the concentrates that are having other milk components like carbohydrates and fats up to certain levels. The spoilage of food items is an issue of major concern to ensure overall food quality. The microbial spoilage often leads to major food losses besides derogatory effects on health on their consumption. The coatings having moisture barrier properties are often found effective in controlling microbial activity. Besides, incorporation of antimicrobial compounds in the coating material can also be beneficial (Henriques et al. 2016; Pereira et al. 2018; Shendurse et al. 2018). Some recent applications of whey protein in the edible coatings are listed in Table 11.2.

Min et al. (2008) used lysozyme and lactoperoxidase while preparation of films based on WPI to coat smoked salmon that resulted in its extended shelf. Zinoviadou et al. (2009) incorporated oregano oil into sorbitol-plasticised films based on WPI to coat beef meat. The study reported effective increase in shelf life with minimal changes in beef meat colour. Whey protein concentrates (WPC) have been used to prepare coatings as well. Wang et al. (2010) used WPC to prepare a film with improved protein concentration and mechanical properties. However, such film was having decreased barrier properties. Addition of glycerol further reduced barrier properties but the extensibility was improved to a greater extent. The optimised parameters using response surface methodology (RSM) indicated WPC (10.2%), glycerol (2.7%) and heating at 82 °C for optimal film development. Javanmard (2009) produced WPC-based films using emulsification. In addition, polyethylenglycol (PEG), olive oil and glycerol were incorporated into the film making.

Table 11.2 Applications of dairy whey protein in the formation of coating on various food products

Food product	Coating material	Observed functions	References
Refrigerated *Scomberoides commersonnianus* fillets	WPI-chitosan active coatings with *Artemisia dracunculus* essential oil	Antimicrobial activity against mesophilic and psychrotrophic bacteria. Reduced deterioration of chilled seafood	Farsanipoor et al. (2020)
Strawberries (*Fragaria ananassa*)	WPI-chitosan conjugate	Enhanced shelf life with improved antioxidant activity and reduced weight loss, pH and titrable acidity	Muley and Singhal (2020)
Chestnuts (*Castanea sativa*)	WPI, chitosan and alginate	Effective against control of microbial growth	Fernandes et al. (2020)
Fresh-cut cheese	WPI nanofibrils with carvacrol	Improved antioxidant and antimicrobial activity with better textural properties and lower weight losses	Wang et al. (2019)
Chilled meat	WP nanofibrils with TiO_2 nanotubes	Surface smoothness and homogeneity of films with improved mechanical properties. Improved antioxidant activity, reduced lipid peroxidation and weight loss. Antimicrobial activity with extended shelf life	Feng et al. (2019)
Cheese slices	WP films treated with heat, ultrasound and/or transglutaminase	Lower WVP, raised tensile strength and puncture strength due to ultrasound treatment. Heat treatment resulted in better mechanical strength. Transglutaminase increased puncture deformation values and affected appearance in terms of colour	Cruz-Diaz et al. (2019)
Fresh-cut turkey pieces	WP-chitosan films	Antimicrobial properties against *S. typhimurium*, *E. coli* and *C. jejuni*	Brink et al. (2019)
Sliced ham	WP coating with *Bifidobacterium* and *Lactobacillus*	Reduction in water and weigh loss. Inhibition of *Staphylococcus* spp., *Pseudomonas* spp., *Enterobacteriaceae* and yeasts/moulds (45 days, 4 °C). Probiotic count maintained at 10^8 CFU/g	Pereira et al. (2018)

(continued)

Table 11.2 (continued)

Food product	Coating material	Observed functions	References
Cooked meatballs in frozen storage	WPI-based edible films incorporated with natural antioxidant extracts from laurel (*Laurus nobilis*) or sage (*Salvia officinalis*)	Higher antioxidant activity and phenolic compound content were observed leading to oxidative stability of frozen meatballs	Akcan et al. (2017)
Processed meat products	WP active edible coating incorporated with *Origanum virens* essential oils	Protection against colour fading. Reduced lipid peroxidation. Antimicrobial effects	Catarino et al. (2017)
Refrigerated pike-perch fillets	WP edible coating with lactoperoxidaseand α-tocopherol	Improved antibacterial and antioxidant properties	Shokri and Ehsani (2017)
Fresh-cut fruit and vegetable	WP/pectin edible films cross linked with transglutaminase	Reduced weight loss (~80% in apple, ~100% in carrot and potato samples). Preservation of total phenolics and carotenoids in carrots	Marquez et al. (2017)
Cottage cheese (Paneer)	WP-iron based edible coating	Supplementation of iron (93.5 ppm) into paneer (1.5 cm cube size) was successfully achieved with dipping volume of 100 mL at a dipping time of 10 min	Jotarkar et al. (2018)

WP whey protein, *WPI* whey protein isolates, *WVP* water vapour permeability; TiO_2 titanium oxide, *CFU* colony forming units

The films were prepared with 10% WPC and varying levels of PEG, olive oil and glycerol as plasticisers. The results revealed decrease in tensile strength and modulus upon increase in glycerol or PEG levels. Henriques et al. (2013) developed WPC-based edible coatings by heat denaturation and UV irradiation. These coatings resulted in effective antimicrobial properties when applied to ripened cheese as compared to existent cheese coatings.

The scalability of any technique poses a challenge in its advancements and applications. No technology can be viable if it cannot be scaled up to industrial production levels. Bugnicourt et al. (2013) demonstrated scalability of whey protein-based coating production. The study reported a novel bio-based coating solution incorporated with WPI and plasticisers which can check brittleness. Schmid et al. (2009) worked on a project WHEYLAYER, in which dairy whey protein-based films, coating and laminates having high oxygen barrier performance were developed. Besides, biodegradable nature of whey protein-based layer imparts enhanced recyclability to these coatings and films. Wagh et al. (2014) worked on protecting the cheddar cheese using dairy whey protein-based films. The study suggests no alterations with respect to the aspect and sensory characteristics of the coated foods.

Galus and Kadzińska (2016) developed WPI-based edible films modified with walnut and almond oils at low concentrations of 1% and 0.5% respectively. An increase in opacity of the films was observed with increase in the oil content which resulted in a heterogeneous microstructure. Moreover, reduction in water vapour permeability and swelling with increase in hydrophobicity was observed. Almond oil possesses higher plasticising effect; thus, having the ability to modify the properties of the films formed.

Silva et al. (2018) studied the formation of WPI base edible films by adding pectin to enhance its physiochemical and microstructural properties. The interaction between whey proteins and pectin resulted in increased opacity, WVP and solubility. However, the flexibility and tensile strength of the films were reduced significantly.

Feng et al. (2018) worked on the development of whey protein-based nanofibrils as an ingredient for edible film formation. The study incorporated glycerol and trehalose as plasticisers. WPI (5%) can self-assemble into nanofibrils at 80 °C for 10 h under 220 rpm constant magnetic stirring. The observations suggested increase in surface homogeneity, smoothness, transparency and hydrophobicity with decrease in moisture content and low water solubility of the WP nanofibril-based edible films. Glycerine and trehalose helped reduce enzymatic browning and provided protective action against product weight loss.

Chakravartula et al. (2019) developed composite edible films using WPC and pectin/alginate. The study was conducted on different proportions of WPC, pectin and alginate to prepare edible films formulated using simplex-centroid mixture design. Films having higher whey protein content presented lower viscosity, low affinity for water, reduced mechanical strength, lower gas barrier values, favoured higher opacity and formed dense structures owing to protein-polysaccharide aggregates. Thermal stability of the films was high with degradation onset temperature at >170 °C.

Jiang et al. (2019) studied the preparation of WPC-based edible films added with nanocrystalline cellulose and transglutaminase to enhance the barrier ability and mechanical properties of the whey protein edible films. The results suggested improved tensile strength and elongation properties.

Abdalrazeq et al. (2019) worked on the development of whey protein edible films plasticised with glycerol under alkaline conditions without using any heat treatment. The casting was performed at pH 12 using unheated whey protein with glycerine (40–50%). The films formed by this method were more flexible with respect to the whey protein based films. The moisture content of the films reduced with decrease in glycerine content.

11.5 Applications of Casein Protein-Based Edible Coating

Casein and caseinates are biopolymer powders that are obtained as by-product during commercial production of dairy whey protein isolates/concentrates (Badem and Uçar 2017). The heat stability of casein protein makes it an ideal candidate for edible film manufacturing. However, inability to form gel is one of the drawbacks of casein as the film produced lacks elasticity (Apriliyani et al. 2020). The physical properties of edible films include tensile strength, elasticity, thickness, water vapour permeability, solubility, transparency and surface appearance of the film (Warkoyo et al. 2014). Casein possesses ability to melt at high temperatures which is good for film formation. Moreover, it absorbs substantial water serving as matrix former producing specialised plastic materials. Casein-based edible films and coatings are easy to form because of their open secondary structure (Shendurse et al. 2018). Caseinates have ability to form films from their aqueous solutions owing to their randomly coiled structure and capability of forming intermolecular bonds (hydrogen, hydrostatic and electrostatic) resulting in increased cohesion among its chains. Thus, casein and caseinates are good coating-forming materials (Sarode et al. 2016; Shendurse et al. 2018).

As discussed above, the property of low flexibility and readiness to gain moisture poses challenge to use casein as film-forming material. On the other side, good mechanical strength and low permeability to oxygen make it an ideal ingredient for film formation. The shortcomings can be overcome by the addition of certain other ingredients (Shendurse et al. 2018; Apriliyani et al. 2020). Chambi and Grosso (2006) prepared casein-based films blended with gelatine at varying levels. Transglutaminase was used for cross-linking casein and gelatine, which stimulated the formation of high molecular mass polymers. Owing to the action of enzyme, decrease in water vapour permeability and increase in mechanical properties were observed. A synergistic effect was observed in the films produced using casein-gelatine formulation; the greatest effect being produced at 75:25 casein-gelatine mixture.

Apriliyani et al. (2019) studied the preparation of antimicrobial edible films-based casein incorporated with chitosan at varying levels. The casein: chitosan ratio of 1:4 was chosen best with greatest water vapour permeability (WVP) observed as 0.1140 g mm/m^2 h kPa, water activity (a_w) as 0.709 and 32.81% film solubility. The casein-chitosan edible film showed solution ratio-dependent antimicrobial activity against *E. coli, Staphylococcus aureus, Lactobacillus bulgaricus* and *Salmonella* sp.

Calcium caseinate (Ca-Cas) can be isolated from skim milk (non-fat milk) and effectively be used in edible film and coating preparations (Akkurt 2015). Serife and Yam (2015) prepared Ca-Cas-based edible films along with high methoxyl pectin. Pectin is used as an additive to enhance the mechanical strength of the films by inducing cross linkages in the molecular structure within the film. The methoxyl and carboxyl groups of the pectin form cross links with the amino groups of the Ca-Cas. Kozempel and Tomasula (Kozempel et al. 2003) worked on the scalability of the film development using caseinates. They developed a semicontinuous process of Ca-Cas–based film preparation using glycerol as plasticiser. The wetted solutions of Ca-Cas/glycerol were spread using Meier rod on the polyethylene/Mylar belts and were removed readily after the film formation. The films were formed at a feed rate of 3.3 and 5 g/cm having a thickness of 0.14 mm and tensile strength of 5 MPa with 30% elongation. Some food applications of casein and caseinates in the area of edible coatings are illustrated in the Table 11.3.

Table 11.3 Application of casein- and caseinates-based edible coatings on food products (Shendurse et al. 2018)

Food product	Coating material	Observed functions
Peeled carrots	Casein-stearic acid	More moisture retention
Frozen fish	Casein-acetylated monoglyceride	Decreased water loss
Frozen fish fillets	Potassium caseinate-rennet casein	Enhanced sensorial properties
Cherries	Sodium caseinate	Reduced moisture loss
Cut potatoes	Calcium caseinate	Lesser browning
Potato slices	Calcium caseinate	Reduced oxidative browning
Cut carrots	Calcium caseinate	Reduction in moisture losses
Peanut	Calcium caseinate	Oil migration reduced

11.6 Summary

Edible coatings are technology-driven modern packaging solutions ranging from whole food to individual component, and further to ingredient-level packaging needs. These packaging systems are not only consumable but also add no waste to the environment; thus, are eco-friendly. Moreover, most of the time, the materials used for the manufacture of edible coatings are made from by-products. Dairy whey and edible casein are such by-products of the dairy industry produced during commercial production of casein coagulation and whey protein respectively. Their cross-linking properties and high heat stability have been found suitable for the film and coating formation. Plasticisers are added to overcome the lack of flexibility and resistance in the films formed using whey and casein proteins. Additives are often added to achieve maximum food safety, quality and shelf life. The edible films and coatings, thus, produced are more versatile having improved mechanical strength, better barrier properties and other active properties because of the additives present. Thus, they pave a new era for processing ready-to-eat foods and better conditions of their end-use enabling varied range of commercial applications.

References

Abdalrazeq M, Giosafatto CVL, Esposito M et al (2019) Glycerol-plasticized films obtained from whey proteins denatured at alkaline pH. Coatings 9(5):322

Akcan T, Estévez M, Serdaroğlu M (2017) Antioxidant protection of cooked meatballs during frozen storage by whey protein edible films with phytochemicals from *Laurus nobilis* L. and *Salvia officinalis*. LWT 77:323–331

Akkurt S (2015) Edible films and coatings from calcium caseinate and their applications. Doctoral dissertation. Rutgers University-Graduate School, New Brunswick

Apriliyani MW, Purwadi, Manab A, Ahmad BM, Uula LM (2019) Physico-chemical and antimicrobial properties of casein-chitosan edible films as food quality and food safety. Int Conf Food Bio-Ind. https://doi.org/10.1088/1755-1315/443/1/012018

Apriliyani MW, Purwadi AM, Ahmad BM et al (2020) Physico-chemical and antimicrobial properties of casein-chitosan edible films as food quality and food safety. IOP Conf Ser: Earth Environ Sci 443(1):012018

Atamer Z, Post AE, Schubert T et al (2017) Bovine β-casein: isolation, properties and functionality. A review. Int Dairy J 66:115–125

Badem A, Uçar G (2017) Production of caseins and their usages. Int J Food Sci Nutr 2:4–9

Barone JR, Schmidt WF (2006) Nonfood applications of proteinaceous renewable materials. J Chem Educ 83:1003

Belitz HD, Grosch W, Schieberle P (2008) Lehrbuch der Lebensmittelchemie. Springer, Berlin

Bodnar I, Alting AC, Verschueren M (2007) Structural effects on the permeability of whey protein films in an aqueous environment. Food Hydrocoll 21:889–895

Brink I, Šipailienė A, Leskauskaitė D (2019) Antimicrobial properties of chitosan and whey protein films applied on fresh cut turkey pieces. Int J Biol Macromol 130:810–817

Bugnicourt E, Schmid M, Nerney OM et al (2013) Processing and validation of whey-protein-coated films and laminates at semiindustrial scale as novel recyclable food packaging materials with excellent barrier properties. Adv Mater Sci Eng 2013:496207

Catarino MD, Alves-Silva JM, Fernandes RP et al (2017) Development and performance of whey protein active coatings with *Origanum virens* essential oils in the quality and shelf life improvement of processed meat products. Food Control 80:273–280

Chakravartula SSN, Soccio M, Lotti N et al (2019) Characterization of composite edible films based on pectin/alginate/whey protein concentrate. Materials 12(15):2454

Chambi H, Grosso C (2006) Edible films produced with gelatin and casein cross-linked with transglutaminase. Food Res Int 39(4):458–466

Chen H, Wang J, Cheng Y et al (2019) Application of protein-based films and coatings for food packaging: a review. Polymers 11(12):2039

Coltelli MB, Wild F, Bugnicourt E et al (2016) State of the art in the development and properties of protein-based films and coatings and their applicability to cellulose based products: an extensive review. Coatings 6:1–59

Cruz-Diaz K, Cobos Á, Fernández-Valle ME et al (2019) Characterization of edible films from whey proteins treated with heat, ultrasounds and/or transglutaminase. Application in cheese slices packaging. Food Packag Shelf Life 22:100397

Di Pierro P, Mariniello L, Giosafatto VL et al (2018) Dairy whey protein-based edible films and coatings for food preservation. In: Grumezescu AM, Holban AM (eds) Food packaging and preservation, vol 9. Academic Press, 456, p 439

Dinika I, Verma DK, Balia R et al (2020) Potential of cheese whey bioactive proteins and peptides in the development of antimicrobial edible film composite: a review of recent trends. Trends Food Sci Tech 103:57–67

Domininghaus H, Eyerer P, Elsner P et al (2008) Kunststoffe: Eigenschaften und Anwendungen; mit 240 Tabellen. Springer-Verlag GmbH, New York

Embuscado ME, Huber KC (2009) Edible films and coatings for food applications. Springer, New York

Farsanipoor A, Khodanazary A, Hosseini SM (2020) Effect of chitosan-whey protein isolated coatings incorporated with tarragon *Artemisia dracunculus* essential oil on the quality of *Nemipterus japonicus* fillets at refrigerated condition. Int J Biol Macromol 155:766–771

Feng Z, Li L, Wang Q (2019) Effect of antioxidant and antimicrobial coating based on whey protein nanofibrils with TiO_2 nanotubes on the quality and shelf life of chilled meat. Int J Mol Sci 20(5):1184

Feng Z, Wu G, Liu C et al (2018) Edible coating based on whey protein isolate nanofibrils for antioxidation and inhibition of product browning. Food Hydrocoll 79:179–188

Fernandes L, Pereira EL, do Céu Fidalgo M et al (2020) Physicochemical properties and microbial control of chestnuts (*Castanea sativa*) coated with whey protein isolate, chitosan and alginate during storage. Sci Hortic 263:109105

Galus S, Kadzińska J (2016) Whey protein edible films modified with almond and walnut oils. Food Hydrocoll 52:78–86. https://doi.org/10.1016/j.foodhyd.2015.06.013

Gupta VK (2008) Technological advances in the utilisation of dairy by-products. National Dairy Research Institute, Karnal

Han JH (2014) Edible films and coatings: a review. In: Han JH (ed) Innovations in food packaging, 2nd edn. Academic Press, New York, pp 213–255

Henriques M, Gomes D, Pereira C (2016) Whey protein edible coatings: recent developments and applications. In: Nedović V, Raspor P, Lević J et al (eds) Emerging and traditional technologies for safe, healthy, and quality food. Springer, Cham, pp 177–196

Henriques MHF, Santos G, Rodrigues A et al (2013) Replacement of conventional cheese coating by natural whey protein edible coatings with antimicrobial activity. J Hyg Eng Des 3:34–47

Hernandez VM, McHugh TH, Berrios J, et al (2006) Glycerol content effect on the tensile properties of whey protein sheets formed by twin-screw extrusion. In: IFT annual meeting and food expo, Orlando

Hernandez VM, Reid DS, McHugh TH et al (2005) Thermal transitions and extrusion of glycerol-plasticized whey protein mixtures. In: IFT annual meeting and food expo, New Orleans

Jaganmai G, Jinka R (2017) Production of lipases from dairy industry wastes and its applications. Int J Curr Microbiol Appl Sci 5:67–73

Javanmard M (2009) Biodegradable whey protein edible films as a new biomaterials for food and drug packaging. Iran J Pharm Sci 5(3):129–134

Jiang SJ, Zhang T, Song Y et al (2019) Mechanical properties of whey protein concentrate based film improved by the coexistence of nanocrystalline cellulose and transglutaminase. Int J Biol Macromol 126:1266–1272

Jotarkar PS, Panjagari NR, Singh AK, Arora S (2018) Effect of whey protein-iron based edible coating on the quality of Paneer and process optimisation. Int J Dairy Technol 71(2):395–407

Kozempel M, Tomasula PM, McAloon A (2003) Drying kinetics of calcium casemate. J Agric Food Chem 51(3):773–776

Lacroix M, Vu Khanh D (2014) Edible coatings and film materials: proteins. In: Han JH (ed) Innovations in food packaging, 2nd edn. Academic Press, New York, pp 277–304

Limsawat P, Pruksasri S (2010) Separation of lactose from milk by ultrafiltration. As J Food Ag-Ind 3:236–243

Marquez GR, Di Pierro P, Mariniello L et al (2017) Fresh-cut fruit and vegetable coatings by transglutaminase-crosslinked whey protein/pectin edible films. LWT 75:124–130

Min S, Rumsey TR, Krochta JM (2008) Diffusion of the antimicrobial lysozyme from a whey protein coating on smoked salmon. J Food Eng 84:39–47

Mollea C, Marmo L, Bosco F (2013) Valorisation of cheese whey, a by-product from the dairy industry. In: Muzzalupo I (ed) Food industry. IntechOpen, London, pp 549–588

Muley AB, Singhal RS (2020) Extension of post harvest shelf life of strawberries (*Fragaria ananassa*) using a coating of chitosan-whey protein isolate conjugate. Food Chem 329, 127213

Ncama K, Magwaza LS, Mditshwa A (2018) Plant-based edible coatings for managing postharvest quality of fresh horticultural produce: a review. Food Packag Shelf Life 16:157–167

Paviath AE, Orts W (2009) Edible films and coatings: why, what, and how? In: Embuscado ME, Huber KC (eds) . Edible films and coatings for food applications, Springer Science Business Media, pp 1–23

Pereira JO, Soares J, Monteiro MJ et al (2018) Impact of whey protein coating incorporated with Bifidobacterium and Lactobacillus on sliced ham properties. Meat Sci 139:125–133

Ramos OS (2011) Development and characterization of bioactive edible, whey protein films and coatings, to improve quality and safety of food products. Ph.D. thesis, Instituto de Tecnologia Química e Biológica—Universidade Nova de Lisboa, Lisboa

Sarode AR, Sawale PD, Khedkar CD et al (2016) Casein and caseinate: methods of manufacture. Ency Food Health 1:676–682

Schmid M, Bischur G, Wild F, Noller K (2009) Whey coated plastic films to replace expensive polymers and increase recyclability. Paper presented at the 12th Tappi European Place Conference, Budapest, 18–20 May 2009

Schmid M, Herbst C, Müller K et al (2016) Effect of potato pulp filler on the mechanical properties and water vapor transmission rate of thermoplastic WPI/PBS blends. Polym-Plast Technol Eng 55:510–517

Schmid M, Müller K (2019) Whey protein-based packaging films and coatings. In: Deeth HC, Bansal N (eds) Whey proteins: from milk to medicine. Academic Press, pp 407–437

Schmid M, Müller K, Sängerlaub S et al (2014a) Mechanical and barrier properties of thermoplastic whey protein isolate/ethylene vinyl acetate blends. J Appl Polym Sci 131:41172

Schmid M, Noller K, Wild F et al (2013) Whey protein coated films WO/ 2013/014493

Schmid M, Sängerlaub S, Wege L, Stäbler A (2014b) Properties of transglutaminase crosslinked whey protein isolate coatings and cast films. Packag Technol Sci 27:799–817

Schmid M, Zillinger W, Müller K et al (2015) Permeation of water vapour, nitrogen, oxygen and carbon dioxide through whey protein isolate based films and coatings 2 permselectivity and activation energy. Food Packag Shelf Life 6:21–29

Serife A, Yam KL (2015) Edible films and coatings from calcium caseinate and their applications. https://rucore.libraries.rutgers.edu/rutgers-lib/48380/. Accessed 16 Nov 2020

Shendurse AM, Gopikrishna G, Patel AC et al (2018) Milk protein based edible films and coatings–preparation, properties, and food applications. J Nutr Health Food Eng 8(2):219–226

Shokri S, Ehsani A (2017) Efficacy of whey protein coating incorporated with lactoperoxidase and α-tocopherol in shelf life extension of Pike-Perch fillets during refrigeration. LWT Food Sci Technol 85:225–231

Silva KS, Fonseca TMR, Amado LR, Mauro MA (2018) Physicochemical and microstructural properties of whey protein isolate-based films with addition of pectin. Food Packag Shelf Life 16:122–128

Sothornvit R, Krochta JM (2005) Plasticizers in edible films and coatings. In: Han JH (ed) Innovations in food packaging. Academic, London. ISBN 0-12-311632-5

Southward CR (2016) Casein products. Consumer and applications. Dissertation, New Zealand Dairy Research Institute

Suput DZ, Lazic VL, Popovic SZ (2015) Edible films and coatings–sources, properties, and application. Food Feed Res 42(1):11–22

United Sates Dairy Export Council (2006) Reference manual for US whey and lactose products. Arlington, USA

Verbeek CJR, van den Berg LE (2010) Extrusion processing and properties of protein-based thermoplastics. Macromol Mater Eng 295:10–21

Wagh YR, Pushpadass HA, Emerald FME et al (2014) Preparation and characterization of milk protein films and their application for packaging of Cheddar cheese. J Food Sci Technol 51 (12):3767–3775

Wang HJ, Sun C, Huang LQ (2010) Preparation and properties of whey protein packaging film. In: Proceedings of the 17th IAPRI world conference on packaging, pp. 259–264

Wang Q, Yu H, Tian B et al (2019) Novel edible coating with antioxidant and antimicrobial activities based on whey protein isolate nanofibrils and carvacrol and its application on fresh-cut cheese. Coatings 9(9):583

Warkoyo RB, Marseno DW, Karyadi JNW (2014) Sifat Fisik Mekanik dan Barrier Edible Film Berbasis Pati Umbi Kimpul (*Xanthosoma sagittifolium*) yang Diinkorporasidengan Kalium Sorbat. Agritech 34(1):72–81

Yadav JSS, Yan S, Pilli S et al (2015) Cheese whey: a potential resource to transform into bioprotein, functional/nutritional proteins and bioactive peptides. Biotechnol Adv 33:756–774

Zinoviadou KG, Koutsoumanis KP, Biliaderis CG (2009) Physico-chemical properties of whey protein isolate films containing oregano oil and their antimicrobial action against spoilage flora of fresh beef. Meat Sci 82:338–345

Seaweeds: New Source of Packaging Edibles

12

Prabhjot Kaur Sabharwal, Vandana, and Para Dholakia

Abstract

Biopolymers are playing a pivotal role in recent arenas, especially in addressing the environmental concerns caused due to biostable polymers. Edible films and coatings especially from marine macroalgae (seaweed) have gained popularity in food packaging and as active carriers of ingredients. Seaweed-based biopolymers like agar, carrageenan and algae have varied applications in the food industry including gelation, emulsification, water-holding capacity, etc. Traditionally, seaweeds have been used as sushi wrappings, condiments, vegetables, etc. They are present in abundance, economic and are sustainable. Recent research supports the role of edible seaweeds as antioxidants. Seaweeds have found an important niche of application in food packaging.

Keywords

Seaweed · Marine polysaccharides · Active packaging · Edible film · Biodegradable film

12.1 Introduction

Biostable polymers, commonly known as plastics, are playing havoc to the environment. They are non-biodegradable, non-toxic, renewable and biocompatible. The rampant usage of packaging is burdening our environment. Choosing other alternatives can lessen the environmental impact due to petroleum-based materials. Only less than 5% of petroleum-based materials get recycled, the majority goes into

P. K. Sabharwal (✉) · Vandana · P. Dholakia
Department of Food Technology, Shaheed Rajguru College of Applied Sciences for Women, University of Delhi, New Delhi, New Delhi, India

A. Poonia, T. Dhewa (eds.), *Edible Food Packaging*,
https://doi.org/10.1007/978-981-16-2383-7_12

landfills (Mostafavi and Zaeim 2020). Biodegradable packaging is gaining popularity in today's time. Most researchers have turned to edible packaging as an alternative. Seaweeds are a good runner in edible packaging material. They are natural polysaccharides and used for wide application such as in the field of biomedical sciences (Gade et al. 2013), packaging, hydrocolloids and therapeutic benefits.

12.2 Seaweeds

Traditionally, seaweeds are defined as multicellular and macroscopic brown, red and green algae. Seaweeds or marine algae are a budding renewable resource derived from the marine domain (Syad et al. 2013). They have been used for varied applications in the past, especially in food industry as phycocolloids (Bixler and Porse 2011), gelling and thickening agent and as nutritional additive (Holdt and Kraan 2011). Seaweeds contain large amount of polysaccharides, high dietary fibre and mineral content (Mabeau and Fleurence 1993). Apart from polysaccharides, seaweeds are valuable source of vitamins, minerals, polyphenols and peptides, offering nutraceutical properties (Hentati et al. 2020). There are various types of seaweed used in industry including brown, red and green algae. The major application of seaweed in India is in the production of agar and alginates (Kahlon and Malhotra 1986). Seaweeds offer varied advantages as bioplastic since they can be easily grown even under harsh conditions and yield high biomass. Seaweeds are also cost-effective in comparison to other microbial sources when used as bioplastic (Pathak et al. 2014).

12.3 Seaweeds as Bioplastic

Seaweeds are biodegradable and, thus, are quite popular in the application of edible bioplastics in packaging zone (Gade et al. 2013). Polysaccharide is the main component in seaweed exploited as reservoir for synthesis of bioplastic (Rajendran et al. 2012). Polysaccharide content in seaweed ranges from 4 to 76% (dry weight) (Holdt and Kraan 2011). Carrageenan, agar and alginate are the main polysaccharides extracted from marine seaweeds. The polysaccharides extracted from seaweed are categorized into sulphated and non-sulphated. Sulphated polysaccharides include galactans, agars, carrageenans and fucoidans. Non-sulphated includes alginates and laminaran (Hentati et al. 2020).

12.4 Polysaccharides

Polysaccharides are high molecular weight biopolymers. Seaweeds produce them as primary and secondary metabolites. Polysaccharides from microalgae have been employed for varied application. They are used as novel foods, natural pigment in food industry, cosmetic industry and in pharmaceutical industry as texturizing agent, carriers and antioxidant agent and as biomedical device (Pierre et al. 2019). Major polysaccharides from seaweeds extracted include carrageenan, agar, floridean, starch and alginate. There are various methods available for extraction of polysaccharides from seaweeds. Conventional techniques of extraction lead to low yield of biomass and also have impact on bioactivity of the extract. Novel techniques like ultrasound-assisted, supercritical fluid, microwave-assisted, hydrothermal-assisted (Rajauria et al. 2021) and enzyme-assisted (Wang et al. 2020) extraction have been employed to protect the bioactivity of biomolecules and increase biomass yield.

12.5 Brown Seaweed Polysaccharides

Brown seaweeds polysaccharides constitute majorly of alginate, fucoidan and laminaran (Rioux et al. 2007). Laminaran is made up of (1,3)-β-D-glucan (Kadam et al. 2015). Laminaran is a proven potent antioxidant and antimicrobial (Rajauria et al. 2021; Wang et al. 2020). Fucoidan is made up of fucose and sulphated ester groups. Research has found fucoidan as an anticoagulant with antimicrobial properties against virus, bacteria, etc. Fucoidan has potent antioxidant activity and is used in varied biomedical applications like anti-inflammatory, anticomplementary, antitumor, gastric protection and as therapeutic agent in surgery (Li et al. 2008). Alginate is a major polysaccharide obtained from brown seaweed, and mainly made up of (1,4) β-D-mannuronic acid and α-L-guluronic acid. Alginates are used in food industry as thickening agents (Fenoradosoa et al. 2010).

12.6 Red Seaweed Polysaccharides

There are more than 4000 species of red seaweeds with diverse polysaccharide compositions including agarose, carrageenan, agar, cellulose, floridean and xylan (Usov 1992). Red seaweed carrageenan is sulphated polysaccharides made up of repeated units of three-linked β-D-galactopyranose and four-linked α-D-galactopyranose (Youssouf et al. 2017). Carrageenan is commonly used in the food industry as a thickening and gelling agent because of its water-holding capacity. Souza et al. (2007) in their studies concluded direct correlation between sulphur content and antioxidant activity.

Many species of red seaweeds synthesize agar (sulphated galactan) in the cell wall. Agar is also the major polysaccharide of red seaweeds made up of D- and L-galactose. There are different agar components sectioned on the basis of side-chain substitutions and confirmations as agarose, agaroids, agarans and agropectin (Lee et al. 2017). Agars are high gelling agent, whereas agaroids are weak gelling agents (Hentati et al. 2020). Agaroids are categorized into funorans and porphyrans.

12.7 Green Seaweed Polysaccharides

Chlorophyceae family of seaweeds synthesizes sulphated complex polyholosides, also referred to as green seaweed. Ulvan is water-soluble polysaccharide from the cell wall of algae and majorly comprised of rhamnose, xylose, uronic acids, glucose and mannose (Paulert et al. 2009). Ulvan has proven to be an important biomass with applications in food, pharmacy and chemical industrial applications (Lahaye and Robic 2007).

12.8 Polysaccharide Extraction from Seaweeds

The extraction of polysaccharides from seaweeds requires precision so as to ensure an optimum-quality grade is obtained. The process of harvesting begins followed by drying, grinding and sieving. The seaweed should be free from any impurities (sand, salt, etc.). Step clarification process is employed to separate polysaccharides by hot extraction. The polysaccharides mixture is dissolved in water and centrifuged to remove heavy cellulose particles. Lighter particles are removed by filtration. Excess water is evaporated from the mixture. The filtrate is gelled by adding potassium chloride and then frozen and compressed to trench excess water. The filtrate can also be precipitated in alcohol, further compressed and vacuum dried to eliminate alcohol. The process is finished after filtrate is dried (Rajendran et al. 2012).

12.9 Bioplastics from Alginate

Alginates have promising potential as bioplastic. Alginates offer various advantages like they are non-toxic, renewable, biocompatible, etc. (Cheng et al. 2012; Farias et al. 2021). However, their applications are limited due to low mechanical strength. Alginate films are water soluble and brittle. Thus, alginates are often used in combination with different plasticizers, cross-linking agents and other ingredients to be used as packaging material or in biomedical applications. Sirviö et al. (2014) in their study prepared alginate- and cellulose-based biocomposite film. Mechanical property of the films was remarkably improved due to cross linking between cellulose and alginate. There have been various studies reported in literature to improvise the properties of alginate-based films (Table 12.1).

12.10 Bioplastics from Fucoidan

Fucoidan in sulphated polysaccharide obtained from brown algae lacking gel or film formation properties. Fucoidan has been blended with other polysaccharides and polymers to form film. In one of the studies, fucoidan was blended with alginate to form edible film. Gomma et al. (2018) in their study worked on extraction of alginate and fucoidan from *Sargassum latifolium*. The blend films showed good antioxidant

Table 12.1 Alginate-based films

Category of film	Functional component	Technology	Reference
Biocomposite film	Montmorillonite clay	Solvent casting	Alboofetileh et al. (2013)
Biocomposite film	Sodium montmorillonite clay, benzethonium chloride, polyethyleneimine and urea	Solvent casting	Tezcan et al. (2012)
Film	Barbatimão extract, xanthan gum and β-cyclodextrin	Microencapsulation	Nascimento et al. (2021)
Bionanocomposite film	Myrtle berries extract	Solution casting method	Cheikh et al. (2020)
Active antioxidant films	Yerba mate (*Ilex paraguariensis*)	Solution casting	Farias et al. (2021)
Film	Glycerol and calcium chloride	Cross linking	Giz et al. (2020)
Edible film	Pectin, *Lactobacillus plantarum* KMC 45, glycerol and sorbitol	Gel formation and solution casting	Shahrampour et al. (2020)
Hydrogel film	Carbon nanofibre	Solution casting	Llorens-Gámez et al. (2020)
Blend film	Deacetylated konjac glucomannan	Ca^{2+} cross linking and deacetylation	Li et al. (2018a, b)

properties. In another study, fucoidan and collagen blend films were synthesized for tissue regeneration purpose (Perumal et al. 2018). Fucoidan-based films have been very popularly used for bone tissue engineering (Jeong et al. 2013; Lu et al. 2018). Fucoidan-based nanoparticles have been used for various biomedical applications (Huang et al. 2016; Karunanithi et al. 2016; Tae Young et al. 2016).

12.11 Bioplastics from Laminaran

Lot of studies have been conducted in isolation and characterization of laminaran from brown seaweeds. Laminaran has been reported for various biomedical applications like anticancer and radiosensitizing agent (Malyarenko et al. 2020), pre-biotic (Seong et al. 2019), radioprotector in melanoma therapy (Malyarenko et al. 2019) and in vivo wound healing properties (Sellimi et al. 2018).

12.12 Bioplastics from Agarose

Agarose is one of the most exploited natural biopolymers used for biomedical and packaging applications due to its excellent biocompatibility and good mechanical strength (Vivcharenko et al. 2020). Agarose has been used very popularly in packaging, biomedical and pharmaceutical applications. A limitation of agarose-based films is their hydrophobicity, which can be improvised by

Table 12.2 Agarose-based packaging films

Component	Component	Application	Reference
Hydrogel film	Carrageenan and natural plant extract	Treatment of cutaneous wound	Ditta et al. (2020)
Film	Hydroxyapatite and gelatin	IR window and microelectronic applications	Ramya et al. (2020)
Chips coated with film	3D polystyrene chips	Capture, recovery and culture of cancer cells	Jeong et al. (2018)
Flexible nanocomposite film	Cellulose nanowhisker	Packaging and biomedical application	Felfel et al. (2018)
Hydrogel	Konjac glucomannan	In vitro drug release	Yuan et al. (2018)
Nanoparticle embedded in film	Silver nanoparticle	Antimicrobial packaging	Onofre-Cordeiro et al. (2018)
Blend films	Chitosan	Biodegradable films	Cao et al. (2018)

incorporating different polymers. Table 12.2 shows some of the agarose-based packaging films.

12.13 Bioplastics from Carrageenan

Carrageenan has been used as a biopolymer in industry due to its superior gelation and viscosity properties. Carrageenan tends to swell in aqueous medium and, thus, limits its application as bioplastic (Hamdan et al. 2020). FDA (Food and Drug Administration) has approved use of starch and carrageenan in food packs. A film with low water vapour permeability was obtained using carrageenan, cassava starch, polyvinyl alcohol and glycerol (de Lima Barizão et al. 2020). Carrageenan has also been used to reduce surface roughness and hydrophobicity (Zhuikova et al. 2020). Carrageenan-based antioxidant films were prepared by incorporating prunus maackii pomace extract and cork bark extract with hydroxypropyl cellulose (Zhang et al. 2021). Carrageenan-based intelligent films have been prepared with carboxymethyl starch and mulberry anthocyanin extract. The films showed high pH responsiveness (Zhang et al. 2020). In another study, curcumin was incorporated in carrageenan. The films were used for monitoring freshness in shrimps and prawns (Liu et al. 2018). In another study, carrageenan-based active and intelligent films have been prepared by incorporating pomegranate peel extract. The film showed high UV light barrier and antimicrobial properties (Liu et al. 2020a, b). In a recent study, copper sulphide nanoparticles were incorporated into carrageenan to obtain broad-spectrum antibacterial films. The film showed high transparency and exhibited good mechanical properties (Li et al. 2020). Thus, carrageenan-based films are very commonly used in green packaging.

12.14 Bioplastics from Agar

Agar-based films have been extensively developed as green packaging alternatives. Pure agar lacks mechanical and barrier properties (Mostafavi and Zaeim 2020). Agar has good gelling properties. Various plasticizers, biopolymers (polysaccharides, protein and lipids), nanoparticles and essential oils have been incorporated to obtain agar-based films. Sugar polyols have been extensively used as plasticizers. Active and intelligent agar-based packaging systems have been developed. Davidović et al. (2019) developed agar-silver nanocomposite films with magnesium ions. The films showed improved thermal and mechanical properties and also posed good biocidal efficacy. Nanoparticles have been very commonly used in literature with agar films. In another study, melanin nanoparticles have been incorporated into agar films for higher antioxidant activity and improved mechanical and water vapour barrier properties (Roy and Rhim 2019). The shelf life of green grapes was extended by using agar-based bionanocomposite films. Bionanocomposite films were used as an active packaging system by incorporating zinc oxide nanoparticles (Kumar et al. 2019). Table 12.3 shows bioplastics from agar.

12.15 Bioplastics from Cellulose

Cellulose is one of the most abundant biomasses available and, thus, has been exploited extensively in the field of food packaging, optical devices and electronics (Zhou et al. 2021). Cellulose exhibits good tensile strength, water-holding capacity and is non-allergenic and non-toxic. Because of these properties, it has been exploited in edible packaging films. Cellulose lacks antimicrobial, antioxidative and stress resistance properties, which limits its application in packaging. Pure cellulose films also lack optical transparency (Liu et al. 2020a, b). Lot of work has been done on development of biodegradable and edible cellulose-based films. Li et al. (2021) synthesized carboxymethylcellulose-based edible films with *Dioscorea opposita* mucilage, glycerol and ZnO nanoparticles. The film possessed good

Table 12.3 Bioplastics from agar

S. no.	Category	Functional component	Reference
1	Edible film	Maltodextrin and starch	Wongphan and Harnkarnsujarit (2020)
2	Nanocomposite film	Nanocrystalline cellulose	Atef et al. (2014)
3	Blend films	Soy protein and glycerol	Tian et al. (2011)
4	Blend films	Fish gelatin	Mohajer et al. (2017)
5	Composite film	Lignin	Mohajer et al. (2017)
6	Thermoplastic blend	Sugar palm starch	Jumaidin et al. (2016)
7	Biocomposite film	Carboxymethyl cellulose and summer savoury essential oil	Abdollahi et al. (2019)

antibacterial activity and a potential option towards green packaging. In another study, an edible active packaging system was developed using chitosan biguanidine and frankincense oil. The novel edible films showed excellent physical and antibacterial properties (Salama et al. 2019).

Cellulose nanoparticles have also been successfully incorporated into composite films. Jancy et al. (2020) synthesized cellulose nanoparticles from non-edible parts of jackfruit. Film was prepared using cellulose nanoparticle, polyvinyl alcohol and fennel seed essential oil. Cellulose nanocrystals improved the mechanical, thermal and barrier properties of soy protein isolate-based active nanocomposite films (Xiao et al. 2020). Cellulose nanofibres (CNF) are known to enhance the mechanical properties of biopolymer-based films. Starch-based films when incorporated with cellulose nanofibers (5–20% loading) showed significant improvement in thermal stability, tensile strength and elasticity modulus (Li et al. 2018a, b). CNF significantly improved the barrier properties of bionanocomposite films made from whey protein isolate, titanium oxide nanoparticles and rosemary essential oil (Alizadeh-Sani et al. 2018). Thus, plant and bacterial cellulose have potential application in packaging as they open the avenues for biodegradable films with improved physical and mechanical strength in comparison to films obtained from pure biopolymers.

12.16 Bioplastics from Xylan

Xylan has excellent oxygen barrier properties. Pure xylan films have poor mechanical and film-forming properties. Pure xylan films are hydrophilic that further limits their application in packaging. Chemical modification and addition of biopolymers and plasticizers have improved the film-forming properties of xylan film (Rao et al. 2021). Cellulose in combination with xylan has been used to obtain transparent composite films using facile solution casting methods. The surface of the film was modified by using enzyme xylanase-permitting, nanoscale roughing (Long et al. 2019). Xylan-based composite films with UV shielding performance have been synthesized using polyvinyl alcohol, nano-ZnO and nano-SiO_2 particles. Nano-ZnO and nano-SiO_2 particles improved the mechanical strength and barrier properties of the composite films (Xinxin Liu et al. 2019). Xylan-based films have been used to mitigate growth of food-borne pathogens, *Enterococcus faecalis* and *Listeria monocytogenes,* by incorporating licorice essential oil (Luís et al. 2019). In another study, β-cyclodextrin/sodium benzoate complex was incorporated in xylan composite film to impart antimicrobial activity (Yang et al. 2019). Xylans as biopolymer have promising potential to be used as biodegradable and active packaging material.

12.17 Bioplastics from Ulvan

Ulvan in combination with plasticizers and biopolymers has shown film with good optical, mechanical and barrier properties. Glycerol and sorbitol when incorporated into ulvan gave films with good optical, thermal and antioxidant properties suitable for food packaging (Guidara et al. 2020; Yang et al. 2019). Ulvan-based edible films have also been synthesized in combination with semi-refined carrageenan and glycerol as plasticizer. The films possessed strong antioxidant activity (Ramu Ganesan et al. 2018). Much work is needed to investigate and exploit ulvan as a biopolymer for packaging applications.

References

Abdollahi M, Damirchi S, Shafafi M, Rezaei M, Ariaii P (2019) Carboxymethyl cellulose-agar biocomposite film activated with summer savory essential oil as an antimicrobial agent. Int J Biol Macromol 126:561–568

Alboofetileh M, Rezaei M, Hosseini H, Abdollahi M (2013) Effect of montmorillonite clay and biopolymer concentration on the physical and mechanical properties of alginate nanocomposite films. J Food Eng 117(1):26–33

Alizadeh-Sani M, Khezerlou A, Ehsani A (2018) Fabrication and characterization of the bionanocomposite film based on whey protein biopolymer loaded with TiO2 nanoparticles, cellulose nanofibers and rosemary essential oil. Ind Crop Prod 124(July):300–315

Atef M, Rezaei M, Behrooz R (2014) Preparation and characterization agar-based nanocomposite film reinforced by nanocrystalline cellulose. Int J Biol Macromol 70:537–544

Bixler HJ, Porse H (2011) A decade of change in the seaweed hydrocolloids industry. J Appl Phycol 23(3):321–335

Cao Q, Zhang Y, Chen W, Meng X, Liu B (2018) Hydrophobicity and physicochemical properties of agarose film as affected by chitosan addition. Int J Biol Macromol 106:1307–1313

Cheikh D, Martín-Sampedro R, Majdoub H, Darder M (2020) Alginate bionanocomposite films containing sepiolite modified with polyphenols from myrtle berries extract. Int J Biol Macromol 165:2079–2088

Cheng Y, Lu L, Zhang W, Shi J, Cao Y (2012) Reinforced low density alginate-based aerogels: preparation, hydrophobic modification and characterization. Carbohydr Polym 88(3):1093–1099

Davidović S, Lazić V, Miljković M, Gordić M, Sekulić M, Marinović-Cincović M, Ratnayakc IS, Ahrenkiel SP, Nedeljković JM (2019) Antibacterial ability of immobilized silver nanoparticles in agar-agar films co-doped with magnesium ions. Carbohydr Polym:224

de Farias NS, Silva B, de Oliveira Costa AC, Müller CM (2021) Alginate based antioxidant films with yerba mate (Ilex paraguariensis St. Hil.): characterization and kinetics of phenolic compounds release. Food Packag Shelf Life 28:100548

de Lima Barizão C, Crepaldi MI, Oscar de Oliveira S, de Oliveira AC, Martins AF, Garcia PS, Bonafé EG (2020) Biodegradable films based on commercial κ-carrageenan and cassava starch to achieve low production costs. Int J Biol Macromol 165:582–590

de Souza MC, Marques CT, Dore CM, da Silva FR, Rocha HA, Leite EL (2007) Antioxidant activities of sulfated polysaccharides from brown and red seaweeds. J Appl Phycol 19(2):153–160

Ditta LA, Rao E, Provenzano F, Sánchez JL, Santonocito R, Passantino R, Costa MA, Sabatino MA, Dispenza C, Giacomazza D, San Biagio PL (2020) Agarose/κ-carrageenan-based hydrogel film enriched with natural plant extracts for the treatment of cutaneous wounds. Int J Biol Macromol 164:2818–2830

Felfel RM, Hossain KM, Kabir SF, Liew SY, Ahmed I, Grant DM (2018) Flexible and transparent films produced from cellulose nanowhisker reinforced agarose. Carbohydr Polym 194:328–338

Fenoradosoa TA, Ali G, Delattre C, Laroche C, Petit E, Wadouachi A, Michaud P (2010) Extraction and characterization of an alginate from the brown seaweed *Sargassum turbinarioides* Grunow. J Appl Phycol 22(2):131–137

Gade R, Tulasi MS, Bhai VA (2013) Seaweeds: a novel biomaterial. Int J Pharm Pharm Sci 5:40–44

Ganesan AR, Shanmugam M, Bhat R (2018) Producing novel edible films from semi refined carrageenan (SRC) and ulvan polysaccharides for potential food applications. Int J Biol Macromol 112:1164–1170

Giz AS, Berberoglu M, Bener S, Aydelik-Ayazoglu S, Bayraktar H, Alaca BE, Catalgil-Giz H (2020) A detailed investigation of the effect of calcium crosslinking and glycerol plasticizing on the physical properties of alginate films. Int J Biol Macromol 148:49–55

Gomaa M, Fawzy MA, Hifney AF, Abdel-Gawad KM (2018) Use of the brown seaweed Sargassum latifolium in the design of alginate-fucoidan based films with natural antioxidant properties and kinetic modeling of moisture sorption and polyphenolic release. Food Hydrocoll 82:64–72

Guidara M, Yaich H, Benelhadj S, Adjouman YD, Richel A, Blecker C, Sindic M, Boufi S, Attia H, Garna H (2020) Smart ulvan films responsive to stimuli of plasticizer and extraction condition in physico-chemical, optical, barrier and mechanical properties. Int J Biol Macromol 150:714–726

Hamdan MA, Ramli NA, Othman NA, Amin KN, Adam F (2020) Characterization and property investigation of microcrystalline cellulose (MCC) and carboxymethyl cellulose (CMC) filler on the carrageenan-based biocomposite film. Mater Today: Proc 42:56–62. https://doi.org/10.1016/j.matpr.2020.09.304

Hentati F, Tounsi L, Djomdi D, Pierre G, Delattre C, Ursu AV, Fendri I, Abdelkafi S, Michaud P (2020) Bioactive polysaccharides from seaweeds. Molecules 25(14):1–29

Holdt SL, Kraan S (2011) Bioactive compounds in seaweed: functional food applications and legislation. J Appl Phycol 23(3):543–597

Huang YC, Li RY, Chen JY, Chen JK (2016) Biphasic release of gentamicin from chitosan/fucoidan nanoparticles for pulmonary delivery. Carbohydr Polym 138:114–122

Jancy S, Shruthy R, Preetha RR (2020) Fabrication of packaging film reinforced with cellulose nanoparticles synthesised from jack fruit non-edible part using response surface methodology. Int J Biol Macromol 142:63–72

Jeong HS, Venkatesan J, Kim SK (2013) Hydroxyapatite-fucoidan nanocomposites for bone tissue engineering. Int J Biol Macromol 57:138–141

Jeong J, Lee Y, Yoo Y, Lee MK (2018) Specific capture, recovery and culture of cancer cells using oriented antibody-modified polystyrene chips coated with agarose film. Colloids Surf B: Biointerfaces 162:306–315

Jumaidin R, Sapuan SM, Jawaid M, Ishak MR, Sahari J (2016) Characteristics of thermoplastic sugar palm starch/agar blend: thermal, tensile, and physical properties. Int J Biol Macromol 89: 575–581

Kadam SU, Tiwari BK, O'Donnell CP (2015) Extraction, structure and biofunctional activities of laminarin from brown algae. Int J Food Sci Technol 50(1):24–31

Kahlon SS, Malhotra S (1986) Production of gibberellic acid by fungal mycelium immobilized in sodium alginate. Enzym Microb Technol 8(10):613–616

Karunanithi P, Murali MR, Samuel S, Raghavendran HR, Abbas AA, Kamarul T (2016) Three dimensional alginate-fucoidan composite hydrogel augments the chondrogenic differentiation of mesenchymal stromal cells. Carbohydr Polym 147:294–303

Kumar S, Boro JC, Ray D, Mukherjee A, Dutta J (2019) Bionanocomposite films of agar incorporated with ZnO nanoparticles as an active packaging material for shelf life extension of green grape. Heliyon 5(6):e01867

Lahaye M, Robic A (2007) Structure and function properties of Ulvan, a polysaccharide from green seaweeds. Biomacromolecules 8(6):1765–1774

Lee WK, Lim YY, Leow AT, Namasivayam P, Abdullah JO, Ho CL (2017) Biosynthesis of agar in red seaweeds: a review. Carbohydr Polym 164:23–30

Li B, Lu F, Wei X, Zhao R (2008) Fucoidan: structure and bioactivity. Molecules 13(8):1671–1695
Li J, Ma J, Chen S, He J, Huang Y (2018a) Characterization of calcium alginate/ deacetylated konjac glucomannan blend films prepared by Ca^{2+} crosslinking and deacetylation. Food Hydrocoll 82:363–369
Li M, Tian X, Jin R, Li D (2018b) Preparation and characterization of nanocomposite films containing starch and cellulose nanofibers. Ind Crop Prod 123:654–660
Li F, Liu Y, Cao Y, Zhang Y, Zhe T, Guo Z, Sun X, Wang Q, Wang L (2020) Copper sulfide nanoparticle-carrageenan films for packaging application. Food Hydrocoll 109:106094
Li X, Ren Z, Wang R, Liu L, Zhang J, Ma F, Khan MZ, Zhao D, Liu X (2021) Characterization and antibacterial activity of edible films based on carboxymethyl cellulose, Dioscorea opposita mucilage, glycerol and ZnO nanoparticles. Food Chem 349:129208
Liu J, Wang H, Wang P, Guo M, Jiang S, Li X, Jiang S (2018) Films based on κ-carrageenan incorporated with curcumin for freshness monitoring. Food Hydrocoll 83:134–142
Liu X, Chen X, Ren J, Chang M, He B, Zhang C (2019) Effects of nano-ZnO and nano-SiO_2 particles on properties of PVA/xylan composite films. Int J Biol Macromol 132:978–986
Liu X, Xu Y, Zhan X, Xie W, Yang X, Cui SW, Xia W (2020a) Development and properties of new kojic acid and chitosan composite biodegradable films for active packaging materials. Int J Biol Macromol 144:483–490
Liu Y, Zhang X, Li C, Qin Y, Xiao L, Liu J (2020b) Comparison of the structural, physical and functional properties of κ-carrageenan films incorporated with pomegranate flesh and peel extracts. Int J Biol Macromol 147:1076–1088
Llorens-Gámez M, Salesa B, Serrano-Aroca Á (2020) Physical and biological properties of alginate/carbon nanofibers hydrogel films. Int J Biol Macromol 151:499–507
Long L, Shen F, Wang F, Tian D, Hu J (2019) Synthesis, characterization and enzymatic surface roughing of cellulose/xylan composite films. Carbohydr Polym 213:121–127
Lu HT, Lu TW, Chen CH, Lu KY, Mi FL (2018) Development of nanocomposite scaffolds based on biomineralization of N,O-carboxymethyl chitosan/fucoidan conjugates for bone tissue engineering. Int J Biol Macromol 120:2335–2345
Luís Â, Pereira L, Domingues F, Ramos A (2019) Development of a carboxymethyl xylan film containing licorice essential oil with antioxidant properties to inhibit the growth of foodborne pathogens. LWT 111:218–225
Mabeau S, Fleurence J (1993) Seaweed in food products: biochemical and nutritional aspects. Trends Food Sci Technol 4(4):103–107
Malyarenko OS, Usoltseva RV, Zvyagintseva TN, Ermakova SP (2019) Laminaran from brown alga *Dictyota dichotoma* and its sulfated derivative as radioprotectors and radiosensitizers in melanoma therapy. Carbohydr Polym 206:539–547
Malyarenko OS, Usoltseva RV, Silchenko AS, Ermakova SP (2020) Aminated laminaran from brown alga *Saccharina cichorioides*: synthesis, structure, anticancer, and radiosensitizing potential in vitro. Carbohydr Polym 250:117007
Mohajer S, Rezaei M, Hosseini SF (2017) Physico-chemical and microstructural properties of fish gelatin/agar bio-based blend films. Carbohydr Polym 157:784–793
Mostafavi FS, Zaeim D (2020) Agar-based edible films for food packaging applications: a review. Int J Biol Macromol 159:1165–1176
Nascimento KM, Cavalheiro JB, Netto AÁ, da Silva Scapim MR, de Cássia Bergamasco R (2021) Properties of alginate films incorporated with free and microencapsulated *Stryphnodendron adstringens* extract (barbatimão). Food Packag Shelf Life 28:100637
Onofre-Cordeiro NA, Silva YE, Solidônio EG, de Sena KX, Silva WE, Santos BS, Aquino KA, Lima CS, Yara R (2018) Agarose-silver particles films: effect of calcium ascorbate in nanoparticles synthesis and film properties. Int J Biol Macromol 119:701–707
Pathak S, Sneha CL, Mathew BB (2014) Bioplastics: its timeline based scenario & challenges. J Polym Biopolym Phys Chem 2(4):84–90
Paulert R, Talamini V, Cassolato JE, Duarte ME, Noseda MD, Smania A, Stadnik MJ (2009) Effects of sulfated polysaccharide and alcoholic extracts from green seaweed *Ulva fasciata* on

anthracnose severity and growth of common bean (*Phaseolus vulgaris* L.) Die Wirkung von sulfonierten Polysacchariden und alkoholischen Extrakten aus der Grünalge U. J Plant Dis Protect 116(6):263–270
Perumal RK, Perumal S, Thangam R, Gopinath A, Ramadass SK, Madhan B, Sivasubramanian S (2018) Collagen-fucoidan blend film with the potential to induce fibroblast proliferation for regenerative applications. Int J Biol Macromol 106:1032–1040
Pierre G, Delattre C, Dubessay P, Jubeau S, Vialleix C, Cadoret JP, Probert I, Michaud P (2019) What is in store for EPS microalgae in the next decade? Molecules 24(23):1–25
Rajauria G, Ravindran R, Garcia-Vaquero M, Rai DK, Sweeney T, O'Doherty J (2021) Molecular characteristics and antioxidant activity of laminarin extracted from the seaweed species *Laminaria hyperborea*, using hydrothermal-assisted extraction and a multi-step purification procedure. Food Hydrocoll 112:106332
Rajendran N, Puppala S, Sneha Raj M, Ruth Angeeleena B, Rajam C (2012) Seaweeds can be a new source for bioplastics. J Pharm Res 5(3):1476–1479
Ramya JR, Arul KT, Asokan K, Ilangovan R (2020) Enhanced microporous structure of gamma irradiated agarose-gelatin-HAp flexible films for IR window and microelectronic applications. Mater Today Commun 24:101215
Rao J, Lv Z, Chen G, Hao X, Guan Y, Peng F (2021) Fabrication of flexible composite film based on xylan from pulping process for packaging application. Int J Biol Macromol 173:285–292
Rioux LE, Turgeon SL, Beaulieu M (2007) Characterization of polysaccharides extracted from brown seaweeds. Carbohydr Polym 69(3):530–537
Roy S, Rhim JW (2019) Agar-based antioxidant composite films incorporated with melanin nanoparticles. Food Hydrocoll 94:391–398
Salama HE, Aziz MS, Sabaa MW (2019) Development of antibacterial carboxymethyl cellulose/chitosan biguanidine hydrochloride edible films activated with frankincense essential oil. Int J Biol Macromol 139:1162–1167
Sellimi S, Maalej H, Rekik DM, Benslima A, Ksouda G, Hamdi M, Sahnoun Z, Li S, Nasri M, Hajji M (2018) Antioxidant, antibacterial and in vivo wound healing properties of laminaran purified from *Cystoseira barbata* seaweed. Int J Biol Macromol 119:633–644
Seong H, Bae JH, Seo JS, Kim SA, Kim TJ, Han NS (2019) Comparative analysis of prebiotic effects of seaweed polysaccharides laminaran, porphyran, and ulvan using in vitro human fecal fermentation. J Funct Foods 57:408–416
Shahrampour D, Khomeiri M, Razavi SM, Kashiri M (2020) Development and characterization of alginate/pectin edible films containing *Lactobacillus plantarum* KMC 45. LWT 118:108758
Sirviö JA, Kolehmainen A, Liimatainen H, Niinimäki J, Hormi OE (2014) Biocomposite cellulose-alginate films: promising packaging materials. Food Chem 151:343–351
Syad AN, Shunmugiah KP, Kasi PD (2013) Seaweeds as nutritional supplements: analysis of nutritional profile, physicochemical properties and proximate composition of G. Acerosa and S. Wightii Biomed Prev Nutr 3(2):139–144
Tezcan F, Günister E, Özen G, Erim FB (2012) Biocomposite films based on alginate and organically modified clay. Int J Biol Macromol 50(4):1165–1168
Tian H, Xu G, Yang B, Guo G (2011) Microstructure and mechanical properties of soy protein/agar blend films: effect of composition and processing methods. J Food Eng 107(1):21–26
Usov AI (1992) Sulfated polysaccharides of the red seaweeds. Food Hydrocoll 6(1):9–23
Vivcharenko V, Benko A, Palka K, Wojcik M, Przekora A (2020) Elastic and biodegradable chitosan/agarose film revealing slightly acidic pH for potential applications in regenerative medicine as artificial skin graft. Int J Biol Macromol 164:172–183
Wang L, Jayawardena TU, Yang HW, Lee HG, Kang MC, Sanjeewa KK, Oh JY, Jeon YJ (2020) Isolation, characterization, and antioxidant activity evaluation of a fucoidan from an enzymatic digest of the edible seaweed, hizikia fusiforme. Antioxidants 9(5):363
Wongphan P, Harnkarnsujarit N (2020) Characterization of starch, agar and maltodextrin blends for controlled dissolution of edible films. Int J Biol Macromol 156:80–93

Xiao Y, Liu Y, Kang S, Wang K, Xu H (2020) Development and evaluation of soy protein isolate-based antibacterial nanocomposite films containing cellulose nanocrystals and zinc oxide nanoparticles. Food Hydrocoll 106:105898
Yang YC, Mei XW, Hu YJ, Su LY, Bian J, Li MF, Peng F, Sun RC (2019) Fabrication of antimicrobial composite films based on xylan from pulping process for food packaging. Int J Biol Macromol 134:122–130
Young AT, Kang JH, Kang DJ, Venkatesan J, Chang HK, Bhatnagar I, Chang KY, Hwang JH, Salameh Z, Kim SK, Kim HT (2016) Interaction of stem cells with nano hydroxyapatite-fucoidan bionanocomposites for bone tissue regeneration. Int J Biol Macromol 93:1488–1491
Youssouf L, Lallemand L, Giraud P, Soulé F, Bhaw-Luximon A, Meilhac O, D'hellencourt CL, Jhurry D, Couprie J (2017) Ultrasound-assisted extraction and structural characterization by NMR of alginates and carrageenans from seaweeds. Carbohydr Polym 166:55–63
Yuan Y, Wang L, Mu RJ, Gong J, Wang Y, Li Y, Ma J, Pang J, Wu C (2018) Effects of konjac glucomannan on the structure, properties, and drug release characteristics of agarose hydrogels. Carbohydr Polym 190:196–203
Zhang C, Sun G, Cao L, Wang L (2020) Accurately intelligent film made from sodium carboxymethyl starch/κ-carrageenan reinforced by mulberry anthocyanins as an indicator. Food Hydrocoll 108:106012
Zhang C, Sun G, Li J, Wang L (2021) A green strategy for maintaining intelligent response and improving antioxidant properties of κ-carrageenan-based film via cork bark extractive addition. Food Hydrocoll 113:106470
Zhou Z, Wang L, Hu Y, Song R, Mei N, Chen T, Tang S (2021) Preparation of AAEK-functionalized cellulose film with antibacterial and anti-adhesion activities. Int J Biol Macromol 167:66–75
Zhuikova YV, Zhuikov VA, Zubareva AA, Akhmedova SA, Sviridova IK, Sergeeva NS, Varlamov VP (2020) Physicochemical and biological characteristics of chitosan/κ-carrageenan thin layer-by-layer films for surface modification of nitinol. Micron 138:102922

13 By-Products of Oilseeds Industry (Defatted Cake Waste): Biodegradable Alternatives for Plastics

Vedpal Yadav and Usman Ali

Abstract

In the modern world, most of our needs completely relay on the plastic material. These plastic materials are developed from the petrochemical-based materials. The disposability of such type of plastic materials is a serious environmental issue after their use when they enter into the mainstream. Therefore, sustainability, eco-efficiency and green chemistry are enforcing the researcher to develop new-generation biodegradable plastic material. The latest environmental regulation and environmental awareness motivate the researcher for the development of biodegradable/environmentally friendly plastic from natural resources. India is the second largest producer of the oilseeds. After extraction of oil from oilseeds (soybean, cotton seed, peanut, rapeseed, mustard and sunflower seeds), abundant quantity of oilseeds cake is produced as one of the major by-product of oilseeds industries. These oilseeds cake is a good source of protein and some polysaccharide, and can be used for the development of biodegradable plastic. Most of these oilseeds cakes (such as mustard, cotton seed and sunflower) are underutilized and generally used for the animal feed. These industrial defatted oilseed cakes are a good source of raw material for the development of biodegradable plastic and are rapidly being developed as new source of bio-economy.

Keywords

Oilseeds · Biodegradable · Plastics · Biodegradable

V. Yadav (✉)
Food Technology, Government Polytechnic, Mandi Adampur, Hisar, Haryana, India

U. Ali
Department of Food & Nutritional Biotechnology, National Agri-Food Biotechnology Institute, Sector-81 (Knowledge City), Mohali, Punjab, India

A. Poonia, T. Dhewa (eds.), *Edible Food Packaging*,
https://doi.org/10.1007/978-981-16-2383-7_13

13.1 Introduction

The environmental pollution caused by the plastic polymer is a global issue of concern. Due to the rapid development in the polymer industries, a number of products have been developed with modernization of the technology. This progress in the modern technology greatly improved the living quality of human being. Most of the polymer materials used now a days are synthetic and obtained from the non-renewable sources. High production and non-biodegradability draw public attention due to the huge amount of environmental pollution created by these plastics (Patel et al. 2016). The durability of petrochemical-based plastic makes it ideal for a number of applications such as packaging and hygiene of the products, commodities and building material. But the disposal of such type of polymer is a major problem that affect the ecosystem and persist for centuries. Such types of plastic polymers are resistant to the microbial degradation; therefore, accumulate in the environment (Vroman and Tighzert 2009).

Currently, the production of plastic accounts approximately 4–8% consumption of oil globally and expected to reach up to 20% by 2050 globally. Due to the reduced cost and wide range of application and properties, the production of plastic increased globally from 15 million metric tonnes (MMT) in 1964 to 359 MMT in 2018 and projected to twofold increase within the next 20 years (Ellen MacArthur Foundation 2016; Narancic et al. 2020). According to the survey conducted by the Central Pollution Control Board (CPCB, India), in 60 different cities, 56 lakh tonnes of plastic waste are generated annually in India and approximately 6137 tonnes of plastic waste are remaining uncollected per day. This remaining or uncollected plastic waste may end up in the water or landfill, and thus, develop toxic and greenhouse gases (Patel et al. 2016).

The term 'bio-plastic' is generally used confusingly. The bio-plastics consist of either biodegradable plastic (i.e. the plastics developed from the fossil materials) or bio-based plastics (i.e. plastic that is synthesized from renewable resources and biomass) (Tokiwa et al. 2009). According to the ASTM, a biodegradable plastic is the plastic in which the process of degradation takes place by the action of naturally occurring microbes such as bacteria, fungi and algae and converts into biomass, carbon dioxide, water and methane (Patel et al. 2016).

A process in which an organic material (such as polymer) is converted into inorganic substance (such as carbon dioxide) is called mineralization. There are a number of factors that are involved in the biodegradation process such as polymer structure, variable reaction condition and microorganism (Krzan 1994).

In this concern, researchers are using renewable resources like protein and polysaccharide (starch, cellulose and cellulose derivatives) for the development of biodegradable plastic because of their low cost and abundant presence in nature. But the reduced elasticity of the developed material is a major drawback that limits their field of application (Ayhllon-Meixueiro et al. 2000; Patel et al. 2016).

Plant oil can be a suitable alternative to the fossil oil. Generally, plant oils are produced for human nutrition from plants' seeds such as soy, sesame seed, sunflower seed, mustard, rapeseed and groundnut. The processing of the oilseed generates the residual meal after the extraction of oil and is generally used for animal feed; however, the residual meal from some of the oilseed is also used to meet the need of human consumption. Development of bio-based plastic is a promising application of these protein-rich feedstocks (Leckband et al. 2002; Rouilly and Rigal 2002; Domenek et al. 2004; Carlsson 2009; Carlsson et al. 2011; Newson 2012; Zhang and Mittal 2014). The attractive gas barrier properties of these protein make them possible alternative for conventional polymer. However, the high cost, brittleness, poor processability and reduced strength and stiffness are the main hurdle in these polymer-based materials. Most of the plant proteins showed glassy state in pure form and materials developed from these proteins are brittle. Therefore, in order to reduce the glass transition temperature and to reduce the brittleness of the material, plasticizers are used to induce the flexibility in the material (Newson 2012; Zhang and Mittal 2014).

Protein is generally considered as thermoplastic hetropolymers and contains both polar and non-polar α-amino acids. The amino acids in the protein have the ability to develop intramolecular linkage resulting in multiple interactions and offer a number of functional properties. The protein-based bio-plastics in a classical manner are developed by mixing protein with plasticizers and the properties of the products are improved by the use of plasticizers (Vroman and Tighzert 2009).

According to the German Plastic Museum, the application of protein was first mention in 1530 for the development of biodegradable plastic from milk protein (casein). There are a number of natural resources available as a source of protein for the development of biodegrade material.

The deoiled cake obtained after the extraction of oil from oilseed (such as groundnut deoiled cake, mustard deoiled cake, soybean deoiled cake, cotton seed deoiled cake, sesame seed deoiled cake and rapeseed deoiled cake) is an abundant source of protein for the development of biodegradable plastic and films (Patel et al. 2016). The protein in combination with the polysaccharides such as starch have been used as a filler with conventional petroleum-based thermoplastic to decrease the cost of the raw material and to impart biodegradability in the plastic from petroleum sources. A number of processing methods (extrusion, film casting, injection moulding and compression moulding) have been used for the development of biodegradable (Zárate-Ramírez et al. 2014; Patel 2015).

13.2 Development of Biodegradable Polymer/Plastic

In the recent years, biodegradable plastics have been studied intensively. The biodegradable plastics are commercialized for the manufacturing of a number of products, such as compost bags, garbage bags, agricultural mulch films, etc. (Faris et al. 2014).

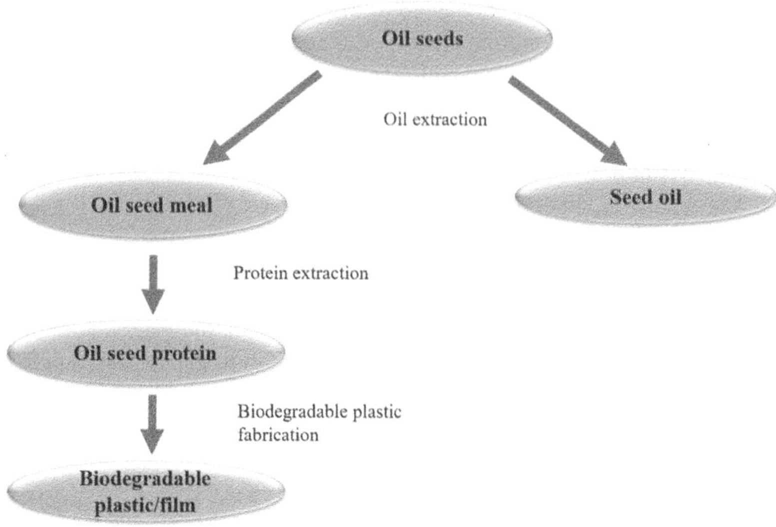

Fig. 13.1 General process for the development of biodegradable plastic from oilseed meal

Since the production cost of bio-plastic at a large scale is very high, it has not been extensively used. During the twentieth century, the production of bio-plastic was mainly dominated by the countries like North America, Western Europe, Japan, etc. In 2008, biodegradable plastics such as polylactic acid, polyester and resin account approximately 90% of the total bio-plastic demand (Shivam 2016). The polylactic acid (PLA) and polyhydroxyalkanoate (PHA) are the biodegradable polymer. The PLA is obtained from the renewable sources such as corn and starch, and offers multiple applications. The PHA comes under the categories of the green plastic (obtained from the renewable resources) via fermentation process. But, the high production coat is a major drawback of PHA (Faris et al. 2014).

For the development of biodegradable plastic from the deoiled cake of plant seeds (cotton seed, mustard seed, sesame seed, soybean seed, etc.), firstly, the seed protein is isolated from the deoiled cake. The isolated protein is further used for the development of biodegradable plastic/film by using different methods. The functionality of the material such as biodegradable film can be modified by using plasticizer, antioxidant, antimicrobial compounds, etc. (Fig. 13.1)

13.3 Production of Oilseeds (Tables 13.1 and 13.2)

Table 13.1 Production of oilseeds crops in India (with three major producing states) during 2017–2018

Sr. no.	Oilseed	Major producing states	Production (million tonnes)
1	Mustard and Rapeseed	Rajasthan	3.40
		Haryana	1.11
		Madhya Pradesh	0.98
		All India	8.32
2	Groundnut	Gujarat	3.94
		Rajasthan	1.26
		Andhra Pradesh	1.04
		All India	9.18
3	Soybean	Madhya Pradesh	5.32
		Maharashtra	3.89
		Rajasthan	1.07
		All India	10.98
4	Sunflower seeds	Karnataka	0.10
		Bihar	0.02
		Odisha	0.02
		All India	0.21
5	Total oilseeds	All India	31.31

Reference: Agricultural Statistics (2018)

Table 13.2 Conversion factors between primary and secondary agriculture commodities

Sr. no.	Commodity	Conversion	Conversion factor
1	Rapeseed and mustard seeds	Oil to seeds crushed	33%
		Cake to seeds crushed	67%
2	Groundnut	Oil to kernels crushed	40%
		Cake to kernels crushed	60%
3	Soybean seeds	Oil to soybean seed crushed	18%
		Meal to soybean seed crushed	73%
		Wastage from soybean seed crushed	1%
4	Sesamum seeds	Oil to seeds crushed	40%
		Cake to seeds crushed	60%
5	Cotton seeds	Oil to seeds crushed	14–18%
		Cake to seeds crushed	82–86%

Reference: Agricultural Statistics (2018)

13.4 Composition of Oilseed Cake

The chemical composition provides the critical information for the development of a product with good quality and also helps in the exploration of products for new applications. The oilseed cake obtained after the extraction of oil is a rich source of protein. The concentrations of the protein in oilseed cake/meal depend upon the type of the source (oilseed). The chemical composition of the oilseed cake obtained from the different oilseeds is presented in Tables 13.3–13.6.

Table 13.3 Composition of sunflower oilseed cake and protein isolate

Sr. no.	Components	Sunflower oilseed cake	Isolate of sunflower proteins
1	Protein	34.4 ± 1.0	90 ± 1.0
2	Lipid	1 ± 0.5	0.6 ± 0.4
3	Cellulose	22.3 ± 2.0	0
4	Lignin	5.2 ± 1.0	1.7 ± 1.0
5	Hemicellulose	9.8 ± 1.0	0
6	Phenolic compounds	5.7 ± 0.80	4.8 ± 0.8
7	Moisture	10 ± 2.0	6 ± 1.0
8	Ash	7.6 ± 0.4	2.4 ± 0.4

Reference: Ayhllon-Meixueiro et al. (2000)

Table 13.4 Chemical composition of cottonseed and soybean meal

Sr. no.	Components	Soybean meal (g/kg)	Cottonseed meal (g/kg)
1	Crude protein	470.1	465.2
2	Ether extract	15.1	10.8
3	Crude fibre	62.1	102.1
4	Ash	54.3	60.2
5	Major amino acids		
5.1	Leucine	35.2	26.8
5.2	Arginine	34.3	49.8
5.3	Lysine	29.2	21.3
5.4	Valine	22.6	21.5
5.5	Threonine	18.2	14.5
5.6	Phenylalanine	23.3	24.3
5.7	Histidine	12.1	12.6
6	Total phosphorus	5.60	11.1
7	Calcium	2.70	2.50
8	Free gossypol	ND	0.82

ND Not detected. Reference: Tang et al. (2012)

Table 13.5 Chemical composition of semi-defatted sesame cake and peanut deoiled cake

Sr. no.	Components	Semi-defatted sesame cake (g/100 g)	Peanut deoiled cake (%)
1	Moisture	8.1 ± 0.1	11.26
2	Protein	35.0 ± 0.3	38.11
3	Fat	11.2 ± 0.2	1.25
4	Dietary fibre	22.7 ± 0.0	10.54
5	Ash	8.6 ± 0.0	4.71
6	Carbohydrate	14.4	34.13
7	Amino acid profile (g/100 g) of protein		
7.1	Threonine	4.95 ± 0.04	–
7.2	Methionine	NA	–
7.3	Valine	7.75 ± 0.06	–
7.4	Leucine	11.85 ± 0.16	–
7.5	Isoleucine	6.70 ± 0.10	–
7.6	Phenylalanine	10.70 ± 0.14	–
7.7	Lysine	4.25 ± 0.06	–
7.8	Histidine	3.95 ± 0.04	–
7.9	Tryptophan	3.50 ± 0.05	–

References: Nascimento et al. (2012), Patel et al. (2016)

Table 13.6 Chemical composition of black and yellow mustard cake and rapeseed cake

Sr. no.	Components	Black mustard cake (%)	Rapeseed cake (%)
1	Moisture	9.20 ± 0.5	–
2	Minerals	7.10 ± 0.3	–
3	Acid insoluble ash	1.93 ± 0.4	–
4	Oil	8.70 ± 0.8	–
5	Crude fibre	12.17 ± 1.3	10.35 ± 0.65
6	Crude protein	38.17 ± 1.0	34.78 ± 0.55
7	Allyl isothiocyanate	0.086 ± 0.009	–
8		Amino acids profile (precipitated protein isolate obtained from cake (%))	Amino acids content in rapeseed meal (%)
8.1	Threonine	19.17	1.40
8.2	Lysine	4.55	1.33
8.3	Isoleucine	5.57	1.24
8.4	Tryptophan	1.96	0.42
8.5	Valine	1.20	1.62
8.6	Methionine	2.25	0.60
8.7	Alanine	3.56	–
8.8	Arginine	2.74	1.82
8.9	Aspartate	4.49	–
8.10	Glycine	2.54	–

References: Dingyuan and Jianjun (2007), Sarker et al. (2015)

13.5 Oilseed Cake-Based Biodegradable Polymer

13.5.1 Cotton Seed

More than 30 countries are producing cotton as a major source of fibre for the textile industries. Harvesting and ginning of the cotton produce lint and seed as two marketable products. The residual obtained after oil extraction is called the defatted cottonseed meal. But, this defatted meal is generally used as an animal feed or fertilizer. Although this is a potential value-added product for the cotton grower and the processor, the defatted cottonseed meal can be used for the development of a number of value-added products such as bio-plastics and films, antioxidant meal hydrolysates, wood adhesives, superabsorbent hydrogel and bio-oil and bio-char (He et al. 2015).

Protein from cottonseed flour in combination with urea, aldehydes and glycerol was successfully used for the development of bio-plastic by using hot-press moulding. The effect of cross linking and plasticizer on the properties (thermal stability, mechanical strength and moisture barrier property) of the bio-plastic was investigated. The study showed that cross linking improved the thermal stability, mechanical strength and moisture resistance properties of the bio-plastic due to Maillard-driven formation. The improved mechanical strength and moisture barrier properties make these bio-plastic application in low load bearing and environmentally sensitive industries (Yue et al. 2012) (Fig. 13.2).

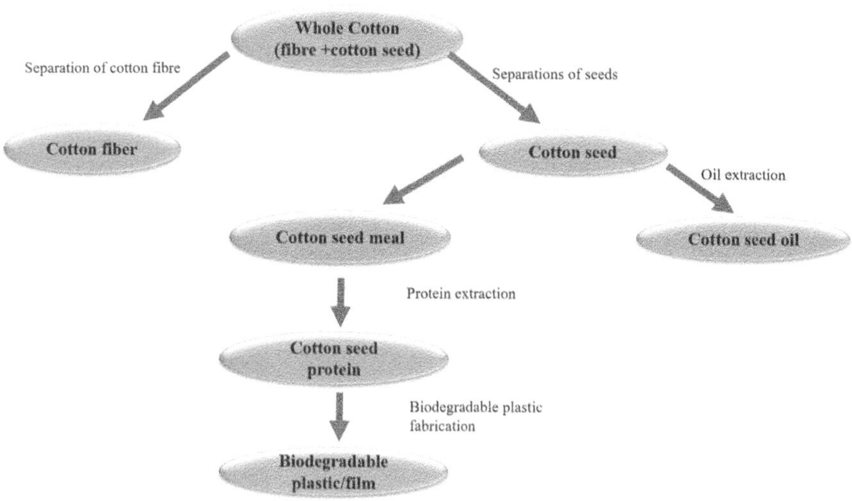

Fig. 13.2 General process for the development of biodegradable plastic from cottonseed meal

13.5.2 Mustard Seed

India is the leading producer of the mustard and rapeseed along with other countries (USA, Canada and other European Countries). According to the Agricultural Statistics (2018), 8.32 million tonnes of mustard and rapeseed were produced in 2018. The mustard seed contains oil (30–35%) and protein (34–39%) (Anil Kumar et al. 2002). After oil extraction from the mustard seed, approximately 60% of the by-product is produced as a mustard seed cake and it is generally used for the animal feed. This mustard seed cake obtained after oil extraction is a rich source of protein and contains good quantity of the essential amino acid and is relatively rich in methionine. However, the reduced palatability of the mustard seed cake is a major problem for its utilization as animal feed due to the presence of glucosinates. These glucosinates produce pungent and hot metabolite upon hydrolysis by the endogenous enzymes. This mustard seed cake is also used as a suitable raw material for development of biodegradable film and plastic generation (Anil Kumar et al. 2002; Hendrix et al. 2012). The developed films possess good mechanical and moisture barrier properties that are important for application in food commodities. Further, the properties of the film can be improved by optimizing film-forming process variables, type and concentrations of plasticizer and emulsifier for its commercial applications in food industry (Anil Kumar et al. 2002; Hendrix et al. 2012).

The biodegradable films were developed from the mustard seed cake and the properties of the films such as tensile strength and elongation and water vapour were determined in the range of 1.3–5.5 MPa, 0.9–18.1% and 3.4–5.0 g mm/m^2 kPa h respectively. The study also showed that coating containing mustard cake as a component effectively decrease the lipid oxidation during storage in the smoked salmon (Hendrix et al. 2012; Swati and Das 2015).

13.5.3 Rapeseed

Rapeseed is the second largest oilseed crop after soybean around the world. The rapeseed contains approximately 40–48% of oil content and the by-product generated after the oil extraction contains 35% of protein, which is approximately 55–60% of its initial weight. However, the presence of antinutritional factors (glucosinolates and phytates) makes it unsuitable for human consumptions; therefore, it is generally used for low value applications. The rapeseed cake is used as an ingredient for the development of bio-plastics. The cake contains 20% of napin and 60% of cruciferin protein which allow the cross linking of protein. It has been reported that rapeseed meal was used for the development of bio-plastic by injection mouldings at different moulding temperature (Anil Kumar et al. 2002; Delgado et al. 2018).

13.5.4 Peanut

Peanut deoiled cake is also obtained as by-product from peanut oil processing industry. The deoiled cake contains 35–40% of the protein and can be extracted by using alkali and acid extraction. The protein extracted from the peanut deoiled cake was used for the development of the bio-plastic. Recent study showed that the peanut deoiled cake protein isolate was used for the development of biodegradable film in combination with chitosan. The effect of different concentrations of protein on functional, structural and biodegradable properties was determined. The results showed that the developed film has potential as a packaging material (Patel et al. 2016).

13.5.5 Sesame Seed

Sesame is an important oilseed crop in India, which is cultivated in numerous tropical countries. The sesame seed cake obtained after the extraction of oil is a good source of protein; generally, contains 35–40% of protein. The sesame seed cake is generally considered as by-products of the sesame seed oil industry, and is generally used as animal feed as a source of protein (Sharma and Singh 2016). The high heat stability of the sesame seed protein allows for multiple industrial applications. This sesame seed cake is further defatted and used to obtain protein isolate or concentrate. The protein isolate from the cake is obtained either by isoelectric precipitation or salt precipitation. This protein isolate is rich in methionine (2.4–4.0%) and sulphur-containing amino acids (3.8–5.5%) (Onsaard et al. 2010; Sandhu et al. 2017). The protein isolate from the cake can be used for the development of biodegradable or edible film. The films developed from the sesame seed cake protein isolate have good barrier properties as compare to the films obtained from the peanut protein, soya protein and whey protein film. The edible film developed from the sesame seed protein isolate exhibited better thermal and moisture barrier properties compared to the faba bean protein, peanut protein and lentil protein. In addition to this, the films also unveil good mechanical properties and can be used for multiple packaging and coating application (Sharma and Singh 2016).

13.5.6 Sunflower Seed Cake

The sunflower seed cake is generally used for the animal feed but it contains reduced nutritional value as compared to the other oilseed cake such as soybean, canola and rapeseed.

Sunflower oilseed cake was used for the extraction of sunflower protein isolation by alkali extraction method and the isolated protein is used for the development of biodegradable film.

Different methods such as casting, thermos moulding, injection moulding and extrusion moulding have been used for the development of the film from sunflower seed cake protein isolate (Ayhllon-Meixueiro et al. 2000; Geneau-Sbartaï et al. 2008). The films were developed by the casting method and the effect of plasticizer and dissolving solvent on mechanical properties of the films was determined. The use of glycerol and 1,3-propanediol resulted in highest tensile strength (27.1 MPa) and greatest elongation at break (251%) respectively (Ayhllon-Meixueiro et al. 2000).

The sunflower protein isolate was used for the development of film by thermo-moulded method and the effect of different additives on the film properties (mechanical, hydrophobicity and water uptake) was studied. The use of octanoic acid and octanol as additives provides film with high tensile strength (7Mpa) and tensile elongation (54%). Finally, the octanoic acid also reduces the films' water uptake (Orliac et al. 2002).

The injection-moulded plastic was developed from the sunflower protein insolate and the mechanical properties were determined to study the effect of glycerol. The increase in the glycerol content causes decrease in the tensile strength. The study explains the thermoplastic behaviour of sunflower protein isolate mixture and their suitability for the development of injection-moulded objects (Orliac et al. 2003).

The extrusion method of film formation was used for the development of film from sunflower protein isolate. The effect of plasticizer and temperature on the film properties (appearance, thermomechanical and mechanical properties and film swelling behaviour) was studied. A mixture of water (20 parts), glycerol (70 parts) and sunflower protein isolate (100 parts) at 160 °C die temperature and 20 rpm of screw speed was the best condition to obtain a homogenous and smooth film (Antoine Rouilly et al. 2006).

13.6 Properties

The application of a biodegradable material for their end use depends upon the properties of the material such as barrier (moisture barrier and gas barrier), mechanical and thermal properties.

13.6.1 Barrier Properties

13.6.1.1 Moisture Barrier Properties

The moisture barrier property of biodegradable plastic is the ability of the material to resist the migration of moisture/water vapours and also the affinity towards the moisture. This is determined by the water vapour transmission rate (WVTR). The undesirable effect due to the moisture absorption by the material resulted in the regain of the moisture in the packed material. This property is very important when we use the packing material to pack the dry food material. Therefore, the moisture barrier properties can be developed by the external coating of water-resistant material or hydrophobic material, cross linking with inorganic filler and blending with hydrophobic material (Bourtoom and Chinnan 2008; Kumar et al. 2011; Ashok and Rejeesh 2018; Mangaraj et al. 2019).

13.6.1.2 Gas Barrier Property

The gas barrier property is also a critical factor to determine the application of a material especially in the food packaging industries. The material must have the good gas barrier properties (Kumar et al. 2011; Ashok and Rejeesh 2018).

However, the gas barrier property is also dependent upon their end use. The biodegradable films having high gas permeability than the oxygen vapour pressure inside packaging lead to the oxidation in the products. Further, the high gas permeability also causes the quality deterioration due to respiration in case of fruits and vegetables after packaging (Mangaraj et al. 2019).

13.6.1.3 Mechanical Property

The mechanical property of biodegradable material is a critical property that helps to determine the suitability of the material for their wide applications. Therefore, the biodegradable material must have a good mechanical property against the mechanical damage. The application of material especially in the automobile industry requires a competitive mechanical property (Ashok and Rejeesh 2018).

It is well known that the architecture of the polymer plays an important role in their mechanical properties and their methods of processing/production (i.e. blow moulding, film forming, injection moulding, extrusion, etc.). In addition to this, the commercially used packaging container is operated below room temperature. Therefore, the mechanical performance must be assessed on different temperature range (Mangaraj et al. 2019).

13.6.1.4 Other Property

The thermal property of a biodegradable material is important for the protection of the product against thermal damage. Different technique like plasticization is used to improve the thermal stability of the material by increasing the temperature of degradation (Ashok and Rejeesh 2018).

13.7 Applications

The decomposition of the biodegradable plastics with the passage of time is a major reason for their wider applications in different sectors. The market growth of the bio-plastic is driven by continuous research and development, environmental awareness and regulation. However, the high production cost is a major hurdle. The main factors that drive the growth of biodegradable plastic are as follow: biodegradability, disposable houseware, consumer preference, government policies, etc. A number of companies are manufacturing products from biodegradable polymers (Ashter 2016).

These plastics have found their use in a number of application as given below.

13.7.1 Packaging

Biodegradable material offers a wide range of application in packaging. A number of companies are using biodegradable polymers for bags development. Such products must be compostable and should confirm to ASTM standard D6400-99 (Ashter 2016). These biodegradable bags decompose when in the presence of water, air and sunlight (Ashter 2016).

13.7.2 Food Packaging

Modified atmospheric packaging is a technique of food preservation to increase the shelf life of the products by reducing the metabolic activity (by reducing O_2 concentrations and increasing the CO_2 Concentrations). Generally, petroleum-based films are used to increase the shelf life of different commodities.

The utilization of the biopolymer for smart and active polymer system in the food industry has emerged from the last decade. These polymers have been used to entrap the micronutrient or active component in the matrix based on biopolymers. The technologies associated with these polymers have the potential for the development of new-generation active and intelligent packaging (Mangaraj et al. 2019).

13.7.3 Disposable Housewares

The disposable housewares that are developed from the biodegradable polymer are now replacing houseware that are developed from the traditional plastics (polystyrene and polyolefin) (Ashter 2016).

13.7.4 Agriculture and Horticulture

The biodegradable plastic also has a strong potential market in agriculture and horticulture. A number of agriculture products such as agricultural mulches, tapes, foil, nets and the seed strips are developed from the biodegradable polymer. As the germination of the seeds takes place the tape biodegrades in the soil (Corbin 2012; Ashter 2016; Narancic et al. 2020).

13.7.5 Medical

Since last 50 years, the biodegradable polymers have been the main focus of research for their application in biomedical (Narancic et al. 2020).

Sutures developed from the non-toxic biodegradable polymer have been used by the surgeons in life-saving operation (heart). These sutures remain intact until the healing of surrounding tissue takes place and dissolve, and these sutures are readily metabolized in the body with the passage of time (Ashter 2016; Narancic et al. 2020).

TYRX absorbable antibacterial envelope is developed from tyrosine-derived polyarylate, which is an important component of hernia repair device by TyRx Pharma. In addition to this, the biodegradable plastic pins, screws and tacks have been developed from the non-toxic biodegradable polymer for application in scattered bones. The dental implants developed from the biodegradable polymer (porous polymer particles) are being used for quick hole filling after a tooth has been extracted. These polymers are also used as delivery vehicles for the controlled release of drugs (Vert 2009; Ashter 2016; Prakasam et al. 2017).

13.8 Future Development

The development of the biodegradable plastic is an open and wide area of research. The deoiled cake is cheap and an abundant source of the protein and can be utilized for the development of a number of products. These biopolymers can be modifying (chemical and enzymatic modification) for the development of the products having the desired characteristics. Further, lots of research and development need to be done by using multiple (analytical and biotechnological) tools to make the product and technology more commercially viable. In addition, advance technology such as active packaging and nanotechnology can further be used to modify the functional properties of the biodegradable plastic. Finally, the future of the technology or material also depends upon its cost and functional properties. Therefore, in future development, the researches must focus to improve the functional properties for their wider application.

13.9 Conclusion

Biodegradable polymer received more attention due to their application in multiple fields (environment, health, medical, etc.). The utilization of the by-products for the development of value-added products such as bio-plastic also helps to add economic value to these by-products and reduce the problem of pollution caused by the non-biodegradable plastic.

References

Agricultural Statistics (2018) agristatglance2018.pdf https://agricoop.gov.in/sites/default/files/agristatglance2018.pdf (agricoop.gov.in)

Anil Kumar GK, Panwar VS, Yadav KR, Sihag S (2002) Mustard cake as a source of dietary protein for growing lambs. Small Rumin Res 44:47–51. https://doi.org/10.1016/S0921-4488(02)00034-2

Ashok A, Rejeesh RR (2018) Biodegradable polymers for sustainable packaging applications. Int J Mech Eng Technol 9:293–303

Ashter SA (2016) Commercial applications of bioplastics. Introd to Bioplastics Eng:227–249. https://doi.org/10.1016/b978-0-323-39396-6.00009-9

Ayhllon-Meixueiro F, Vaca-Garcia C, Silvestre F (2000) Biodegradable films from isolate of sunflower (*Helianthus annuus*) proteins. J Agric Food Chem 48:3032–3036. https://doi.org/10.1021/jf9907485

Bourtoom T, Chinnan MS (2008) Preparation and properties of rice starch-chitosan blend biodegradable film. LWT—Food Sci Technol 41:1633–1641. https://doi.org/10.1016/j.lwt.2007.10.014

Carlsson AS (2009) Plant oils as feedstock alternatives to petroleum: a short survey of potential oil crop platforms. Biochimie 91:665–670. https://doi.org/10.1016/j.biochi.2009.03.021

Carlsson AS, Yilmaz JL, Green AG et al (2011) Replacing fossil oil with fresh oil—with what and for what? Eur J Lipid Sci Technol 113:812–831. https://doi.org/10.1002/ejlt.201100032

Corbin A (2012) Using biodegradable plastics as agricultural mulches. Washingt State Univ Ext:1–6

Delgado M, Felix M, Bengoechea C (2018) Development of bioplastic materials: from rapeseed oil industry by products to added-value biodegradable biocomposite materials. Ind Crops Prod 125:401–407. https://doi.org/10.1016/j.indcrop.2018.09.013

Dingyuan F, Jianjun Z (2007) Nutritional and anti-nutritional composition of rapeseed meal and its utilization as a feed ingredient for animal. Int Consult Gr Res Rapeseed:265–271

Domenek S, Feuilloley P, Gratraud J et al (2004) Biodegradability of wheat gluten based bioplastics. Chemosphere 54:551–559. https://doi.org/10.1016/S0045-6535(03)00760-4

Ellen MacArthur Foundation (2016) The new plastics economy: rethinking the future of plastics. Ellen MacArthur Found, p 120

Faris NA, Noriman NZ, Sam ST et al (2014) Current research in biodegradable plastics. Appl Mech Mater 679:273–280. https://doi.org/10.4028/www.scientific.net/AMM.679.273

Geneau-Sbartaï Ć, Leyris J, Silvestre F, Rigal L (2008) Sunflower cake as a natural composite: Composition and plastic properties. J Agric Food Chem 56:11198–11208. https://doi.org/10.1021/jf8011536

He Z, Zhang H, Olk DC (2015) Chemical composition of defatted cottonseed and soy meal products. PLoS One:10. https://doi.org/10.1371/journal.pone.0129933

Hendrix KM, Morra MJ, Lee HB, Min SC (2012) Defatted mustard seed meal-based biopolymer film development. Food Hydrocoll 26:118–125. https://doi.org/10.1016/j.foodhyd.2011.04.013

Krzan A (1994) Biodegradable plastics and polymers. J Pestic Sci 19:1–8. https://doi.org/10.1584/jpestics.19.S11

Kumar MNS, Yaakob Z, Siddaramaiah (2011) Biobased materials in food packaging applications. Handb Bioplastics Biocomp Eng Appl:121–159. https://doi.org/10.1002/9781118203699.ch5

Leckband G, Frauen M, Friedt W (2002) NAPUS 2000. Rapeseed (*Brassica napus*) breeding for improved human nutrition. Food Res Int 35:273–278. https://doi.org/10.1016/S0963-9969(01)00196-X

Mangaraj S, Yadav A, Bal LM et al (2019) Application of biodegradable polymers in food packaging industry: a comprehensive review. J Packag Technol Res 3:77–96. https://doi.org/10.1007/s41783-018-0049-y

Narancic T, Cerrone F, Beagan N, O'Connor KE (2020) Recent advances in bioplastics: application and biodegradation. Polymers (Basel) 12. https://doi.org/10.3390/POLYM12040920

Nascimento EM d GC d, CWP C, Takeiti CY et al (2012) Use of sesame oil cake (*Sesamum indicum* L.) on corn expanded extrudates. Food Res Int 45:434–443. https://doi.org/10.1016/j.foodres.2011.11.009

Newson WR (2012) Protein based plastics from the residuals of industrial oil crops protein based plastics from the residuals of industrial oil crops

Onsaard E, Pomsamud P, Audtum P (2010) Functional properties of sesame protein concentrates from sesame meal. Asian J Food Agro-Industry 3:420–431

Orliac O, Rouilly A, Silvestre F, Rigal L (2002) Effects of additives on the mechanical properties, hydrophobicity and water uptake of thermo-moulded films produced from sunflower protein isolate. Polymer (Guildf) 43:5417–5425. https://doi.org/10.1016/S0032-3861(02)00434-2
Orliac O, Silvestre F, Rouilly A, Rigal L (2003) Rheological studies, production, and characterization of injection-molded plastics from sunflower protein isolate. Ind Eng Chem Res 42:1674–1680. https://doi.org/10.1021/ie020913x
Patel A (2015) MRAGJ. Bio-plastics: a value added products from renewable resources. A Peer Rev Int J Asian Acad Res Assoc:166–184
Patel AV, Panchal T, Thomas M (2016) Preparation and characterization of biodegradable packaging film using groundnut protein isolate. In: 24th European biomass conference and exhibition, Amsterdam, The Netherlands. https://doi.org/10.5071/24thEUBCE20163BO.CITATIONS
Prakasam M, Locs J, Salma-Ancane K et al (2017) Biodegradable materials and metallic implants: a review. J Funct Biomater 8:1–15. https://doi.org/10.3390/jfb8040044
Rouilly A, Rigal L (2002) Agro-materials: a bibliographic review. J Macromol Sci—Polym Rev 42:441–479. https://doi.org/10.1081/MC-120015987
Rouilly A, Mériaux A, Geneau C, Silvestre F, Rigal L (2006) Film extrusion of sunflower protein isolate antoine. Wiley Intersci:1635–1640. https://doi.org/10.1002/pen
Sandhu KS, Sharma L, Charanjiv Singh A (2017) Plant biotechnology: recent advancements and developments. Plant Biotechnol Recent Adv Dev:1–390. https://doi.org/10.1007/978-981-10-4732-9
Sarker AK, Saha D, Begum H et al (2015) Comparison of cake compositions, pepsin digestibility and amino acids concentration of proteins isolated from black mustard and yellow mustard cakes. AMB Express:5. https://doi.org/10.1186/s13568-015-0110-y
Sharma L, Singh C (2016) Sesame protein based edible films: development and characterization. Food Hydrocoll 61:139–147. https://doi.org/10.1016/j.foodhyd.2016.05.007
Shivam P (2016) Recent Developments on biodegradable polymers and their future trends. Int Res J Sci Eng Int Res J Sci Eng 4:17–26
Swati SS, Das M (2015) A brief overview: present status on utilization of mustard oil and cake. Indian J Tradit Knowl 14:244–250
Tang JW, Sun H, Yao XH et al (2012) Effects of replacement of soybean meal by fermented cottonseed meal on growth performance, serum biochemical parameters and immune function of yellow-feathered broilers. Asian-Australasian J Anim Sci 25:393–400. https://doi.org/10.5713/ajas.2011.11381
Tokiwa Y, Calabia BP, Ugwu CU, Aiba S (2009) Biodegradability of plastics. Int J Mol Sci 10:3722–3742. https://doi.org/10.3390/ijms10093722
Vert M (2009) Degradable and bioresorbable polymers in surgery and in pharmacology: beliefs and facts. J Mater Sci Mater Med 20:437–446. https://doi.org/10.1007/s10856-008-3581-4
Vroman I, Tighzert L (2009) Biodegradable polymers. Materials (Basel) 2:307–344. https://doi.org/10.3390/ma2020307
Yue HB, Cui YD, Shuttleworth PS, Clark JH (2012) Preparation and characterisation of bioplastics made from cottonseed protein. Green Chem 14:2009–2016. https://doi.org/10.1039/c2gc35509d
Zárate-Ramírez LS, Romero A, Bengoechea C et al (2014) Thermo-mechanical and hydrophilic properties of polysaccharide/gluten-based bioplastics. Carbohydr Polym 112:16–23. https://doi.org/10.1016/j.carbpol.2014.05.055
Zhang H, Mittal G (2014) Effect of hydrothermal carbonization reaction parameters on. Environ Prog Sustain Energy 33:676–680. https://doi.org/10.1002/ep.11974

Seed Gums: Sources, Applications, and Recent Trends in Edible Films

14

Para Dholakia, Prabhjot Kaur Sabharwal, and Vandana

Abstract

Natural gums are essentially polysaccharides which can be extracted from their source and used for various functions in food. They tend to attract water molecules and form gels; thus, they increase the viscosity of the solution, act as thickeners, emulsifiers, encapsulate flavors, colors, vitamins, and antioxidants, and have wide pharmaceutical applications. A relatively recent development is their use in the formulation of edible films for packaging and coating fresh foods like fruits and vegetables. In the present scenario of global burden of plastics, the need of the hour is to produce safe, biodegradable food packaging; this demand is met by gums. Seed gums are basically galactomannans that are obtained from seed endosperm. Guar gum and locust beans are well-known sources; however, a large variety of seed gums are fast emerging. Many of these are used for edible film production and studies have been done on their effect on shelf life, quality retention, and sensory properties of fresh fruits and vegetables and other foods. Inclusion of plasticizers like glycerol, essential oils, and antimicrobial compounds in the films made from seed gums has shown an encouraging effect on the shelf life of fresh food commodities. With a lot of research going on in this area, seed gums are surely going to lead the quest for an environmentally safe, toxin-free food packaging.

Keywords

Seed gums · Guar gum · Locust bean · Biodegradable · Glycerol · Edible films

P. Dholakia (✉) · P. K. Sabharwal · Vandana
Department of Food Technology, Shaheed Rajguru College of Applied Sciences for Women, University of Delhi, New Delhi, New Delhi, India

A. Poonia, T. Dhewa (eds.), *Edible Food Packaging*,
https://doi.org/10.1007/978-981-16-2383-7_14

We all have heard about gum arabic, carob bean gum, and guar gum. These are food gums that are in use for hundreds of years. Trees, seeds, roots, and seaweed are the natural sources of these gums.

Natural gums are essentially polysaccharides, which can cause an increase in a solution's viscosity, even at a minute concentration. They are nontoxic, odorless, less expensive, inert, and available widely. Mostly, they are botanical gums found in the woody portion of plants or in seed coatings. The name "gum" generally represents a group of polysaccharides that are found naturally. They form gels or a viscous or thick solution with water, and act as stabilizers in the emulsion system. They do not dissolve in oils or organic solvents like ether and alcohol. Sugars such as arabinose, galactose, mannose, and glucuronic acid are obtained when these composite mixtures are hydrolyzed (Coppen 1995). They are also known as hydrocolloids as they are water soluble.

Their ability to act as emulsifiers, stabilizers, thickeners, and fiber in a number of products makes them stand out. They have shown a lot of promise in the development of edible coatings for foods. This is because of their power to attract water molecules, unique film-forming abilities, and encapsulation of a variety of substances like flavors, aromas, and nutraceutical or medicinal compounds. Food gums may be primarily classified into four: seed gums, plant exudate gums, tree or shrub exudates and microbial fermentation gums, on the basis of their source.

In this chapter, we will discuss seed gums and their applications in the food industry, particularly in the formulation of edible coatings and coatings for packaging edibles.

14.1 Seed Gums

Seed galactomannans are commercially known as seed gums. Galactomannan polysaccharides are obtained from mainly two sources: plant seed endosperm and microbes, in particular, yeast and other fungi. However, seed gums are preferred over microbial sources as their production can be customized as per the demand of the industry (Kapoor and Srivastava 2005). These are usually obtained from leguminous plants; however, there are also some nonleguminous sources of galactomannans.

Seed gums come under the category of hydrocolloids; their role in the seed is to strengthen the matrix. They are extensively distributed in nature. These are linear, substituted gums or polysaccharides with linear backbones and sugar substituents that occur mainly on C-6; the sugar substituent in seed gums is galactose, a monosaccharide present on 16–50% of backbone sugar units (Nieto 2009). They comprise of "a linear β-(1 → 4)-mannane backbone attached to single D-galactopyranosyl residues via α-(1 → 6) linkages together with side chains of galactose" (Mirhosseini and Amid 2012a).

The ratio of mannose to galactose and the percentage of galactose affect the structure of the galactomannans, which in turn is responsible for their properties and interaction with other polysaccharides. This ensures their wide industrial applications in food, pharmaceuticals, and cosmetics. They are used in tablet coatings, microcapsules, thickeners, and stabilizers in emulsions, gelling agents, as dietary fiber, and fat replacers in foods, in bakery foods, icings, coffee whiteners, soups and sauces, and in frozen and cured meat products.

Most commonly used seed gums are as follows: guar, carob or locust bean, and cassia, but a number of nontraditional sources of gums are also emerging and there is a lot of research going on in this area. We will discuss these one by one.

14.1.1 Traditional Sources of Seed Gums

Guar gum, locust bean gum, and cassia seed gum are the most common or traditional sources of seed gums. These come under the family Leguminosae.

14.1.1.1 Guar Gum

Guar gum is obtained from the plant *Cyamopsis tetragonoloba*, a plant that is resistant to drought. Indian cluster bean, guar, and guaran are the names by which the bean, flour, and the galactomannan fraction are known (Mudgil and Barak, 2014). Gaur gum is manufactured by separating the endosperm from the hull and germ. After the separation, the dried and powdered endosperm is sold as guar gum. The large amount of branching in its structure is responsible for its ability to hydrate easily and form hydrogen bonds with other polymers. These properties make it useful as a thickener, emulsifier, and binder in food industries. The composition of guar gum is given in Table 14.1.

14.1.1.2 Locust Bean Gum

The locust or carob tree (*Ceratonia siliqua* L.) is grown in Mediterranean countries. The gum obtained is a "heteropolysaccharide galactomannan extracted by mashing the seed endosperm of the fruit pod of the carob tree (*C. siliqua* L.)" (Dakia et al. 2008).

The composition of locust bean gum is given in Table 14.2.

The structure of the bean is similar to guar gum but with less numbers of galactose units. The basic structure of the galactomannan extracted from carob bean is a linear chain of 1-4 ß-D-mannopyranosyl units associated with side chains of 1-6-linked α-D-galactopyranosyl (Barak and Mudgil 2014).

Table 14.1 Composition of guar gum

Component	Amount (%)
Galactomannan	75–85
Protein	5.0–6.0
Moisture	8.0–14
Fiber	2.0–3.0
Ash	0.5–1.0

Source: Chudzikowski (1971)

Table 14.2 Composition of commercial carob or locust bean gum

Component	Amount (%)
Galactomannan	80–85
Moisture	10–12
Protein	5.0
Acid insoluble ash	1.0
Crude fiber	1.0

Source: Barak and Mudgil (2014)

Rheological properties of locust bean gum significantly depend on its chemical and molecular structure (Kök 2007). The water solubility of the gum is affected by the presence of galactose units on the mannose chain. Therefore, the solubility of locust bean gum is less than guar gum as the galactose:mannose ratio is high (3.5), whereas for guar gum it is 1.8. Guar gum can be dissolved in cold water, whereas locust beans have to be slightly heated to achieve full solubility (Maier et al. 1993). Locust or carob bean gum is used in various food products for different reasons. It imparts richness to cream cheese spread, smoothness to sauces, reduces ice crystal formation in ice creams, and prevents syneresis in yoghurt and gels (Ünal et al. 2003). Its use in edible films has been studied (Cerqueira et al. 2011). There are some nonfood uses also like in cosmetics, medicinal products, and paper and textile manufacturing (Dakia et al. 2008).

14.1.1.3 Cassia Seed Gum

Cassia is thought to be a "non-conventional" resource for seed gums. *Cassia tora* and *Cassia obtusifolia,* of the legume family, are the sources of the cassia gum (Mahungu and Meyland 2008). With a mannose: galactose ratio of 3:1, the solubility of the gum in water is lesser than that of carob and other gums (Saha et al. 2011). There are similarities between carob and cassia gum with respect to some properties. They are a little expensive in terms of production cost and not as easily available as locust bean and guar gums. Cassia seed gum may be used in food products as a thickening and gelling agent (Embuscado and Huber 2009). There are nonfood uses also in the cosmetic, pharmaceutical, and other industrial products such as paint, paper, adhesive, textile, and explosives (Singh et al. 2009). The gum has a possibility of use in edible films with incorporation of plasticizers like glycerol, sorbitol, oleic acid, linoleic, and linolenic acid, which improve film properties.

14.1.2 Nontraditional Sources of Seed Gums

There are some nontraditional sources of galactomannans from the seeds of other legume plants, which can be substituted for conventional galactomannan sources. Some of these are listed in Table 14.3.

Table 14.3 Seed gums from nontraditional sources and their uses in foods

Seed gum	Source	Uses	Source
Tara Gum	Tara tree	Thickening and stabilizing agent	Antoniou et al. (2015)
Mesquite seed gum	Mesquite pods	Binder and emulsifier, confectionery, and microencapsulated products	Mirhosseini and Amid (2012b) and Saha et al. (2017)
Fenugreek seed gum	Fenugreek seeds	Less used in food industry	Saha et al. (2017)
Tamarind gum	Tamarind seed	Stabilizer, binder, thickener, and gelling agent	Mirhosseini and Amid (2012b)
Sesbania seed gum	Sesbania seed	Suspending agent, stabilizer, and thickener	Patel et al. (2009)
Psyllium seed gum	Psyllium seeds	Stabilizer, gelling agent, and medicinal uses	Saha et al.(2017)

14.1.3 Other Seed Gums

Various seeds are being explored for their gums, such as durian seed, epsina corona, cress, sage, malva nut, chia, flax, basil, corn, marigold flower, quince, and soybean. Some of these have gelling and thickening properties and, therefore, used as a substitute of locust bean gum and guar gum. Many of them have shown good potential in developing edible coatings. In the next section, we will discuss the use and potential of some of the seed gums in the development of edible coatings for foods.

14.2 Edible Coatings from Seed Gums

Fresh fruits and vegetables lose moisture and firmness quickly during handling, transport, and storage. This seriously damages their marketability and affects shelf life. A lot of research is going on in the use of gums in developing edible coatings having bioactive agents to improve the organoleptic properties of fresh food like fruits and vegetables, shrimps, beef, etc. as well as augment their nutritional value and keeping quality.

Edible coatings are used in fruits and vegetables to lengthen their shelf life. Gums have appeared on the forefront as an ingredient in the formulations for coatings that are edible. This is especially useful for use in fresh produce like fruits and vegetables. They have shown good barrier properties in low humidity environments. These gums are superior to plastics as their availability is good, they are relatively cheaper, and the most important of all, they are biodegradable and, thus, environment friendly. The films synthesized from natural gums could have many advantages over the traditional films, like better shelf life, delay in the ripening process, decreased respiration rate, biodegradability, economic affordability, better moisture barrier properties, and ability to act as vehicles for additives such as vitamins, antioxidants, nutraceuticals, flavors, colors, and antimicrobial compounds. These substances get adhered to the food surface when they are applied through these films.

Thus, not only they prolong the shelf life but also enhance the organoleptic quality and nutritive value.

> Edible coatings represent a thin layer of edible material, which forms a protective film/cover/ layer on the surface of fresh fruits and vegetables to extend their post-harvest quality and shelf life—(Saha et al. 2017 p. 960).

Three significant characteristics of gums make them very useful in edible coating applications: first—their hygroscopicity, which is evident in their strong ability to bind water and ability to make a viscous solution; second—less number of calories that they contribute in the diet because of the restricted digestion and assimilation in the body; and third—and right now our main concern, their excellent film-forming ability. The capacity to form a film comes due to the presence of a huge number of hydrophilic units such as hydroxyl functional groups and other polar groups in their structure which form hydrogen bonds (Janjarasskul and Krochta, 2010). The rheological properties of the gums are affected by various factors like molecular size, shape, water-binding property, conformation, etc. They also depend on the viscosity of the dispersing medium (Hershko et al. 1996).

14.2.1 Guar Gum and Locust Bean Gum in Edible Coating Formulations

Guar gum has shown to delay ripening and extend shelf life in tomatoes, cucumber, prickly pear, and pumpkin in studies (Saha et al. 2016; Ghosh et al. 2014). Guar gum films incorporated with spice extract have shown to delay postharvest changes and also provided bactericidal properties (Naeem et al. 2019). Fruit firmness was maintained and other postharvest changes were delayed in cherries by using a coating of guar gum and the extract of ginseng (Dong and Wang 2018). Several other researchers have developed coatings of guar gum with essential oils, sesame protein, etc. Guar gum coatings have also shown reduced oil absorption in banana chips making it healthier (Sothornvit 2011). The unique physicochemical properties make guar gum a common choice for developing edible films. Best thing about guar gum is its easy availability and biodegradability. It is also less costly than other gums. Its ability to form extensive cross linkages also makes it an ideal vehicle for supplementing the edible film with antimicrobials, essential oils, etc. as mentioned before. However, low mechanical strength and higher permeability to water vapor are some of the drawbacks. Many researchers have worked on treatments like UV curing and thermal processing (Micard et al. 2000), gamma irradiation (Ibrahim 2011), and addition of various additives (Rhim et al. 2006) to overcome these drawbacks. One percent solution of guar gum with inclusion of 1% potassium sorbate solution showed antifungal activity and prolonged shelf life in cut and fresh tomato and cucumber (Mehyar et al. 2014).

Locust bean or carob gum has shown better mechanical and film-forming properties than guar gum due to its lower galactose substitution or higher mannose-to-galactose ratio (Liu et al. 2020). It shows synergistic effect when used along with guar gum. It has also shown synergistic effect with whey protein isolate films as shown in the study done by Silva et al. (2016), in which incorporation of locust bean gum in whey protein isolate film increased the barrier to oxygen and carbon dioxide and increased the resistance to deformation. Similar effects have been shown with locust bean gum and κ-carrageenan (Martins et al. 2012). Sensory quality and shelf life have been shown to improve markedly in apples and mandarins with locust bean-based films. Incorporation of locust bean gum with lipids enhances film transport properties significantly. Also, these films can incorporate antioxidants, which also improve the color, barrier properties, and antioxidant capacity of edible films (Saha et al. 2017). Locust bean gum and tragacanth-based films exhibited better transparency, moisture barrier, and mechanical properties than tragacanth alone in a study (Mostafavi et al. 2016).

14.2.2 Emerging or Nontraditional Gums in Edible Film Formulations

Various gums other than guar gum and carob gum, discussed in this chapter, have shown good potential in their use as edible films, individually and in combination. Table 14.4 lists some of these seed gums and the studies done on edible film for foods.

Let us discuss some of these and more in the following sections.

14.2.2.1 Tara Gum

Tara gum is extracted from *Caesalpinia spinosa*, a tree which is locally grown in Peru and is commonly cultivated in many countries (Pizato et al. 2013). Edible films

Table 14.4 Seed gums in edible coatings

Seed gum	Edible coating use	Source
Tara gum	Peaches	Pizato et al. (2013)
Basil seed gum	Cheese, shrimp, and Peaches	Khazaei et al. (2016), Mazidi and Hosseini Ghaboos (2020)
Almond gum	Sweet cherry, tomato, and banana slices	Mahfoudhi and Hamdi (2015), Mahfoudhi et al. (2014), and Farahmandfar et al. (2017)
Psyllium gum	Papaya, strawberry, and apple	Yousuf and Srivastava (2015) and Banasaz et al. (2013)
Mesquite gum	Persian lime and guava	Tomás et al. (2005) and Bosquez-Molina et al. (2003)
Fenugreek seed gum	Occasionally in meat products	Embuscado and Huber (2009)
Soybean gum	Apple	El-Anany et al. (2009)
Balangu seed gum	Peaches	Mazidi and Hosseini Ghaboos (2020)

have been developed from Tara gum with inclusion of oleic acid with good water vapor barrier properties. Tara gum in combination with carob gum formed tough, flexible films with "high elastic modulus, tensile strength and elongation at break values" (Liu et al. 2020). Incorporation of polyols like glycerol and sorbitol in Tara gum affected the thermomechanical and barrier properties showing effectiveness of Tara gum in edible films (Antoniou et al. 2014).

14.2.2.2 Basil Seed Gum

Biodegradable films of basil seed gum have shown a significant possibility of use in food packaging applications. Incorporating essential oils of *Zataria multiflora* and oregano have shown to improve mechanical and barrier properties of basil seed gum (Gahruie et al. 2017, Hashemi and Khaneghah 2017). Using glycerol solutions at various concentrations and oleic acid incorporation in basil seed gum film has increased moisture barrier and mechanical properties (Salehi 2019). Microbial growth in shrimps in cold storage was reduced by using different concentrations of thymol as plasticizer in basil seed gum film (Khazaei et al. 2016).

14.2.2.3 Almond Gum

Almond gum is exuded from the trunk, branches, and fruits of *Prunusdulcis* trees after a microbial infection or injury. Used as coatings for sweet cherry fruit, almond gum delayed postharvest ripening and ethylene production in the coated fruits (Mahfoudhi and Hamdi 2015). Almond gum coatings in tomatoes delayed changes in color, firmness, titratable acidity, ascorbic acid, and soluble solids content and maintained sensory quality even after 20 days of storage (Mahfoudhi et al. 2014). Almond gum film coating reduced browning index in banana slices after drying (Farahmandfar et al. 2017). Oil absorption was found to be reduced by 34% and sensory characteristics were significantly improved in almond seed gum-coated fried potato chips as compared to the uncoated chips (Bouaziz et al. 2016).

14.2.2.4 Psyllium Gum

Psyllium is produced from the genus *Plantago*. Extensively grown in India and Iran, it has been in use for the management of constipation, colon cancer, diarrhea, high cholesterol, diabetes, and ulcerative colitis (Singh 2007; Ahmadi et al. 2012). High tensile strength, surface hydrophobicity and glass transition point, high flexibility, low water vapor permeability, elongation, and water solubility—all features required for an effective coating film—were found in a film produced from psyllium (Ahmadi et al. 2012). High ascorbic acid retention, firmness, and microbial inhibition in freshly cut stored apples treated with psyllium seed gum were seen (Noshad et al. 2019). Many studies are being carried out on the efficiency of psyllium gum as films for foods, and the future prospects look good.

14.2.2.5 Mesquite Gum

Mesquite is the common name of the gum derived from the *Prosopis* species. With a high amount of protein and polysaccharides, it is a good source of mucilages for making edible films. They are used as emulsifiers for colors and flavors, microencapsulating agents, coating agents in emulsions, stabilizers in colloids, and binders in tablets (Salehi 2019). Studies have been done on mesquite gums to check their film-forming and emulsifying capacity for orange peel oil (Román-Guerrero et al. 2009). Film structure of whey protein isolates with addition of mesquite gums was studied (Osés et al. 2009). Effectiveness of mesquite gum was demonstrated in both studies. Reduced ethylene production and weight loss were seen in stored guava fruit having a film of mesquite and candellila wax (Tomás et al. 2005).

14.2.2.6 Fenugreek Seed Gum

There is a widespread cultivation of fenugreek seeds in Egypt, India, and Middle Eastern countries. It also belongs to the legume family. Its use as stabilizer, emulsifier, and thickener is well documented. However, the films made from fenugreek were not stable; thus, its use in packaging films is limited. Studies have been done on fenugreek gum-based films with incorporation on different types of nanoclays. Good antimicrobial properties in the developed films and up to 5% inclusion of nanoclays improved the mechanical properties (Memiş et al. 2017). Some researchers have used fenugreek-based edible coatings in meat products (Salehi 2019).

14.2.2.7 Soybean Gum

Researchers have extracted soluble soybean polysaccharides from the soybean protein isolate having components: D-galactose, L-arabinose, D-galacturonic, and L-rhamnose. The prominent features of soybean gum are relatively low viscosity and good stability in water-based solutions. They can be used as stabilizers, emulsifiers, and for adhesion in foods. Films made from a mixture of soybean polysaccharides with tragacanth gums gave promising results in extending the sensory and keeping quality of fresh apple slices (Jafari et al. 2018; El-Anany et al. 2009).

14.2.2.8 Balangu Seed Gum

Balangu plant (*Lallemantia iberica* F. & C. M.) from the mint family is originally from the Caucasus and the Middle East and widely available in European and eastern countries. The plant is well known for its use in traditional medicines. As compared to other polymers, this plant is economically affordable, with better efficiency and medicinal properties. Films developed from Balangu seed mucilage with glycerol have superior thermal and mechanical properties (Sadeghi-Varkani et al. 2018). Balangu seed gum films with desired levels of glycerol have shown good potential in edible film production. Coating formulated from Balangu seed gum and basil seed gum was able to delay the ripening process in peaches (Mazidi and Hosseini Ghaboos 2020).

14.2.2.9 Cress Seed Gum

Cress (*Lepidium sativum*) belongs to the Cruciferae family and grows commonly in the USA, Asia, and Europe. It shows a quick water adsorption on soaking in water (Sahraiyan et al. 2013). Edible films made from cress seed gums with varied concentration of glycerol as plasticizer have shown that it can be effectively used to produce films for packaging foods.

14.2.2.10 Flax Seed Gum

Flax seed is a major crop of Canada and cultivated mainly for its oil. The flax seed meal obtained after oil extraction is still rich in essential fatty acids and its mucilaginous material can be extracted for edible film formation. Edible films with good mechanical properties and transparency with inclusion of up to 5% glycerol have been developed showing its potential in this area (Tee et al. 2016). Flax seed gums with lemongrass essential oils have been used for coating pomegranate arils and their sensory and microbial quality studied during storage. The film reduced total plate count and yeast and mold population significantly as compared to controls. Increasing concentration of essential oil delayed the ripening process and reduced weight loss during storage (Yousuf and Srivastava 2017). Flax seed gum-cellulose nanocrystal nanocomposites have shown to be promising materials for use as a sustainable biopolymer for application as bioplastics (Prado et al. 2018).

14.2.2.11 Chia Seed Gum

The polysaccharides of chia seed have shown excellent mucilaginous properties in water. This results in their use as a stabilizer, coating agent, emulsifier, suspending agent, adhesive, and binder. Studies have been done on blending chia seed gum with whey protein concentrate to develop films (Munoz et al. 2012). Also, glycerol at varying levels was used as plasticizer to formulate edible films. These films showed excellent thermal stability (Dick et al. 2015). Inclusion of chia mucilage loaded with oregano essential oil in gelatin films synthesized edible films with better functional and antimicrobial properties (Luo et al. 2019).

14.2.2.12 Lepidium perfoliatum Seed Gum

Lepidium perfoliatum is a flowering plant belonging to *Brassica* (mustard) family, known commonly as clasping pepperweed. It is known as "Quodumeshahri" in Iran where it has been used in traditional medicine for hundreds of years. It is native to Asia and Europe. It can provide viscosity to the solutions and texture to foods. It has thickening, stabilizing, or emulsifying action in foods. Therefore, *Lepidium perfoliatum* is an emerging source of gums having an ability to produce films with superior thermal and mechanical properties. Coatings made from this seed with addition of glycerol at different levels have given promising results in edible film formation (Seyedi et al. 2014).

14.2.2.13 Sage Seed Gum

Sage (*Salvia macrosiphon*) is a plant from the genus *Salvia*. The gum which is extracted from sage seeds is a galactomannan just like the gums obtained from other seeds. Good yield, easy and economical extraction, and comparable functional and rheological characteristics of sage seed gums make them a good choice for use in foods as thickeners and film formers (Amini and Razavi 2020). Razavi et al. (2015) have shown that incorporation of plasticizers like glycerol and sorbitol had resulted in edible films with improved moisture content, flexibility, moisture barrier, and water solubility. In another study, they also showed an improvement in barrier properties of the film made from sage seed gum by inclusion of fatty acids.

14.3 Conclusion

Natural gums obtained from seeds have an edge over the traditional packaging materials that are based on plastics because of their easy accessibility, better economy, and biodegradability. Therefore, with their use, it is easier to meet the consumer expectation for food produce having better keeping quality and freshness. Seed gums are emerging as a very significant source of edible coatings because of their good barrier properties, ability to extensively form cross linkages, and molecular interactions. It has been studied that the ratio of galactose to mannose in the galactomannan can significantly alter the mechanical properties of these films. The research in this area is ever growing as there is a possibility of including bioactive agents in these coatings and improving the organoleptic properties of fresh foods with an added advantage of nutritional enhancement and better shelf life. Studies have been done with various seed gums incorporated with essential oils, cellulose nanocrystals, fatty acids, glycerol, and sorbitols and have shown to improve the mechanical and thermal properties, provide improved barrier to water vapor, and in some cases, better transparency. These films have clearly shown increased shelf life in foods with better retention of sensory quality, delayed ripening in fruits and vegetables, and lowered microbial activity. The use of seed gums is being studied at a large scale and new information is being generated. The ease of using seed gums in edible films, their easy availability and variety, low cost, and their unique ability to make interactions with other materials like bactericidal compounds, spice extracts, proteins, and lipids make them exceptional in the area of edible coatings. It can be said that the future prospects of seed gums as an environment-friendly ingredient of edible films are very high.

References

Ahmadi R, Kalbasi-Ashtari A, Oromiehie A, Yarmand MS, Jahandideh F (2012) Development and characterization of a novel biodegradable edible film obtained from psyllium seed (PlantagoovataForsk). J Food Eng 109(4):745–751

Amini AM, Razavi SMA (2020) Physicochemical characterisation of Salvia macrosiphon gum based edible films incorporated with various fatty acids. Int J Bio Macromol 162:1494–1499

Antoniou J, Liu F, Majeed H, Qazi HJ, Zhong F (2014) Physicochemical and thermomechanical characterization of tara gum edible films: effect of polyols as plasticizers. Carbohydr Polym 111: 359–365. https://doi.org/10.1016/j.carbpol.2014.04.005

Antoniou J, Liu F, Majeed H, Zhong F (2015) Characterization of tara gum edible films incorporated with bulk chitosan and chitosan nanoparticles: a comparative study. Food Hydrocolloids 44:309–319

Banasaz S, Hojatoleslami M, Razavi SH, Hosseini E, Shariaty MA (2013) The effect of psyllium seed gum as an edible coating and in comparison to Chitosan on the textural properties and color changes of Red Delicious Apple. Int J Farm Allied Sci 18:651–657

Barak S, Mudgil D (2014) Locust bean gum: processing, properties and food applications—a review. Int J Bio Macromol 66:74–80. https://doi.org/10.1016/j.ijbiomac.2014.02.017

Bosquez-Molina E, Guerrero-Legarreta I, Vernon-Carter EJ (2003) Moisture barrier properties and morphology of mesquite gum-candelilla wax based edible emulsion coatings. Food Res Int 36 (9–10):885–893

Bouaziz F, Koubaa M, Neifar M, Zouari-Ellouzi S, Besbes S, Chaari F, Ghorbel RE (2016) Feasibility of using almond gum as coating agent to improve the quality of fried potato chips: evaluation of sensorial properties. LWT-Food Sci Technol 65:800–807
Cerqueira MA, Bourbon AI, Pinheiro AC, Martins JT, Souza BWS, Teixeira JA, Vicente AA (2011) Galactomannans use in the development of edible films/coatings for food applications. Trends Food Sci Technol 22(12):662–671
Chudzikowski RJ (1971) Guar gum and its applications. J Soc Cosmet Chem 22(1):43
Coppen JJW (1995) Gums, resins and latexes of plant origin and latexes. Food and Agriculture Organization of the United Nations, Rome
Dakia PA, Blecker C, Robert C, Wathelet B, Paquot M (2008) Composition and physicochemical properties of locust bean gum extracted from whole seeds by acid or water dehulling pre-treatment. Food Hydrocolloids 22(5):807–818
Dick M, Costa TMH, Gomaa A, Subirade M, de Oliveira RA, Flôres SH (2015) Edible film production from chia seed mucilage: effect of glycerol concentration on its physicochemical and mechanical properties. Carbohydr Polym 130:198–205
Dong F, Wang X (2018) Guar gum and ginseng extract coatings maintain the quality of sweet cherry. LWT 89:117–122
El-Anany AM, Hassan GFA, Ali FMR et al (2009) Effects of edible coatings on the shelf-life and quality of Anna apple (Malusdomestica Borkh) during cold storage. J Food Technol 7(1):5–11
Embuscado ME, Huber KC (2009) Edible films and coatings for food applications, vol 9. Springer
Farahmandfar R, Mohseni M, Asnaashari M (2017) Effects of quince seed, almond, and tragacanth gum coating on the banana slices properties during the process of hot air drying. Food Sci Nut 5(6):1057–1064
Gahruie HH, Ziaee E, Eskandari MH, Hosseini SMH (2017) Characterization of basil seed gum-based edible films incorporated with Zatariamultiflora essential oil nanoemulsion. Carbohydr Polym 166:93–103
Ghosh A, Dey K, Bhowmick N, Medda PS, Dutta P (2014) Effect of guar gum as an edible coating to improve shelf life and quality of tomato (Solanumlycopersicum L.) fruits during storage. Ecoscan 6:201–207
Hashemi SMB, Khaneghah AM (2017) Characterization of novel basil-seed gum active edible films and coatings containing oregano essential oil. Prog Org Coatings 110:35–41
Hershko V, Klein E, Nussinovitch A (1996) Relationships between edible coatings and garlic skin. J Food Sci 61(4):769–777
Ibrahim SM (2011) Characterization, mechanical, and thermal properties of gamma irradiated starch films reinforced with mineral clay. J Appl Polym Sci 119(2):685–692
Jafari S, Hojjati M, Noshad M (2018) Influence of soluble soybean polysaccharide and tragacanth gum based edible coating to improve the quality of fresh-cut apple slices. J Food Pro Pres 42(6): e13638
Janjarasskul T, Krochta JM (2010) Edible packaging materials. Ann Rev Food Sci Technol 1:415–448
Kapoor VP, Srivastava M (2005) Seed galactomannans: an overview. Chem Biodiv 2:295–317
Khazaei N, Esmaiili M, Emam-Djomeh Z (2016) Effect of active edible coatings made by basil seed gum and thymol on oil uptake and oxidation in shrimp during deep-fat frying. Carbohydr Polym 137:249–254
Kök MS (2007) A comparative study on the compositions of crude and refined locust bean gum: in relation to rheological properties. Carbohydr Polym 70(1):68–76
Liu F, Chang W, Chen M, Xu F, Ma J, Zhong F (2020) Film-forming properties of guar gum, tara gum and locust bean gum. Food Hydrocolloids 98:105007. https://doi.org/10.1016/j.foodhyd.2019.03.028
Luo M, Cao Y, Wang W, Chen X, Cai J, Wang L, Xiao J (2019) Sustained-release antimicrobial gelatin film: effect of chia mucilage on physicochemical and antimicrobial properties. Food Hydrocolloids 87:783–791

Mahfoudhi N, Hamdi S (2015) Use of almond gum and gum arabic as novel edible coating to delay postharvest ripening and to maintain sweet cherry (P runusavium) quality during storage. J Food Proc Preser 39(6):1499–1508
Mahfoudhi N, Chouaibi M, Hamdi S (2014) Effectiveness of almond gum trees exudate as a novel edible coating for improving postharvest quality of tomato (Solanumlycopersicum L.) fruits. Food Sci Technol Int 20(1):33–43
Mahungu SM, Meyland I (2008) Cassia gum, chemical and technical assessment (CTA). Brussels, Belgium
Maier H, Anderson M, Karl C, Magnuson K, Whistler ROYL (1993) Guar, locust bean, tara, and fenugreek gums. In Ind gums (Elsevier), pp. 181–226
Martins JT, Cerqueira MA, Bourbon AI, Pinheiro AC, Souza BWS, Vicente AA (2012) Synergistic effects between $κ$-carrageenan and locust bean gum on physicochemical properties of edible films made thereof. Food Hydrocolloids 29(2):280–289
Mazidi D, Hosseini Ghaboos SH (2020) Effect of Basil and Balangu seeds gum on sensorial properties, quality and shelf life of peaches. J Food Res 30(3):137–149
Mehyar GF, Al-Qadiri HM, Swanson BG (2014) Edible coatings and retention of potassium sorbate on apples, tomatoes and cucumbers to improve antifungal activity during refrigerated storage. J Food Proc Pres 38(1):175–182
Memiş S, Tornuk F, Bozkurt F, Durak MZ (2017) Production and characterization of a new biodegradable fenugreek seed gum based active nanocomposite film reinforced with nanoclays. Int J Bio Macromol 103:669–675
Micard V, Belamri R, Morel MH, Guilbert S (2000) Properties of chemically and physically treated wheat gluten films. J Agric Food Chem 48(7):2948–2953
Mirhosseini H, Amid BT (2012a) A review study on chemical composition and molecular structure of newly plant gum exudates and seed gums. FRIN 46(1):387–398. https://doi.org/10.1016/j.foodres.2011.11.017
Mirhosseini H, Amid BT (2012b) A review study on chemical composition and molecular structure of newly plant gum exudates and seed gums. Food Res Int 46(1):387–398
Mostafavi FS, Kadkhodaee R, Emadzadeh B, Koocheki A (2016) Preparation and characterization of tragacanth--locust bean gum edible blend films. Carbohydr Polym 139:20–27
Mudgil D, Barak S (2014) Guar gum: processing, properties and food applications—a review. J Food Sci Technol 51:409–418. https://doi.org/10.1007/s13197-011-0522-x
Munoz LA, Aguilera JM, Rodriguez-Turienzo L, Cobos A, Diaz O (2012) Characterization and microstructure of films made from mucilage of salvia hispanica and whey protein concentrate. J Food Eng 111(3):511–518
Naeem A, Abbas T, Ali TM, Hasnain A (2019) Application of guar gum-based edible coatings supplemented with spice extracts to extend post-harvest shelf life of lemon (*Citrus limon*). Qual Assur Safety Crops Foods 11(3):241–250
Nieto MB (2009) Structure and function of polysaccharide gum-based edible films and coatings. In: Embuscado ME, Huber KC (eds) Edible films and coatings for food applications, vol 9. Springer, New York, NY. 57:112
Noshad M, Rahmati-Joneidabad M, Badvi Z (2019) Effects of natural mucilage as an edible coating on quality improvement of freshly-cut apples. Nut Food Sci Res 6(2):21–27
Osés J, Fabregat-Vázquez M, Pedroza-Islas R, Tomás SA, Cruz-Orea A, Maté JI (2009) Development and characterization of composite edible films based on whey protein isolate and mesquite gum. J Food Eng 92(1):56–62
Patel GC, Patel MM et al (2009) Preliminary evaluation of sesbania seed gum mucilage as gelling agent. Int Pharm Tech Res 1:840–843
Pizato S, Cortez-Vega WR, de Souza JTA, Prentice-Hernández C, Borges CD (2013) Effects of different edible coatings in physical, chemical and microbiological characteristics of minimally processed peaches (P runuspersica L. B atsch). J Food Safety 33(1):30–39

Prado NS, Silva ISV, da Silva TAL, de Oliveira WJ, Motta LA d C, Pasquini D, Otaguro H (2018) Nanocomposite films based on flaxseed gum and cellulose nanocrystals. Mater Res 21(6). https://doi.org/10.1590/1980-5373-MR-2018-0134

Razavi SMA, Amini AM, Zahedi Y (2015) Characterisation of a new biodegradable edible film based on sage seed gum: influence of plasticiser type and concentration. Food Hydrocolloids 43: 290–298

Rhim JW, Hong SI, Park HM, Ng PKW (2006) Preparation and characterization of chitosan-based nanocomposite films with antimicrobial activity. J Agric Food Chem 54(16):5814–5822

Román-Guerrero A, Orozco-Villafuerte J, Pérez-Orozco JP, Cruz-Sosa F, Jiménez-Alvarado R, Vernon-Carter EJ (2009) Application and evaluation of mesquite gum and its fractions as interfacial film formers and emulsifiers of orange peel-oil. Food Hydrocolloids 23(3):708–713

Sadeghi-Varkani A, Emam-Djomeh Z, Askari G (2018) Physicochemical and microstructural properties of a novel edible film synthesized from Balangu seed mucilage. Int J Bio Macromol 108:1110–1119

Saha S, Gupta A, Singh SRK, Bharti N, Singh KP, Mahajan V, Gupta HS (2011) Compositional and varietal influence of finger millet flour on rheological properties of dough and quality of biscuit. LWT-Food Sci Technol 44(3):616–621. https://doi.org/10.1016/j.lwt.2010.08.009

Saha A, Tyagi S, Gupta RK, Tyagi YK (2016) Guar gum based edible coating on cucumber (*Cucumis sativus* L.). Eur J Pharm Med Res 3(9):558–570

Saha A, Tyagi S, Gupta RK, Tyagi YK (2017) Natural gums of plant origin as edible coatings for food industry applications. Crit Rev Biotechnol:8551. https://doi.org/10.1080/07388551.2017.1286449

Sahraiyan B, Naghipour F, Karimi M, Davoodi MG (2013) Evaluation of Lepidium sativum seed and guar gum to improve dough rheology and quality parameters in composite rice--wheat bread. Food Hydrocolloids 30(2):698–703

Salehi F (2019) Characterization of new biodegradable edible films and coatings based on seeds gum : a review. J Packaging Technol Res 3(2):193–201. https://doi.org/10.1007/s41783-019-00061-0

Seyedi S, Koocheki A, Mohebbi M, Zahedi Y (2014) Lepidiumperfoliatum seed gum: a new source of carbohydrate to make a biodegradable film. Carbohydr Polym 101:349–358

Silva KS, Mauro MA, Gonçalves MP, Rocha CMR (2016) Synergistic interactions of locust bean gum with whey proteins: effect on physicochemical and microstructural properties of whey protein-based films. Food Hydrocolloids 54:179–188. https://doi.org/10.1016/j.foodhyd.2015.09.028

Singh B (2007) Psyllium as therapeutic and drug delivery agent. Int J Pharm 334(1–2):1–14

Singh V, Srivastava A, Tiwari A (2009) Structural elucidation, modification and characterization of seed gum from Cassia javahikai seeds: a non-traditional source of industrial gums. Int J Bio Macromol 45(3):293–297

Sothornvit R (2011) Edible coating and post-frying centrifuge step effect on quality of vacuum-fried banana chips. J Food Eng 107(3–4):319–325

Tee YB, Wong J, Tan MC, Talib RA (2016) Developent of edible film from flaxseed mucilage. Bio Res 11(4):10286–10295

Tomás SA, Bosquez-Molina E, Stolik S, Sánchez F (2005) Effects of mesquite gum-candelilla wax based edible coatings on the quality of guava fruit (Psidiumguajava L.). J de Phys IV (Proc) 125: 889–892

Ünal B, Metin S, Işıklı ND (2003) Use of response surface methodology to describe the combined effect of storage time, locust bean gum and dry matter of milk on the physical properties of low-fat set yoghurt. Int Dairy J 13(11):909–916

Yousuf B, Srivastava AK (2015) Psyllium (Plantago) gum as an effective edible coating to improve quality and shelf life of fresh-cut papaya (*Carica papaya*). Int J Biol Biomol Agric Food Biotech Eng 9(7):702–707

Yousuf B, Srivastava AK (2017) Flaxseed gum in combination with lemongrass essential oil as an effective edible coating for ready-to-eat pomegranate arils. Int J Bio Macromol 104:1030–1038

Part III

Shelf-Life, Safety and Improvement Packaging Edibles

Edible Packaging: A Review on Packaging Strategy Through Novel Sources, Processing Techniques and Its Sustainability

15

Pratibha Tiwari, Vatsala Sharma, and Monika Thakur

Abstract

There is an increase in the growth of edible films and coatings because of rising awareness among consumers and also because of progress in material science and processing technology. Additionally, it is potential to reduce environmental impact by diminishing the complexity, more sustainability and thus improving the recyclability of materials, compared to the more traditional non-environmentally friendly materials, and may be able to substitute such synthetic polymers. Due to the favourable potential as contemporary food packaging systems, unconventional sources of edible materials, as well as the new processing techniques, it is a matter of great interest. A large number of bio-based polymers have been utilized in the development of edible films and coatings. Furthermore, it is observed that for each kind of foods, edible coatings need to be tailored considering its effects and implication on several food models. Thus, this chapter also aims to explore the concept of edible packaging for food, their sustainability and industrial applications.

Keywords

Edible films · Edible coatings · Sustainability · Packaging

P. Tiwari
Maharshi Markandeshwar University, Maulana, Ambala, Haryana, India

V. Sharma · M. Thakur (✉)
Amity Institute of Food Technology, Amity University, Noida, Uttar Pradesh, India
e-mail: vsharma6@amity.edu; mthakur1@amity.edu

A. Poonia, T. Dhewa (eds.), *Edible Food Packaging*,
https://doi.org/10.1007/978-981-16-2383-7_15

15.1 Introduction

As a substitute of conventional packaging, where the problems like recycling, disposal, non-renewal sources, non-biodegradable are on the peak, the emerging technologies like organic packaging, MAP and CAP have received significant attention as good alternatives to food quality optimization. Moreover, nowadays consumers are demanding more natural, high quality food packages which do not increase pollution, that too in an inexpensive way. In order to fulfil these demands, observing different aspects of innovative packaging, researchers are now more focussed on developing sustainable, biodegradable substance that can provide food safety along with quality and productivity. In this sense, biopolymers provide an effective alternative to traditional packaging by fabricating biodegradable and edible films (Mahalik and Nambiar 2010).

To address key challenges of conventional packaging, innovative sustainable packaging has received substantial recognition in recent years. Among potential solution, emerging concepts of comestible coating or films provide numerous solutions for long-term crucial issue of environmental impact, food contamination and food borne disease. Edible packaging has been recognized as a possibility to take the edge off the waste and to construct new applications for enhancing parameters in the products such as stability, quality, safety, variety and comfort for consumers.

Edible films and coatings can be understood as fine covering of comestible material applied on a food surface as a coating, or placed (pre-formed) between food components. Moreover, their usefulness as a carrier of anti-fungal and anti-microbials medium to enhance the shelf life of foods products and as carriers of nutrients to increase the nutritional value of final processed food products has received an excellent attention of researcher and consumers. Proving to this statement in recent researches, it has been manifested that edible wrapping provides a replacement and/or fortifies the natural layers at the outer surfaces of food product to prevent moisture losses, gas aromas and solute movements out of the food, while selectively allowing for administered exchange of essential gases, namely, carbon dioxide, oxygen and ethylene, which are involved in food product respiration (Embuscado and Huber 2009). Thus, organoleptic characteristics of packaged foods can be enhanced by supplying various components (flavourings, colourings, sweeteners) (Bourtoom 2008). The need for food protection can further be improved by combining primary edible-packaging with secondary non-edible packaging. The purpose of secondary packaging is usually for controlling and for maintaining and cleanliness. Many new innovative technologies have also been augmented by various researchers in developing world (Thakur and Modi 2020; Thakur et al. 2020).

Edible packaging generally comprises of edible films, pouches, coatings and sheets. Since the films or coatings are produced completely from renewable, edible elements therefore are expected to degenerate more promptly than polymeric materials. Although edible coatings and films have seemed to be similar in their definition, there are dissimilarities. Edible coatings are formed directly onto the food surfaces, while edible films are prepared separately and then applied to the exterior of the food. Therefore, coatings are put in application in liquid forms, while films are acquired as solid laminates and then put onto food stuff (Falguera 2011). Therefore, in this chapter, description of multiplicity of polysaccharides (starch and hydrocolloids), proteins (whey proteins, soybean proteins and fish proteins) and lipids as the important types either independently or in mixtures along with new tendencies based on the incorporation of diverse active compounds (antimicrobials, antioxidants, nutraceuticals), that are applied on fruits and vegetables, are mentioned as well. Furthermore, synergistical approach of active packaging, intelligent packaging and green packaging technologies yielding a multipurpose food-packaging structure with no counteracting interactions between components have also been included as superlative future goal for food packaging technology. In continuation with the new technologies to use edible coatings as porter of functional ingredients, the development of new technologies to improve the delivery attributes of edible films and coatings is one of the issues requiring future research. Thus, future trends on cost cutback and manufacturing in larger scales and on stability and safety necessary for promoting the feasibility of commercialized edible coated fruits with vegetables are discussed as well.

15.2 Salient Feature of Edible Materials

Based on the course of action and the purposive function, substantially, filmogenic solutions or suspensions are used to obtain coatings or films. In order to be adequate to the application coating manner, the rheological behaviour of the filmogenic solution has to be controlled, that is, spraying requires a low viscosity, while a higher viscosity is required for immersion. Lopez et al. (2010) grouped the edible packaging systems based on presence of different sources (Fig. 15.1):

Coating integrity is a critical factor that depends on surface tension, adhesion to food substrate and pliability of the coating itself, and the plasticizer addition. Various characteristics that are associated with biodegradable films are mentioned in Fig.15.2:

15.3 Classification of Edible Packaging Material

Polysaccharides, proteins and lipids are the three types of an edible packaging on the basis of the materials used (Fig. 15.3). The material used for producing edible packaging must meet two criteria: edible and the potentiality to form a continuous layer or film. Proteins, polysaccharides, lipids belong to this group by creating a continuous film or coating.

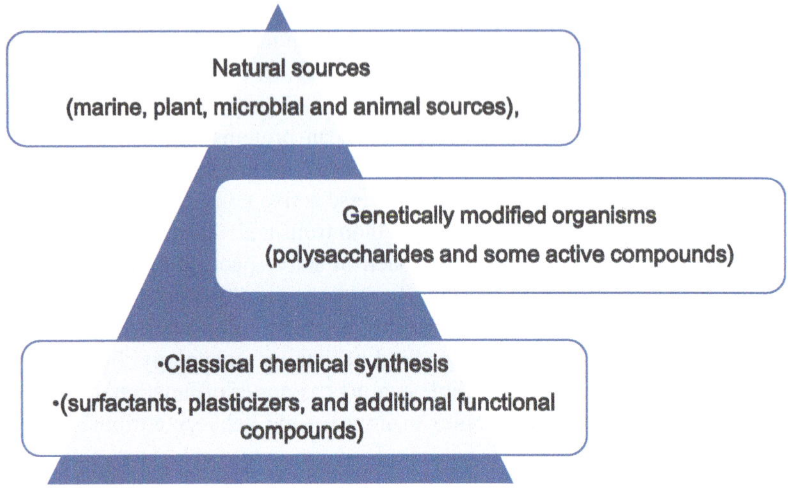

Fig. 15.1 Categorization of edible packaging systems based on sources

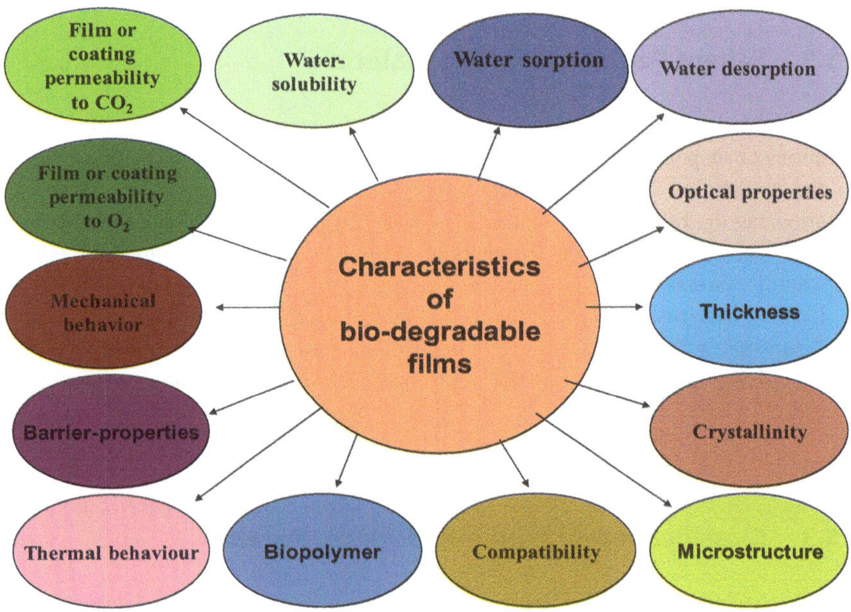

Fig. 15.2 Characteristics which are associated with biodegradable films

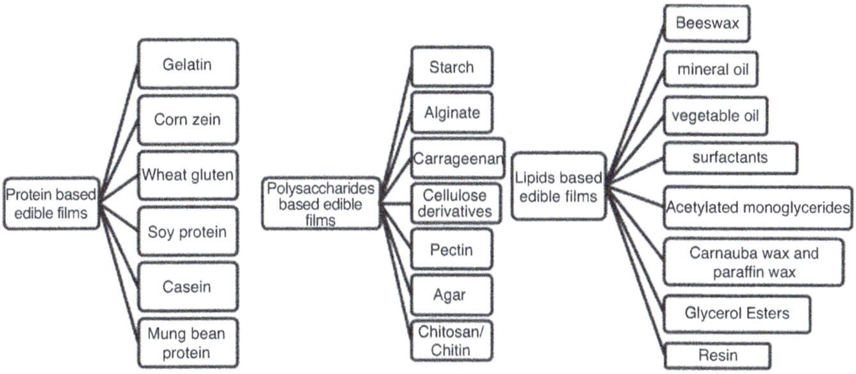

Fig. 15.3 Edible food packaging material classification (Saklani et al. 2019)

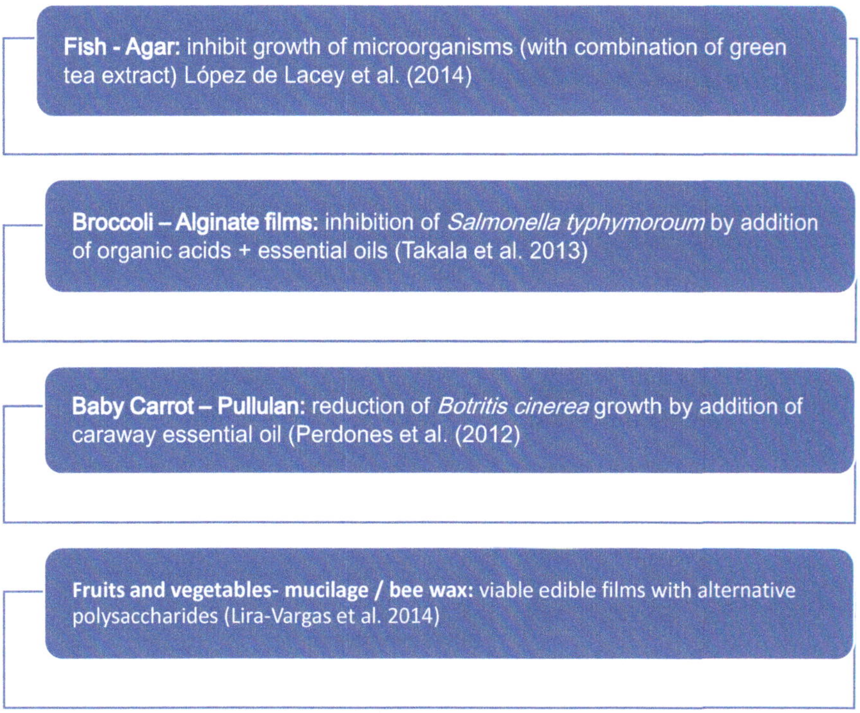

Fig. 15.4 Polysaccharides-based edible films

15.3.1 Polysaccharide-Based Edible Films

Polysaccharide-based films have various specific features like excellent aroma, O_2, oil-barrier properties, strength and also structural integrity to the food but they provide very little resistance to water migration. Various examples in relation with food application are cited below in Fig. 15.4.

15.3.2 Lipid-Based Films

Lipid coating on the food products has barrier properties for water vapour and O_2. Usually, wax films have been much resistant to moisture migration (Raajeswari and Pragatheeswari 2019).

Various combinations in the formulations of lipid-based films have been researched:

- Beeswax or glycerol.
- Palm fruit oil/tween 80.
- Citrus essential oils.
- Glycerol–Chia seed mucilage.
- Shellac/oleic acid–Aloe gel.
- Rice bran oil/glycerin–Whey protein.

15.3.3 Protein-Based Edible Films

These films are used for different food products to minimize the loss of moisture, restrict the absorption of oxygen, reduce the migration of lipids, improve mechanical handling properties, provide physical protection and suggest an alternative to synthetic packaging materials. Collagen (fibrous proteins) has received the most attention in the production of edible films. Few examples of protein-based film applications are cited below in Fig. 15.5.

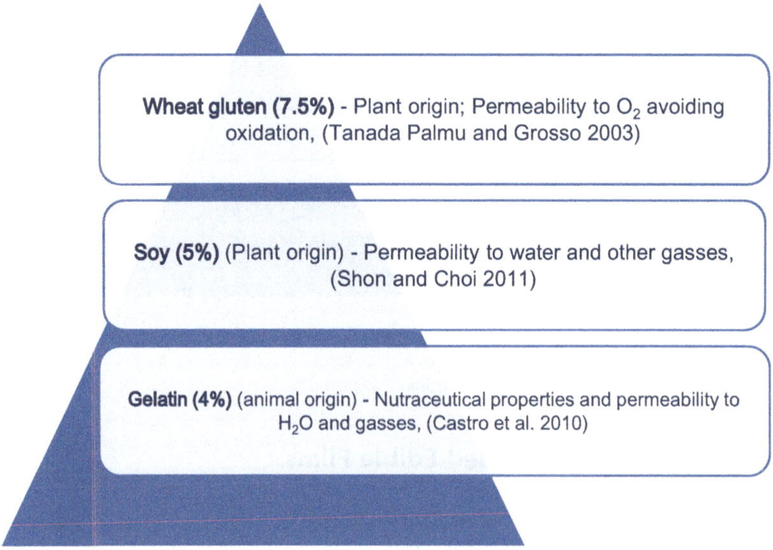

Fig. 15.5 Protein-based edible films

15.4 Antimicrobial Packaging

Food safety has been the prime concern globally and one of the significant goals of the contemporary food legislation. In this sense, edible packaging integrated with antimicrobial composites has the potency to restraint food-borne pathogens activities and reduce the microbiological risks of food products.

Various examples for the use of the anti-microbial packaging are mentioned:

- **Clove essential oil** (Matrix: Glycerol + SPC).
- **Lemongrass + Rosemary + Clove bud essential oils** (Matrix: PSM + red algae).
- **Clove Bud oil** (Matrix:CFP + glycerol + tween 20).
- **Oregano + Cinnamon + Star-anise essential oils** (Matrix: Protein + SCSG + chitosan).

15.5 Novel Sources of Edible Packaging

Nowadays, lot many research efforts are more focused on evolving and conducting novel sources as latest new edible films and coatings (Fig. 15.6). Additionally, the new processing systems are tested in terms of their optimization of the structure and the active components. These materials may exist in the midst of the available food components but have not been analysed as edible packaging compounds or by-products from food industries and waste finding a sustainable application.

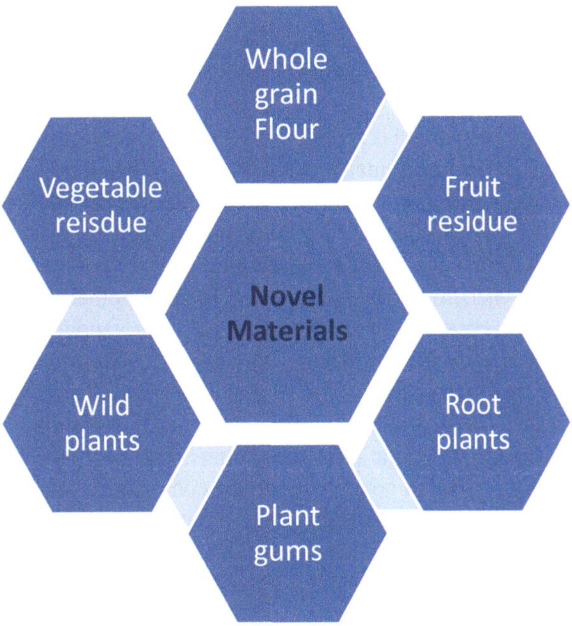

Fig. 15.6 Novel Materials that can be applied as edible packaging and films

López et al. (2010) also reviewed new edible films and coatings, mainly having multi-components, with or without lipids, as well as those included with different active compounds. Various novel sources and their comparison with traditional ones are described below in flow chart (Fig. 15.7).

15.6 Trends and Sustainability in Food Packaging

Nowadays consumers are more aware in terms of expecting high value foods and ready to eat, ready to serve foods with proper food safety concerns and less environmental impacts as well as new manufacturing trends (like moderately preserved, new, palatable and favourable food products with enhanced shelf-life and controlled quality) are the chief forces prompting the unfolding of modern and ingenious packaging approaches which sustain and observe food safety and quality, extend shelf-life and bring down the environmental problem of food packaging (Dainelli et al. 2008).

Consequently, to minimize the use of ordinary, imperishable plastics, the development of edible or biodegradable films or coatings for effective food packaging has aroused considerable interest in recent years. Additionally, the replacement with alternative biodegradable forms fulfilling the food manufacturers requirements, for packaging materials to be food grade, to maintain and enhance product shelf-life stability and safety and to utilize nominal values of packaging would significantly allow improvement in overall operating costs while reducing the waste streams.

15.7 Future Trends

A lot of research is currently on going on the production of edible and biodegradable, but the profit-oriented use of films is still limited at present. However, due to its natural protection to foods according to specific packaging requirements, it is expected that the use of edible/biodegradable packaging substances in foods has a promising and sustainable future. Despite various advantages as presented by researcher, a number of hurdles in development potential have to be overruled, such as cost effectiveness, improved water vapour permeability barrier and technological application methods. Moreover, it should not be utilized singly where there are unsanitary conditions during food handling and development of off flavour further needs greater attention. There have been many hurdles in industrial implementation of edible films or coatings, but they have been the most convenient and useful options in comparison to the conventional plastic packages. The commercialization of edible packaging is still in its infancy stage, so the potential of edible and bio-degradable is still underexploited.

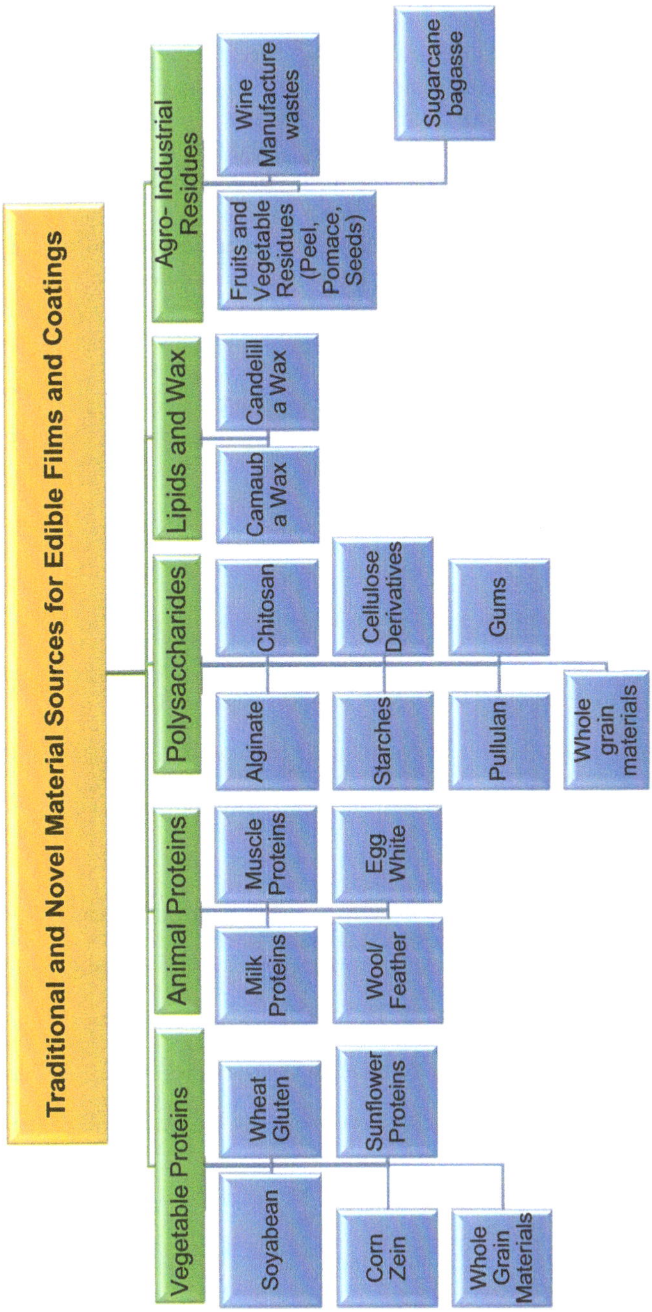

Fig. 15.7 Flow chart for the comparison of traditional and modern novel sources

15.8 Conclusion

The edible films and coatings play a pivotal role as porter of bio-active components in the food products and bio-medical industry. Since they have been a sustainable solution, therefore, they contribute to a very important research now days. Researches have already been increased in the preceding years; however, some research problems are yet to be tackled, especially difficulties in processing due to the recent studies application with wet methods. Moreover, very few industrial applications of edible packaging have been evolved. Various new components should also be investigated to provide bioactive components important in both the food and biomedical sector and also have significant role in building a global sustainable economic system. Even though there have been many hurdles, starting from in vitro trials to the final industrial applications and implementations, the research is still having unlimited wings. There have been many underexploited feasible options in active food packaging or its use in biomedical parameters instead of conventional plastic packages. Hence, many new avenues in the sustainable use of edible coatings and films have been explored.

References

Bourtoom T (2008) Edible films and coatings: characteristics and properties. Int Food Res J 15:237–248

Dainelli D, Gontard N, Spyropoulos D, Zondervan-van den Beuken E, Tobback P (2008) Active and intelligent food packaging: legal aspects and safety concerns. Trends Food Sci Technol 19 (1):S103–S112

Embuscado ME, Huber KC (2009) Edible films and coatings for food applications. Springer, New York

Falguera V (2011) Edible films and coatings: Structures, active functions and trends in their use. Trends Food Sci Technol 22(6):292–303

López O, García MA, Zaritzky NE (2010) Novel sources of edible films and coatings. Stewar Posthar Rev 3:1–8

Mahalik NP, Nambiar AN (2010) Trends in food packaging and manufacturing systems and technology. Trends Food Sci Technol 21(3):117–128

Raajeswari P A, Pragatheeswari R (2019) Edible packaging and market overview. Food Marketing & Technology, Indian addition, article

Saklani P, Siddhnath S, Kishor D, Shiv MS (2019) A review of edible packaging for films. Int J Curr Microbiol App Sci 8(7):2885–2895

Thakur M, Modi VK (2020) Emerging Technologies in Food Science- focus on the developing world. Springer Nature

Thakur M, Modi VK, Khedkar R, Singh K (2020) Sustainable food waste management: concepts and innovations. Springer Nature

Nanosized Additives for Enhancing Storage Quality of Horticultural Produce

16

T. S. Hanumesh Gowda, Priya Pal, S. Vijay Rakesh Reddy, B. N. Ashwija, and Shalini Gaur Rudra

Abstract

Use of nanosized materials for preventing spoilage and active packaging of horticultural produce has increased rapidly over the past two decades. Nanoparticles, like metal oxides, nanoemulsions, and nanocomposites, have the ability to enhance mechanical and barrier properties while acting as antimicrobial agent. The presence of higher surface to volume ratio provides better contact with the microorganism and consequently long-lasting antimicrobial efficacy. Present chapter encompasses recent advances in the area of application of nanosized additives for antimicrobial activity, and nanocomposite active and smart packaging of fruits, vegetables, and flowers are presented. The material used, preparation techniques, characterization, efficacy, and degradation are discussed. Moreover, there is little knowledge regarding the pharmacokinetics, toxicity, and pharmacodynamics of nanomaterials in humans, but still there are many conceivable advantages of such technology. Developments on regulatory aspects and specific directives for the same have been included.

Keywords

Active packaging · Biodegradable films · Nanosized materials

T. S. Hanumesh Gowda · P. Pal · B. N. Ashwija · S. G. Rudra (✉)
Division of Food Science and Postharvest Technology, Indian Agricultural Research Institute, New Delhi, India

S. Vijay Rakesh Reddy
ICAR-Indian Agricultural Research Institute, New Delhi, India

A. Poonia, T. Dhewa (eds.), *Edible Food Packaging*,
https://doi.org/10.1007/978-981-16-2383-7_16

16.1 Introduction

Nanotechnology is an emerging approach that refers to understanding, controlling, and mastering the properties of matter at the nanometer scale. Nowadays, nanoparticle science has emerged as one of the most imperative tools in modern agro-horticulture preharvest and postharvest production sectors.

Nanotechnology is an emerging approach in many sectors like medicine, packaging, bioremediation. It comprises of self-assembly or manipulation of discrete molecules or their gatherings for the creation of devices and materials possessing unique properties. This technology has been developed as a result of efforts from the researchers and engineers functioning in separate fields of photonics, microelectronics, biotechnology, and chemistry. It has become an imperative tool in agro-horticulture both in the preharvest and postharvest sectors. The implications of nanotechnology for shelf life extension and management of horticultural can be far-reaching. Nanotechnological developments have resulted from the integration of several branches of science such as biology, chemistry, photonics, biotechnology, and electronic engineering. For example, nanofabrication involves the integration of biology with material science and engineering dc (Opara 2004). Horticulture is a vital part of the wider biological industry. Since the whole biological world resides within the sphere of nanotechnological scale, there is a strong logic in a convergence of nanobiology, biotechnology, and bioengineering to solve the practical problems faced by the horticultural industry.

Nanomaterials have been proven as effective tools for enhancing the shelf life of horticultural crops since they have low dosage requirements, exhibit target specificity, offer controlled or slow-release, and have precise action on active sites. Nanoparticles produced for effective action have other advantages of no pollutant release, less space requirements, and energy efficient.

When the particle size of a substance is compacted under its threshold level, the consequential material would exhibit significantly different physicochemical properties from those of the macroscale substances. Nanosized particles that exhibit greater surface area per unit mass are considered to be biologically more active as compared to their respective larger-sized particles of similar chemistry (Oberdorster et al. 2005). With a reduction in particle size, the water solubility, thermal stability, and overall bioavailability of functional composites can be increased through the improvement of their solubility, delivery properties, and the extended residence time in the gastro intestinal tract and its absorption by cells (Li et al. 2009; Hellmann et al. 2011).

Proteins and sugars are considered as main target-recognition groups by nanoparticles. These can be used to increase the keeping quality of foods in the form of nanosensors. They are also used as protective agents against environmental factors that may damage and decrease the shelf life of concerning foods. These nanoparticles can be used in designing of flavors and antioxidants (Li et al. 2009). Most commonly employed nanoparticles in agriculture are silver ions, gold nanoparticles, polymeric compounds, etc. Sulfur-based nanoparticles have been studied for organic farming to prevent the growth of fungus from apples and grapes

(Gupta et al. 2013). Other potential applications of nanotechnology in horticultural systems include nanocoatings, nucleic acid bio-engineering, bio-analytical nanosensors, and bio-selective surfaces, etc. Low-cost nanosensors are used for detecting horticulturally important pathogens and food-borne pathogens and nanofilters for exclusion of undesirable compounds from the foods and drinks. Micronutrient deficiency and mineral toxicity were found to be the major limitation in horticultural production. Input use efficiency can be attained through the use of nanocarriers for smart delivery to the targeted sites and nanoformulations for controlled chemical release. Nanosensors can be utilized to study the soil nutrient status as well as the soil microflora.

In wake of climate change and the growing needs of the human population, natural resources such as water, nutrients, and fertilizers need to be used judiciously and efficiently in the horticulture systems. This can be realized through the use of nano-based smart delivery systems and nano-based sensors. Quality of the horticultural produce can be monitored through nanobarcodes and other nano-based tracking systems. Thus, the infusion of nanotechnology into horticultural science would serve to ensure global food and nutritional security. This includes an early prediction of the biotic stresses followed by their management, enhancement of input use efficiencies, and shelf-life of the perishables viz. fruits, vegetables, and flowers.

In this chapter, recent developments in the area of application, characterization, and preparation of nanomaterials for increasing shelf life, improving the package, designing smart indicators for horticultural produce shall be elaborated. Biodegradation and regulatory aspects shall also be dealt with.

16.2 Techniques of Nanosizing

The nanoparticles can be prepared by size reduction of bigger materials into smaller particles also called as top-down approach or through joining and clustering of atoms and molecules to form particles also referred as bottom-up approach (Ayuk et al. 2017).

The top-down approach mainly includes mechanical milling and mechanochemical, laser ablation, electrospraying, flame pyrolysis, etching, etc., while bottom-up approach comprises mainly vapor-based techniques, chemical vapor condensation, solution-phase techniques, etc. The bottom-up technique is more common and favorable for the preparation of nanostructures (nanoshel.com) as shown in Fig. 16.1.

16.2.1 Physical Methods

16.2.1.1 High Ball Milling Method

It comprises many hard steel balls or tungsten carbide balls inside a rotating stainless-steel container (drum). The raw material is fed into the rotating steel drum and these balls randomly impact the particles making them a smaller size.

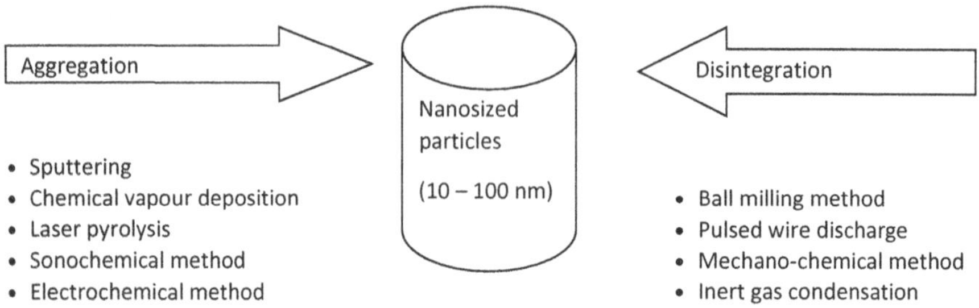

Fig. 16.1 Top-down and bottom-up techniques for Nanosizing

Major factors affecting the final product quality include the type of mill, speed, temperature, size distribution, a weight ratio of ball to powder, time of the process, and extent of filling the vial. However, this process of making nanoparticles is not very efficient, since the production of ultrafine particles is very difficult or takes a very long time. Although if the cost of production and scaling up the process for large-scale production is the ultimate goal, then this technique is advantageous to use (Satyanarayana and Sudhakar 2018).

16.2.1.2 Pulse Laser Ablation

The technique uses a high-power laser beam that is focused inside a vacuum chamber to target the material and strike it, thus creating plasma. This is converted into a colloidal solution of nanoparticles. The second harmonic generation type laser (ND: YAG) is commonly used for the preparation of nanoparticles. It is considered for many technical applications like production of nanomaterial, carbon nanotubes, nanowires, deposition of thin metallic coating, fabrication for superconducting material, dielectric films, etc. Factors affecting the final product of laser ablation include the type of laser, pulsing time and number of pulses, type of solvent (Kim et al. 2017).

16.2.1.3 Pulsed Wire Discharge

This method is employed for generating nanoparticles from metals that are easily available in thin wire form with high electrical conductivity. The pulsed electric field uses the phenomenon of joule heating that arises when a pulse electric current is passed through a thin metal wire, turning it into plasma. On cooling this plasma with ambient gas, a large quantity of fine-sized nanoparticles is generated. This process is not commonly used for industrial purposes due to its relatively high cost and difficulty in using it explicitly for different materials (etigo.nagaokaut.ac.jp).

16.2.1.4 Mechano-Chemical Method

The combination of mechanical and chemical processes encompasses chemical reaction during milling and subsequent heat treatment for nanosizing. This produces ultrafine particles which are recovered by suitable solvents (Yevale et al. 2019). This process is used to synthesize a range of metal nanoparticles like Ag, Cu, Cr, Co, etc., and also other compounds such as oxides and sulfides. During the mechanochemical process, pulverization of reagents and chemical interactions between components results in the formation of disperse particles (Namita 2015).

16.2.1.5 Inert Gas Condensation

This process involves two primary steps: first, evaporation of material and second, rapid condensation under controlled conditions to produce particles of the required size (nanoshel.com). The material is evaporated in the chamber in the presence of pressurized inert gas. The evaporated atoms bombard with gas atoms forming small, discrete crystals by a subsequent decrease in their kinetic energy. The gas atoms heated by evaporation sources result in convection currents and the fine powder is condensed using liquid nitrogen inside the chamber. The size of the crystal obtained is dependent upon evaporation rate, gas pressure, and gas composition. The light inert gases can be used in place of heavy gases for the chemical vapor deposition process to obtain extremely fine particles or a decrease in evaporation rate and gas pressure can aid in the production of particles having a very fine size (Vozga et al. 2020).

16.2.1.6 Chemical Vapor Deposition (CVD)

The process involves the coating of a fine solid film upon a heated surface via vapor or gas phase as a result of a chemical reaction (Creighton and Ho 2001). The deposition of a thin layer on the hot surface is the result of a chemical reaction occurring on or near the hot surface. It is followed by the formation of chemical by-products which are released out of the chamber with unreacted gases. These CVD reaction systems are very complex considering a series of gas-solid phase surface reactions. There can be both vertical and horizontal type CVD to produce materials of the desired quality. This technique produces high-quality and high-performance-based nanoparticles. The reaction is activated by a high temperature of 300 °C to above 900 °C depending upon the type of chemical vapors used, which may be attained by thermal, laser, photo laser, metal-organic, etc. (Namita 2015).

16.2.1.7 Sputtering

It is a physical vapor vacuum deposition process wherein atoms are ejected from the material onto which deposition takes place and condenses ejected atoms on the surface under high vacuum (angstromsciences.com). The sputtering displaces atoms of specific material in the presence of an ionized gas molecule. Prominent methods of sputtering include ion beam, magnetron, and diode sputtering (nanografi.com).

16.2.1.8 Laser Pyrolysis

Laser pyrolysis makes use of a laser beam to selectively heat the gas stream containing nanoparticle precursors thereby decomposing them and inducing nucleation of nanoparticles (Singaravelan and Alwar 2015). The production of nanoparticles occurs abruptly when sufficient supersaturation of condensable products occurs in a vapor phase. Once nucleation sets in, there is rapid particle growth with coagulation/coalescence instead of further nucleation (nanoshel.com). Due to the flexibility of synthesizing particles of diverse materials, this process finds more use. One of the major advantages of this technique over other heating methods is that only the gas portion of the system is heated with the use of less energy for rapid heating and cooling simply mixing heated gas with unheated gas (Malekzadeh et al. 2018).

16.2.2 Chemical Methods

16.2.2.1 Sonochemical Method

Ultrasound radiations (20 kHz–10 MHz) are applied in the sonochemical process to hasten the process and aid in a chemical reaction. The sonochemical method was initially aimed at producing iron nanoparticles but is now used in nanosizing different metals and oxides. The technique involves passing ultrasounds of fixed frequency causing acoustic cavitation through the solution or slurry of a selected metal precursor (Satyanarayana and Sudhakar 2018). The use of volatile organometallics with the sonochemical process is exploited and used as a common process for the production of different nanophase materials by replacing the reaction medium. The main advantage of the sonochemical process involves the simple operating conditions and easy control of nanoparticles with respect to their size by making use of precursors of different concentrations. The process synthesizes nanoparticles by making use of both ultrasonic pulses and electrolytes (Boddolla and Thodeti 2018).

16.2.2.2 Electrochemical Method

The process of electrochemical deposition happens in electrolyte solution at the interface that is mixed with desired metal that is to be deposited and a conductive site for metal substrate. The method has advantages over other bottom-up approaches because it produces nanoparticles of high purity. The size of nanoparticles can be enhanced by controlling the electrolytic concentration and current density applied making it a flexible method for nanoparticle production (Estrin et al. 2009). The method has a disadvantage of excess deposition of metal on the cathode during the electrochemical process, reducing the surface available for particle formation. The particle formation altogether comes to halt as the entire cathode surface gets covered with the metal electrodeposits (Salloom 2017).

16.2.2.3 Chemical Reduction Method

Owing to process simplicity, this is the most commonly used method for nanoparticles. In this method, metal precursors are reduced by making use of reducing agents. Choice of a suitable reducing agent is important as many parameters such as the shape and size of a particle depend on the nature of the reducing agent. This technique provides control over particle size by providing variation in feed rate, the molar concentration of reactant, dispersant. Reaction rate is important during synthesis process. Higher reaction rate leads to speedy formation of a large number of nuclei, resulting in the production of very small particles and vice versa. The copper metal salts are the easiest to produce nanoparticles, with the use of controlled morphology and sizes (nanoshel.com).

16.2.3 Biological Method

The use of bacteria, fungi, plant proteins, yeast, potential viruses as a medium for the production of nanoparticles has been exploited extensively in making a green process for the synthesis of nanomaterials.

Researchers have shown the ability of bacteria to consistently generate nanoparticles with their fabrication extracellularly. Bacillus species have been commonly used to synthesize metal nanoparticles. Besides, bioproduction of nanoparticles with the use of fungi is being investigated due to their higher bioaccumulation, receptivity towards toxicity, simple downstream processing, effortless synthesis, and easy biomass handling. The use of *Aspergillus niger*, *Fusarium*, and *Aspergillus oryzae* has been documented for their extracellular synthesis. Yeasts like *Candida glabrata* and *Schizosaccharomyces pombe* have been used for the production of cadmium nanoparticles. Silver and gold nanoparticles were biosynthesized from extremophilic yeast strain obtained through acid mine drainage. Biological particles such as viruses, peptides, and enzymes could be exploited for the production of nanoparticles. The nucleated nanocrystal growth has been done by peptides (Lade and Shanware 2020). Figure 16.2 depicts various green synthesis approaches for the preparation of metallic nanoparticles.

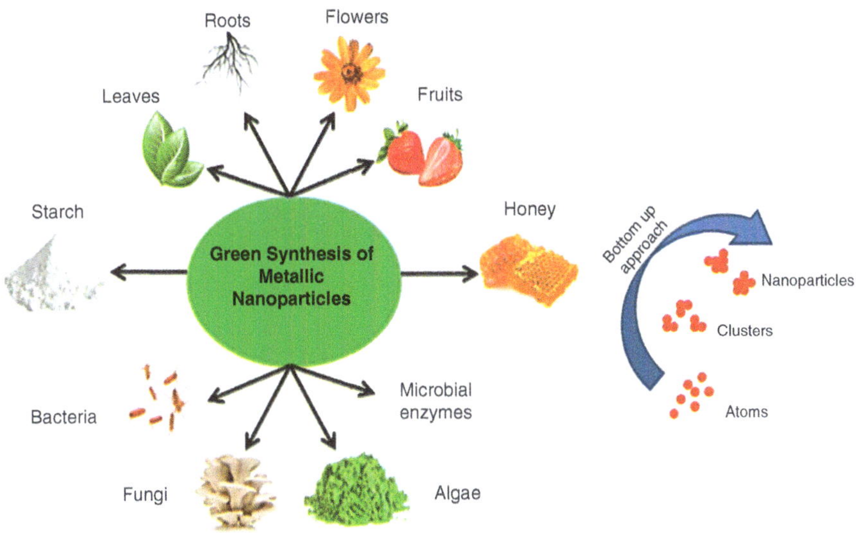

Fig. 16.2 Various green synthesis approaches for preparation of metallic nanoparticles*. Kumar et al. (2020)

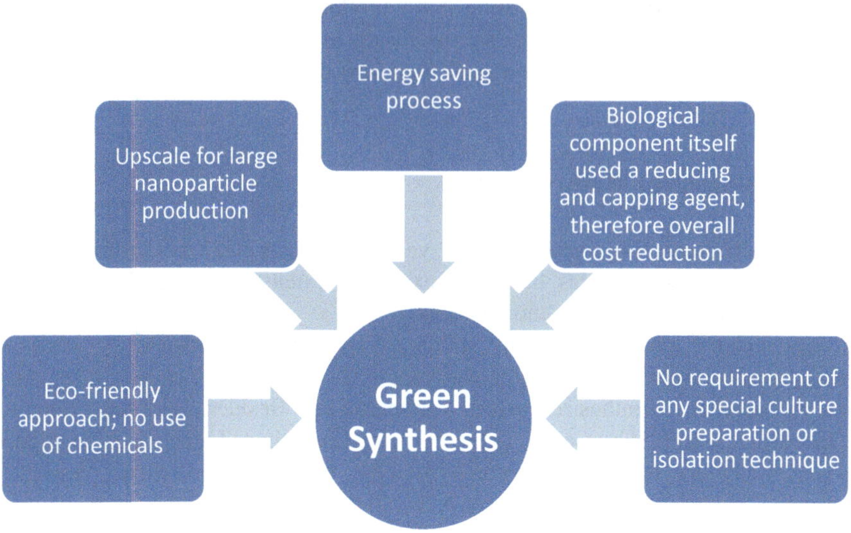

Fig. 16.3 Merits of green synthesis

Silver and gold nanoparticles have been produced from geranium, aloe vera, sundried *Cinnamomum camphora*, and extracts of different flowers due to the toxicant-free nature (Lade and Shanware 2020). Figure 16.3. shows merits of green synthesis.

Biological preparations used for capping and stabilizing are microorganisms, plants parts, enzymes, insects, amino acids, vitamins, tea, winery waste, red grape, and banana.

Biosynthesis of nanoparticles from plant sources has advantages over other processes due to their ecofriendly nature, cost-effectiveness, easy extraction, processing, reproducible, and toxins-free nature (Kumar et al. 2020). A general procedure of plant-based synthesis involves the determined weight of plant part along with extracts preparation in solvents, followed by dilution of filtrate with sterile water and this filtrate acts as reducing agent for the preparation of nanoparticles (Siddiqi and Husen 2017a). Table 16.1 shows biosynthesis of nanomaterials using plants.

16.3 Nanostructures

Synthetically engineered nanomaterials are also available for a variety of horticultural applications such as nanocomposites, nanoemulsions, nanocarriers (nanocapsules), nanosensors, nanobots, nanocides, nanosensors, nanobiosensors, and nanofiltration. Categorization scheme of nanomaterials is shown in Fig. 16.4.

Nanotubes: One of the most researched nanomaterials is nanotubes that too made from carbon due to their mechanical, thermal, and electrical properties (Husen and

Table 16.1 Biosynthesis of nanomaterials using plants

S. no.	Plant species	Plant part	Nanoparticle size (nm)
1.	*Cocos nucifera*	Inflorescence	22
2.	Olive	Extract 1 mL	30
3.	*Allium sativum*	Sucrose and fructose	4–22
4.	*Aloe vera*	Leaves	50–350
5.	*Citrus sinensis*	Peel	10–35
6.	*Acorus calamus*	Rhizome	31.83
7.	*Calotropis procera*	Plant	19–45
8.	*Argyreia nervosa*	Seed	20–50

[a]Source: Lade and Shanware (2020)

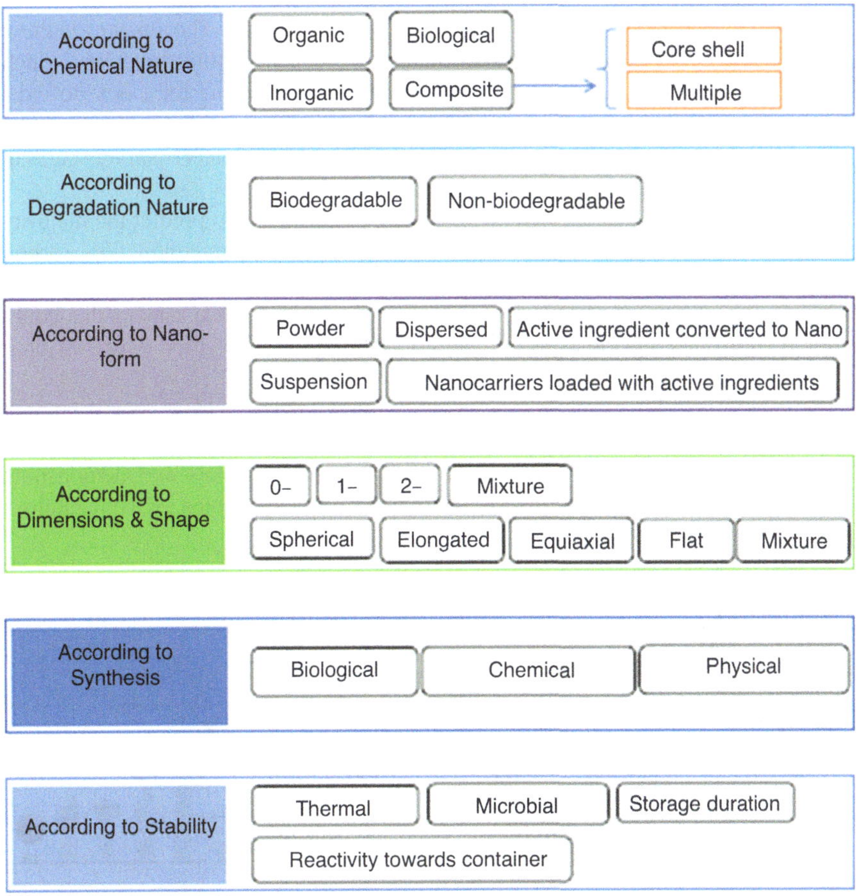

Fig. 16.4 Categorization scheme of nanomaterials. *As per the directions of Government of India (2019) for evaluation of nano-agri input and nano-agriproducts

Siddiqi 2014). The nanotubes are a more advanced generation of nanomaterials and are used by Sumio Lijima of NEC Corporation, Japan. These are prepared by "winding" sheets into very long, thin tubes that exhibit a flexible, stable, strong properties (Siddiqi and Husen 2017b). The carbon is bonded covalently in the lattice. The carbon nanotubes might be single-walled tubes or multiwalled tubes, former being lighter yet stronger than steel, making them the strongest material is known (Rudakiya et al. 2019). Nanotubes made from carbon are well utilized in agriculture due to their number of advantages, namely, water solubility, mechanical strength, biocompatibility, large surface to volume ratio, higher immobilization efficiency, etc. (Husen and Siddiqi 2014).

Carbon multiwalled nanotubes used are absorbed by tomato seedling through seeds and roots. These nanoparticles enter the seed coat by piercing and rapid water absorption. These nanoparticles are distributed in aerial parts via root. The interaction of chemicals present on the root surface is not the driving force of nanoparticles transport. Capillary action takes place in the wider passage. The uptake of these carbon nanotubes by plants can show changes in their metabolic functions which help in the increase in biomass and yield. However, concerns for cases do exist whether these carbon nanotubes become phytotoxic so their concentration must be controlled to prevent damage (Liu et al. 2019). Figure 16.5 depicts ripening stages in different fruits were analyzed by the sensor.

Carbon nanotubes using ethylene sensors have been used to predict the maturity and ripeness of the fruit effectively. Developed from semiconducting carbon nanotubes, these can detect ethylene emitted when fruits ripe and flowers bloom in very low concentrations of 15 ppb. The researchers at MIT made sensors that could monitor fruits and vegetables as they are transported to reduce food waste (Sridhar et al. 2020).

Nanowhiskers: These are a type of filamentary crystal with a cross-sectional diameter of 1–100 nm with distinct anisotropy of properties and a dislocation-free structure. They are several times stronger than regular crystals and demonstrate flexibility and resistance to corrosion (Hernandez and Dos Santos 2016). Cellulose

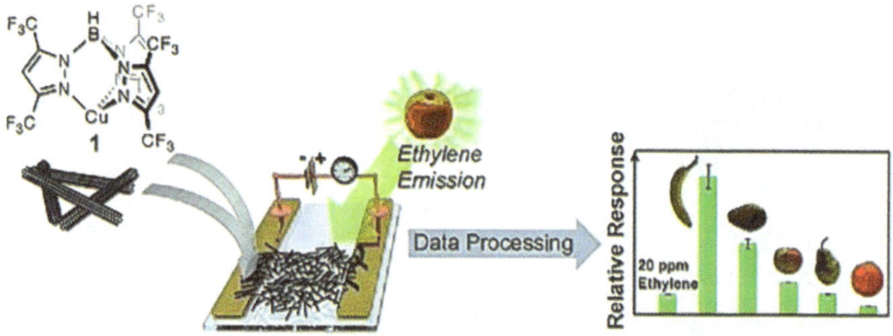

Fig. 16.5 Copper complex 1 mixed with single walled carbon nanotubes making chemoresistive sensor. The device represents sub ppm sensitivity and higher selectivity towards ethylene. The ripening stages in different fruits were analyzed by the sensor. Source nanowerk.com

is the most widely used polymer in food industry. Cellulose nanowhiskers are obtained through biological sources like sea creatures, plants, and bacteria. Carbon nanofillers like graphene and carbon nanotubes are used in cellulose material due to their exceptional material properties including mechanical strength, electrical conductivity, electromagnetic shielding effectiveness, and capacity (Shadia 2014).

Cellulose nanowhiskers (CNW) have a shortcoming of difficulty in controlling the dispersion level by the virtue of crystallite isolation and their incorporation in a matrix. One of the techniques involves exposure to mechanical dispersion or ultrasonication which would aid in permitting dispersion of CNW aggregates and produce a stable colloidal suspension. Appropriate microcrystalline cellulose (MCC) treatment with H_2SO_4 will result in forming a stable colloidal system by providing a negatively charged surface on which esterification of hydroxyl groups occurs by sulfate ions (Shadia 2014). The blends of lignocelluloses exhibit positive effects like thermal stability and mechanical strength since it has high durability and strength. Cassava bagasse with thermoplastic starch and polylactic acid, banana fiber with polyvinyl alcohol, extracted cellulose whiskers fiber obtained from ramie with polyethylene showed that a variety of combinations can be applied for reinforcement (Sridhar et al. 2020).

Nanoplatelets: Another type of nanomaterial used for the shelf life extension is nanoplatelets wherein making graphite nanoplatelets and applying them to pesticides, etc., are most common. Fin like cadmium-selenium-sulfur (CdSeS) nanoplatelets has been developed from selenium to sulfur solution by adjusting the molar ratio. These nanoplatelets were used to detect the popular organophosphorus pesticide in fruits and vegetables by fluorescence emission from these nanoplatelets (Carrillo-López et al. 2016). An interesting study conducted by Lancaster University has strongly advocated the use of nanoplatelets developed from vegetables mainly root vegetables for enhancing the strength and durability of buildings and bridges. Concrete mixtures of nanoparticles from sugar beet or carrot significantly improved the mechanical properties of concrete (Fruit&veggie.com).

16.4 Characterization of Nanoparticles

Nanomaterials have distinct properties which are size-dependent in 1–100 nm involving quantum phenomenon. The approaches relating to particle radius, Bohr radius, make quantum confinement apparent. The large surface area of these nanomaterials is the reason behind their novel properties. The characterization of nanoparticles takes two main parameters into consideration, that is, size and shape. The degree of accumulation, surface area and charge, size distribution, and surface chemistry are also important to some extent. The application of nanoparticles and other properties are affected by the size, size distribution, and organic ligands (Marian et al. 2019).

The techniques to characterize these nanomaterials include atomic force microscopy, x-ray scattering, scanning probe microscopes, electron microscopy, acoustic wave technique, contact angle measurements, and other spectroscopies (Yadollahi et al. 2009).

16.5 Applications in Horticultural Crops

16.5.1 Nanomaterials for Antimicrobial Activity

Nanoparticles are known to exhibit inherent antimicrobial potential. The food-borne microbes as well as spoilage pathogens are developing resistance for most of the available antimicrobial agents. Thus, there is a need for the development of new and novel antimicrobial agents that could help in combating spoilage microbes and achieving food safety. Use of metal nanoparticles (NPs) such as titanium (Ti), zinc (Zn), copper (Cu), gold (Au), and silver (Ag) are one such option found to possess antimicrobial properties, considering their spectra of activity and potential. The antimicrobial efficacy and effectiveness are mostly dependent on composition and particle size. Moreover, antimicrobial mechanisms of nanoparticles were not established clearly; there are two possible theories like (a) oxidative stress through formation of reactive oxygen species (ROS) on nanoparticles surface and (b) free metal ion toxicity obtained from the dissolution of metals from nanoparticles surface (Besinis et al. 2014). Table 16.2 shows the selected recent studies on nanomaterials employed for antimicrobial action in horticultural crops.

Antimicrobial action of the nanoparticles also differs with their shape, interaction with proteins, surface charge, and receptors, etc. Antimicrobial activities of the most commonly used nanoparticles are discussed hereunder.

16.5.1.1 Nano-Silver

Silver nanoparticles are the most commonly used inorganic nanomaterial for inducing antimicrobial activity especially in plastics, biopolymers, textiles, and coating industries (Egger et al. 2009). Silver nanoparticles exhibit a high surface area to volume ratio and unique physical and chemical properties that are gaining importance as an excellent antimicrobial agent against various spoilage-causing microorganisms. These particles damage DNA by rupturing and destroying the cell membrane of spoilage causing microbes. Khezerlou et al. (2018) have revealed that silver nanoparticles exhibit greater antimicrobial activity compared to their ionic forms against a wide range of bacteria. The antibacterial action of silver nanoparticles is largely because of their ability to damage bacterial outer membranes by inducing pits and gaps and thus eventually fragmenting the cells. In addition, silver ions interact with sulfhydryl and disulfide groups of enzymes that lead to the destruction of metabolic processes and ultimately, death of cell. European Union has declared a safety limit in foods up to 0.05 mg/kg. Besides reduction in spoilage causing microbes, it also helps in inhibition of ethylene biosynthesis by enhancing postharvest quality and longevity of horticultural produce (Naing et al. 2017; Park et al. 2017). Owing to these factors, usage of nano-silver particles up to 0.06 mg/l is

Table 16.2 Selected recent studies on nanomaterials employed for antimicrobial action in horticultural crops

Crops	Treatments	Nanosizing technique	Effects	References
Nanosilver				
Flower crops				
Anthurium	Nanosilver (10 mg L^{-1})	–	Bacterial growth was reduced and water uptake improved	Amin (2017)
Dianthus caryophyllus	Nanosilver (0.23 mM) and sucrose (58 mM)	–	Reduced bacterial growth and ethylene content	Park et al. (2017)
Orchid	Nanosilver (5 mg L^{-1})	Biosynthesis of plant extract	Bacterial growth was suppressed	Rahman et al. (2019)
Gladiolus	Biologically synthesized nanosilver (4 ppm) and sucrose (4%)	Biosynthesis of plant extract	Reduced bacterial growth and nutrient, as well as water uptake, was enhanced in the floral stalk	Maity et al. (2019)
Rosa hybrida cv. "Black magic"	Preservative solution(100 mL) containing silver nitrate ($AgNO_3$), 8-hydroxyquinoline sulfate (8-HQS), silver nanoparticles, chitosan, and distilled water as a control	–	$AgNO_3$ and silver nanoparticles were effective in suppressing bacterial growth and led to more water uptake by cut stem ends	Abdel-Kader et al. (2017)
Vegetables				
Fresh cut capsicum	Nanosilver (NS) *with* PVP-based glycosomes	*Biosynthesis* of plant extract	*Increased shelf life up to* 12 days at 4 °C without causing any changes in cellular and physiochemical property	Saravanakumar et al. (2020a, b)
Tomato	NS (100 ppm)	–	NS treated fruits had a lesser incidence of gray mold (*Botrytis cineraria*), highest TSS, highest pH, and highest firmness were maintained as compared to control samples	Salem et al. (2019)
Fruits				
Strawberry	NS encapsulation of *Thymus daenensis* and *Anethum graveolens* essential oils	Encapsulation method	More than 80% inhibition of anthracnose (*Colletotrichum nymphaea*)	Weisany et al. (2019)

(continued)

Table 16.2 (continued)

Crops	Treatments	Nanosizing technique	Effects	References
Blueberry and blackberry	NS particles with plant extracts like lemon peel (*Citrus limon*), Calendula flowers, and French marigold flowers (*Tagetes patula*)	Biosynthesis of plant extract	Odor, color, texture, and overall acceptability of blueberry and blackberry fruits were better in NS of citrus lemon peel extract	Rodino et al. (2019)
Kinnow fruit (*Citrus reticulata*)	NS particles with carboxy methyl cellulose (CMC) and guar gum-based coatings stability of the kinnow mandarin for period of 4 months (85–90% RH) at 4 and 10 °C	–	Combined use of silver nanoparticle coating along with low-temperature storage (4 °C) maintained excellent preservation extending shelf life up to 120 days as compared kinnow fruits with 10 °C storage temperature up to 60 days shelf life	Shah et al. (2015)
Zinc				
Flowers				
Lisianthus	Zinc oxide (ZnO) nanoparticles with zinc oxide/graphene (ZnO/G)	–	Quality of flowers was enhanced up to 16 days through generating greater turgor, greater water absorption weight gain in leaves, firmness of the pedicel, and stimulating the opening of the flowers, as well as more green coloration in leaves, was maintained	Soriano Melgar et al. (2018)
Gerbera	ZnO nanoparticles + different capping agents (citric acid, starch, chitosan, xylan)	Plant-based capping agents	Starch capped ZnO nanoparticles found to be the best for shelf life enhancement	Deepshikha et al. (2018)
Vegetables				
Tomato	Carboxy methyl cellulose (CMC) with ZnO nanoparticles from pineapple peel extracts	Biosynthesis of plant extract	Respiration rate was reduced, maintained fruit firmness, antioxidant activity, reduction of weight loss, and reduction of black spot (*Alternaria alternata*) by 20% was noticed	Saekow et al. (2019)

Okra	Nanocomposite zinc oxide-chitosan coatings	–	63% reduction in fungal and bacterial spores	Al-Naamani et al. (2018)
Fruits				
Guava	Chitosan-based nanocomposite film containing leaf extract of *Urtica dioica* obtained from copper oxide and zinc oxide nanoparticles	Biosynthesis of plant extract	CuO NPs were found to be the best than ZnO NPs and CuO NPs-chitosan film inhibited *E. cloacae* MTCC 509 (15 mm), *Salmonella* (13 mm), *S. aureus* (12 mm), and *campylobacter* (11 mm) and ZnO NPs inhibited *Salmonella* (11 mm)	Kalia et al. (2021)
Persimmon	Carboxy methyl cellulose (CMC) with ZnO nanoparticles from pineapple peel extracts	Biosynthesis of plant extract	Respiration rate was reduced, maintained fruit firmness, reduction of weight loss, and reduction of black spot (*Alternaria alternata*) by 20%	Saekow et al. (2019)
Chitosan				
Fruits				
Cherry	*Eryngium Campestre* essential oil (ECEO) encapsulated chitosan nanoparticles	Ion gelation method	Reduced microbial counts up to 21 days storage at 4 °C	Arabpoor et al. (2021)
Loquat	Chitosan or nanosilica coating	–	Browning was delayed, inhibited decrease of titratable acidity, activities of phenylalanine ammonia lyase, lipoxidase, polyphenol oxidase were inhibited	Song et al. (2016)
Carambola	Chitosan (CH), gum arabic (GA), and alginate (AL) coating	Deacetylation	GA 1% and CH 0.3% coating maintained better quality and inhibition of cell wall degrading enzyme activities and exhibited good appearance even after 12 days of storage	Gol et al. (2015)
Vegetables				
Cucumber	Chitosan coating with cinnamomum essential oil	Ionic gelation technique	Better quality, firmness, color was maintained, microbial growth was suppressed and extended shelf life up to 21 days at 10 °C	Mohammadi et al. (2015)

(continued)

Table 16.2 (continued)

Crops	Treatments	Nanosizing technique	Effects	References
Capsicum	Chitosan coating with *Cymbopogon citratus* essential oil (EO)	Biosynthesis of plant extract	Maintained the quality of capsicum throughout the storage period	Ali et al. (2015)
Button mushroom (*Agaricus bisporus*)	*Citrus aurantium* essential oil with nanoparticles of chitosan	Ionic gelation technique	Lowered the microbial growth, decreased activity of peroxidase, glutathione reductase, and ascorbate peroxidase	Karimirad et al. (2020)
Titanium dioxide				
Vegetables				
White button mushroom	Chitosan, nano-titanium dioxide + chitosan, thymol + tween-80 + nano-titanium dioxide + chitosan coating treatment	Ionic gelation technique	Chitosan + nano-titanium dioxide coating showed lower PPO activity, lower microbial contamination, lower respiration rate, and lowest weight loss	Sami et al. (2020)
Fresh cut cantaloupe	Chitosan+nano-silicon dioxide+nisin coating	Centrifugation method	Combination treatment of chitosan+nano-silicon dioxide+nisin maintained good color, sensory attributes, vitamin-c, decreased peroxide activity, and microbial contamination at 4 °C up to 1 week of storage	Sami et al. (2021)
Fresh cut capsicum	The antibacterial polymeric film was formed by using sodium alginate 3% + cellulose nanowhisker 0.5% + copper oxide nanoparticles 5 mM	Sonication	Suppressed the growth of *S. aureus* by 27.49 ± 0.91 mm, *E. coli* by 12.12 ± 0.58 mm, *Salmonella sp.*by 25.21 ± 1.05 mm and *Trichoderma spp.* by 5.31 ± 1.16 mm	Saravanakumar et al. (2020a, b)
Mung bean (*Vigna radiata*)	Sulfur nanoparticles (SNPs) and ZnO nanoparticles (ZNPs)	Liquid-phase precipitation method	Both nanoparticles maintained good plant nutrition	Patra et al. (2013)

Fruit crops				
Blueberry	Chitosan, chitosan or nano-titanium dioxide and chitosan or nano-titanium dioxide with tween and thymol essential oil	Blending and centrifugation of chemicals	Chitosan or nano-titanium dioxide with tween and thymol essential oil coating proved to be the best by shelf life extension and reduced shrinkage rate, decay rate, weight loss, mesophilic aerobic bacteria, yeast, and mold counts as compared to other coatings	Rokayya et al. (2021)
Loquat cv. "Qingzhong" and "Dawuxing"	0.10% of nano-SiO_2 in the range of 40–60 nm	–	Inhibited decay, maintained quality and expanding shelf life	Wang et al. (2020)
Mango	Chitosan with nano-TiO_2 modified by sodium laurate	Ultrasonication	Decay index and malonaldehyde (MDA) content was lower, maintained firmness, POD, PPO activity, total phenols, and flavonoid content was higher in composite coated fruits than the control	Xing et al. (2020)
Orange	Copper (cu) and copper oxide (CuO) nanoparticles	Precipitation method and chemical reduction	Decreased *Penicillium italicum* and *Penicillium digitatum by* increasing nanoparticles concentration to 15%	Chalandar et al. (2017)
Mango	Polylactic acid film, Polylactic acid and bergamont essential oils film, polylactic acid nanocomposite film (polylactic acid/bergamont/nano-TiO2)	Solvent evaporation method	Polylactic acid nanocomposite films reduced microbial count and extended postharvest shelf life up to 15 days	Chi et al. (2019)
Fresh cut apples	ZnO nanoparticles with poly-lactic acid (PLA) matrix + cinnamaldehyde	–	Firmness, total phenolic content, color, the sensory quality was maintained and inhibited the growth of microorganisms	Li et al. (2017)

acceptable as a carrier in the edible coating of horticulture commodity (Shah et al. 2015).

Zarei et al. (2014) studied the antibacterial effects of silver nanoparticles against four pathogenic foodborne bacteria viz. *S. typhimurium*, *V. parahaemolyticu*, *E. coli*, and *L. monocytogenes*. Results showed antibacterial properties of silver nanoparticles on the above-mentioned bacteria and thus declared as alternative ecofriendly method for disinfecting equipment and their cleaning. A study showed that a low concentration of silver nanoparticles enhanced inflorescence opening and leaf yellowing as compared to control which showed low inflorescence opening but remained green. The high concentration also caused early senescence. Longer vase life and less weight loss resulted when exposed to a lower concentration of silver nanoparticles.

16.5.1.2 Zinc Oxide Nanoparticles

Zinc oxide (ZnO) has the ability to inhibit cell growth and cause cell death by creating mitochondrial weakness and creating stress by gene regulation (Sirelkhatim et al. 2015). The ZnO nanoparticles are known to exhibit antibacterial properties against spores resistant to high temperatures, pressures, as well as Gram-positive and Gram-negative bacteria (Li et al. 2013). These ZnO particles act as highly effective antimicrobial agents and are used in food packaging which helps in synthesizing reactive oxygen species (ROS). These are used in edible coatings because of their nontoxic effects, antibacterial properties (Almoudi et al. 2018), biosafe, recognized as GRAS substance, provide a good barrier and mechanical properties and can be used as Zinc fortificant to foods (Anugrah et al. 2020).

Akbar and Anal (2014) examined zinc oxide nanoparticles for its antibacterial properties produced using hydro-thermal synthesis against *S. typhimurium* and *S. aureus*. It was demonstrated that film loaded with ZnO nanoparticles effectively inhibited the growth of foodborne pathogens. Espitia et al. (2013) studied ZnO nanoparticles for their antimicrobial activity against spoilage microorganisms and foodborne pathogens. Based on their studies, no antimicrobial activity was seen against *L. plantarum*, *P. aeruginosa*, and *L. monocytogenes*. But showed antimicrobial activity against *S. aureus*, *E. coli*, *S. choleraesuis*, *A. niger*, and *Saccharomyces cerevisiae*.

16.5.1.3 Chitosan Nanoparticles

Chitosan is an antibrowning and antimicrobial agent used for the storage of horticultural commodity. Moreover, it acts as a good barrier to loss of aroma, moisture, safer for human consumption, bio-compatibility and has the ability to form a layer of a thin film on fruit and vegetable surfaces which acts as a barrier and all these properties step ahead towards controlling spoilage-causing microbes like fungi, molds, and bacteria (Xing et al. 2016). Mechanisms responsible for antimicrobial activities of chitosan is occurred by causing cell death due to leakage of intracellular content and alteration in the membrane barrier properties due to interaction of amine group (+ve charged) with microbial cell membranes (−ve charged) (Hosseinnejad and Jafari 2016). Another study showed the use of chitosan coating delayed the discoloration associated with enzymes like polyphenol oxidase, peroxidase, phenylalanine, laccase, and a lower value of total phenolics on mushrooms. Increasing the concentration of this chitosan in the coating base greatly enhanced the quality and shelf life of fresh-cut mushrooms.

16.5.1.4 Titanium Dioxide/Silicon Dioxide/Copper/Sulfur/PLA Nanoparticles

Their antimicrobial effects are attributed mainly to their size, shape, and crystalline structure (Haghighi et al. 2013). They kill resting spores of microbes by degrading their cytoplasmic membrane and cell wall by synthesizing reactive oxygen species (ROS) like hydrogen peroxide and hydroxyl radicals. Initially, leakage of cellular contents occurs and subsequently leads to lysis of cell. The photocatalytic damage of microorganism involves several steps. The antibacterial properties and efficacy of TiO_2 photocatalytic properties depend on its abilities like (i) to make stable TiO_2 nanoparticles or nanocomposites, (ii) to produce electron-hole pairs by extending wavelength to the visible light region, and (iii) to achieve a newly created electron-hole carrier with reduced recombination rate. Maneerat and Hayata (2006) observed TiO_2 photocatalytic reaction for its antifungal activity by using TiO_2 powder and TiO_2 coated plastic film against fruit tests and microorganism *P. expansum* in vitro. They found that "TiO_2 photocatalytic reaction" indicates antifungal effects against *P. expansum* indicating its potential use in controlling postharvest diseases. The use of nano-titanium dioxide along with chitosan and thymol as shelf life enhancers in mushrooms acted as antimicrobial (41).

16.5.1.5 Copper Nanoparticles

Copper nanoparticles draw a great degree of interest from researchers due to their exclusive physicochemical, biological properties and their antimicrobial activities. Yoon et al. (2007) investigated copper and silver nanoparticles for their susceptibility constants for *B. subtilis* and *E. coli* and found that *B. subtilis* exhibited greater susceptibility to Cu nanoparticles, while *E. coli* exhibited the lowest susceptibility to silver nanoparticles. Copper-chitosan nanoparticles were studied for their antimicrobial properties on *P. aeruginosa*, *S. choleraesuis,* methicillin-resistant *S. aureus*, *B. subtilis*, and *C. albicans* were studied by Usman et al. (2013). They introduced these nanoparticles as antimicrobial agents; nevertheless, Cu nanoparticles act at high concentrations as a pro-oxidant, while their rapid oxidation limits their application. Azam et al. (2012) revealed that CuO nanoparticles have a size-dependent antibacterial activity against Gram-positive bacteria (*B. subtilis* and *S. aureus*) and Gram-negative bacteria (*E. coli and P. aeruginosa*). They found that the CuO nanoparticles inhibited both types of bacteria while the bactericidal activity depends on their growth medium, stability, concentration, and its size.

16.5.1.6 Gold (Au) Nanoparticles

Gold nanoparticles are considered crucial for their antibacterial properties due to polyvalent effects, ease of detection, nontoxicity, high ability functional ability, and photothermal activities (Lima et al. 2013). Though most of the nanoparticles function by generating ROS, gold nanoparticles differ in their mechanism for antibacterial activities. The small size of the gold nanoparticles makes them favorable to penetrate the cell through the endosome vesicle (Verma and Stellacci 2010). Many studies showed little or no cytotoxicity from gold nanoparticles related tom their surface charge, shape, or size (Simpson et al. 2013). Cui et al. (2012) found that the antibacterial property of gold nanoparticles is related to (i) modification of membrane potential and subsequent decrease in ATP level (ii) suppressing tRNA binding to protein factory (ribosome). The antimicrobial property of gold nanoparticles against *E. coli* and *S. typhi* was reported by Lima et al. (2013). These nanoparticles decreased their colony count by 90–95% and the main factors responsible for antibacterial properties were dispersion of gold nanoparticles and its roughness over the medium. This makes gold nanoparticles to act against targeted microbial cells.

16.5.1.7 Silica (SiO_2) Nanoparticles

Silica particles are known to exhibit antimicrobial activity at a nanoscale due to the enhancement of their surface areas (Dhapte et al. 2014). Cousins et al. (2007) discovered that Silica nanoparticles suppressed adherence of bacterial cells to oral biofilms and their mixture with other biocidal nanoparticles such as silver have been studied widely since decades. Kim et al. (2006) observed the antibacterial effects of the Copper/Silicon dioxide nanocomposite by adopting *P. citrinum*, *E.cloacae*, *C. albicans*, *E. coli*, and *S. aureus* as indicators and through disk diffusion methods. Results showed were clear indicating their antibacterial activities against Gram-negative, Gram-positive bacteria, and fungi. The efficacy of the nanocomposite could be due to Cu nanoparticles which would have formed well over the surface of SiO_2 nanoparticles without aggregation and thus resulting in a large surface area. Fellahi et al. (2013) evaluated the antibacterial properties of Silica nanowire substrates with silver or copper nanoparticles and revealed their strong antibacterial effects to suppress *E. coli.* Silver decorated silicon nanowires were found to be biocompatible with human cells while copper decorated silicon nanowires exhibited greater cytotoxicity. Silicon nanoparticles and nanocomposites with other metal nanoparticles with novel properties have been developed recently by researchers (Potemkin et al. 2018; Kumar et al. 2021).

16.5.1.8 Magnesium Oxide and Calcium Oxide Nanoparticles

Strong antimicrobial activity of the Ca and Mg oxides is due to their alkalinity and strong antibacterial activity. It is widely accepted that CaO and MgO nanoparticles generate superoxide on their surfaces and also increase their pH with their hydration (Yamamoto et al. 2010). Jin and He (2011) have revealed that MgO nanoparticles led to death of microbial cell through leakage of intracellular contents. Vidic et al. (2013) reported the antimicrobial effects of mixed nanostructures of zinc oxide-magnesium oxide with pure ones. Among them, zinc oxide nanocrystals showed enhanced antimicrobial action against both Gram-negative and Gram-positive bacteria. Cells of *B. subtilis* were damaged by zinc oxide-magnesium oxide nanoparticles was shown by microscopic analysis and hence suggesting them as a safe and new therapeutic agent to avoid bacteria. These nanoparticles also are inexpensive, available in abundance, and biocompatible, hence considered to be vital antibacterial agent. They can also be effectively utilized in food processing, pharmaceuticals, and in environmental preservation (Wang et al. 2017).

16.5.1.9 Aluminum Oxide Nanoparticles

Aluminum nanoparticles have a corundum-like structure and are thermodynamically stable over a wide range of temperatures (Martınez-Flores et al. 2003). These particles carry a positive charge on their surface at close neutral pH and they attract the negatively charged microorganisms viz. *E. coli* with electrostatic interaction. Increasing nanoparticles concentration in the suspension solution led to an increase in reaction and adhesion levels (Li and Logan 2004). The adhesion of bacterial cell on nanoparticles is usually directed by electrostatic interactions between surface of nanoparticles and bacteria cells, along with cross linking of polymers and hydrophobic interactions. The antimicrobial effects of these nanoparticles are relation with reactive oxygen species (ROS) which leads to ultimate cell death by degradation of cell membrane (Ruparelia et al. 2008). Nanoparticles can protect cells from induced oxidative stress cell death (Mohammad et al. 2008). Ansari et al. (2013) reported the antibacterial nature of aluminum nanoparticles against methicillin-resistant coagulase-negative *S. aureus* by serial dilution method.

16.5.1.10 Clay Nanoparticles

These are the layered mineral silicate groups into various classes viz. halloysite, montmorillonite, hectorite, kaolinite, and bentonite based on their morphology and nanoparticle composition (Mierzwa et al. 2013). Also, there are a new class of organically modified nanoclays popularized these days. These are hybrid organic-inorganic nanoparticles used in different polymer nanocomposites such as gas absorbents, rheological modifiers, drug delivery carriers, and antimicrobial agents (Busolo et al. 2010). The most commonly used nanoclay in various nanomaterials applications is montmorillonite (MMT). This plate-like MMT could be dispersed in a matrix of polymer to form nanocomposite (Shameli et al. 2011). Costa et al. (2011) examined the antimicrobial property of silver/montmorillonite nanoparticles for enhancing the shelf life of fresh fruit salad. They found that these nanoparticles suppressed the growth of coliforms, yeasts, psychrotrophic bacteria, and mesophilic, lactic acid bacteria and molds). Kanmani and Rhim (2014) examined the antimicrobial effects of gelatin laced active nanocomposite films, silver nanoparticles against *L. monocytogenes* and *E. coli* and found that silver nanoparticles possessed greater antimicrobial action against both the bacteria, but the clay nanoparticles showed antimicrobial activity against Gram-positive (*L. monocytogenes*) bacteria only. This property is usually related to alkyl quaternary group of ammonium salts (de Azeredo 2013).

Flowers: The flower of plants is loaded with diverse secondary compounds which include pigments, phenolics, and volatiles contributing to the fragrance which plays a major role in ethnobotanical relevance or floral therapy. These floral components are used as reducing agents for a variety of nanoparticles such as Au, Ag, Zn, Fe, etc. These are also used as stabilizing agents.

16.5.2 Nanopackaging

The nanopackaging application is predicted to have a maximum share of >25% among all the nano-enabled products in the current market. Incorporating nanomaterials into the plastic polymers has led to the development of innovative packaging materials such as (i) polymer nanocomposites with improved properties, (ii) polymers incorporating antimicrobial nanomaterials, (iii) active nanocoatings for sterile food contact surfaces, (iv) hydrophobic nanocoatings for self-cleaning surfaces, and (v) nanobiosensors for smart/intelligent packaging systems (Chaudhry and Laurence 2011).

Nanopackaging offers greater hopes for horticultural crops through safer packaging, extending shelf-life, better traceability of horticultural products, and healthier food. Packaging material incorporated with nanocomposites may enhance material properties, such as strength, thermal stability, gas barrier, and antimicrobial properties, without affecting their toughness and transparency. Most of the horticultural crops are seasonal in nature, and to extend their availability over the off-season, nanopackaging is one of the best options available as it serves numerous functions such as protection from moisture, dust, oxygen, light, harmful microorganisms. For effective utilization, the packaging material should also be safe to use, inert, eco-friendly, economic, lightweight, easy to dispose of, reusable, good stake strength, ability to withstand extreme conditions like filling, sealing, and transportation. Fresh horticultural produce (flowers, vegetables, fruits) is metabolically active and needs oxygen permeable materials with a good transmission rate, while the value-added horticultural products do not require such gaseous transfer. The critical issue relating to packaging material for extending the shelf-life of perishable products is their poor barrier properties to water vapor and other gases as the fruits and vegetables respire continuously by taking in oxygen and giving out CO_2. In the case of apple, the oxygen causes apple scab on its surface. Another major critical issue in the packaging of processed foods is that of moisture and gaseous migration and permeability: no packaging material is fully resistant to water vapor, atmospheric gases, and natural substances of the contained food or even the packaging material itself.

Certain polymer films are crammed with "silica nanoparticles" to substantially decrease the O_2 in-flow and moisture out-flow to keep the produce fresh and appealing to the consumers. The plastic material used in packaging of fruit and vegetable-based beverages must have high $O_2$and CO_2 barrier properties to check decarbonation of juices and oxidation of its contents. As a result of these complexities, processed horticultural products require sophisticated packaging material for extending the product shelf-life and prevention spoilage. New ecofriendly materials have been exploited for the development of edible films for extending the shelf-life and quality improvement in fruit and vegetable products Tharanathan

2003). However, the use of edible polymers has been limited due to their brittleness, poor barrier (gas and moisture) properties, less thermal distortion temperature, and high cost. The application of nanoscience to these polymers may open up new prospects for refining the barrier properties as well as improving the cost-price efficiency (Sorrentino et al. 2007).

16.5.3 Nanocomposites

Nanocomposite technology has proved to be suitable for improving the physical properties of biopolymers and man-made thermoplastics. Packaging material incorporated with nanocomposites may enhance physical properties viz. strength, thermal stability, antimicrobial, and gas barrier properties without affecting their inherent toughness and transparency. Such packaging may also serve as smart packaging by regulating the oxygen levels for enhancing the quality and safety of horticulture products as well as for interacting with consumers in case of spoilage or pathogen detection (Yam et al. 2005). Usage of nanoparticles can enhance the mechanical and heat resistance properties of packaging material and thereby extend their shelf-life, by hindering gas or water vapor permeability. The use of bio-polymers in the preparation of novel nanocomposites is expected to increase since they offer the rare opportunity of carbon-neutral biodegradable films for packing horticultural produce (Reddy et al. In press; Table 16.2). This also offers a great opportunity for the developing countries in the best utilization of their agricultural and forestry by-products and waste materials for producing low-cost bio-polymer nanocomposites. Table 16.3 presents nanocomposites used in packaging of horticultural produce.

Nanolaminates consist of two or more bonded (physical/chemical) layers of nanomaterial and are well suited for packaging fruits and vegetables. They are also helpful in formulating edible coatings, films, and foaming are that are current needs

Table 16.3 Nanocomposites used in packaging of horticultural produce

Composite	Function
Silicate	Mechanical strength, improved gas barrier, heat resistance
Organic substance (protein, peptide, or lipid)	Increased toughness of a packaging material
Potato starch and calcium carbonate	Improves thermal stability and also biodegradable
Clay nanocomposite	Lower water vapor transmission rate (WVTR) and higher tensile properties
PET coated with silica and alumina	Improved moisture and gas barrier, transparent
Montmorillonite 3–5%	Lightweight, increased barrier properties, and stronger with higher thermal stability
Ethylene vinyl alcohol copolymer containing nanoclays	Increases barrier properties and shelf-life of fruits and vegetables

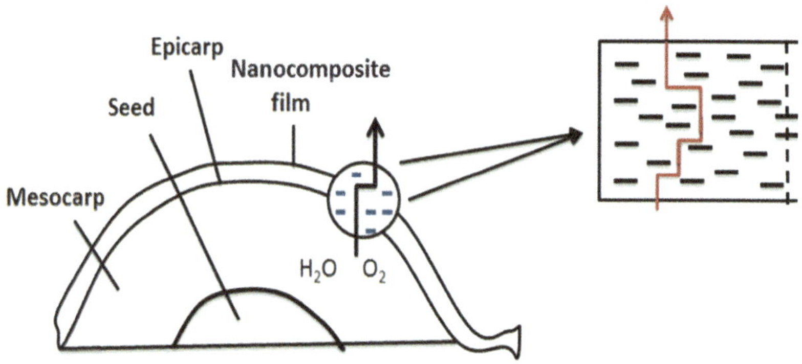

Fig. 16.6 Tortuous path of water movement from a nanocomposite film or coating in fruits and vegetables

of the fruit and vegetable industry (Table 16.3). These coatings or films would act as a barrier to fat, vapor or gases and enhance the food textural properties or act as carriers of functional agents including antioxidants, nutrients, and antimicrobials (Morillon et al. 2002). Currently proteins, polysaccharides, and lipids are being used for fabricating these types of films and coatings. Fig. 16.6 shows tortuous path of water movement from a nanocomposite film or coating in fruits and vegetables.

Composite materials are those which comprise more than one phase, wherein one is a polymer that is the continuous phase and filler or composite material is a dispersed phase. In general, nanocomposites are referred to as those comprising of a single or a mixture of polymers with at least one organic or inorganic filler with less than 100 nm dimensions. Several nanocomposites are incorporated into the packaging film such as SiO_2, clay, $KMnO_4$, nanocellulose, SiC, nano-fibrillated cellulose (NFC), carbon nanotubes that could enhance their mechanical properties (shear strength toughness, tensile strength, delamination resistance, fatigue, mechanical stiffness thermal stability) (Reddy et al. In press; Table 16.3). Table 16.4 shows application of various nanocomposites to the horticultural crops.

16.5.4 Nano-Biosensors

Nano-biosensors are analytical devices employing a biological material as recognition molecule integrated within a physicochemical transducer or in intimate contact with the transducing microsystems, while chemical sensors translate chemical data into signals that contain analytical information. The sensing capacity of chemical sensors depends on specific molecule concentrations in the sample. The components of sensors include a receptor-recognition system and a transducer. However, biosensors contain an optoelectronic recognition system which utilizes cells, enzymes, or organelles mediated biochemical reactions (Nič et al. 2020).

Table 16.4 Application of various nanocomposites to the horticultural crops

Crop	Polymer composite used	Function
Strawberry	Nano-silver and nano-silicate composites are incorporated into polyethylene and polypropylene	• Retains high ascorbic acid and phenolic content • Reduces fruit decay and weight loss
Orange juice	LDPE films with ag and ZnO nanoparticles	• Reduced microbial load below the threshold level up to 28 days • Reduced ascorbic acid degradation and Brown pigmentation
Mushroom-*Flammulinavelutipes*	Nano-ag, nano-TiO_2, attapulgite, nano-SiO_2 incorporated into polyethylene and LDPE	• Reduced weight loss and nutrient loss • Limits the ROS accumulation and lipid peroxidation
Kiwifruit	Nano-ag, nano-TiO_2, and montmorillonite incorporated in polyethylene film	• Decreased oxygen and water vapor permeability • Reduced weight loss, spore germination
Acerola fruits	Fruit purees and alginate incorporated into cellulose whiskers or MMT as edible film	• Reduced weight loss, decay incidence, and ripening rates of fruits • Retention of higher ascorbic acid content
Cherry tomatoes, lychee, and grapes	Carvacrol and aluminosilicate incorporated into polyamide (nylon 6) film	• Broad-spectrum antifungal activity against *Alternaria alternata, Botrytis cinerea, Penicillium digitatum,* and *Aspergillus niger*
Fresh apples, fresh carrot, fresh orange juice	Ag, TiO_2 incorporated into polyethylene	• Inhibition of *E. coli, Listeria* growth
Apple and lettuce	Ag nanoparticles incorporated into ethylene-vinyl alcohol copolymer (EVOH) films	• Inhibits the growth of L. monocytogenes
Barberry	AgNPs + LDPE	• Reduced microbial spoilage
Fresh-cut melon	Cellulose + AgNPs	• Reduced total mesophilic aerobic bacteria, psychrotrophic bacteria, yeasts, molds
Pears, carrots	Sodium alginate + AgNPs	• Reduced spoilage from *E. coli, S. aureus*
Fresh-cut carrots	Calcium alginate + Ag-montmorillonite	• Reduced decay or rot

Nano-biosensors-equipped packages are also intended for tracking of the internal and external conditions of food containers and pallets throughout the supply chain and these sensors allow the detection of contaminants, spoilage microbes, pathogens, and temperature abuse of the product during cold chain supply (Jerish and Dhinesh

2015). Ethylene efflux was used for measuring the fruit metabolism in a respiration chamber and also used for the detection of ethylene (for apple, avocado, pear, and kiwi) by Blanke and Shekarriz (2010). For example, these smart packaging provides relevant information on humidity and temperature by changing color. Nanosensors incorporated into the packaging films can alarm the consumers regarding food spoilage by detecting the gases given off by the spoiled food or by changing its own color. The use of nanoparticles to develop nanosensors for the detection of food contaminants, toxins, and pathogens in the food system is getting much popularity nowadays. Table 16.5 presents recent developments in the area of biosensors using nanomaterials.

Table 16.5 Recent developments in the area of biosensors using nanomaterials

Products	Nanocomponents	Findings	Reference
Time-temperature indicators	Silver nanoparticles	Silver nanoparticles did not show optical changes when exposed to 4 °C	Lanza et al. (2019)
Intelligent pH sensing, antimicrobial, UV barrier wraps	Zinc nanoparticles	Starch-PVA composite films +ZnO nanoparticles and phytochemicals were found effective as a UV barrier wrap and pH sensor	Jayakumar et al. (2019)
Beta glucosidase immobilization	SiO_2 nanoparticles	The purified β-glucosidase was immobilized on silicon dioxide nanoparticles to improve the thermostability and Michaelis constant value from 0.9 mM to 1.1 mM. The immobilized enzyme showed enhanced storage stability and was reusable	Agrawal et al. (2016)
Oxygen scavenger	Nanoscale zero-valent iron particles.	Nanoscale iron acts as active oxygen scavenging composites for dried products. Moreover, the reaction rate was much higher at 100% relative humidity	Foltynowicz et al. (2017)
UV barrier	Poly(L-lactide) (PLLA)/ZnO nanoparticles	UV–vis spectroscopy confirmed the UV-shielding effect in poly (L-Lactide) or ZnO nanocomposites, where the UV spectrum is effectively blocked by 61.2% for a concentration as low as 0.45 vol %, while 95.9% of the visible radiation passes through the material	Lizundia et al. (2016)
Determination of ripeness of fruits.	Carbon nanotube-based device	Carbon nanotubes added with copper atoms and polystyrene beads can detect the ethylene concentration as low as 0.5 ppm.	Trafton (2012)

16.6 Biodegradation and Toxicity

Assessing the biodegradability of nanomaterials is of utmost importance for their development and application in food particularly fruits and vegetables. Studying the biodegradation of nanomaterials is of critical importance in view of their structural variations and design for practical applications, so as to meet future challenges related to nanomaterials emitted into the environment. However, there are not many studies investigating the removal of nanomaterials from the environment. On release into the environment, nanomaterials may enter cells through penetration or endocytosis pathways. Inside the body, they have potential to cause damage to cell membranes, stimulate oxidative stress, and disrupt DNA (Kord and Roohani 2016). Issues which need to be looked into to determine the safety of nanocomposites in food packaging include:

(*i*) Physicochemical properties like size, shape, crystallinity, doping, surface tailoring, aggregation; (*ii*) reliable methodologies to identify and evaluate the possibilities and modes of migration; (*iii*) reliable and suitable methodologies for identification, characterization, and quantification of nanomaterials in sophisticated food matrices; (*iv*) interrelationships between inherent properties of nanoparticles and their associated toxicity; and (*v*) toxico-kinetic properties post-oral consumption and dose-response relationships. Maisanaba et al. (2018) have reported the effect of clay montmorillonite (Mt) quite frequently used nanofiller for food packaging applications on Caco-2 cells. Two silane-modified clays based on Mt., with 3-aminopropyltriethoxysilane (3-APTES), and vinyltrimethoxysilane (VTMS: 0–250 μg/mL), were assessed for genotoxicity, cytotoxicity, cell death, and oxidative stress up to 48 h of exposure. Of these, VTMS silane modified clay induced cytotoxicity, showing cell viability reduction to 63% of the control, besides oxidative stress proportional to concentration. Significant differences in the presence of Ca, Mg, and Si were obtained, although migration levels were within the regulatory range for food contact materials. A recent review on safety, recyclability, and properties of recycled polymers reinforced with nanoadditives by Acevedo-Del-Castillo et al. (2021) opines that nanoclays can decrease the rate of release of migrants because of three aspects: (i) the tortuous path inside polymer matrix lengthens the components' diffusion path, (ii) nucleating effect of the nanoclays increase crystalline regions within the polymer, which are impervious to migration of solutes, and (iii) the interactions between solute and clay by absorption which hinder the diffusion of solute inside the polymer.

Cwiek-Ludwicka and Ludwicki (2017) have reviewed the criteria for risk assessment of nanomaterials in food contact materials. Some important excerpts from their review are presented here. Due to the presence of actively charged surfaces which can absorb biomolecules during transit through the GI tract, the presence of nanomaterials in food enhanced the bioavailability of other substances usually found in the diet (Govers et al. 1994). These so-called "Trojan horses" may possess the ability to carry any toxins from food into the intestinal mucosa that may be associated with *Crohn's* disease (Lomer et al. 2002). Another possible scenario is that active NM surfaces adsorb beneficial food nutrients leading to decline in their bioavailability. It is therefore advisable to consider the nutritional implications when NMs are present or expected in food (Bouwmeester et al. 2009) on case to case basis. There is still lack of available information on toxic implications following low dose chronic or acute oral exposure, which presents serious concerns for assessing impact on health arising from prolonged exposure to NMs. Their effects on the cardiovascular and immune systems and other organs, for example, liver, lungs, and brain, remain unexplored. Besides, there is sufficient data needs to be generated on neurotoxicity, genotoxicity, carcinogenicity, reproduction toxicology teratogenicity nor the endocrine-disrupting potential.

In a study by Krug and Wick (2011), 50- to 100 nm-sized polystyrene particles were orally administered to test animals. While 98% of the nanoparticles got excreted, approximately 80% material which was intravenously administered got accumulated in liver. Therefore, nanoparticles uptake by the GI tract would not be of much significance. Another concern is the complex nature of food and therefore, NM should be characterized within the food matrix administered to test systems. Such studies would reflect the consumer's exposure to the NM-matrix complex, which might not be aptly reflected by experimentation on animals.

Specific NMs features need to be considered like their unique biological characteristics arising from chemical composition, surface/body-receptors interactions, appropriate physicochemical properties, and tissue interaction. As per EFSA, identification and characterization of hazard resulting from NM exposure needs to include specific attributes, with comparison on data for the nonnanoform of the same chemical under test.

Jain et al. (2018) have also emphasized in-depth knowledge of various mechanisms involved in the action of nanomaterials and their toxicity. Nanotoxins may lead to convoluted effects on human health, since they may easily enter humans via respiratory, dermal, or gastrointestinal routes. Their unique physicochemical properties often make investigation of their toxic consequences highly challenging and intricate.

Ashammakhi et al. (2020) have reviewed the developments in new testing methods for toxicology assessment of nanomaterials using microphysiological systems, also known as organ-on-a-chip platforms. They suggest that these can be more accurate than both conventional in vitro cultures and expensive animal studies. The development of more intricate microphysiological systems may be able to mimic human physiology better and rapidly highlight systemic effects of different drugs and materials. However, they might not be able to bring about secondary and systemic toxicity. Thus, the utilization of multiorgan-on-a-chip systems for advancing and assessing nanotechnology will emerge in the upcoming toxicology studies, drug development, and precision medicine.

In horticultural produce, particularly the use of biodegradable nanoparticles is emphasized. Some of the biodegradable NMs include dextran, alginate, carrageenan, polyhydroxybutyrate, silk protein, micelles and emulsions (based on biodegradable surfactants/emulsifiers), PEG, albumin, Polylactides PLA, PLGA poly(lactic-co-glycolic acid), chitosan, gelatin, polycaprolactone, poly(alkyl cyanoacrylates), and nanoparticles of bioactive and nanoclay.

Besides, the nanomaterials generally used are GRAS in nature with demonstrated safety on long-term consumption. Biodegradable nanoparticles may be categorized based on the structure and arrangement of the nanomaterials, with nanocapsules formed by encapsulation or incorporation into nanosphere. Some examples of classic nanocapsules include micelles and liposomes, while dendrimers are an example of nanospheres. Irrespective of the structure or how the mode of incorporation of target compound, biodegradable nanoparticles provide advantages in application, such as slow and/or controlled release and targeted delivery (Chen et al. 2017). In general, polylactic acid is degraded to lactic acid, poly (lactic-co-glycolic acid) hydrates yield lactic acid, and glycolic acid. Polycaprolactone also degrades monomer 6-hydroxy caproic acid. Chitosan also degrades to chitosan oligosaccharide.

Migration from packaging materials may lead to direct contact of nanomaterials with food. Because of general concerns of consumers over the possible adverse health implication of migrating food packaging components, this is a critical factor in their risk assessment.

From commercially available polyethylene plastic bags, Huang et al. (2015) investigated the migration of nanosilver into various food stimulants (Table 16.6). Although the least migration was observed into 95% ethanol, no significant differences in migration for other food simulants were observed.

Not only reduction of migration to zero, but also reliable data on the nanomaterial effects on customer safety after exposure must remain the most important criterion for optimization of design and manufacturing of packaging material (Huang et al. 2015).

Biodegradation of CNTs, GRA, and their derivatives has been studied using various microbes and enzymes. Zhang et al. (2013) have reported three bacteria (*Burkholderia kururiensis*, *Delftia acidovorans*, and *Stenotrophomonas maltophilia*) to constitute a community of potential multiwalled CNT (MWCNT) degraders. They were able to decompose MWCNTs into CO_2 along with several intermediate products such as 2-naphthol, 2-methoxy naphthalene, isophthalic acid, and cinnamaldehyde. These bacteria are commonly found in the soil rhizosphere, surface water, and groundwater. Although individual bacteria could only weakly degrade MWCNTs, in combination they could degrade with much higher efficiency. Recently, Chouhan et al. (2016) sourced soil bacteria (*Trabusiella guamensis*) from a goldsmith site contaminated with nanomaterials and demonstrated that the bacteria had adapted and could tolerate the nanomaterials. These bacteria biotransformed MWCNTs through oxidation. Besides bacteria, fungi have also been reported to decompose nanomaterials. For example, the *Sparassis latifolia* mushroom can secrete lignin peroxidase (LiP) to degrade both thermally treated and raw-grade carboxylated SWCNTs (Chandrasekaran et al. 2014). Andón et al. (2013) have reported biodegradation of SWCNT on incubation with human eosinophil peroxidase and H_2O_2.

Biodegradation and migration behavior of PLA and PLA-based nanocomposite films having different loads of cellulose nanocrystal and nanoclay has been studied through biodegradation, burial in soil, and compost, enzymatic degradation, and overall migration by Kord and Roohani (2016). They have reported higher

Table 16.6 Summary of studies pertaining to migration from food nanocomposite packaging

Nanomigrant	Main packaging component	Food/simulant	Storage temperature	Storage time	Key observation
Clay	Starch	Lettuce, spinach, water	40 °C	10 d	The films' contact with the vegetables did not lead to increase of iron and magnesium in the vegetables, while higher silica contents was observed
Carbon black	LDPE, PS	3% acetic acid, 95% ethanol	40, 60 °C	1–10 d	No migration of carbon black was recorded from the packaging material into food simulants
Fe, clay	HDPE, LLDPE	Water, isooctane	20, 40 °C	2, 10 d	Very small or negligible migration levels of iron and aluminum from active composites
Ag	PLA	3% acetic acid	Room temperature	1–8 d	Partial hydrolysis of PLA from the contact surfaces could potentially facilitate the migration process
Cu	PLA	Saline solution	Not available	4–24 h	First-order equation explained the ion release process during the first hours
Ag	PVC	Chicken meat	5, 20 °C	1–4 d	Quantity of silver migration from the nanocomposite was influenced by fill, time, and temperature and not the size of the nanoparticle
Ag, cu	PE	Chicken breasts	8.13, 21.8 °C	1.1, 3.1 d	Time and temperature were not significant on the migration
Ag	PE	Water, 3% acetic acid	40 °C	10 d	Percentage fill rate composite and simulant type were significant on silver migration
Ag	Commercial plastic food containers	50% ethanol, 3% acetic acid	40, 70 °C	2 h, 10 d	Heating in a microwave oven increased the migration values over heating in conventional oven
Ag, ZnO	LDPE	Orange juice	4 °C	28 d	Compared to silver, zinc migrated with a higher rate
Clay	PET	3% acetic acid	25, 45 °C	7–90 d	Dissolved silicon concentration was always greater than dissolved aluminum

MMT	PP	Water, 3% acetic acid, 15% ethanol, olive oil, grape seed oil, coconut oil	20, 40, 70 °C	2 h–14 d	3% acetic acid showed the highest migration among aqueous food simulants. Migration into fatty simulants was higher than those into aqueous ones but independent of fatty acid composition
Ag	PE	Water, 4% acetic acid, 95% ethanol, hexane	25, 40, 50 °C	3–15 d	Migrated silver was within 300 nm
TiO_2	PE	3% acetic acid, 50% ethanol	25, 70, 100 °C	1–8 h	Higher migration of Ti occurred in 3% acetic acid than into 50% ethanol. Increase in additives in the film enhanced the migration
MMT	Wheat gluten	Water, 3% acetic acid, 15% ethanol, olive oil, Tenax, agar gel	40 °C	10 d	High pressure/temperature treatment lead to higher amount of silicon migration in all the simulants
ZnO, ag	LDPE	Aqueous solution	40 °C	10 d	ZnO migration was higher than ag migration
Ag	PE	3% acetic acid, 95% ethanol	20, 40, 70 °C	1–9 h	The amount of ag migration was higher in 3% acetic acid than that in 95% ethanol
Ag	Commercial plastic food containers	Water, 3% acetic acid, 10% ethanol, olive oil	20 °C	1 h–10 d	The silver released was in both ionic and nanoparticle form
Ag_2O	LDPE	Apple	5, 15 °C	2–24 d	The PE/Ag_2O bag was found to have acceptable safety for food packaging

LDH layered double hydroxide, *HDPE* high-density PE, *LLDPE* linear LDPE, *PLA* polylactic acid, *MALLS* multi-angle laser light-scattering spectrometry, *ASV* anodic stripping voltammetry, *EDX* energy-dispersive X-ray, *LPSA* laser particle size analysis biodegradability of the composites with increase of nanoparticles in the enzymatic, soil, and compost media. Besides, they reported that the addition of cellulose nanocrystal and nanoclay to the polymer matrix decreased the overall migration of composites. This was attributed to higher adhesion of the nanoparticles to the polymer matrix and the tortuosity of their path.

Alaraby et al. (2020) investigated the impact of differently shaped titanium dioxide nanomaterials in Drosophila. The journey of the TiNMs through the midgut of larvae was carefully followed using TEM. Nano-titanium, regardless of its shape, was found to get eroded and degraded just after entering the gut lumen of the larvae. The peritrophic membrane, as the first defense line of the intestinal barrier, succeeded in reserving the NMs; however, the perpendicular particles of nanowires were able stab it, make pores, and therefore permitted translocation into intestinal cells. On the other hand, exposure to TiO_2 nanoparticles did not decrease viability of egg-to-adult, but all different shapes, especially nanowires, mediated a wide molecular response including changes in expression of genes involved in antioxidant, stress, repair, and physical interaction responses. All these changes point to concerns about their ability to elevate ROS levels, ultimately leading to potential genotoxicity.

Suravajhala et al. (2020) have reported bioaccumulation of green synthesized silver nanoparticles in various biological systems. The silver nanoparticles have been extensively explored to treat potential antibacterial, antimicrobial, antifungal agents in food crops and health applications. They have reviewed the impact of these particles on the environment and biological systems and deliberated the toxicity associated with it.

López de Dicastillo et al. (2020) have reviewed the development of recycled polymer nanocomposites reinforced with nanoadditives, their recyclability, and safety assessment. They investigated the effect of incorporation of nanoparticles on different postconsumer recyclable polymers while projecting nanotechnology as an alternative to contribute to the recovery of plastic waste. Deterioration of plastics by reprocessing, such as cross-linking, chain breaking, and generation of secondary products, was found to vary depending on structure and interactions between the polymer and nanoparticles. Further, plastics may also get contaminated with non-volatile and volatile compounds due to contact with other plastics. Therefore, although the inclusion of nanoparticles improves the barrier and mechanical properties of packaging developed from postconsumer plastic, their effect on the chemical safety and physical properties should be investigated to address in favor of a circular economy.

16.7 Regulatory Aspects

The use of nanomaterial directly overproduce is still in its infancy and lack of clarity over regulation for clinical use is greatly hindering their translation. Great anticipation surrounds the field of nanotechnology and its influence on the industry, regulatory guidance in this area is urgently required, which is critical to provide legal certainty to manufacturers, policymakers, healthcare providers, and the general public.

In the area of food contact materials (FCMs), definition for the term "nanomaterial" (NM) was adopted in 2011 which states that "nanomaterial means a natural, incidental or manufactured material containing particle, in an unbound state or as an aggregate or as an agglomerate and where, for 50% or more of the particles in the number size distribution, one or more external dimensions is in the size range from 1 to 100 nm. In specific cases and where warranted by concerns for the environment, health, safety, or competitiveness, the number size distribution threshold of 50% may be replaced by a threshold between 1 and 50%." This definition is as a reference for determining whether a material should be considered as a "nanomaterial" for legislative and policy purposes in the European Union.

The principle underlying the European Regulation (EC) 1935/2004 is that any material or article intended to come into contact with food must be sufficiently inert to preclude substances from being transferred to the food in quantities large enough to endanger human health or to bring about an unacceptable change in the composition of the food or a deterioration in its organoleptic properties. The European Union Regulation on plastic food contact materials and articles emphasizes that because the nanosized engineered substances possess chemical and physical properties significantly different from those at a larger scale may lead to generation of different toxicological attributes, these substances should be assessed for health-risk on a case-by-case basis by EFSA. It should therefore be made clear that any authorizations based on risk assessment of conventional particle sizes of a given substance do not cover nanoparticles.

At present, the number of nanoparticles allowed for use in food contact materials in EU is small (Table 16.7). Any application for authorization of a nanoparticle substance is subject to a risk assessment by EFSA. This is also when the nanomaterial is used behind the functional barrier. A coordinated approach to assess

Table 16.7 Nanomaterials authorized for use in plastic food contact materials (FCMs) according to EU legislation

Substance	Restrictions and specifications
Silicon dioxide, silanated (produced using primary particles in nanoform in the final material)	For synthetic amorphous silicon dioxide, silanated: Primary particles of 1–100 nm which are aggregated to a size of 0.1–1 μm and may form agglomerates within the size distribution of 0.3 μm to the mm size
Titanium nitride (TiN), nanoparticles	No migration of titanium nitride nanoparticles Only to be used in PET bottles up to 20 mg/kg. In the PET, the agglomerates have a diameter of 100–500 nm consisting of primary titanium nitride nanoparticles; primary particles have a diameter of approximately 20 nm
Zinc oxide, nanoparticles, uncoated	Only to be used in unplasticized polymers.
Zinc oxide, nanoparticles, coated with [3-(methacryloxy)propyl] trimethoxysilane (FCM no. 788)	Only to be used in unplasticized polymers. The restrictions and specifications specified for FCM substance no 788 shall be respected (0.05 mg/kg)

the human health risks of micro- and nanoplastics in food was planned for 2020 by the EFSA, but has been postponed due to the COVID-19. This scientific colloquium shall forge a path for the introduction and assessment of many new materials, ways, and insights for safety assessment and approvals.

In India, the Department of Science and Technology and the Government of India created a group to regulate nanotechnology and draft a set of guidelines creating a three-tiered governance framework that has been implemented to assist policymakers in developing a pathway for regulation of nanomedicine. Department of Science and Technology Nano's mission came up with safe handling of NMs in the workplace and industry in 2019. But there is no specific legislation/regulation for agriculture.

As per guidelines, novel food or novel food ingredients or processed with the use of novel technology New additive New processing aids including enzymes, articles of food and food ingredients consisting of, or isolated from microorganisms, bacteria, yeast, fungi or algae, enzymes, flavoring, and additives, health supplements, nutraceuticals, food for special dietary use, food for special medical purpose, functional food, novel food, and food contact material are legislated by the Food Safety Authority of India under the Ministry of Health and Family Welfare, Government of India. Insecticides and Biocides by Central Insecticide Board and Registration Committee (CIB&RC) under the Directorate of Plant Protection, Quarantine & Storage, Department of Agriculture & Cooperation is set up by the Ministry of Agriculture. It has been suggested to obtain preliminary toxicity analysis data on cytotoxicity through ATP Cell Titer-Glo, neutral red uptake, the release of lactate dehydrogenase, MTT assay [3-(4,5-Dimethylthiazol-2-yl)-2,5-Diphenyltetrazolium Bromide]; XTT assay (2,3-Bis-(2-Methoxy-4-Nitro-5-Sulfophenyl)-2H-Tetrazolium-5-Carboxanilide); cell impedance, trypan blue, BrdU (Bromodeoxyuridine/5-bromo-2′-deoxyuridine), Alamar Blue, WST-1 (water-soluble tetrazolium salts), live/dead cell counting, colony-forming efficiency, and genotoxicity: through Organisation for Economic Co-operation and Development (OECD) Technical guidelines TG 471, 473, 476, 482, and 487 methods. The confirmatory toxicity analysis should be performed by Central Insecticide Board.

There are no unanimously acceptable international guidelines for nano-agriproducts (NAPs). A few provisions that are in place globally for nanomaterials include REACH, EPA, AVMPA, OECD, and FAO/WHO with certain specific guidelines for quality, safety, and efficacy. However, innovations with alteration of the functionality of nanosystems make it difficult to apply a universal set of evaluation parameters for different nanoproducts with different applications. Many a time the case-by-case basis evaluation approach is advocated for NAPs.

16.8 Conclusions

Nanoscience has emerged as an important domain with enhanced scientific and technological prospects and novel applications with a plethora of applications. Recent years have witnessed its multidisciplinary nature and its rapidly increasing scope for the development of commercially viable applications. Various researchers have demonstrated the beneficial aspects of quality and efficacy of nanosized materials in horticulture to encourage the commercialization of nanotechnology-based innovations. It is expected to reduce the postharvest losses, improve the product quality, and simultaneously increase the competitiveness of the horticultural products. R&D in the postharvest nanotechnology could help in the long term retention of freshness and quality along with the prevention of postharvest diseases in a safer way. The application of green chemistry in the synthesis of nanoparticles using plant extracts and living cells could reduce the use of toxic solvents to a greater extent. However, safety of novel products needs to be complied with while emphasizing the high benefit to low-risk ratio compared to bulk counterparts. Modification of the existing policies and new standard guidelines for the evaluation of novel products on the basis of current scientific understanding shall prove an important milestone in the applications of nanoscience to prevent spoilage and meet global sustainable guidelines.

References

Abdel-Kader HH, Hamza AM, Elbaz TT, Eissa SM (2017) Effects of some chemicals on vase life of some cut flowers I. Effect of 8-Hydroxyquinoline sulfate, silver nitrate, silver nano particles and chitosan on vase life and quality of cut rose flowers (Rosa hybrida. Cv."Black Magic"). J Plant Prodn 8(1):49–53

Acevedo-Del-Castillo A, Águila-Toledo E, Maldonado-Magnere S, Aguilar-Bolados H (2021) A brief review on the high-energy electromagnetic radiation-shielding materials based on polymer nanocomposites. Int J Mol Sci 22(16):9079

Agrawal R, Verma AK, Satlewal A (2016) Application of nanoparticle-immobilized thermostable β-glucosidase for improving the sugarcane juice properties. Innovative Food Sci Emerg Technol 33:472–482

Akbar A, Anal AK (2014) Zinc oxide nanoparticles loaded active packaging a challenge study against Salmonella typhimurium and Staphylococcus aureus in ready-to-eat poultry meat. Food Control 38:88–95

Alaraby M, Hernández A, Marcos R (2020) Novel insights into biodegradation, interaction, internalization and impacts of high-aspect-ratio TiO_2 nanomaterials: a systematic in vivo study using Drosophila melanogaster. J Hazard Mater 124474

Ali A, Noh NM, Mustafa MA (2015) Antimicrobial activity of chitosan enriched with lemongrass oil against anthracnose of bell pepper. Food Pack Shelf Life 3:56–61

Almoudi M, Hussein AS, Hassan MIA, Zain NM (2018) A systematic review on antibacterial activity of zinc against Streptococcus mutans. Saudi Dental J 30(4):283–291

Al-Naamani L, Dutta J, Dobretsov S (2018) Nanocomposite zinc oxide-chitosan coatings on polyethylene films for extending storage life of Okra (*Abelmoschus esculentus*). Nanomaterials 8(7):479

Amin OA (2017) Influence of nanosilver and stevia extract on cut Anthurium inflorescences. Middle East J Appl Sci 7(2):299–313

Andón FT, Kapralov A, Yanamala N, Feng W, Baygan A, Chambers BJ, Kagan VE (2013) Biodegradation of single walled carbon nanotubes by eosinophil peroxidase. Small 9(16):2721–2729

Ansari MA, Khan HM, Khan A, Pal R, Cameotra S (2013) Antibacterial potential of Al_2O_3 nanoparticles against multidrug resistance strains of *Staphylococcus aureus* isolated from skin exudates. J Nanopart Res 15(10):1–12

Anugrah DSB, Alexander H, Pramitasari R, Hudiyanti D, Sagita CP (2020) A review of polysaccharide-zinc oxide nanocomposites as safe coating for fruits preservation. Coatings 10(10):988

Arabpoor B, Yousefi S, Weisany W, Ghasemlou M (2021) Multifunctional coating composed of Eryngium campestre essential oil encapsulated in nano-chitosan to prolong the shelf-life of fresh cherry fruits. Food Hydrocoll 111:106394

Ashammakhi N, Darabi MA, Clebi Saltik B, Tutar R, Hartel MC, Lee J, Khademhosseini A (2020) Microphysiological systems: next generation systems for assessing toxicity and therapeutic effects of nanomaterials. Small Methods 4(1):1900589

Ayuk EL, Ugwu MO, Aronimo SB (2017) A review on synthetic methods of nanostructured materials. Chem Res J 2(5):97–123

Azam A, Ahmed AS, Oves M, Khan MS, Memic A (2012) Size-dependent antimicrobial properties of CuO nanoparticles against gram-positive and-negative bacterial strains. Int J Nanomedicine 7:3527

Besinis A, De Peralta T, Handy RD (2014) The antibacterial effects of silver, titanium dioxide and silica dioxide nanoparticles compared to the dental disinfectant chlorhexidine on Streptococcus mutans using a suite of bioassays. Nanotoxicol 8(1):1–16

Blanke M, Shekarriz R (2010) Gold nanoparticles and sensor technology for sensitive ethylene detection. In XXVIII International Horticultural Congress on Science and Horticulture for People (IHC2010) 934:255–262

Boddolla S, Thodeti S (2018) A review on characterization techniques of nanomaterials. Int J Eng Sci Maths 7(1):169–175

Bouwmeester H, Dekkers S, Noordam MY, Hagens WI, Bulder AS, De Heer C, Ten Voorde SCG, Susan WP, Wijnhoven SWP, Marvin HJP, Sips AJAM (2009) Review of health safety aspects of nanotechnologies in food production. Regul Toxicol Pharmacol 53:52–62

Busolo MA, Fernandez P, Ocio MJ, Lagaron JM (2010) Novel silver-based nanoclay as an antimicrobial in polylactic acid food packaging coatings. Food Addit Contam 27:1617–1626

Carrillo-López LM, Morgado-González A, Morgado-González A (2016) Biosynthesized silver nanoparticles used in preservative solutions for Chrysanthemum cv. puma. J Nanomat 2016: 1769250

Chalandar HE, Ghorbani HR, Attar H, Alavi SA (2017) Antifungal effect of copper and copper oxide nanoparticles against Penicillium on orange fruit. Biosci Biotechnol Res Asia 14(1):279–284

Chandrasekaran G, Choi SK, Lee YC, Kim GJ, Shin HJ (2014) Oxidative biodegradation of single-walled carbon nanotubes by partially purified lignin peroxidase from Sparassis latifolia mushroom. J Indus Engg Chem 20(5):3367–3374

Chaudhry Q, Laurence C (2011) Food applications of nanotechnologies: an overview of opportunities and challenges for developing countries. Trends Food Sci Technol 22(11):595–603

Chen M, Qin X, Zeng G (2017) Biodegradation of carbon nanotubes, graphene, and their derivatives-review. Trends Biotechnol 35(9):836

Chi H, Song S, Luo M, Zhang C, Li W, Li L, Qin Y (2019) Effect of PLA nanocomposite films containing bergamot essential oil, TiO_2 nanoparticles, and Ag nanoparticles on shelf life of mangoes. Sci Hortic 249:192–198

Chouhan RS, Qureshi A, Yagci B, Gülgün MA, Ozguz V, Niazi JH (2016) Biotransformation of multi-walled carbon nanotubes mediated by nanomaterial resistant soil bacteria. Chem Eng J 298:1–9

Costa CA, Buonocore G, Del Nobile MA (2011) Antimicrobial silver-montmorillonite nanoparticles to prolong the shelf life of fresh fruit salad. Int J Food Microbiol 148(3):164–167

Cousins BG, Allison HE, Doherty PJ, Edwards C, Garvey MJ, Martin DS, Williams RL (2007) Effects of a nanoparticulate silica substrate on cell attachment of Candida albicans. J Appl Microbiol 102(3):757–765

Creighton JR, Ho P (2001) Introduction to chemical vapor deposition (CVD). Chem Vap Depos 2:1–22

Cui L, Pan G, Li L, Yan J, Zhang A, Bian R, Chang A (2012) The reduction of wheat Cd uptake in contaminated soil via biochar amendment: a two-year field experiment. Bioresources 7(4):5666–5676

Cwiek-Ludwicka K, Ludwicki JK (2017) Nanomaterials in food contact materials; considerations for risk assessment. Rocz Panstw Zakl Hig 68(4)

De Azeredo HM (2013) Antimicrobial nanostructures in food packaging. Trends Food Sci Technol 30(1):56–69

Deepshikha G, Verma AL, Monika T (2018) Application of ZnO nanoparticles in enhancing shelf life of cut flowers with special reference to Gerbera jamesonii. Res J Chem Environ 22(8)

Dhapte V, Kadam S, Moghe A, Pokharkar V (2014) Probing the wound healing potential of biogenic silver nanoparticles. J Wound Care 23(9):431–441

Egger S, Lehmann RP, Height MJ, Loessner MJ, Schuppler M (2009) Antimicrobial properties of a novel silver-silica nanocomposite material. Appl Environ Microbiol 75(9):2973–2976

Espitia PJP, Soares NDF, Teófilo RF, dos Reis Coimbra JS, Vitor DM, Batista RA, Medeiros EA (2013) Physical–mechanical and antimicrobial properties of nanocomposite films with pediocin and ZnO nanoparticles. Carbohydr Polym 94(1):199–208

Fellahi O, Sarma RK, Das MR, Saikia R, Marcon L, Coffinier Y, Boukherroub R (2013) The antimicrobial effect of silicon nanowires decorated with silver and copper nanoparticles. Nanotechnology 24(49):495101

Foltynowicz Z, Bardenshtein A, Sängerlaub S, Antvorskov H, Kozak W (2017) Nanoscale, zero-valent iron particles for application as an oxygen scavenger in food packaging. Food Packag Shelf Life 11:74–83

Gol NB, Chaudhari ML, Rao TR (2015) Effect of edible coatings on quality and shelf life of carambola (Averrhoa carambola L.) fruit during storage. J Food Sci Technol 52(1):78–91

Govers M, Termont D, Vanaken GA, Vandermeer R (1994) Characterization of the adsorption of conjugated and unconjugated bile-acids to insoluble amorphous calcium phosphate. J Lipid Res 35(5):741–748

Gupta S, Rajesh K, Abdelwahab O (2013) Formulation strategies to improve the bioavailability of poorly absorbed drugs with special emphasis on self-emulsifying systems. Int School Res Not

Haghighi F, RoudbarMohammadi S, Mohammadi P, Hosseinkhani S, Shipour R (2013) Antifungal activity of TiO2 nanoparticles and EDTA on Candida albicans biofilms. Infect Epidemiol Microbiol 1(1):33–38

Hellmann C, Andreas G, Joachim WH (2011) Design of pheromone releasing nanofibers for plant protection. Poly Adv Technol 22(4):407–413

Hernandez CC, Dos Santos RD (2016) Extraction of cellulose Nanowhiskers: natural fibers source, methodology and application. Matrix 3:16

Hosseinnejad M, Jafari SM (2016) Evaluation of different factors affecting antimicrobial properties of chitosan. Int J Biol Macromol 85:467–475

Huang JI, Li X, Zhou W (2015) Safety assessment of nanocomposite for food packaging application. Trends Food Sci Technol 45:187–199

Husen A, Siddiqi KS (2014) Carbon and fullerene nanomaterials in plant system. J Nanobiotechnol 12(1):1–0

Jain A, Ranjan S, Dasgupta N, Ramalingam C (2018) Nanomaterials in food and agriculture: an overview on their safety concerns and regulatory issues. Crit Rev Food Sci Nutr 58(2):297–317

Jayakumar A, Heera KV, Sumi TS, Joseph M, Mathew S, Praveen G, Radhakrishnan EK (2019) Starch-PVA composite films with zinc-oxide nanoparticles and phytochemicals as intelligent pH sensing wraps for food packaging application. Int J Biol Macromol 136:395–403

Jerish JJ, Dhinesh V (2015) Nanosensors and their applications in food analysis: a review. Int J Sci Technol 3(4):80

Jin T, He Y (2011) Antibacterial activities of magnesium oxide (MgO) nanoparticles against foodborne pathogens. J Nanopart Res 13(12):6877–6885

Kalia A, Kaur M, Shami A, Jawandha SK, Alghuthaymi MA, Thakur A, Abd-Elsalam KA (2021) Nettle-leaf extract derived zno/cuo nanoparticle-biopolymer-based antioxidant and antimicrobial nanocomposite packaging films and their impact on extending the post-harvest shelf life of guava fruit. Biomol Ther 11(2):224

Kanmani P, Rhim JW (2014) Properties and characterization of bionanocomposite films prepared with various biopolymers and ZnO nanoparticles. Carbohydr Polym 106:190–199
Karimirad R, Behnamian M, Dezhsetan S (2020) Bitter orange oil incorporated into chitosan nanoparticles: preparation, characterization and their potential application on antioxidant and antimicrobial characteristics of white button mushroom. Food Hydrocol 100:105387
Khezerlou A, Alizadeh-Sani M, Azizi-Lalabadi M, Ehsani A (2018) Nanoparticles and their antimicrobial properties against pathogens including bacteria, fungi, parasites and viruses. Microb Pathog 123:505–526
Kim W, Zide J, Gossard A, Klenov D, Stemmer S, Shakouri A, Majumdar A (2006) Thermal conductivity reduction and thermoelectric figure of merit increase by embedding nanoparticles in crystalline semiconductors. Phy Rev Lett 96(4):045901
Kim M, Osone S, Kim T, Higashi H, Seto T (2017) Synthesis of nanoparticles by laser ablation: a review. KONA Powder Part J 2017009
Kord B, Roohani M (2016) Biodegradation and migration behavior of cellulose nanocrystal-nanoclay reinforced PLA composites. Iran J Wood Paper Sci Res 31(1)
Krug HF, Wick P (2011) Nanotoxicology: an interdisciplinary challenge. Angew Chem Int Ed Engl 50:1260–1278
Kumar H, Bhardwaj K, Kuča K, Kalia A, Nepovimova E, Verma R, Kumar D (2020) Flower-based green synthesis of metallic nanoparticles: applications beyond fragrance. Nanomaterials 10(4):766
Kumar A, Choudhary A, Kaur H, Mehta S, Husen A (2021) Metal-based nanoparticles, sensors, and their multifaceted application in food packaging. J Nanobiotechnol 19(1):1–25
Lade BD, Shanware AS (2020) Phytonanofabrication: methodology and factors affecting biosynthesis of nanoparticles. Smart Nanosystems for Biomed, Optoelectronics and Catalysis, Intech open publications
Lanza GA, Perez-Taborda JA, Avila A (2019) Time-temperature indicators (Ttis) based on silver nanoparticles for monitoring of perishables products. J Phys Conf Ser 1247(1):012055
Li B, Logan BE (2004) Bacterial adhesion to glass and metal-oxide surfaces. Collds Surf B Biointerfaces 36(2):81–90
Li H, Feng L, Lin W, Jianchun S, Zhihong X, Liyan Z, Hongmei X, Yonghua Z, Qiuhui H (2009) Effect of nano-packing on preservation quality of Chinese jujube (Ziziphus jujuba Mill. Var. Inermis (Bunge) Rehd). Food Chem 114(2):547–552
Li M, Lin D, Zhu L (2013) Effects of water chemistry on the dissolution of ZnO nanoparticles and their toxicity to Escherichia coli. Environ Pollut 173:97–102
Li W, Li L, Cao Y, Lan T, Chen H, Qin Y (2017) Effects of PLA film incorporated with ZnO nanoparticle on the quality attributes of fresh-cut apple. Nanomaterials 7(8):207
Lima E, Guerra R, Lara V, Guzmán A (2013) Gold nanoparticles as efficient antimicrobial agents for Escherichia coli and Salmonella typhi. Chem Cent J 7(1):1–7
Liu D, Dong Y, Liu Y, Ma N, Sui G (2019) Cellulose Nanowhisker (Cnw)/graphene Nanoplatelet (Gn) composite films with simultaneously enhanced thermal, electrical and mechanical properties. Front Mater 6:235
Lizundia E, Ruiz Rubio L, Vilasjl LLM (2016) Poly (L-Lactide)/Zno nanocomposites as efficient UV-shielding coatings for packaging applications. J Appl Pol Sci 133(2)
Lomer MC, Thompson RP, Powell JJ (2002) Fine and ultrafine particles of the diet: influence on the mucosal immune response and association with crohn's disease. Proc Nutr Soc 61(1):123–130
López De Dicastillo C, Velásquez E, Rojas A, Guarda A, Galotto MJ (2020) The use of nanoadditives within recycled polymers for food packaging: properties, recyclability and safety. Compr Rev Food Sci Food Saf 19(4):1760–1776
Maisanaba S, Guzmán-Guillén R, Puerto M, Gutiérrez-Praena D, Ortuño N, Jos Á (2018) In vitro toxicity evaluation of new silane-modified clays and the migration extract from a derived polymer-clay nanocomposite intended to food packaging applications. J Hazard Mater 341: 313–320

Maity TR, Samanta A, Saha B, Datta S (2019) Evaluation of Piper betle mediated silver nanoparticle in post-harvest physiology in relation to vase life of cut spike of Gladiolus. Bull Natl Res Centre 43(1):1–11

Malekzadeh M, Rohani P, Keskar M, Swihart MT (2018) Laser pyrolysis synthesis of novel nanoparticles using spray-based precursor delivery. In: AIChE Annual Meeting. AIChE. https://www.aiche.org/conferences/aiche-annual-meeting/2018/proceeding/paper/375i-laser-pyrolysis-synthesis-novel-nanoparticles-using-spray-based-precursor-delivery

Maneerat C, Hayata Y (2006) Antifungal activity of TiO_2 photocatalysis against Penicillium expansum in vitro and in fruit tests. Int J Food Microbiol 107:99–103

Marian B, Steliana R, Alina B (2019) Application of silver nanoparticles on fresh fruits preservation, proceedings of the 5th World Congress on New Technologies (Newtech'19) Lisbon, Portugal Paper No. ICNFA 156

Martınez-Flores E, Negrete J, Villasenor GT (2003) Structure and properties of Zn–Al–cu alloy reinforced with alumina particles. Mater Des 24(4):281–286

Mierzwa JC, Arieta V, Verlage M, Carvalho J, Vecitis CD (2013) Effect of clay nanoparticles on the structure and performance of polyethersulfone ultrafiltration membranes. Desalination 314:147–158

Mohammad G, Mishra VK, Pandey HP (2008) Antioxidant properties of some nanoparticle may enhance wound healing in T2DM patient. Dig J Nanomat Biostruct 3(4):159–162

Mohammadi A, Hashemi M, Hosseini SM (2015) Chitosan nanoparticles loaded with Cinnamomum zeylanicum essential oil enhance the shelf life of cucumber during cold storage. Postharvest Biol Tech 110:203–213

Morillon V, Debeaufort F, Blond G, Capelle M, Voilley A (2002) Factors affecting the moisture permeability of lipid-based edible films: a review. Crit Rev Food Sci Nutr 42(1):67–89

Naing AH, Win NM, Han JS, Lim KB, Kim CK (2017) Role of nano-silver and the bacterial strain Enterobacter cloacae in increasing vase life of cut carnation 'Omea'. Front Plant Sci 8:1590

Namita R (2015) Methods of preparation of nanoparticles – a review. Int J Adv Eng Technol 7(4):1806–1811

Nič M, Jirát J, Košata B, Jenkins A, McNaught A (2020) IUPAC compendium of chemical terminology. IUPAC, Research Triangle Park; 2009

Oberdorster G, Oberdorster E, Oberdorster J (2005) Nanotoxicology: an emerging discipline evolving from studies of ultrafine particles. Environ Health Perspect:823–839

Opara L (2004) Emerging technological innovation triad for smart agriculture in the 21st century. Part I. Prospects and Impacts of Nanotechnology in Agriculture

Park DY, Naing AH, Ai TN, Han JS, Kang IK, Kim CK (2017) Synergistic effect of nano-sliver with sucrose on extending vase life of the carnation cv. Edun. Front Plant Sci 8:1601

Patra P, Choudhury SR, Mandal S, Basu A, Goswami A, Gogoi R, Gopal M (2013) Effect sulfur and ZnO nanoparticles on stress physiology and plant (*Vigna radiata*) nutrition. Adv Nanomater Nanotechnol:301–309

Potemkin DI, Maslov DK, Loponov K, Snytnikov PV, Shubin YV, Plyusnin PE et al (2018) Porous nanocrystalline silicon supported bimetallic Pd-Au catalysts: preparation, characterization, and direct hydrogen peroxide synthesis. Front Chem 6:85

Rahman MM, Ahmad SH, Mohamed MTM, Ab Rahman MZ (2019) Improving the vase life of cut Mokara red orchid flower using leaf extracts with silver nanoparticles. Proc Natl Acad Sci India Sect B Biol Sci 89(4):1343–1350

Rodino S, Butu M, Butu A (2019) Application of biogenic silver nanoparticles for berries preservation. Dig J Nanomater Biostruct 14:601–606

Rokayya S, Jia F, Li Y, Nie X, Xu J, Han R, Helal M (2021) Application of nano-titanum dioxide coating on fresh highbush blueberries shelf life stored under ambient temperature. LWT 137: 110422

Rudakiya D, Patel Y, Chhaya U, Gupte A (2019) Carbon nanotubes in agriculture: production, potential, and prospects. In: Nnanotechnology for agriculture. Springer, Singapore, pp 121–130

Ruparelia JP, Chatterjee AK, Duttagupta SP, Mukherji S (2008) Strain specificity in antimicrobial activity of silver and copper nanoparticles. Acta Biomater 4(3):707–716

Saekow M, Naradisorn M, Tongdeesoontorn W, Hamauzu Y (2019) Effect of carboxymethyl cellulose coating containing ZnO-nanoparticles for prolonging shelf life of persimmon and tomato fruit. J Food Sci Agric Technol 5:41–48
Salem EA, Nawito MAS, Abd El-Raouf AER (2019) Effect of silver nano-particles on gray mold of tomato fruits. J Nanotechnol Res 1(4):108–118
Salloom HT (2017) Third order optical nonlinearity of silver nanoparticles prepared by chemical reduction method. Al-Nahrain J Sci 20(3):99–104
Siddiqi KS, Husen A (2017a) Plant response to engineered metal oxide nanoparticles. Nanoscale Res Lett 12(1):1–18
Siddiqi KS, Husen A (2017b) Recent advances in plant-mediated engineered gold nanoparticles and their application in biological system. J Trace Elem Med Biol 40:10–23
Simpson CA, Salleng KJ, Cliffel DE, Feldheim DL (2013) In vivo toxicity, biodistribution, and clearance of glutathione-coated gold nanoparticles. Nanomedicine 9(2):257–263
Sami R, Elhakem A, Alharbi M, Benajiba N, Almatrafi M, Jing J, Helal M (2020) Effect of titanium dioxide nanocomposite material and antimicrobial agents on mushrooms shelf-life preservation. PRO 8(12):1632
Sami R, Almatrafi M, Elhakem A, Alharbi M, Benajiba N, Helal M (2021) Effect of nano silicon dioxide coating films on the quality characteristics of fresh-cut cantaloupe. Membranes 11(2):140
Saravanakumar K, Hu X, Chelliah R, Oh DH, Kathiresan K, Wang MH (2020a) Biogenic silver nanoparticles-polyvinylpyrrolidone based glycerosomes coating to expand the shelf life of fresh-cut bell pepper (Capsicum annuum L. var. grossum (L.) Sendt). Postharvest Biol Technol 160:111039
Saravanakumar K, Sathiyaseelan A, Mariadoss AVA, Xiaowen H, Wang MH (2020b) Physical and bioactivities of biopolymeric films incorporated with cellulose, sodium alginate and copper oxide nanoparticles for food packaging application. Int J Biol Macromol 153:207–214
Satyanarayana T, Sudhakar RS (2018) Methods of nanomaterials. IJRASET 6(1):2321–8653
Shadia J (2014) Characterization of nanomaterials. Springer J Mineral 16:28–29
Shah SWA, Jahangir M, Qaisar M, Khan SA, Mahmood T, Saeed M, Liaquat M (2015) Storage stability of kinnow fruit (Citrus reticulata) as affected by CMC and guar gum-based silver nanoparticle coatings. Molecules 20(12):22645–22661
Shameli K, Ahmad MB, Zargar M, Wan Y, Ibrahim NA, Sha-banzadeh P, Ghaffari-Moghadam M (2011) Synthesis and characterization of silver /montmorillonite/chitosan bionanocomposites by chemical reduction method and their antibacterial activity. Intt J Nanomed 6:271–284
Simpson CA, Salleng KJ, Cliffel DE, Feldheim DL (2013) In vivo toxicity, biodistribution, and clearance of glutathione-coated gold nanoparticles. Nanomedicine 9(2):257–263
Singaravelan R, Alwar SB (2015) Electrochemical synthesis, characterisation and phytogenic properties of silver nanoparticles. Appl Nano Sci 5(8):983–991
Sirelkhatim A, Mahmud S, Seeni A, Kaus NHM, Ann LC, Bakhori SKM, Mohamad D (2015) Review on zinc oxide nanoparticles: antibacterial activity and toxicity mechanism. Nano-Micro Lett 7(3):219–242
Song H, Yuan W, Jin P, Wang W, Wang X, Yang L, Zhang Y (2016) Effects of chitosan/nano-silica on postharvest quality and antioxidant capacity of loquat fruit during cold storage. Postharvest Biol Technol 119:41–48
Soriano Melgar LAA, López-Guerrero AG, Cortéz-Mazatan G, Mendoza-Mendoza E, Peralta-Rodríguez RD (2018) Nanopartículas de óxido de zinc y óxido de zinc/grafeno empleadas en soluciones florero durante la vida poscosecha de lisianthus (Eustoma grandiflorum). Agroproductividad 11(8)
Sorrentino A, Gorrasi G, Vittoria V (2007) Potential perspectives of bio-nanocomposites for food packaging applications. Trends Food Sci Technol 18:84–95
Sridhar A, Ponnuchamy M, Kumar PS, Kapoor A (2020) Food preservation techniques and nanotechnology for increased shelf life of fruits, vegetables, beverages and spices: a review. Environ Chem Lett 9:1–21
Suravajhala R, Bhagat M, Malik B (2020) Emerging toxicity aspects of silver nanoparticles: an overview. Cure Nanomater 5(2):158–164

Tharanathan RN (2003) Biodegradable films and composite coatings: past, present and future. Trends Food Sci Technol 4(3):71–78

Trafton A (2012) Comparing apples and oranges–new sensor can accurately measure fruits ripeness, helping prevent loss of produce from spoilage. Sens Rev

Usman MS, El Zowalaty ME, Shameli K, Zainuddin N, Salama M, Ibrahim NA (2013) Synthesis, characterization, and antimicrobial properties of copper nanoparticles. Int J Nanomedicine 8: 4467

Verma A, Stellacci F (2010) Effect of surface properties on nanoparticle–cell interactions. Small 6(1):12–21

Vidic J, Stankic S, Haque F, Ciric D, Le Goffic R, Vidy A, Delmas B (2013) Selective antibacterial effects of mixed ZnMgO nanoparticles. J Nanopart Res 15(5):1–10

Vozga I, Kacani J (2020) Application of inert gas condensation, review Article. World J Engg Res Tech 6(2):11–22. ISSN 2454-695X. https://www.wjert.org/download/article/37022020/1582957855.pdf

Wang L, Hu C, Shao L (2017) The antimicrobial activity of nanoparticles: present situation and prospects for the future. Int J Nanomed 12:1227

Wang L, Shao S, Madebo MP, Hou Y, Zheng Y, Jin P (2020) Effect of nano-SiO_2 packing on postharvest quality and antioxidant capacity of loquat fruit under ambient temperature storage. Food Chem 315:126295

Weisany W, Amini J, Samadi S, Hossaini S, Yousefi S, Struik PC (2019) Nano silver-encapsulation of Thymus daenensis and Anethumgraveolens essential oils enhances antifungal potential against strawberry anthracnose. Ind Crop Prod 141:111808

Xing Y, Xu Q, Li X, Chen C, Ma L, Li S, Lin H (2016) Chitosan-based coating with antimicrobial agents: preparation, property, mechanism, and application effectiveness on fruits and vegetables. Int J Polymer Sci

Xing Y, Yang H, Guo X, Bi X, Liu X, Xu Q, Zheng Y (2020) Effect of chitosan/Nano-TiO_2 composite coatings on the postharvest quality and physicochemical characteristics of mango fruits. Sci Hortic 263:109135

Yadollahi A, Arzani K, Khoshghalb H (2009) The role of nanotechnology in horticultural crops postharvest management. Insoutheast Asia Symposium on Quality and Safety of Fresh and Fresh-Cut Produce 875:49–56

Yam KL, Takhistov PT, Joseph M (2005) Intelligent packaging: concepts and applications. J Food Sci 70(1):1–10

Yamamoto D, Watanabe S, Miyahara MT (2010) Coordination and reduction processes in the synthesis of dendrimer-encapsulated Pt nanoparticles. Langmuir 26(4):2339–2345

Yevale R, Khan N, Bhadane S (2019) An overview on nanoparticles. J Drug Del Therap 9(5):181–184

Yoon KY, Byeon JH, Park JH, Hwang J (2007) Susceptibility constants of Escherichia coli and Bacillus subtilis to silver and copper nanoparticles. Sci Total Environ 373(2–3):572–575

Zarei M, Jamnejad A, Khajehali E (2014) Antibacterial effect of silver nanoparticles against four foodborne pathogens. Jundishapur J Microbiol 7(1):8720

Zhang L, Petersen EJ, Habteselassie MY, Mao L, Huang Q (2013) Degradation of multiwall carbon nanotubes by Bacteria. Environ Pollut 181:335–339

Edible Packaging: Mechanical Properties and Testing Methods

17

Yogesh Kumar, Soumen Roy, Anjali Sharma, Dev Kumar Yadav, Anand Kishore, Nitin Kumar, and Murlidhar Meghwal

Abstract

Traditionally people use food packaging materials from nonrenewable sources such as plastics obtained from petroleum. The traditional food packaging materials play a significant role in the total plastics consumption of the world. They are nonbiodegradable, cause environmental pollution, and large carbon adds to water footprint in their manufacturing process. Subsequently, this leads to a heavy dependence on nonrenewable resources. Reducing the amount of nonrenewable material by biodegradable and renewable sources of packaging holds the potential to improve waste disposal and decrease the cost of packaging and overall product. The need for alternatives to petrochemical-based plastics is more than ever. The current trend in food packaging calls for the utilization of natural, "environment-friendly" materials, which also have some additional

Y. Kumar
Department of Agronomy, Food, Natural Resources, Animals and Environment (DAFNAE), University of Padova, Agripolis Avenue of the University, Legnaro (PD), Padua, Italy

S. Roy · M. Meghwal (✉)
Department of Food Science and Technology, National Institute of Food Technology Entrepreneurship & Management, Kundli, HR, India

A. Sharma
IGNOU Study Centre, National Institute of Food Technology Entrepreneurship & Management Kundli, Sonipat, Haryana, India

D. K. Yadav · A. Kishore
Defence Food Research Laboratory, Grain Science & Technology (GST) Division, DRDO, Ministry of Defence, Government of India, Mysore, India

N. Kumar
Department of Food Engineering, National Institute of Food Technology Entrepreneurship & Management Kundli, Sonipat, Haryana, India

A. Poonia, T. Dhewa (eds.), *Edible Food Packaging*,
https://doi.org/10.1007/978-981-16-2383-7_17

functional properties wherever possible and with some cost benefit. Biodegradable packaging is a very good alternative packaging and a specialized type of biodegradable packaging, that is, the "edible packaging," is a very good alternative as they also have some other benefits like enhancing shelf life, maintaining the freshness of foods, and can be eaten along with the food without altering the original organoleptic characteristics of the food. The two main considerations with the use of edible packaging can be understood from its name itself, which are firstly "edible," it means the package is an integral part of the food and may or may not be consumed together with the product and hence should have all the necessary properties. At present, 12 types of edible films are approved by FDA and commercially used, which are made up of materials like polysaccharides, lipids, cellulose, chitosan, starch (potato, corn, wheat, rice), hemicellulose, hydrocolloids like gums (guar, pectin) and alginates, and plant and animal proteins. Secondly, the material chosen should have all the properties to protect the food from the outer environment, provide a good barrier against light, water vapor, other gases, and also have the required mechanical strength to protect the food against damages incurred during storage, transportation, and handling.

Keywords

Biodegradable packaging · Edible packaging · Hydrocolloids · Mechanical properties

17.1 Introduction

Packaging material plays a pivotal role in acceptability and safety of the processed food material. It not only acts as a barrier, but also as a carrier of important information for the consumers. As a primary barrier it must be of food grade. These materials have varying properties which affects the keeping and eating quality of the products. Food packaging imparts a significant role in food chain supplies. Consumable films/coatings are among emerging strategies for food quality assurance (Banerjee et al. 2017). Their application is based on the capacity to maintain the quality, to extend the shelf life, and to contribute to the cost efficiency of it. Preferentially, consumers demand for natural ingredients based high-quality products. Since decades, efforts have been made and several techniques have been used to preserve and package the food items using edible and biodegradable films and coatings. A package has to assure various necessities efficiently and inexpensively; for this rationale, a contemporary food package should be optimized and integrated efficiently with the food supply chain.

Biological material from various sources was used to develop and exemplify, but these packaging materials are not fit for universal uses in a variety of products (Banerjee et al. 2017). Such implication poses a challenge for the development of new and specific coatings and films for different foods. The distinctiveness mandatory for edible films and coatings depends principally on the product which is to be packed/coated. Consequently, low oxygen permeability is logically required for oxygen-sensitive food products. While there is a requirement of selectively moisture permeable packing based on biological material for various fruits and vegetables so as to ensure their freshness and extended shelf life.

Recent advances in food technology have made our lives virtually easier and efficient. Changing lifestyle has led to the emergence of packaged food that provides convenience to the consumers. The authentic packaging of food implements consumers with the role of communication, containment, convenience, and protection. Any typical food package provides us with the information about the product and its nutritional fact, thus an efficient way of marketing/branding. It protects the food product from the external environment and adulteration, thus keeps it safe. It allows the consumers to relish the food as per their convenience (reheating and direct consumption) and holds the product of any size, shape, and restraints during handling and transportation (Robertson 2006). Food packages can be altered or modified according to one's requirement with features like single-serving dishes and portability. However, traditional packaging is lacking adequacy to meet unending consumer demands and product intricacies. Therefore, a contemporary concept of packaging with higher proficiency is needed to meet various consumer requirements. For instance, a packaging that can enhance product's shelf life with lesser or no preservatives, that maintains freshness, that meets permissible regulatory requisites, and that enables tracking throughout the product life cycle (Suhag et al. 2020). Emerging technologies like smart packaging have the expertise to invigilate a product and its milieu and pursue any modifications into these. Moreover, smart packaging helps in gaining global market recognition and meets international food safety standards. In recent years, there has been rapid development in food packaging that includes active packaging, intelligent packaging, or smart packaging. These are the technologies that deal with the packaging of food, beverages, cosmetics, and pharmaceutical products (Banerjee et al. 2017). Moreover, these terminologies are apparently identical, but each has a different function to deal with and is as follows.

Active packaging is "packaging in which subsidiary constituents have been deliberately included in or on either the packaging material or the package headspace to enhance the performance of the package system" (Robertson 2006). *Intelligent packaging* is "a packaging system that is capable of carrying out intelligent functions (like detecting, sensing, recording, tracking, communicating, and applying scientific logic) to facilitate decision making, to extend shelf life, enhance safety, improve quality, provide information, and warn about possible problems" (LaCoste et al. 2005). *Smart packaging* is "one that possesses the capabilities of both intelligent and active packaging. Smart packaging provides a total packaging solution that on the one hand monitors changes in the product or the environment (intelligent) and on the other hand acts upon these changes (active)" (Vanderroost et al. 2014). Smart packaging is captivating industrialists and researchers due to its high impact on food packaging and supply chain. This chapter mainly deals with recent advances in smart packaging, its trends, and applications in industries.

17.2 Edible Packaging Materials

17.2.1 Definition

Presently, food is much more than a commodity to assure nourishment of a common being. Edible packaging is a type of packaging material that may be consumed or has the capability to biodegrade competently alike the food packed within it. Such packaging is prepared using various biopolymers and is continuously being improved and innovated with respect to its composition. An edible thin layer can be created into a stand-alone configuration and be used as a food wrap/cover. Edible packaging materials can be classified into two main groups' viz. coatings and films. Coatings are basically used in liquid form, while films are obtained as stand-alone configurations and then used as a packaging which can be further discriminated based on their thickness. Stand-alone layers of material which are thin (0.050–0.250 mm) are defined as films and the one's which are thicker (>0.250 mm) are named sheets. There are numerous advantages associated with natural polymers as coatings and films that is, biodegradability, edibility, biocompatibility, barrier properties, etc. On the other hand, biobased, nanostructured materials are additionally endowed with advanced surface area to volume ratio, in comparison to their micronized counterparts (Kadzińska et al. 2019).

17.2.2 Currently Used Packaging Materials and Their Sources

Diverse materials are being used in formulations of edible packaging viz. proteins, lipids, polysaccharides, resins, etc. To increase flexibility a plasticizer is the most common choice. New additives can be pooled to amend and improve physical properties or functionality of these films (Sakkara et al. 2020).

An imperative research inclination has been exploring food industry by-products and waste as possible edible packaging entity, such as starch from potato chip waste, chitosan from crustacean shells, whey from cheese production, zein (corn protein) from ethanol production, fish proteins from surimi wash water (Bourtoom et al. 2006; Nataraj et al. 2018; Sakkara et al. 2020), mung bean protein and starch (Bourtoom 2008), and pomace from beverage production from fruits (Park and Zhao 2006). It is also possible to produce films from fruit and vegetable purees (Rojas-Graü et al. 2006; Du et al. 2008; Sothornvit and Rodsamran 2008; Azeredo et al. 2009). Descriptions of general materials that form films are mentioned next.

Proteins

1. Wheat gluten.
2. Corn zein.
3. Soy protein isolate.
4. Collagen and gelatin.
5. Milk protein.

(a) Caseins.
(b) Whey proteins.
(c) Other proteins.

Polysaccharides
1. Cellulose derivatives.
2. Starch.
3. Chitosan.
4. Pectin.
5. Alginate.
6. Carrageenan.
7. Exudate gum.
8. Seed gum.
9. Microbial polysaccharides.

Lipids
1. Glycerol esters.
2. Waxes.
3. Resin.

Composite Materials
1. Film additives.

17.2.2.1 Polysaccharide as a Source for Edible Biofilms

An excellent example of biological polysaccharide is bacterial cellulose which is extruded by *Gluconacetobacter xylinus* and yields a 3D nanofibrillar pure cellulosic network (Rühs et al. 2018). The properties which are associated with bacterial cellulose and make it suitable for wrap/film are in situ moldability, high tensile strength, water holding capacity, biocompatibility, and biodegradability. In the case of biomedical fields, the bacterial cellulose is potentially being used for the preparation of temporary skin substitutes and artificial blood vessels owing to its novel properties. "Nata de coco" is a high fiber food and is used in low calorie preparations especially in Asian countries. This is obtained by fermenting various substrates in natural or controlled fermentation processes. A variety of fruits and vegetables such as purees, residues, extracts, and juices have been explored in terms of their matrix-forming ability to produce edible packaging materials to be applied to food products influencing their shelf life, quality, and improving the efficiency of synthetic packaging. These packaging can be useful for a variety of foods in order to improve their quality and appeal. Furthermore, some portions/forms of fruits and vegetables may be included into the matrix-forming solution so as to reduce costs and time of the production process. To improve global food quality, sensory perceptions, and convenience, fruits and vegetables based packaging materials can also serve as carriers of aromas, colorants, flavors, spices, nutrients, and antibrowning and antimicrobial agents.

17.2.2.2 Lipid as a Source for Edible Biofilms

A range of compounds is available for mounting the hydrophobicity of lipid-based edible films and coatings. The most common hydrophobic substance potentially used includes biowaxes (rice bran wax, carnauba wax, candelilla wax, and beeswax); petroleum-based waxes (such as paraffin and polyethylene wax); petroleum-based oils, mineral oils, and vegetable oils; and acetoglycerides and fatty acids (Rhim and Shellhammer 2005). To impart gloss to the commodity mostly resins are applied. The edible biofilms have excellent water vapor barrier properties, but generally are opaque and relatively inflexible. Edible films and coatings are prepared from dispersion or solution of the film-forming agents, followed by film-forming operations such as casting, spraying, dipping, extrusion, and falling film enrobing. Film formulation methods and film preparation methods decide various film properties such as barrier, mechanical, thermal, and optical properties. Other external conditions such as relative humidity differentials across the film, the temperature, and the packaging methods used also play an important role in establishing film properties. Lipid-based edible films and coatings are very useful for packing fresh and processed food products, minimally processed fruits and vegetables, and in designing active edible packaging.

17.2.2.3 Composite Edible Films

The choice of active substances that can be included in the coatings is affected by two main factors:

1. Edible compounds.
2. Consumer perception of the ingredients mentioned on the product label.

Each biomaterial such as polysaccharide, starch and its derivatives, alginates, chitosan, pullulan, and pectin, lipid (waxes, acetyl glycerides, and shellac), and protein (corn zein, gelatin, whey protein, and wheat gluten) possess unique functionality and used as a base for films and edible coatings (Nataraj et al. 2018; Sakkara et al. 2020). The addition of any bio ingredients often with bioactive properties results in the enhancement of the adhesiveness of the edible packaging and the durability on the food facade during processing, storage, and transport (Meghwal and Goswami 2014). Therefore, it is possible to develop contemporary formulations of packaging materials, such as composites or microemulsions, with high barrier properties for moisture retention and longer adhesion to the facade. For example, the incorporation of lipids and fatty acids can improve barriers to the moisture, vapor, gas, and solute. Wettability of the surface is also imperative for the appropriate adhesion of the coating due to the surface tension of the solid body being greater than or equal to the surface of the liquid. This can be overcome by adding a suitable surfactant to the coating solution.

17.2.3 Challenges Associated with Edible Packaging

There are certain desired functional attributes, viz. water or lipid solubility, color and appearance, mechanical and rheological characteristics, nontoxicity, solute and/or gas, moisture barrier for various films and wraps of edible and biodegradable nature. These traits are majorly governed by the material type, its formation, and application. These characteristics of most biofilms depend on several parameters related to the coating and film composition, such as preparation conditions (solvent, pH, components concentration, and temperature) and type of added additives (cross-linking agents, antimicrobials, plasticizers, and emulsifiers). These properties are strengthened further by the use of various plasticizers, cross-linking agents, antimicrobial agents, antioxygen agents, and texture agents.

The governing forces in any polymeric packaging film or coating are grouped as:

1. Cohesion: between the film-forming polymer molecules for all polymeric films or coatings.
2. Adhesion: between the film and the substrate for coatings only.

Various film properties viz. resistance, flexibility, permeability, etc., are affected by the degree of cohesion. The degree of cohesion in material development depends on the chemistry of biopolymer and structure, the fabrication procedure and parameters, the presence of plasticizers and cross-linking additives, and on the final thickness of the film, and cohesion is favored by high chain order polymers. Strong cohesion reduces flexibility, gas, and solute barrier properties and increases porosity. Industrial production employs the use of excessive solvent evaporation or cooling may sometimes produce noncohesive films due to impulsive immobilization of the polymer molecule. There are a number of attributes which are associated with the limited use of these biofilms, such as mechanical (flexibility and tension), thermal, optical (brightness and opacity), wettability, and morphological properties of edible films. These properties govern the surrounding atmosphere that influences the transfer of gas and further becomes a barrier for the transfer of aromatic compounds.

Mechanical Properties: Various natural phytochemicals such as coloring pigments (anthocyanins and carotenoids), phenolic compounds, and ascorbic acid aside from their nutritional and antioxidant value were found to improve polymer matrix organization by falling free volume during controlled dispersion. One of the studies pertaining to influence of the mango leaf extract on the optical character, morphological traits, water contact, and mechanical characteristics of composite chitosan films revealed that the extract concentration was directly proportional to the thickness of the film and resulted into reduced moisture content in edible packaging materials. It was also found that because of mango leaf extract inclusion water solubility, contact angle, and vapor permeability established reduced hydrophilicity and water vapor penetrability of the developed films. It was noticed that biofilm thus developed using mango leaf extract possessed reduced elongation ratio but enhanced tensile strength in comparison to the packaging material based on pure chitosan (Nataraj et al. 2018).

Edible Packaging and Their Barrier Properties

- Barrier to atmosphere.
- Barrier to mass transfers within foods.
- Barrier to moisture (water vapor permeability).
- Barrier to gases (oxygen and carbon dioxide permeability).

The most common deterioration of food products is due to oxidation of lipids in case of fat-rich products and food ingredients, discoloration of myoglobin in fresh meat, and enzymatic browning in case of fresh fruits and vegetables produce. The desirable trait to pack fat-rich and oxidation prone commodities is low oxygen-permeable biofilms, therefore by employing edible packaging with low oxygen permeability (OP) not only ensures better food quality, but also extends shelf life in addition to decreasing practice of using costly nonrecyclable O_2 barrier plastics. Baldwin et al. (2011) mentioned advanced development of edible films which were selectively permeable with defined gas permeability under set of storage conditions. This is to ensure that there is a favorable atmosphere within the pack, restraining the respiration rate of horticultural harvest and/or the ethylene production of physiologically active climacteric produce especially during storage and distribution. At lower pH most of hydrocolloid-based films usually have impressive gas barrier properties. While during high humidity storage (RH), it was found that hydrocolloid-based films and hydrophilic EVOH films are plasticized by absorbed moisture, leading to compromised barrier properties. It is desired that the packing material not only ensures retention of volatile compounds to preserve its characteristic flavor or aroma, but also at the same time the immigration of undesirable off flavors into packaged food during food supply chain (storage and distribution) (Meghwal and Goswami 2014). To ensure barrier properties of any biological packing material for a flavoring compound, it is needed to design the process and develop a product in such a way that the migrating compound has low affinity to film materials and low diffusivity through the polymer matrix. The water loving nature of protein- and polysaccharide-based edible films compose exceptional barriers to nonpolar aroma compounds. And therefore, for encapsulating flavor and aroma, starch and protein-based emulsion biofilms are always preferred and employed (Rosenberg and Lee 2004; Pegg et al. 2007; Fabra et al. 2009; Hambleton et al. 2009; Sakkara et al. 2020). The rationale behind this approach is to retain the hydrophobic organic flavoring compounds or active ingredients in a nonpolar lipid dispersed phase. On the other hand, the matrix made of hydrophilic polymer preserves both aroma loss to the surrounding or oxidation. Debeaufort and Voilley (1994) and Miller and Krochta (1998) have proposed a feasible method of determining flavor permeability of biofilms, which was comparatively tedious in comparison to the measurement of moisture and gas transport.

Similar to other desired properties of edible films, it is also imperative to map their thermal stability at varied temperature ranges. This not only enables to understand their process ability, but also enables to assess the feasibility of their industrial scale production and their potential applications. The previous research finding suggests that cellulose and lignin fibers have positive effects on thermal stability of edible packing. While incorporating a higher amount of grape pomace adversely affected thermal stability of edible starch packaging materials. The reason behind this attribute is its plasticizer nature which weakens intermolecular interactions resulting in lowering the values of thermal decomposition of various biofilms. Thermal stability was found to be increased by the addition of guava puree to hydroxypropyl methylcellulose biofilms as it increases its glass transition temperature (Sakkara et al. 2020). Very low values of thermal stability were recorded in case of mango puree films developed by the inclusion of cellulose nanofibers. This resulted in poor chemical stability of the material owing to high molecular mobility and high reactivity of its components. But at the same time, a promising trait of having good flexibility at lower temperature was one of the desirable traits for the products, especially at sub-zero temperature storage as desired in case of frozen processed foods (Meghwal and Goswami 2014).

17.3 Factor Affecting Mechanical Properties

17.3.1 Mechanical Properties Depend on the Following Factors

17.3.1.1 Molecular Weight

Mechanical properties are significantly affected by the molecular chain length, molecular weight, and the molecular weight distribution. Weak van der Waals forces bind the polymer chains loosely at lower molecular weight and the chain can shift easily, which is responsible for lower strength. However, the chain becomes large in case of large molecular weight and hence the chain is tightly bonded, which gives strength to the material. Molecular weight of a polymer molecule is equal to the degree of polymerization (DP) multiplied by the molecular weight (MW) of the repeating unit (monomer). The average number of repeating units in a single molecule of a polymer is known as the degree of polymerization (DP). Thus, PE with a DP of 10,000 would have a MW of 280,000, since the MW of the repeating unit [–CH2CH2] is 28. Typically, the range of molecular weight of polymer is between 50 and 200 kDa. As the polymer molecules are made by joining many monomers of different sizes during the polymerization reaction, it is necessary to know not only the MW but also the relative proportions of monomers and arrangement of the two structural units in the chain. The proportion of different sizes of repeating units in the chain may be characterized by molecular weight distribution. Two average molecular weights, number-average MW (M_n) and the weight-average MW (M_w), help to quantify the frequency of polymer chains with different lengths. The number-average MW (M_n) and the weight-average MW (M_w) are defined as:

$$M_n = \frac{\sum_i N_i M_i}{\sum_i N_i} \tag{17.1}$$

$$M_w = \frac{\sum_i w_i M_i}{\sum_i w_i} = \frac{N_i M_i 2}{\sum_i N_i M_i} \tag{17.2}$$

where N_i, M_i, and w_i are the number, molecular weight, and weight of a polymer molecule.

If the polymer molecules are averaged in terms of number fractions of various lengths, the M_n is obtained. The number-average MW (M_n) values does not depend on molecular size but is significantly affected by small molecules present in the mixture. However, many bulk properties such as viscosity and toughness depend as much on the size of the molecules as on their number. In this case, the appropriate function is the M_w, which is more dependent on the number of heavier molecules than M_n. As the molecular weight distribution (MWD) becomes increasingly broad, the weight average assumes numerically larger values than the number average, so that the ratio M_w:M_n (also known as the polymer dispersity index, PDI) measures the spread of molecular chain lengths. The value of PDI is always ≥ 1. If the molecular weight of all the polymers is the same, the value of PDI would be one. While if the weight of polymer molecules varies much, the value of PDI will be higher. A higher PDI tends to decrease the impact and tensile strength, but increase its processability. For most commercial polymers, PDI falls between 2 and 8.

17.3.1.2 Extent of Crystallization

The extent of crystallinity is an important factor affecting mechanical properties. In the crystalline region, chains are arranged in orderly fashion as the chains are parallel and closely packed, whereas in the amorphous region, the molecular chains are randomly arranged. The structure of chain (linear and branch) affects the possibility of an ordered arrangement and hence reduces the crystallinity. For example, HDPE (high-density polyethylene) has high crystallinity (75–90%) due to its linear polymer chain, whereas LDPE (low-density polyethylene) has lower crystallinity (usually between 55% and 70%) due to the presence of chain branching. Hardness, stiffness, and modulus of material increase with an increase in crystallinity due to the close order packing between chains. However, because of the substantial interstitial space between the chains, the polymer is able to absorb more energy by deformation and thus contributes to the toughness and flexibility. Oriented films are generally tougher than either amorphous or unoriented crystalline materials because in oriented film the film has stretch and realigns other molecules. Most of the packaging materials are semicrystalline, having a controlled proportion of amorphous and crystalline structures, and thus can attain a material with superior characteristics of stiffness and strength. The extent of crystallinity affects the physical nature of the polymer at above glass transition temperature which is shown in Table 17.1.

Normally, an increase in crystalline brings increased density, tensile, and compression strength, while decreased impact strength, tearing resistance, toughness, and elongation.

Table 17.1 Effect of crystallinity on the physical nature

Degree of crystallinity	Physical nature
Low amorphous (<10%)	Rubbery
Medium (20–70%)	Tough, leathery
High crystalline (>70%)	Hard, stiff

17.3.1.3 Composition

The composition is a primary factor controlling the properties of packaging materials. Composition is determined by the chemical structure and molecular weight. The additives used to enhance the functionality of packaging polymers also greatly influence the properties. The composition may affect the morphology which in turn significantly influences the mechanical properties of the product. The materials are composed of several molecules and the intermolecular force between these molecules affect the material's mechanical properties. The low strength of the material is mainly due to the weak intermolecular forces. The covalent and metallic bonds are stronger than the intermolecular forces, so the glass and metal are stronger and stiffer than plastics and paper. Bond polarity also affects the properties of material. The more polar the molecules, the stronger are the intermolecular force and the stronger molecular forces responsible for higher mechanical properties.

17.3.1.4 Temperature of Application

The mechanical properties of the polymer are greatly affected by the temperature; the tensile strength and elastic modulus are inversely related to the temperature, while the ductility improves with temperature. The relaxation modulus is found to decrease with increase in temperature and time. The mechanical properties of plastics are also affected by the temperatures at which they undergo physical transitions. The important transition temperature for the packaging materials are usually melting point, glass transition, and the crystallization temperature; the crystallites disappear when the polymer is heated and the temperature at which crystalline regions vanish is called the crystalline melting point. At normal temperatures, the cross-links make the solid quite rigid. If the temperature is raised to the point where the cross-links are broken, then irreversible chemical processes also occur that destroy the mechanical properties of the packaging materials. The glass transition temperature (T_g) is the property of the nanocrystalline (amorphous) polymers, whereas the crystalline polymers are depicted by the crystalline melting temperature (T_m). As temperature increases, the stiffness reduces and amorphous material becomes easier to shape, however, the crystalline material is strict in the extent of deformation until the melting temperature is reached. The physical appearance of a polymer depends on thc values of crystalline melting temperature and glass transition temperature relative to room temperature which is shown in Table 17.2. Other properties such as heat capacity, stiffness (modulus), gas permeability, dielectric properties, and refractive index also change at glass transition temperature. T_m and T_g are important parameters that define the upper and lower temperature limits for numerous applications.

Table 17.2 Effect of temperature on physical appearance

T_m and T_g lie below room temperature	Liquid
Room temperature lies between T_m and T_g	Viscous supercooled liquid or a crystalline solid
T_m and T_g are above room temperature	Glassy, brittle

The crystalline melting and glass transition temperatures are also governed by the strength of the intermolecular forces, degree of flexibility, and length of the chains. Thus, polar side groups, such as chloride and hydroxyl groups, favor higher melting and glass transition temperatures because they enhance the strength of the intermolecular bonds.

17.4 Mechanical Properties and its Test

17.4.1 Introduction

The major function of any packaging material includes protecting the contents inside it against various stresses like shock, compression, vibration, atmospheric conditions like humidity, temperature that can damage or reduce the quality of the product. To ensure that a packaging material is strong enough to provide protection, it is tested on various parameters in a laboratory to know its suitability for use. There are many ways in which the same test can be performed to know a particular property of the packaging material and hence yield different results. Therefore, to maintain uniformity in the testing method and procedure, we follow the procedures set-up by local governing authority or internationally accepted standards. The American Society for Testing Materials (ASTM) is widely used in the United States, and standards developed by International Standards Organization (ISO) are internationally recognized and followed. Aside from these two standards, International Safe Transit Association (ISTA) is also used in many tests. Before choosing a standard for testing the material (to be tested), the method and the regulatory guidelines should be considered.

17.4.2 Tensile Strength Test/Tensile Properties

17.4.2.1 Definition and Importance

"Tensile" basically means ability to be drawn out or stretched. Tensile properties are one of the most important properties for packaging materials, as based on these properties their use can be determined. Generally, four separate mechanical properties, which are tensile strength (measured in MPa), elongation at break (%), yield stress (MPa), and elasticity, are categorized under the heading of *tensile properties* as the equipment used for testing all of them is the same.

Tensile property values can be obtained directly by performing individual tests or can be calculated using an equation from the "stress-strain" graph produced by

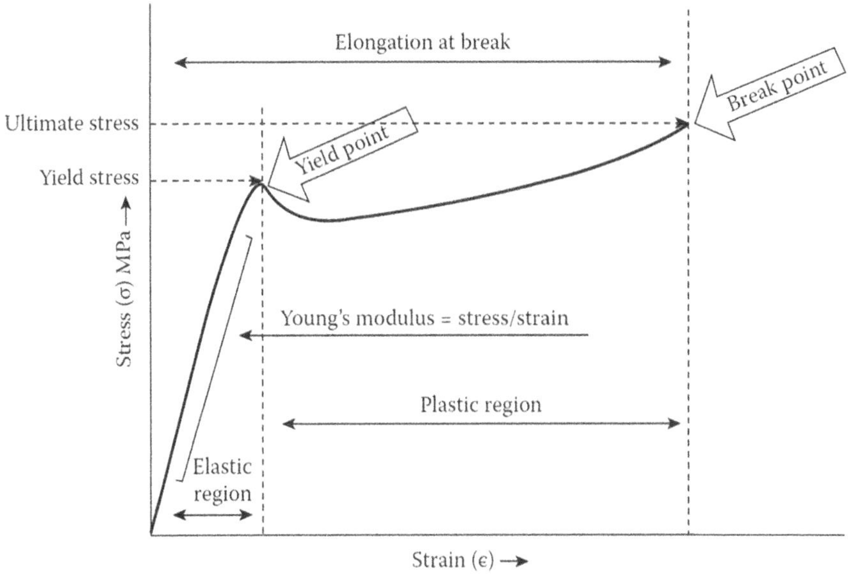

Fig. 17.1 Stress-strain graph (food packaging materials: testing and quality assurance)

varying the stress or load and keeping the rate of crosshead travel constant. The graph is shown in Fig. 17.1.

Before moving forward, it is important to discuss "stress" and "strain." Stress generally represented by 'σ' is defined as *force* applied per unit cross-sectional area of a material, it is often confused with pressure, but the key difference between pressure and stress is that in pressure we consider only the normal force that is the force applied perpendicular to the cross-sectional area not in any other direction. Thus 'σ' is given as (Eq. 17.3):

$$\sigma = \frac{F}{A}\ \mathrm{N\,m^{-2}(Pa)} \quad (17.3)$$

Strain represented by 'ε' is defined as the change in length relative to the original length, that is, fractional change in length. It is a dimensionless quantity and is expressed as (Eq. 17.4):

$$\varepsilon = \frac{\Delta L}{L} \quad (17.4)$$

Tensile properties of materials depend on their rigidity, thickness, test conditions, and sample preparation mainly. Based on these parameters, the testing method to be used is determined.

17.4.2.2 Testing Methods

According to ASTM D638, Standard Test Method for Tensile Properties of Plastics, the samples to be tested can be classified based on their thickness and rigidity into four categories, and each category has its requirements of seed and starting strain rate which are shown in Table 17.3.

Generally followed standards for tensile testing are given as follows:

- ASTM D618, Standard Practice for Conditioning Plastics for Testing.
- ASTM D638, Standard Test Method for Tensile Properties of Plastics.
- ASTM D6287, Standard Practice for Cutting Film and Sheeting Test Specimens.
- ASTM D6988, Standard Guide for Determination of Thickness of Plastic Film Test Specimens.
- ASTM D882, Standard Test Method for Tensile Properties of Thin Plastic Sheeting.

Generally, for calculating *tensile* properties using the graphical method at least five specimens are required. The individual properties can be found out in the following manner:

1. Tensile strength—It is defined as the maximum amount of load (in Newtons) that a material can withstand before breaking. It is calculated, according to ASTM D638 standard, by dividing stress by strain as given in Eq. 17.1.
2. Yield strength—It can be defined as the stress at which an elastic material undergoes nonelastic deformation. This stress point is also called *yield point*. It is calculated according to ASTM D638 standard, using the following Eq. 17.5:

Table 17.3 ASTM (D638) specifications for speed of testing

ASTM(D638) specifications for speed of testing			
Classification	Specimen type	Speed of testing (mm/min)	Strain rate at beginning (mm/min)
Rigid or semirigid	I, II, III rods and tubes	5 ± 1.25	0.1
		50 ± 0.5	1
		500 ± 0.1	10
	IV	5 ± 1.25	0.15
		50 ± 0.5	1.5
		500 ± 0.1	15
	V	1 ± 0.25	0.1
		10 ± 2.5	1
		100 ± 25	10
Nonrigid	III	50 ± 5	1
		500 ± 50	10
	IV	50 ± 5	1.5
		500 ± 50	15

$$\text{Yield strength} = \frac{\text{Yield stress}(\sigma)}{\text{Cross sectional area}\,(\text{A})} \tag{17.5}$$

Elongation at break—Elongation is an important property for a packaging material as it represents its ability to stretch. Greater ability to stretch before breaking shows a greater ability of the packaging material to absorb greater shock/mechanical damage before breaking, thus providing better protection to the contents inside it.

It is calculated according to ASTM D638 by dividing the length of elongation at break by length of original specimen sample. It can be calculated as Eq. 17.6:

$$\text{Elongation at break}\,(\%) = \frac{L_{\mathrm{B}}}{L_{\mathrm{o}}} * 100 \tag{17.6}$$

where L_{B} = length of stretched test specimen at break, L_{o} = length of original test specimen before stretching.

Elasticity—Elasticity is a measure of the ability of a body to resist a deforming force and return to its original shape once the deforming force is applied. In other words, it gives an idea about the internal stiffness of the film. Elasticity calculated using the stress-strain graph using ASTM D638 is called *Young's Modulus* of Elasticity. It is calculated by finding the slope of the linear portion of the stress-strain curve represented by "E" and having units of Nm^{-2} (Pa) same as that of units of pressure. The larger the value of E, the elastic (stiffer) the material is and more suitable for being used as secondary packaging material.

Industrially there is a more common and widespread method of quantifying *elasticity* called the Secant Modulus of Elasticity. It can be defined as the ratio of nominal stress to strain at any given point on the stress-strain curve or as the slope of a line drawn between a point on a stress-strain graph and the origin. It also has the same units as Young's modulus. Unlike Young's modulus, it can be calculated automatically using computer software, whereas Young's modulus usually needs to be calculated from graph manually, hence it is more preferred in an industrial setup (Eq. 17.7):

$$E = \frac{\sigma}{\varepsilon}\mathrm{Nm}^{-2} \tag{17.7}$$

Some key points to keep in mind in context of the scope of this book also will be the testing of biobased packaging material like biopolymers, biodegradable packaging materials, and edible films. These materials generally have sensitive characteristics and lower mechanical strength, in general, compared to traditional metallic or polymeric materials. It is important to note that the principle of testing remains the same for testing any material, the only slight modifications are made in the sample preparation, testing environmental conditions like temperature and RH, and testing parameters according to individual biobased material to be tested as testing methods are heavily dependent on the material.

17.4.3 Bursting Strength Tests

The bursting strength refers to the resistance of a material against applied pressure and is a measure of the strength of film used. The test method usually employed is the "Jumbo Muller Tester" or "bursting strength tester." Under this method, the sample is fixed between clamps and subjected to unwaveringly increasing pressure hydraulically exerted on a rubber diaphragm (30.48 mm diameter) underneath the sample until it ruptures. The maximum pressure required to rupture the sample is known as bursting strength.

17.4.4 Impact Strength Tests

Impact strength of a material can be defined as the ability to withstand fracture after being subjected to a sudden force. Several test methods have been proposed for assessing the impact properties of polymeric materials such as Izod popularly used for plastic materials and Charpy (pendulum) impact tests for metals. These tests are used primarily to assess material specification and quality control. The Izod pendulum impact toughness test is described in the standard ASTM D256. This method involves the determination of the energy needed to break a test specimen clamped at the ends and then struck in the center by a pendulum weight. The difference in potential energy of the hammer before and after the impact is expressed as:

$$E = mg(hS - hE) \tag{17.8}$$

where h is the height of the hammer, m is the mass, and g is the gravitational acceleration (9.81 m/s^2). The results may be described as energy lost per unit cross-sectional area at the notch ($J\ m^{-2}$) or ($ft \cdot lb./in^2$). It must be noted that the total impact energy is dependent on notch shape and length as well as the test specimen size. ASTM D3420 is the standard test method for Izod pendulum impact resistance of plastic films. As a technique to measure toughness, this test method helps determine parameters of material at strain rates more alike to some end-use applications than given by low-speed uniaxial tensile tests. If the film is used as a packaging material, the dynamic tensile behavior of a film becomes necessary. The impact strength is calculated from the difference between the potential energy of the pendulum at the maximum height of its mechanical, optical and barrier properties of thermoplastic polymers and the potential energy of the pendulum after rupturing the sample. Falling dart impact testing as per ASTM D1709 and ISO 7765 is a conventional method for assessing the impact strength or toughness of a plastic film. The dart is inserted into the testing bracket and released onto the center of the test specimen with samples clamped at the base of the drop tower. The weight of the dart and the pass/fail results are recorded.

When the specimen fails, the drop weight is decreased, and vice versa when it passes. The impact *failure weight* is calculated, that is, calculated weight for which 50% or more of the test specimens will fail under the impact at a given height.

17.4.5 Tear Strength Tests

Tear strength or tearing resistance measures how well a material can withstand the effect of tearing, that is, energy absorbed by the test sample in propagating a tear that has already been initiated by cutting a small nick in the test piece. Tear strength is commonly noted in tearing force in millinewtons (mN) with specimen thickness also reported.

The method used for measuring tear strength is the Elmendorf tear tester (ASTM D1922). It consists of a pendulum and a vertical bracket fixed onto a metallic bare. The test specimen is placed between the two jaw clamps when the pendulum is in a raised position. A knife is mounted on the vertical arm that tears through the test specimen. This machine is most suitable for paper and least suitable for testing rigid materials like nylon, polyester films, and rigid PVC. Tear strength is represented as force per unit of specimen thickness—pounds-force per inch (lbf/in), kilograms force per centimeter (kgf/cm), or kilonewtons per meter (kN/m). Another test method tear-propagation resistance (ASTM D1938) is useful in rating the tear-propagation resistance of various plastic films and thin sheeting of comparable thickness. It takes into consideration the force necessary to propagate a tear in plastic film and thin sheeting (thickness of 1 mm [0.04 in.] or less) by a single-tear method. The method is not applicable for film or sheeting material where brittle failures occur during testing.

17.4.6 Stiffness Tests

Bending stiffness also known as flexural rigidity refers to the resistance offered by material while undergoing bending. It is used to understand the relationship between the load and deflection that may be used to determine the modulus of elasticity or stiffness of a given paper or paperboard. Simply said, bending stiffness is the ability to resist an applied bending force proportional to EI, where E is Young's modulus and I the moment of inertia. The Handle-O-Meter stiffness tester measures the combined effects of flexibility and surface friction of films. Stiff materials offer greater resistance to the motion of the beam as it moves into the slot. Rough materials also exert resistance as they are dragged over the edge of the slot. The combined resistance is the reported result.

17.4.7 Crease or Flex Resistance

Crease or flex resistance is a property measured by repeatedly folding the film backward and forward at a given rate. The number of cycles to failure is recorded as the flex resistance.

The flex durability is measured using ASTM F392 method—Gelbo Flex Tester—where sample size (200 $\times$ 280 mm) is attached to the flex tester mandrels. The flexing action consists of a twisting motion combined with horizontal compression, thus repeatedly twisting and crushing the film. After a predetermined number of strokes, the material is examined. In general, the formation of pinholes is determined by the use of colored turpentine and allowing it to stain through the pinholes onto a white backing. The number of pinholes is a measure for the flexing resistance of the material.

17.4.8 Coefficients of Friction

This property is considered as the "resistance to motion" and is determined by the surface adhesivity (surface tension and crystallinity), additives (slip, pigment, and antiblock agents), and surface finish (Hernandez et al. 2000; Bajpai et al. 2020). The coefficients are calibrated from initial and sliding friction of films. COF measurement is based on observation of the relative motion between two bodies in contact and the coefficient that can be expressed as a ratio of force related to friction from different surfaces. It is expressed as (Eq. 17.9):

$$F_{\mathrm{f}} = \mu * N \tag{17.9}$$

where F_{f} = frictional force (N), μ = static or kinetic frictional coefficient, N = normal force (N).

17.4.9 Blocking

Blocking is the tendency of two adjacent layers of film in intimate contact with nearly complete exclusion of air to stick together, for example, when films are stacked in sheets or compacted rolls. Temperature or pressure, or both, can induce or change the adhesion level of surfaces. It can also make bags made from lay flat film difficult to open. Blocking is also affected by static charges, surface treatment (e.g., printing pretreatment), and storage conditions. The film often adds antiblocking additives to reduce the tendency to block. These functions are carried out by spreading on the surface and forming noncohesive layers. In ASTM D3354 and ISO 11502, when applied force perpendicular to the film surface, the level of blocking is determined by the force required to separate the two layers of the blocking film (Meghwal and Goswami 2014).

17.4.10 Orientation and Shrinkage

Although orientation itself is not of primary practical importance, the way in which it affects the physical and mechanical properties of a film is important. Orientation is an important factor that influences tensile strength, impact strength, stiffness, and tear resistance. The measurement of the compressive stress and the orientation release stress of film is described in ASTM D2838. The appearance and performance of the packaging film is dependent on the "shrink tension" of the film also known as *compressive* stress. There are generally two methods to determine the compressive stress. (1) In the first method the film is heated to a fixed predetermined temperature and the maximum force it exerted without shrinkage is measured. These data are very important in determining the extent of shrinkage on exposure to a certain temperature, degree, and direction of the orientation and orientation release stress. (2) This method is used to measure the maximum force exerted by a film specimen that is allowed to shrink when heated rapidly to a certain temperature. The data obtained are used to estimate the final appearance of the package object and also the force an object receives when it is covered by the test specimen film.

17.4.11 Other General Properties

17.4.11.1 Mass

Mass is used to produce a force in test methods. Mass is measured by weighing the test piece or object in question using an appropriate balance or scales. As the magnitude and the accuracy needed vary, the weighing instrument has to be selected accordingly. Accuracies required are often written in terms such as accurate to 1 mg, whereas balances may be quoted as reading to 1 mg. The two are not the same, and the standards are not always clear. ISO 536:2012 specifies a method for determining the grammage of paper and board.

17.4.11.2 Thickness

Thickness is one of the important parameters of the packaging material. Thickness affects the various properties of packaging material, including stiffness and permeability. The thickness is the perpendicular distance between the two outer surfaces of material. Many physical properties of plastic materials are dependent on thickness. In the case of films, the resistance to gas transmission increases directly with an increase in thickness. Strength properties such as stiffness are also influenced by thickness. The method for determining the thickness is given in TAPPI T411. Generally, thickness is measured at different locations of the sample, and the mean is calculated. The thickness is measured by using a micrometer (Rhim and Ng 2007). Micrometers reading in one-ten-thousandth of an inch (0.0001 in.) or even typical fifty millionth (0.00005 in.) increments are simply not sensitive enough. The thickness can also be measured via a microscope (microtome cut) or also via a scanning electron microscope when it comes to very thin films (1 μm). Mean thickness for control soy protein isolate (SPI) and polylactic acid (PLA) films were 78.8 μm and 89.5 μm, respectively (Rhim and Ng 2007). Thickness values are reported in millimeters (mm).

17.4.11.3 Density

Density is an important parameter for packaging material. Density is defined as the mass per unit volume of the tested material. Its SI unit is kg/m^3, but is also expressed as g/cm^3. Higher density value means higher transport and storage costs. Lighter density materials are also easy to handle and transport and less hazardous. Density is also used to calculate specific properties that is diving a mechanical property by its density, for example, specific tear strength, specific bursting strength, etc. Specific properties present a better understanding of the internal strength of the packaging material that one wants to build (Van de Velde and Kiekens 2002). ISO 1183 and ASTM D792 are the most widely used methods for determining the density of a packaging material.

17.4.11.4 Optical Properties

Unique properties, such as excellent clarity and transparency, good impact strength, moldability, and low cost, have made plastics the number one choice of many design engineers. Successful applications of transparent and translucent plastics include automotive tail-light lenses, safety glasses, window glazing, merchandise display cases, and instrument panels. Plastics are much more resistant to impact than glass; therefore, in applications such as street-lamp globes and high school windows, glass has been replaced by high-impact, vandal-resistant plastic materials such as polycarbonate.

Almost all plastics, below a certain minimum thickness, are translucent. Only a few plastics are transparent. Plastic materials' transparency or translucency depends on their basic polymer structure. Generally, all amorphous plastics are transparent. Crystallinity increases the density of the polymer, which decreases the speed of light passing through it, and this increases the refractive index. When crystals are larger than the wavelength of the visible light, the light passing through many successive crystalline and amorphous areas is scattered, and the clarity of the polymer is decreased. A large single crystal scatters light at wide angles and thus causes haze. As a rule, crystalline plastics are translucent. However, the clarity of crystalline plastics can be improved by quenching or random copolymerization. The primary optical properties are:

1. Refractive index.
2. Light transmittance and haze.
3. Color.
4. Gloss.

17.5 Conclusions

An understanding of the food product to be packed, knowledge of the packaging material, and insights from consumers are crucial in designing packaging. Mechanical properties of packaging materials, like tear strength, elasticity, flexibility, hardness, are an indicator of ability to withstand a drop, vibration compression, the barrier properties of edible packaging (gas/water vapor permeability), indicate the resistance to sorption and diffusion of gases, flavor, and aroma compounds through the packaging material, and are very important to determine the suitability of the material for the intended use. These tests and their importance are discussed in this chapter.

References

Azeredo HM, Mattoso LH, Wood D, Williams TG, Avena-Bustillos RJ, McHugh TH (2009) Nanocomposite edible films from mango puree reinforced with cellulose nanofibers. J Food Sci 74(5):N31–N35

Baldwin EA, Hagenmaier R, Bai J (eds) (2011) Edible coatings and films to improve food quality. CRC Press, Boca Raton, FL

Banerjee S, Meghwal M, Vivek CS, Mitra J (2017) Utilization of used packaging materials for steam pipes insulation. J Packag Technol Res 1(1):45–51

Bajpai A, Kumar Y, Singh H, Prabhakar PK, Meghwal M (2020) Effect of moisture content on the engineering properties of Jamun (Syzgium cuminii) seed. J Food Process Eng 43(2):e13325

Bourtoom T (2008) Factors affecting the properties of edible film prepared from mung bean proteins. Int Food Res J 15(2):167–180

Bourtoom T, Chinnan MS, Jantawat P, Sanguandeekul R (2006) Effect of select parameters on the properties of edible film from water-soluble fish proteins in surimi wash-water. LWT Food Sci Technol 39(4):406–419

Debeaufort F, Voilley A (1994) Aroma compound and water vapor permeability of edible films and polymeric packagings. J Agric Food Chem 42(12):2871–2875

Du WX, Olsen CW, Avena-Bustillos RJ, McHugh TH, Levin CE, Friedman M (2008) Storage stability and antibacterial activity against Escherichia coli O157: H7 of carvacrol in edible apple films made by two different casting methods. J Agric Food Chem 56(9):3082–3088

Fabra MJ, Talens P, Chiralt A (2009) Microstructure and optical properties of sodium caseinate films containing oleic acid–beeswax mixtures. Food Hydrocoll 23(3):676–683

Hambleton A, Debeaufort F, Bonnotte A, Voilley A (2009) Influence of alginate emulsion-based films structure on its barrier properties and on the protection of microencapsulated aroma compound. Food Hydrocoll 23(8):2116–2124

Hernandez RJ, Selke S, Culter JD (2000) Plastics packaging, 1st edn. Hanser Gardner, Cincinnati, OH

Kadzińska J, Janowicz M, Kalisz S, Bryś J, Lenart A (2019) An overview of fruit and vegetable edible packaging materials. Packag Technol Sci 32(10):483–495

LaCoste A, Schaich KM, Zumbrunnen D, Yam KL (2005) Advancing controlled release packaging through smart blending. Packag Technol Sci Int J 18(2):77–87

Meghwal M, Goswami TK (2014) Effect of grinding methods and packaging materials on fenugreek and black pepper powder quality and quantity under normal storage conditions. Int J Agric Biol Eng 7(4):106–113

Miller KS, Krochta JM (1998) Measuring aroma transport in polymer films. Trans ASAE 41(2):427

Nataraj D, Sakkara S, Meghwal M, Reddy N (2018) Crosslinked chitosan films with controllable properties for commercial applications. Int J Biol Macromol 120:1256–1264

Park SI, Zhao Y (2006) Development and characterization of edible films from cranberry pomace extracts. J Food Sci 71(2):E95–E101

Pegg RB, Amarowicz R, Naczk M, Shahidi F. (2007). PHOTOCHEM® for determination of antioxidant capacity of plant extracts

Rhim JW, Ng PK (2007) Natural biopolymer-based nanocomposite films for packaging applications. Crit Rev Food Sci Nutr 47(4):411–433

Rhim JW, Shellhammer TH (2005) Lipid-based edible films and coatings. In: Innovations in food packaging. Academic Press, London, pp 362–383

Robertson GL (2006) Modified atmosphere packaging. Food packaging: principles and practice, pp 313–329

Rojas-Graü MA, Avena-Bustillos RJ, Friedman M, Henika PR, Martín-Belloso O, McHugh TH (2006) Mechanical, barrier, and antimicrobial properties of apple puree edible films containing plant essential oils. J Agric Food Chem 54(24):9262–9267

Rosenberg M, Lee SJ (2004) Water-insoluble, whey protein-based microspheres prepared by an all-aqueous process. J Food Sci 69(1):FEP50–FEP58

Rühs PA, Storz F, Gómez YA, Haug M, Fischer P (2018) 3D bacterial cellulose biofilms formed by foam templating. NPJ Biofilms Microbiomes 4(1):1–6

Suhag R, Kumar N, Petkoska AT, Upadhyay A (2020) Film formation and deposition methods of edible coating on food products: a review. Food Res Int (Ottawa, Ont) 136:109582–109582

Sakkara S, Venkatesh K, Reddy R, Nagananda GS, Meghwal M, Patil JH, Reddy N (2020) Characterization of crosslinked Macrotyloma uniflorum (horse gram) protein films for packaging and medical applications. Polym Test 91:106794

Sothornvit R, Rodsamran P (2008) Effect of a mango film on quality of whole and minimally processed mangoes. Postharvest Biol Technol 47(3):407–415

Van de Velde K, Kiekens P (2002) Biopolymers: overview of several properties and consequences on their applications. Polym Test 21(4):433–442

Vanderroost M, Ragaert P, Devlieghere F, De Meulenaer B (2014) Intelligent food packaging: the next generation. Trends Food Sci Technol 39(1):47–62

Methods for the Improvement of Barrier and Mechanical Properties of Edible Packaging

18

Kulsum Jan, Sidra Kazmi, Aayasee Nanda, Khalid Bashir, Ab Lateef Khan, and Shumaila Jan

Abstract

Packaging is an indispensable fragment to preserve quality characteristics and improve storage, transportation and marketing of food products. Increased health interests in nutrition, safety and environmental issues have led to the development of edible packaging. Edible packaging developed from agro-industrial byproducts and wastes has always been a trend. Recent studies have shown reduction in the environmental wastes caused due to the non-biodegradable food packaging. Diverse edible packaging materials developed have barrier properties towards the transfer of oxygen and moisture and have been successfully used to replace synthetic polymer-based films. Much attention acquired by edible packaging is due to increased shelf life, enhanced mechanical properties, enrobing heterogeneous foods and preserving organoleptic characteristics. Biopolymers like pure starch find their way to be used for the development of films. Films produced from starch are hard with poor mechanical properties and are also difficult to handle due to high water sensitivities. Various methods have been employed to mitigate the shortcoming of films that are based on starch and involve its modification by physical, chemical and biological techniques. Modification changes the cross linkages of starch and addition of certain nontoxic

K. Jan (✉) · S. Kazmi · A. Nanda · K. Bashir
Department of Food Technology, Jamia Hamdard, New Delhi, India
e-mail: kulsumjan@jamiahamdard.ac.in

A. L. Khan
School of Biomedical Engineering, Institute for Personalized Medicine, Med-X Shanghai Jiao Tong University, Shanghai, China

S. Jan
Department of Food Science and Technology, National Institute of Food Technology Entrepreneurship and Management, Sonipat, Haryana, India

A. Poonia, T. Dhewa (eds.), *Edible Food Packaging*,
https://doi.org/10.1007/978-981-16-2383-7_18

functional additives. Research on techniques for the amelioration of mechanical and protective properties of these films which are edible in nature has endured rapid expansion in the past years. The current study envisages the methods and techniques for the improvement of characteristics of edible films.

Keywords

Biodegradable · Edible · Films · Modification · Starch

18.1 Introduction

Edible packaging materials such as films and coatings have gained popularity among food researchers, scientists and industries due to their inherent biodegradability and edibility as they can be directly consumed and thereby reducing waste disposal problems. Furthermore, they provide excellent barrier properties and enhance the overall quality of food products (Bourtoom 2008). Important factors taken in consideration while selecting materials for development of edible food packaging includes; their ability to improve processing and preservation techniques, their performance as a barrier to environment and excellent carrier of bioactive compounds. Generally, edible films work as an effective moisture barrier and they have the ability to prevent moisture exchange between food materials and surroundings, thus inhibiting undesirable enzymatic and chemical reactions, textural changes and microbial growth (Janjarasskul and Krochta 2010). These packaging films are likely to be capable of preserving the quality and enhancing the keeping quality of oxygen-sensitive food, therefore they can also work as a good oxygen barrier. Carrageenan and chitosan based edible film are the effective barrier to non-polar aroma compounds, inhibiting oxidation and loss of aroma (Hambleton et al. 2009a, b). Edible coatings based on hydrocolloid for example carboxymethylcellulose (CMC) and alginate also helps in keeping the internal moisture of the product intact and reducing uptake of fat in fried foodstuffs which have been deep-fried (Dragich and Krochta 2010). Selective permeability of edible materials can help in restricting aroma and flavor loss at the time of freeze-drying operations, whereas during brine-freezing operations, materials having high water vapor permeability can help in reducing salt migration into food (Janjarasskul and Krochta 2010). While selecting the various materials for the formulation of these edible packaging films, an important aspect of it being an excellent carrier of antioxidant or antimicrobial compounds and also its ability to restrict the migration of molecules from the edible package to the food product is highly considered for its various food product applications. There have been various researches in this area of edible materials for the preservation of food due to their good delivery properties. There are various advantages of edible films such as reduction of waste and solid disposal problems, enhancing organoleptic properties like color and sweetness and enhancing the nutritional values of food products by supplementation. Edible films are environment friendly as they can be fully consumed or can be biologically degraded or recycled. These edible films can be used in the form of an interface between the various layers of the food product which are heterogeneous in nature so as to prevent it from being deteriorated like in the case of strawberries. These films can be used as carriers of antimicrobial or antioxidant species and can also be used for the purpose of microencapsulation of flavoring agents. There are also various functional properties of these edible films such as retarding moisture migration in fruits and vegetables which are wax coated, retarding gas permeability of carbon dioxide and oxygen, retarding solute transport as well as oil and fat migration. Edible films also carry food additives, retain volatile flavor compounds, improve mechanical handling properties of food and enhance the nutritive value of food. The desired characteristics for the formulations of these films and coatings which are edible in nature depends majorly on its application on the food product which is getting coated. Therefore, for oxidation sensitive food products, low oxygen permeability is required.

There are various sources of edible films such as:

1. Proteins: Pea protein, Soy protein, wheat gluten, whey protein, casein protein, corn zein, collagen and gelatins, fish myofibrillar protein, egg white protein, peanut protein.
2. Polysaccharides: Starches and modified starches, pectins, kefiran, chitin and chitosan, Galactomannans, Cellulose & modified cellulose, Carrageenan, Xanthan gum, Gellan gum, Alginate, Pullulan.
3. Lipids: Oils, Free fatty acids, Carnauba wax, paraffin, beeswax, Shellac resin, Acetoglycerides, Terepene resin.

18.2 Applications of Edible Packaging Films

18.2.1 Oxygen Barrier

There are various reasons for food deterioration such as lipids oxidation and oxidation of food ingredients, enzymatic browning of fresh-cut produce or myoglobin discoloration in fresh meat cuts. Edible films with low oxygen permeability help in preserving the food quality and extending the keeping quality of oxygen sensitive food products while decreasing the usage of non-recyclable plastics. At low relative humidity (RH), the edible films which are based on the hydrocolloidal property mostly have excellent resistance towards gas. In case of high relative humidity (RH) due to the effect of moisture absorption, hydrophilic EVOH film and hydrocolloid-based films are plasticized, this results in the gradual reduction in the resistance or barrier properties of the edible films (Janjarasskul and Krochta 2010).

18.2.2 Moisture Barrier

The migration of moisture from the atmosphere to the food product can be inhibited by the use of modified edible films. Undesirable alterations in the texture of the food product, microbial growth, deteriorative enzymatic and chemical reactions can occur as a result of the alterations in the water content of the packed foods. It has been observed that edible films which are based on the hydrocolloidal properties have higher permeability of water vapor as compared to films prepared by plastics and edible waxes. WVP of hydrophilic films increases at high plasticizer concentration and high relative humidity (RH) due to their significant polarity. Therefore, only in the case of protection of food from moisture migration of food with low moisture content, these films can be used effectively as protective barriers for a short duration of time. Whereas, lipids or any other hydrophobic compounds have low polarity, dense-structured molecular matrixes and low water affinity, therefore they are generally used to prepare moisture barrier coatings or increase the moisture barrier properties of hydrocolloid-based edible films (Janjarasskul and Krochta 2010).

18.2.3 Aroma Barrier

It is extremely important to restrict the loss of characteristic flavor and aroma of the food product which are volatile in nature and the inclusion of off-flavor from the atmosphere into the packaged food during their distribution as well as the storage. In the case of low diffusivity of the compound that is migrating through the polymer matrix and low affinity to film materials, the barrier property of the packaging film is developed. These edible films having the polysaccharides with hydrophilic nature makes them an excellent barrier to the compounds that cause the aroma of the food product which are non-polar in nature. Therefore, the flavor and aroma encapsulation with the help of carbohydrates and protein-based emulsion films were proposed (Rosenberg and Lee 2004; Pegg and Shahidi 2007; Hambleton et al. 2009a, b). Preservation of the active ingredient in the dispersed phase which is non-polar in nature or the aroma compounds which are organic and hydrophobic in nature is the sole objective of this technology. The oxidation of the food product or its aroma loss is prevented by the matrix which is made up of a polymer having hydrophilic properties. When compared with the calculations and measurements of gas and moisture migration, the permeability of aroma compounds is limited and relatively challenging. Various methodologies for the determination of aroma permeability were also proposed (Debeaufort and Voilley 1994; Miller and Krochta 1998).

18.2.4 Oil Barrier

Grease resistance can be provided by edible packaging to any lipid-containing products. There is limited data to the quantitative information of the permeability of oil (Krochta 2002). The property of grease-resistance is expected to be shown by the inherent hydrophilicity of carbohydrate and protein-based polymer films. Examples include whey protein (De Mulder-Johnston 1999) and zein (Trezza and Vergano 1994) were observed to have excellent grease resistance ability.

18.3 Methods of Improvement

Films made from proteins are promising biomaterials since they are excellent gas barriers. However, protein films have certain limitations like poor resistance to moisture and water vapor due to their moisture absorbent nature and lack of mechanical strength. The method of cross-linking is an effective method to improve cohesion, mechanical strength, rigidity, water resistance and moisture restrictive property. There can be a usage of several functional groups of proteins to accomplish this. A broad spectrum of various active compounds can be interacted using protein networks. The various functional groups present on the side groups of it which are reactive in nature can do this. Functional properties of films are enhanced by modifying it via physical, chemical or enzymatic cross-linking.

18.3.1 Physical Modification

18.3.1.1 Casting

The commonly used technique for formation of a film in laboratory as well as pilot scales is casting method or also known as solvent casting. It involves three steps for the development of a film from biopolymers; firstly, suitable solvent is taken to solubilize biopolymer followed by the solution being casted in the mold itself and finally its drying takes place (Rhim et al. 2006). The specific polymer or the whole polymer mixture is being selected and this marks the initiation of the formulation of these edible films. The polymer which is chosen is dispersed or dissolved in a solvent which is fitting to it; like soy protein isolate polymer is dissolved using ethanol (Jensen et al. 2015); which is termed as solubilization. The polymers ability to solubilize is the main factor on which solvent casting of film formation is dependent upon, rather than its melting (Koide et al. 2013). During casting, glass plates coated with Teflon are used where the obtained solution is poured. The process of drying provides enough duration for the solvent to vaporize which results in the attachment of the polymer film to the mold. Air dryers such as vacuum dryers, tray dryers, microwave and hot air ovens are used for the casting of films so that the solvents can be easily removed (Cha-um et al. 2003). For the casting of edible films, the process of air-drying is a crucial step which helps in formulating an edible film with advantageous microstructure as well as improving the various interactions in the polymer chains (Sherrington 2003). Quick-drying methods applied for casting has resulted in undesirable outcomes on structural and physical properties (Velaga et al. 2018). For the development of edible films, there have been quite some studies on the various relations being the air-drying temperatures and methods (Kaya and Kaya 2000; Tapia-Blacido et al. 2013). Developed edible film should be consistent and does not have any imperfections such as non-consistency, mechanical harm and inclusion. The most essential parameters of edible films include transparency, thickness, opacity, thermal stability, swelling degree, physical strength as well as the ability to transport oxygen and moisture along with its biological characteristics (Skurtys et al. 2011; Kanatt et al. 2012; Khanzadi et al. 2015). The formation of an easily peeled edible film is done with the help of plasticizers and cohesive matrix which results in excellent barrier properties, mechanical strength, uniform microstructure and thermal stability (Park et al. 2008; Fakhouri et al. 2013). Owing to the significant increase in the quantity of plasticizers, there has been a positive impact on the mechanical strength, thermal properties and barrier properties of the edible film (Sothornvit and Krochta 2005; Sanyang et al. 2015).

18.3.1.2 Extrusion

Extrusion method is a crucial technique for the processing of polymer provided it is presently in application at commercial scale for producing polymeric films (Hernandez-Izquierdo et al. 2008). It improves the physicochemical properties and varies the structure of extruded products (Fitch-Vargas et al. 2016). Extrusion is generally classified into three categories: first, the zone in which the materials to be processed is fed, then the zone in which these fed materials are being kneaded and finally the zone in which the kneaded material is heated (Hauck and Huber 1989; Calderon-Castro et al. 2018). When the process initiates, the components of the mix are carried into the first zone and then are further. This process is also called a dry process as it works best with a minimum amount of solvent or moisture content. Plasticizers are needed during this process to enhance and increase the flexibility of the film (Peressini et al. 2003). The plasticizers which are extruded are polyethylene glycol or sorbitol which is generally used in a measure of about 10% to 60% weight for weight. The ingredients are passed into the kneading zone wherein the density, strain, and temperature of the mixture increases. To develop extruder-based edible film, the thermal energy (extruder barrel temperature) and mechanical (specific mechanical energy) are involved in this method (Fitch-Vargas et al. 2016). It was reported by (Wojtowicz et al. 2015) about the effect which the speed of the screw had on the specific mechanical energy. The various properties of starch-based films like shear rate, shear stress, and homogeneity are affected by the variations in the screw speed of the extrusion. They also control the time of residence and facilitate the addition and removal of the additives like stabilizers. Due to the increase in the speed of the screw, the torque value of the edible films which were based on starch was decreased (Su et al. 2009; Calderon-Castro et al. 2018). According to a study, after a duration of about half an hour and 2 min, the required specific mechanical energy for extrusion of cassava starch ranged from 242.73 to 56.81 KJ/Kg respectively at 150 rpm having a moisture content of about 30% (Fayose and Agbetoye 2012). The thermal energy usually is between the ranges of 120–170 °C (Pandit et al. 2018). This technique requires high pressure of about 200 bar for the mixing as well as for the desirable shape of the ingredients (Hanani et al. 2012). Various factors such as reduction in the water level conditions and the thermoplastic behavior of polymers when the glass transition temperature is surpassed during plasticization and heating. Multi-layer films can also be formed by the technique of co-extrusion and can also provide flexibility in order to attain necessary and satisfactory properties of these edible films. The structure of the film with multilayers developed along with the improvement in the functionality and processability of the films are developed by these multilayers in the edible films (Winotapun et al. 2019).

18.3.1.3 Antimicrobial Nanostructure Based Edible Films

The usage of antimicrobial compounds and nanostructures for the formulations of various food packages have been extensively studied upon (de Azeredo 2013; Sung et al. 2013). Major health related concerns are caused due to the inclusion of active compounds which are undesirable in nature from the contact materials of food to the matrix of the food. It is important to consider all of the components of edible packaging materials as a component of the food, given that the entire food can be consumed resulting in no harmful outcomes (de Azeredo 2013). Nano encapsulated antimicrobial peptides and nano-emulsions containing essential oils are the two basic methods of incorporating components having antimicrobial properties into the polymer matrices of the food that is being used to formulate films which can be both edible as well as have antimicrobial properties. Nano-emulsions are majorly used for its property of being antimicrobial and further the encapsulation of essential oils which are of plant origin to enhance shelf life and increase food safety (Pathakoti et al. 2017). Various other bioactive compounds like carotenoids, quinones antioxidants, fatty acids and phytosterols are also incorporated into these edible films (Salvia-Trujillo et al. 2017). Furthermore, they should possess the ability to be included in the edible films so as to produce active food packaging systems. According to a study, the edible based films showing antimicrobial activity based on hydroxypropyl methylcellulose (HPMC) contained as essential oil named *Thymus daenensis* against a spectrum of fungus as well as bacteria. The scientists illustrated that focusing on the origin of these essential oil (in other words, cultivated or wild *T. daenensis*), as a result more efficient antimicrobial films were developed against certain specific microorganisms due to their varying compositions (Moghimi et al. 2017). In a study, edible films were produced based upon less or more methyl ester pectins and cinnamaldehyde incorporated papaya puree with nano-emulsions of different sized droplets. Smaller droplets create great inhibition zones result in increased efficiency of antimicrobial activity against pathogenic and bacteria which cause spoilage of food products because of their high bioavailability as well as surface areas (Otoni et al. 2014a). In continuation with this observation, a study also reported that increase in efficiency of antimicrobial activity of edible films based on methylcellulose when these were incorporated by nano-emulsions of oregano and essential oils from clove bud as compared to that of coarse emulsified films (Otoni et al. 2014b). Similar results were observed for different other materials as well (Gul et al. 2018).

18.3.1.4 Irradiation

There has been an immense usage of irradiation technique in order to improve the properties of edible packaging films. Both barriers as well as mechanical properties of edible films based on proteins were improved effectively by inducing cross-linking through irradiation method. The protein content of these films are generally affected by irradiation which causes rupture of covalent bonds, oxidation of amino acids, formation of protein free radicals, conformational changes and rupture of covalent bonds. Proteins can be converted into higher molecular weight aggregates by the formation of electrostatic, hydrophobic interactions, development of disulphide bonds as well as inter-protein cross-linking reactions (Davies and Delsignore 1987). The molecular properties of proteins can be modified by superoxide and hydroxyl anion radicals which are formed by radiation of film-forming solutions. The cross-linkages which are covalent in nature are developed in the solution of protein due to irradiation and these can further modify and improve the protein films (Garrison 1987).

It has been suggested that the genesis of aggregates of high molecular weights are negligible at the range of low-doses, but gradually increases at higher doses. Improvement in protein films by irradiation treatment like gamma irradiation has been immensely used to improve and develop the proteins. Gamma irradiation cross-linking helps in improving the chemical stability and the water vapor permeability of milk protein based films (Ouattara et al. 2002). There was an increased resistance to enzymatic and microbial biodegradation and significant decrease in the water vapor permeability due to gamma irradiation. It has also been observed that the film forming solution had an increased high molecular weight protein concentration. The researchers pointed out that the effect on gamma irradiation can be explained by two hypotheses. The first hypotheses being the presence of more molecular residues in intermolecular interactions when used in proteins with varying physico-chemical properties and the second being the formation of intra- and/or intermolecular covalent cross-links in the film-forming solutions.

A study investigated that gamma irradiation causes various effects on mechanical properties of milk proteins. The gamma irradiations-initiated formation of a structure of protein gel which was finely stranded. Due to this, the viscosity of proteins film solutions which have undergone the process of irradiation was enhanced compared to that of control films (Ciesla et al. 2004). It was seen that there was significant improvement in the β-conformation as compared to non-irradiated milk protein-based films due to application of gamma irradiations. Furthermore, due to the presence of protein conformations which are correctly ordered in the gels derived from irradiated solutions, the formation of more crystalline films takes place. As a result, these films show excellent mechanical strength, increased rigidity and improved barrier properties when compared to solutions which were not irradiated.

Lee et al. (2005) reported that a significant amount of disruption was found in the protein molecules which had ordered structures when treated with gamma irradiation. It changed the tensile strength as well as the elongation at break. Further it also improved its water vapor permeability. Since there is an increased accumulation of polypeptide chains, the tensile strength also increases significantly. Therefore, it is observed that the reduction in diffusion rate through the film caused the decrease in water vapor permeability of the edible film.

The various properties of mechanical strength and water resistance of corn protein-based films can be modified by the application of gamma irradiation (Soliman and Furuta 2009). The physicochemical properties of zein based edible films are potentially modified with the help of gamma irradiation treatment, specifically the properties of moisture resistance. The absorption of water molecules into the film was done by its treatment with proteins having high molecular weight procured from cleaved polypeptide and disaggregated protein particles and the diffusion through these films can be reduced by the formation of the earlier mentioned linkages.

18.3.2 Chemical Modifications

The film forming properties can also be improved using treatments with chemicals like alkali, acid or cross-linking agents. Less permeability and greater tensile strength were obtained when there is an increase in the protein structure interaction. It was observed by (Gennadios et al. 1993) that when the isolate of soy protein is subjected to an alkaline treatment, the oxygen permeability, tensile strength and water vapor permeability remained unaffected, but appearance of the film was improved by making it more uniform and clear and also reduction in the occurrence of air bubbles along with break at the elongation. Various chemical agents like aldehydes such as glutaraldehyde, formaldehyde or gloxal are used for the purpose of protein's covalent cross-linking.

Formaldehyde has extensive reaction specificity and is considered as the simplest of cross-linking agents. It reacts with the side chains of tyrosine, cysteine, tryptophan, arginine and histidine in addition to amine group of lysine. Formaldehyde can cross-link due to its property to react bi-functionally. It was seen that formaldehyde was less specific than glutaraldehyde. Glutaraldehyde was able to react with cysteine, tyrosine, lysine and histidine (Kim and Tae 1983). Lysine and arginine side chain groups are involved in the protein cross-linking by glyoxal (Marquie 2001) at alkaline pH. In general, there are two step processes taking place during the reaction between protein and formaldehyde. There is the formation of methylol compounds in the first step and formation of bridges of methylene which act as cross-links between the various protein chains in the second step.

The usage of aldehydes and their various effects of cross-linking for the development of glutenin-rich films were studied by Hernández-Muñoz et al. (2004). The water vapor permeability values decreased by around 30% when glutenin rich films were incorporated by cross-linking agents like glutaraldehyde, glyoxal and formaldehyde. With help of formaldehyde and glutaraldehyde, the highest tensile strength values were achieved. In addition, with the usage of cross-linking agents, the glass transition temperature of cross-linked films was seen to shift from its original value. Due to this reason glutaraldehyde and gossypol proved to be less efficient than formaldehyde in cross-linking. While synthesizing biodegradable materials, the tendency of it being toxic must be considered. There should be a permanent protein network formed by cross-linking when any aldehyde is used. Also, the outcome of the aldehyde used during this process on the environment at the termination of the material's life should be considered.

The properties of sunflower protein isolate made thermo-molded films showed various changes due to the effects of cross-linking agents which are found naturally such as tannins and gallic acid. It was studied that higher mechanical properties were resulted by the incorporation of gallic acid and tannins than for control films, but were lower than the films obtained with aldehydes. Weak interactions can cause these contrary to the covalent bonds as in the case of aldehydes (Orliac et al. (2002).

The mechanical properties of gelatin films were improved with the help of tannic acid and ferulic acid, (Cao et al. 2007). It was seen that tannic and ferulic acid had an effect of cross-linking on the gelatin film and therefore acted as natural cross-linking agents. When the pH value of the film-forming solution was 9 for the tannic acid and 7 for ferulic acid, the gelatin-based films showed the maximum mechanical strength. In addition, after the storage for more than 90 days, the properties of gelatin films treated by tannic acid can become better; while the storage time had lesser effect on ferulic acid-modified films.

Hydroxypropyl methylcellulose (HPMC) films are incorporated by preparing nanoparticles of chitosan/tripolyphosphate (CS/TPP)). The film mechanical and barrier properties were improved by the incorporation of nanoparticles of chitosan. The HPMC had various empty spaces in its pores which tend to be occupied by the chitosan nanoparticles, which thereby increases the collapse of the pores and thus improve the film tensile properties and water vapor permeability. With the addition of nanoparticles there is an increase in thermal stability. The study by de Moura et al. (2009) investigated the use of CS-TPP nanoparticles to strengthen the HPMC films. In the last decade, various research programs have concentrated their word on the development of better edible coatings and films (Denavi et al. 2009; Sebti et al. 2007). Amongst these, chitosan and hydroxypropyl methylcellulose (HPMC) which are polysaccharide polymers are closed studied upon (Hernández Muñoz et al. 2008; Perez et al. 2008; Dogan and McHugh 2007). To enhance the quality and shelf life of food products of seafood, agriculture and poultry origin, the various applications of chitosan were discussed in the study by No et al. (2007). However, when solubilized into foods containing high water activity and a reduction in the resistance of these films due to their hydrophilic nature, discourages their applications on an industrial level (Bertuzzi et al. 2007). The amalgam of films with sucrose palmitate, HPMC and sorbitanmonostearate was reported by Villalobos et al. (2006). With a help of a hydrocolloidal to surfactant ratio, the permeability property of moisture for these films was decreased. Syntheses of these films formulated with pectin and polypropylene (PP) was found in a study by Elsabee et al. (2008). Chitosan can form various complex compounds and this property was used to construct a firm multi-layered form on the surface of the PP film, to form a film of increased antimicrobial properties which are used to formulate materials for packing of crops after their harvest. The method was quite successful in which specific polyanions were ionically cross-linked with cationic chitosan. However, it had its complexity with negatively charged polymers. Upon the contact with multivalent polyanions, chitosan has the ability to rapidly form gel caused by the emergence of cross-linkages of inter- and intramolecular type which are arbitrated by these polyanions. Tripolyphosphate is the most popular one among some polyanions that were studied. This is due to its property to be non-toxic and ability to form gel relatively faster. Some interesting features are being exhibited by the CS-TPP nano system which makes them reliable transporters of macromolecules (Gan et al. 2005). Chitosan tripolyphosphate infused hydroxypropyl MC films could be a used material for the application of food packaging applications for the purpose of extension of the storage life of food materials. There can be expansion in the usage of edible films of biodegradable types by the application of nanocomposites (Lagarón et al. 2005; Sinha Ray and Bousmina 2005; Sorrentino et al. 2007). Composites of polymer which have low molecular weight and loadings are fortified due to the restriction of chains within these nanocomposites by the restricted field present among the sheets of the films (Orts et al. 2005; Usuki et al. 1995). Improvements in mechanical properties and the decrease in gas and liquid permeability was resulted by the combination of arrangement of nanoparticle or nanostructure, robust interactions of the surface and chain confinement. The nanoparticles of CS and TPP were formulated and infused into the HPMC films.

CS/TPP nanoparticles in hydroxypropyl methylcellulose edible films ameliorated their barrier and mechanical properties. The thermal characteristics of the films were modified by the nanoparticles in HPMC films (Du et al. 2008). Coatings of chitosan on strawberries reduced the physical damage caused during processing, transportation and finally its storage. However, owing to the sturdy adhesive type density of the energy index of carbohydrate and protein-based edible materials; they tend to have less tensile strength. The flavor and appearance of the food product is also enhanced by these appealing edible coatings. The fruits (such as apple, lemon and oranges) have waxes on them which polishes the surface and gives a glossy appearance. It also decreases the spoilage of a product which is caused by high moisture content by acting as a moisture barrier for it (Lin and Zhao 2007).

In a recent research by (Ghanbarzadeh et al. 2011), the combination of the individual effects of carboxymethyl cellulose and citric acid was used to ameliorate the physical and resistive properties of corn starch-based edible films. The relationship between the carboxyl group of CA and the hydroxyl group on the starch resulted in multi-carboxylic structure. This interaction helps in improving its barrier to moisture because of decreasing the free OH groups of starch. The CA concentration was raised from 0 to 10% weight by weight and this significantly improved the ultimate tensile strength (UTS) and water vapor barrier property ($p < 0.05$).

Carrot basically consists of substances made of cellulose, water, protein and pectin (Bao and Chang 1994) and these earlier mentioned materials may help in providing desirable properties to form biodegradable, renewable and economical packaging films. In a study, gelatin, carboxymethyl cellulose (CMC) and starch were added to carrot-based films which may help in improving their strength by creating a synergistic effect. Gelatin that is derived from partial degradation of collagen is used as edible films due to its biodegradability, myriad and excellent gelling properties (Cao et al. 2007; Dangaran et al. 2009).

18.3.3 Enzymatic Modifications

Amongst the various methods for improving the barrier properties and mechanical strength of the edible films, one effective technique is using enzymatic methods for a cross linking. Enzymes like polyphenol oxidase, transglutaminase, lysyl oxidase, peroxidase and lipoxygenase have been used since a long time for the cross linking of proteins. However, high molecular weight (MW) bio-polymers can be formed by the action of an enzyme called transglutaminase. The transglutaminase enzyme catalyzes the reactions of acyl transfer between the λ-carboxyamide groups of glutamine residues which is and acyl donor and ε-amino groups of lysine residues which is an acyl acceptor and results in emergence of inter and intramolecular cross-linked proteins of ε-(λ-glutaminyl) lysine DeJong and Koppelman (2002).

Various sources of proteins for example gelatin, soy and casein proteins were studied upon for polymerization using transglutaminase where various reactions in the strength of the gel were dependent on the different sources of proteins and the reaction (Sakamoto et al. 1994). The intensity and order of the gel of these proteins influence their strength which helps the enzymes to produce the required cross-links. The occurrence of hindrance to the physical cross-linkages by the new covalent linkages depends on the formation of the triple helix and the process of renaturation during the gel formation (Babin and Dickinson 2001). Films acquired from slightly deamidated gluten had high chances of formation of covalent bonds by transglutaminase Larre et al. (2000). The tensile strength of protein films is improved by the cross-linkage of transglutaminase which decreases the solubility properties and its elongation at break.

It is still undetermined that enzymatic modification by the use of cross-linking can enhance and modify the film forming properties. The types of substrate proteins or amount of enzyme used are the various processing parameters and factors on which the enhancement in the edible films properties which are protein-based. The enzyme concentration is also affected along with the improvement in the properties of edible film. In a study on the properties of soy protein-based films, it was reported that the hydrophobicity and tensile strength can be modified and improved with the usage of transglutaminase (Tang and Jiang 2007).

18.4 Conclusions

The edible films have various advantages over synthetic ones as they can be used as packaging films which are edible in nature. During these recent years edible films have gained popularity due to their biodegradability and edibility. These films are derived from various biopolymer sources such as protein, carbohydrates and lipids and can be further modified using various physical, chemical and enzymatic techniques. Some of these methods include casting, extrusion, irradiation etc. The presence of unique structures is exhibited by proteins which favors the modification and development of these films. The mechanical properties of such films are far better than that of the fat and polysaccharide-based films. However, these edible packaging films when compared with synthetic polymers show poor water vapor resistance and inferior quality of mechanical strength, therefore, limiting their food packaging applications. The use of various physical, chemical and enzymatic methods can be used to improve these properties of films. The properties of the film that are formed by the application of the earlier mentioned methods depend on the various types of conditions and modification undergone. The properties such as water resistance and mechanical strength are efficiently increased by the help of chemical and enzymatic modifications. Although, toxicity can be caused by the use of aldehydes in chemical modifications of these films. This can cause huge matters of concern in application of chemical modification. Therefore, these advancements in modification as well as development of edible packaging films pose to be highly effective and have promising future prospects to it.

References

Babin H, Dickinson E (2001) Influence of transglutaminase treatment on the thermoreversible gelation of gelatin. Food Hydrocoll 15(3):271–276

Bao B, Chang KC (1994) Carrot juice color, carotenoids, and nonstarchy polysaccharides as affected by processing conditions. J Food Sci 59(6):1155–1158

Bertuzzi MA, Armada M, Gottifredi JC (2007) Physicochemical characterization of starch-based films. J Food Eng 82(1):17–25

Bourtoom T (2008) Edible films and coatings: characteristics and properties. Int Food Res J 15(3):237–248

Calderon-Castro A, Vega-Garcia MO, de JesusZazueta-Morales J, Fitch-Vargas PR, Carrillo-Lopez A, Gutierrez-Dorado R, Aguilar-Palazuelos E (2018) Effect of extrusion process on the functional properties of high amylose corn starch edible films and its application in mango (Mangiferaindica L.) cv. Tommy Atkins. J Food Sci Technol 55(3):905–914

Cao N, Fu Y, He J (2007) Mechanical properties of gelatin films cross-linked, respectively, by ferulic acid and tannin acid. Food Hydrocoll 21(4):575–584

Cha-um S, Mosaleeyanon K, Kirdmanee C, Supaibulwatana K (2003) A more efficient transplanting system for Thai neem (Azadirachtasiamensis Val.) by reducing relative humidity. Sci Asia 29:189–196

Ciesla K, Salmieri S, Lacroix M, Le Tien C (2004) Gamma irradiation influence on physical properties of milk proteins. Radiat Phys Chem 71(1–2):95–99

Dangaran K, Tomasula PM, Qi P (2009) Structure and function of protein-based edible films and coatings. In: Edible films and coatings for food applications. Springer, New York, NY, pp 25–56

Davies KJ, Delsignore ME (1987) Protein damage and degradation by oxygen radicals. III Modification of secondary and tertiary structure. J Biol Chem 262(20):9908–9913

de Azeredo HM (2013) Antimicrobial nanostructures in food packaging. Trends Food Sci Technol 30(1):56–69

de Moura MR, Aouada FA, Avena-Bustillos RJ, McHugh TH, Krochta JM, Mattoso LH (2009) Improved barrier and mechanical properties of novel hydroxypropyl methylcellulose edible films with chitosan/tripolyphosphate nanoparticles. J Food Eng 92(4):448–453

De Mulder-Johnston CLC (1999) Thermal analysis of, and oil migration through films from, whey protein isolate. University of California, Davis

Debeaufort F, Voilley A (1994) Aroma compound and water vapor permeability of edible films and polymeric packagings. J Agric Food Chem 42(12):2871–2875

DeJong GAH, Koppelman SJ (2002) Transglutaminase catalyzed reactions: impact on food applications. J Food Sci 67(8):2798–2806
Denavi G, Tapia-Blacido DR, Anon MC, Sobral PJA, Mauri AN, Menegalli FC (2009) Effects of drying conditions on some physical properties of soy protein films. J Food Eng 90(3):341–349
Dogan N, McHugh TH (2007) Effects of microcrystalline cellulose on functional properties of hydroxy propyl methyl cellulose microcomposite films. J Food Sci 72(1):016–022
Dragich AM, Krochta JM (2010) Whey protein solution coating for fat-uptake reduction in deep-fried chicken breast strips. J Food Sci 75(1):43–47
Du WX, Olsen CW, Avena-Bustillos RJ, McHugh TH, Levin CE, Friedman M (2008) Antibacterial activity against E. coli O157:H7, physical properties, and storage stability of novel carvacrol-containing edible tomato films. J Food Sci 73(7):378–383
Elsabee MZ, Abdou ES, Nagy KS, Eweis M (2008) Surface modification of polypropylene films by chitosan and chitosan/pectin multilayer. Carbohydr Polym 71(2):187–195
Fakhouri FM, Costa D, Yamashita F, Martelli SM, Jesus RC, Alganer K, Innocentini-Mei LH (2013) Comparative study of processing methods for starch/gelatin films. Carbohydr Polym 95(2):681–689
Fayose FT, Agbetoye LAS (2012) Characterizing the specific mechanical energy requirement (SME) of extrudates of cassava and cereals from a locally developed extruder. In Rural Development. International Conference of Agricultural Engineering-CIGR-AgEng 2012: Agriculture and engineering for a healthier life, Valencia, Spain, 8–12 July 2012. CIGR-EurAgEng
Fitch-Vargas PR, Aguilar-Palazuelos E, de JesusZazueta-Morales J, Vega-García MO, Valdez-Morales JE, Martínez-Bustos F, Jacobo-Valenzuela N (2016) Physicochemical and microstructural characterization of corn starch edible films obtained by a combination of extrusion technology and casting technique. J Food Sci 81(9):2224–2232
Gan Q, Wang T, Cochrane C, McCarron P (2005) Modulation of surface charge, particle size and morphological properties of chitosan–TPP nanoparticles intended for gene delivery. Colloids Surf B: Biointerfaces 44(2–3):65–73
Garrison WM (1987) Reaction mechanisms in the radiolysis of peptides, polypeptides, and proteins. Chem Rev 87(2):381–398
Gennadios A, Brandenburg AH, Weller CL, Testin RF (1993) Effect of pH on properties of wheat gluten and soy protein isolate films. J Agric Food Chem 41(11):1835–1839
Ghanbarzadeh B, Almasi H, Entezami AA (2011) Improving the barrier and mechanical properties of corn starch-based edible films: effect of citric acid and carboxymethyl cellulose. Ind Crop Prod 33(1):229–235
Gul O, Saricaoglu FT, Besir A, Atalar I, Yazici F (2018) Effect of ultrasound treatment on the properties of nano-emulsion films obtained from hazelnut meal protein and clove essential oil. Ultrason Sonochem 41:466–474
Hambleton A, Debeaufort F, Bonnotte A, Voilley A (2009a) Influence of alginate emulsion-based films structure on its barrier properties and on the protection of microencapsulated aroma compound. Food Hydrocoll 23(8):2116–2124
Hambleton A, Fabra MJ, Debeaufort F, Dury-Brun C, Voilley A (2009b) Interface and aroma barrier properties of iota-carrageenan emulsion–based films used for encapsulation of active food compounds. J Food Eng 93(1):80–88
Hanani ZN, Beatty E, Roos YH, Morris MA, Kerry JP (2012) Manufacture and characterization of gelatin films derived from beef, pork and fish sources using twin screw extrusion. J Food Eng 113(4):606–614
Hauck BW, Huber GR (1989) Single screw vs twin screw extrusion. Cereal Foods World 34(11):930–939
Hernandez-Izquierdo VM, Reid DS, McHugh TH, De J, Berrios J, Krochta JM (2008) Thermal transitions and extrusion of glycerol-plasticized whey protein mixtures. J Food Sci 73(4):169–175
Hernández-Muñoz P, Villalobos R, Chiralt A (2004) Effect of cross-linking using aldehydes on properties of glutenin-rich films. Food Hydrocoll 18(3):403–411
Janjarasskul T, Krochta JM (2010) Edible packaging materials. Annu Rev Food Sci Technol 1:415–448

Jensen A, Lim LT, Barbut S, Marcone M (2015) Development and characterization of soy protein films incorporated with cellulose fibers using a hot surface casting technique. LWT Food Sci Technol 60(1):162–170
Kanatt SR, Rao MS, Chawla SP, Sharma A (2012) Active chitosan–polyvinyl alcohol films with natural extracts. Food Hydrocoll 29(2):290–297
Kaya S, Kaya A (2000) Microwave drying effects on properties of whey protein isolate edible films. J Food Eng 43(2):91–96
Khanzadi M, Jafari SM, Mirzaei H, Chegini FK, Maghsoudlou Y, Dehnad D (2015) Physical and mechanical properties in biodegradable films of whey protein concentrate–pullulan by application of beeswax. Carbohydr Polym 118:24–29
Kim CS, Tae GS (1983) Ecological studies on the Halophyto communities at Western and southern coasts in Korea (IV)-the halophyte communities at the different salt marsh habitats. Korean J Ecol 6(3):167–176
Koide Y, Ikake H, Muroga Y, Shimizu S (2013) Effect of the cast-solvent on the morphology of cast films formed with a mixture of stereoisomeric poly (lactic acids). Polym J 45(6):645–650
Krochta JM (2002) Proteins as raw materials for films and coatings: definitions, current status, and opportunities. Protein Based Films Coatings 1:1–40
Larre C, Denery-Papini S, Popineau Y, Deshayes G, Desserme C, Lefebvre J (2000) Biochemical analysis and rheological properties of gluten modified by transglutaminase. Cereal Chem 77(2):121–127
Lee SL, Lee MS, Song KB (2005) Effect of gamma-irradiation on the physicochemical properties of gluten films. Food Chem 92(4):621–625
Lin D, Zhao Y (2007) Innovations in the development and application of edible coatings for fresh and minimally processed fruits and vegetables. Compr Rev Food Sci Food Saf 6(3):60–75
Marquie C (2001) Chemical reactions in cottonseed protein cross-linking by formaldehyde, glutaraldehyde, and glyoxal for the formation of protein films with enhanced mechanical properties. J Agric Food Chem 49(10):4676–4681
Miller KS, Krochta JM (1998) Measuring aroma transport in polymer films. Trans ASAE 41(2):427
Moghimi R, Aliahmadi A, Rafati H (2017) Antibacterial hydroxypropyl methyl cellulose edible films containing nanoemulsions of Thymus daenensis essential oil for food packaging. Carbohydr Polym 175:241–248
Orliac O, Rouilly A, Silvestre F, Rigal L (2002) Effects of additives on the mechanical properties, hydrophobicity and water uptake of thermo-moulded films produced from sunflower protein isolate. Polymer 43(20):5417–5425
Orts WJ, Shey J, Imam SH, Glenn GM, Guttman ME, Revol JF (2005) Application of cellulose microfibrils in polymer nanocomposites. J Polym Environ 13(4):301–306
Otoni CG, de Moura MR, Aouada FA, Camilloto GP, Cruz RS, Lorevice MV, Mattoso LH (2014a) Antimicrobial and physical-mechanical properties of pectin/papaya puree/ cinnamaldehydenanoemulsion edible composite films. Food Hydrocoll 41:188–194
Otoni CG, Pontes SF, Medeiros EA, Soares NDF (2014b) Edible films from methylcellulose and nanoemulsions of clove bud (Syzygiumaromaticum) and oregano (Origanumvulgare) essential oils as shelf-life extenders for sliced bread. J Agric Food Chem 62(22):5214–5219
Ouattara B, Canh LT, Vachon C, Mateescu MA, Lacroix M (2002) Use of γ-irradiation cross-linking to improve the water vapor permeability and the chemical stability of milk protein films. Radiat Phys Chem 63(3–6):821–825
Pandit P, Nadathur GT, Maiti S, Regubalan B (2018) Functionality and properties of bio-based materials. In: Bio-based materials for food packaging, vol 81–103. Springer, Singapore
Park JW, Whiteside WS, Cho SY (2008) Mechanical and water vapor barrier properties of extruded and heat-pressed gelatin films. LWT Food Sci Technol 41(4):692–700
Pathakoti K, Manubolu M, Hwang HM (2017) Nanostructures: current uses and future applications in food science. J Food Drug Anal 25(2):245–253
Pegg RB, Shahidi F (2007) Encapsulation, stabilization, and controlled release of food ingredients and bioactives. In: Handbook of food preservation. CRC Press, Boca Raton, FL, pp 527–586
Peressini D, Bravin B, Lapasin R, Rizzotti C, Sensidoni A (2003) Starch–methylcellulose based edible films: rheological properties of film-forming dispersions. J Food Eng 59(1):25–32

Perez OE, Sanchez CC, Pilosof AM, Patino JMR (2008) Dynamics of adsorption of hydroxypropyl methylcellulose at the air–water interface. Food Hydrocoll 22(3):387–402

Ray SS, Bousmina M (2005) Biodegradable polymers and their layered silicate nanocomposites: in greening the 21st century materials world. Prog Mater Sci 50(8):962–1079

Rhim JW, Mohanty AK, Singh SP, Ng PK (2006) Effect of the processing methods on the performance of polylactide films: Thermocompression versus solvent casting. J Appl Polym Sci 101(6):3736–3742

Rosenberg M, Lee SJ (2004) Calcium-alginate coated, whey protein-based microspheres: preparation, some properties and opportunities. J Microencapsul 21(3):263–281

Sakamoto H, Kumazawa Y, Motoki M (1994) Strength of protein gels prepared with microbial transglutaminase as related to reaction conditions. J Food Sci 59(4):866–871

Salvia-Trujillo L, Soliva-Fortuny R, Rojas-Grau MA, McClements DJ, Martin-Belloso O (2017) Edible nanoemulsions as carriers of active ingredients: a review. Annu Rev Food Sci Technol 8: 439–466

Sanyang ML, Sapuan SM, Jawaid M, Ishak MR, Sahari J (2015) Effect of plasticizer type and concentration on tensile, thermal and barrier properties of biodegradable films based on sugar palm (Arengapinnata) starch. Polymers 7(6):1106–1124

Sebti I, Chollet E, Degraeve P, Noel C, Peyrol E (2007) Water sensitivity, antimicrobial, and physicochemical analyses of edible films based on HPMC and/or chitosan. J Agric Food Chem 55(3):693–699

Sherrington DC (2003) Introduction to physical polymer science. Reactive polymers, vol 20. Wiley, Hoboken, NJ

Skurtys O, Velásquez P, Henriquez O, Matiacevich S, Enrione J, Osorio F (2011) Wetting behavior of chitosan solutions on blueberry epicarp with or without epicuticular waxes. LWT Food Sci Technol 44(6):1449–1457

Soliman EA, Furuta M (2009) Influence of γ-irradiation on mechanical and water barrier properties of corn protein-based films. Radiat Phys Chem 78(7–8):651–654

Sorrentino A, Gorrasi G, Vittoria V (2007) Potential perspectives of bio-nanocomposites for food packaging applications. Trends Food Sci Technol 18(2):84–95

Sothornvit R, Krochta JM (2005) Plasticizers in edible films and coatings. In: Innovations in food packaging. Academic Press, London, pp 403–433

Su HJ, Dorozhkin DV, Vance JM (2009) A screw theory approach for the conceptual design of flexible joints for compliant mechanisms. J Mech Robot 1(4)

Sung SY, Sin LT, Tee TT, Bee ST, Rahmat AR, Rahman WA. Vikhraman M (2013) Antimicrobial agents for food packaging applications. Trends Food Sci Technol 33(2):110–123

Tang CH, Jiang Y (2007) Modulation of mechanical and surface hydrophobic properties of food protein films by transglutaminase treatment. Food Res Int 40(4):504–509

Tapia-Blacido DR, do AmaralSobral PJ, Menegalli FC (2013) Effect of drying conditions and plasticizer type on some physical and mechanical properties of amaranth flour films. LWT Food Sci Technol 50(2):392–400

Trezza TA, Vergano PJ (1994) Grease resistance of corn zein coated paper. J Food Sci 59(4):912–915

Usuki A, Koiwai A, Kojima Y, Kawasumi M, Okada A, Kurauchi T, Kamigaito O (1995) Interaction of nylon 6-clay surface and mechanical properties of nylon 6-clay hybrid. J Appl Polym Sci 55(1):119–123

Velaga SP, Nikjoo D, Vuddanda PR (2018) Experimental studies and modeling of the drying kinetics of multicomponent polymer films. AAPS PharmSciTech 19(1):425–435

Winotapun C, Phattarateera S, Aontee A, Junsook N, Daud W, Kerddonfag N, Chinsirikul W (2019) Development of multilayer films with improved aroma barrier properties for durian packaging application. Packag Technol Sci 32(8):405–418

Wojtowicz A, Mitrus M, Oniszczuk T, Moscicki L, Kręcisz M, Oniszczuk A (2015) Selected physical properties, texture and sensory characteristics of extruded breakfast cereals based on wholegrain wheat flour. Agric Agric Sci Proc 7:301–308

Multifunctional Edibles and Their Applications in Food Industry

19

Mrinal Samtiya, Sunil Kumar, and Tejpal Dhewa

Abstract

Food packaging industries face enormous challenges regarding the maintenance of foodstuff's safety and quality and demand minimally processed and safe food to develop packaging ideas. Subsequently, active food packaging concepts are useful and provide certain beneficial functions compared to outdated passive materials, which provide limited protection to foodstuffs against external impacts. For meeting the demand for food protection, multifunctional active methods production is considered to be more operational. Some active packaging formulations are considered multifunctional edibles, which are used for food packaging purposes and protect foodstuffs against other stimuli. Examples of active food packaging are films, coatings, blends, etc. Previous evidence reported that due to microbial changes, massive food damages have occurred. Therefore, from several previous years, many physical and chemical methods have been established that could be used to enhance the food product's quality and shelf life. Among such procedures, acceptable packaging is a vital feature in their preservation and promoting stages for food products. Therefore, the packaging of food is essential and, in reality, more significant for the quality maintenance of food. As consumer's demands arc increased for the products free from preservation, processed minimally, and active ingredients must be used for packaging in a very minimal amount that only very little quantity of these preservatives come into food contact. In this chapter, the author discussed some important

M. Samtiya · T. Dhewa (✉)
Department of Nutrition Biology, School of Interdisciplinary and Applied Sciences, Central University of Haryana, Mahendragarh, Haryana, India

S. Kumar
ICMR-RMRC-Regional Medical Research Center, Gorakhpur BRD Medical College, Gorakhpur, UP, India

A. Poonia, T. Dhewa (eds.), *Edible Food Packaging*,
https://doi.org/10.1007/978-981-16-2383-7_19

multifunctional edible packaging such as edible polymers, edible films, edible coatings, edible superhydrophobic surfaces, and edible colloids, which are implemented in food industries for enhancement of foodstuffs shelf life and quality.

Keywords

Multifunctional edibles · Edible superhydrophobic surfaces · Edible colloids · Active food packaging

19.1 Introduction

Throughout the year, people are demanding different types of food. Due to this requirement suitability, they have stimulated the development of innovative food packaging methods to confirm the availability of healthy and safe food. Food packaging's primary purposes are the isolation of foodstuffs from the outer environs and ensure their shelf life betterment by inhibiting spoilage issues comprising the effects of humidity, temperature, oxygen, and microorganisms to reduce nutrition and quality loss (Kumar et al. 2020). There are currently some of the chief global challenges such as aging population, urbanization, sustainability, and food security and safety, which influence the future of food. Demand for the sustainable, abundant, healthy, and safe foods are continually rising, which pressurizing the food researchers to explore new concepts from the other research area to resolve the unsolved challenges related to designing of the food product; it also increased the interdisciplinary investigation opportunities in the food discipline (Patel 2020). Food packaging industries mainly dominated by petroleum-based synthetic plastics, it comprises 37% of food packaging constituents of the total market, which is due to its low cost, heat sealing ability, good barrier properties, high rigidity and mechanical strength, lighter weight, low transportation, and shape versatility (Muller et al. 2017). Alternatively, on the environment, harmful impacts have been caused by these synthetic plastics (Cox et al. 2019). For the food businesses and food manufacturers, it is now one of the biggest challenges to developing safe and healthy food free from synthetic preservatives (chemicals). Due to consumers' increasing worries related to the environment and health, investigators have increased investigators' attention on the bio-based packaging materials used in food packaging as a substitute for synthetic plastic polymers (Rodríguez-Rojas et al. 2019). Some of the recyclable potential alternatives of plastic polymers include proteins, for instance, casein, and polysaccharides, for example, agar, chitosan, starch, and cellulose. (Fabra et al. 2014). In current years, technological substitutes have been searched to decrease the environmental impact and consumption of high energy, in addition to confirm the foodstuff's fresh quality (Kaushik et al. 2014; de São José et al. 2014; Villegas and Albarracín 2016). Edible coatings (ECs) that contain complex materials support the improvement of perishable fruits shelf life through barrier formation, which can be improving the physiological processes of fruits, gaseous exchange (ethylene, O_2, CO_2) and aromas, and water vapor retaining (de Aquino et al. 2015; Oriani et al. 2014; Silva et al. 2015; Thomas et al. 2016).

Consequently, ECs are deliberated as vehicles for transferring bioactive components, which help to defend them microbiologically and also add value to fruits (Boesso-Oriani et al. 2014; Zheng et al. 2019). To protect product stability and quality, edible polymers can be applied directly on the product surface for extra safety. During storage and production of the product, edible polymers that are imposed as requirements were determined by the particular properties of the product and changes in these characteristics. Due to natural polymer decent biodegradability and specific taste, they can be applied as an alternate source of packaging ingredients (Kokoszka and Lenart 2007; Shit and Shah 2014). Consequently, for food surfaces, superhydrophobic materials are used, corresponding legal regulations, additional protections, and material selection need to be followed. Hence, carnauba wax and beeswax, categorized by the U.S. Food and Drug Administration (FDA) as GRAS (generally recognized as safe; 21CFR184.1973 and 21CRF184.1978), were selected as the resources of superhydrophobic surfaces (Wang et al. 2016; Li et al. 2018a, b). Recently, functional and engineered colloids made up of edible constituents gained much of the researcher's interest for advanced applications in the food area for different purposes, from the development of microstructure to health-promoting bioactive delivery to the management of food–body connections (Patel 2020). Coatings and edible films based on chitosan could be one of interest for developing packaging material due to its some particular properties and significant advantage, including good mechanical properties and specific selective permeability to CO_2 and O_2 gases (Fortunati 2016). This chapter aimed to review the uses of food packaging technology and present a general idea of research inventions and developments concerning different food packaging categories. Some interesting concerns must be addressed to improve and maintain the quality and safety of food.

19.2 Multifunctional Edibles and Their Use in Industrial Applications

Even though several techniques are developed for preservation to lower down the postharvest fruit losses, and maximum of these techniques are time-consuming and expensive. Fruit postharvest protection is an international issue (Valero and Serrano 2010; Devlieghere et al. 2004). Low-rate petroleum-based synthetic plastics appear to be a solution; however, their deprived environmental compatibility becomes a chief contaminant (Harding et al. 2007). A promising solution for these concerns is the development of multifunctional packaging constituents, which are environmentally benign, mechanically strong, and have low costing. Besides, it also possesses some other attributes, for example, toxic degrading, antimicrobial, and antioxidant. For the development of biodegradable and multifunctional films most frequently used base materials are lipids, proteins, alginate, chitosan, cellulose, and starches (Li et al. 2018a, b). Because of having a high amount of protein and starch, legume flours could be used as material for film production.

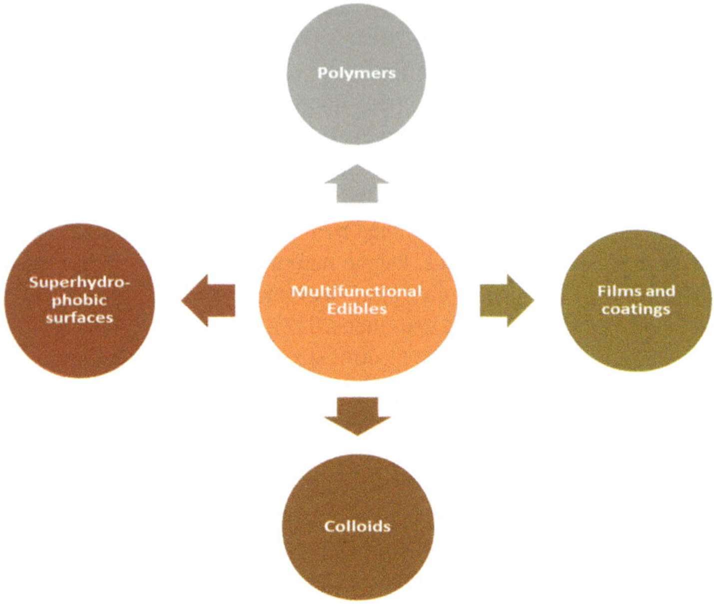

Fig. 19.1 Types of multifunctional edibles

Additionally, numerous other flour films have been produced and evaluated, such as films of flours from mung beans (Bourtoom 2008), chickpeas (Díaz et al. 2019), lentils (Aydogdu et al. 2018), and grass peas (Giosafatto et al. 2018). Previous evidence reported that vegetable and fruit purees were primarily used to make hydrocolloid coatings and films. Pelissari et al. (2013) evaluated the characteristics of plantain bananas-based edible films that are formed from starch and flour. The food industry produced a vast quantity of solid remains annually. Furthermore, these industries' residues are an excellent source of bioactive compounds and nutrients, main biopolymers such as dietary fibers or polysaccharides.

Moreover, higher antioxidant activity is possessed by seed fractions, pomace, and peel of some fruits compared to pulp fractions. For functional food product development, plant residues have been used as components since they are a rich source of nutrients and readily available (Roberta et al. 2014). Recently edible films and coatings have gained much importance prepared by using fruit or vegetable residues. Andrade et al. (2016) reported flour-based edible films composed of diverse residues, containing those from the rocket, cucumber, taro, mint, spinach, carrot, courgette, lettuce, watermelon, passion fruit, and orange. For the development of constant coatings/film structures, several polymers are examined that are bio-based. The most common biopolymers group used in the edible materials production is hydrocolloids, including proteins and polysaccharides. They are easily acquired from different sources, such as microorganisms, animals, and plants (Galus et al. 2020). Figure 19.1 Showing the types of multifunctional edibles.

The different novel film-forming constituents have the utmost potential for promising food preservation applications. In general, to enhance the food shelf life, even though sustaining their sensory and nutritional quality, edible coatings/films are used to cover several food products' surfaces. For the enhancement and improvement of antimicrobial properties of edible packing, considerable deliberation has been given. Consequently, possible findings have been analyzed for the incorporation of the active compound into coatings and films, having a promising biological activity (Kraśniewska and Gniewosz 2012; Atarés and Chiralt 2016). Maqbool et al. (2011) has reported the gum arabic antifungal properties comprising cinnamon and lemongrass oil to regulate papaya and banana postharvest anthracnose. Results of the study concluded that in tropical fruit use of gum arabic containing essential oils efficiently decreased the *Colletotrichum gloeosporioides* and *Colletotrichum musae* growth in reaction to anthracnose. Another storage study of 30 days confirmed that on cold-smoked salmon storage, a film based on peel waste of potato comprising oregano essential oil had efficiently regulated the growth of *Listeria monocytogenes* (Tammineni et al. 2013). Sprays packaging gained much interest in industrial applications not only due to their promising cost reduction although by the excellent quality of the final product. Through this application, although at high pressures, low viscous coating materials can be effortlessly sprayed (Dhanapal et al. 2012; Andrade et al. 2012). The formation of coatings onto meat, vegetables, and fruits, among other foodstuffs dipping methods, has been used. To determine the thickness of films, properties, for instance, surface tension, viscosity, and density of the coating solutions are very important, subsequently dipping techniques are capable of making thick coating layers (Tavassoli-Kafrani et al. 2016; Lu et al. 2010; Dhanapal et al. 2012). The brushing or spreading technique comprises the organized spreading of suspension on the surface of the material to be additionally dried. This technique is deliberated as an adequate substitute for the preparation of the film with dimensions more significant than those formed by casting processes. The thickness of coating suspension could be regulated through a blade connected to the lower part of the spreading device, and drying of the film is detained on the support itself through hot air circulation. This technique can be used for the development of protein and polysaccharides films (Méndez-Vilas 2013). Table 19.1 Representing the studies related to potential applications of edible packaging.

19.3 Types of Multifunctional Edible

19.3.1 Edible Polymers

Edible polymers are polymeric materials that could be simply consumed by lower animals or humans completely or partially through the oral cavity and not provide any harmful effects on the health. Due to several potential reasons, edible polymers would be investigating. One of the reasons was introducing novel food product groups, for example, high quality, convenient, and safe products (Shit and Shah

Table 19.1 Potential applications of edible packaging materials

Material used in edible	Types of edible	Component/ product	Main advantages	References
Almond gum	Edible coatings	Sweet cherries	Reduce ethylene production, increase the duration of color change, weight loss,	Mahfoudhi and Hamdi (2015)
Aloe vera gel	Edible coatings	Apple	Bioactive compounds improvement, decrease the weight loss as well as firmness	Aidilla and Thevinta (2017)
Tragacanth gum, almond gum, quince seed gum	Edible coatings	Bananas (slices)	Decrease color changes and loss of weight	Farahmandfar et al. (2017)
Cajanus cajan seed protein isolate and gum	Edible coatings	Strawberries	Maintained sensory acceptability, reduce total soluble solid content	del C Robles-Flores et al. (2018)
Nanoscale colloidal particles of shellac	Edible colloids	Silibinin	Prevent degradation in stomach and release in the intestine	Patel et al. (2011)
Zein protein particles	Edible colloids	Curcumin	Maximum protection against chemical instability in gastrointestinal tract	Patel et al. (2010)
Mucoadhesive Zein–curcumin composite particles	Edible colloids	Curcumin	Enhance absorption of curcumin due to slower release, prevent degradation, improve solubilization	Zou et al. (2016)
Beeswax nanoparticles	Edible superhydrophobic surface	—	Edible container surface	Wang et al. (2016)
Gelatin, arabic gum, honeycomb wax	Edible superhydrophobic surface	—	Showed high contact angles for a diversity of liquid foods, although liquid foods residue could be decreased efficiently	Li et al. (2018a, b)
Quaternized chitosan films combined with carboxymethyl cellulose	Edible film	Whole banana	Prolong shelf life	Hu et al. (2016)

(continued)

Table 19.1 (continued)

Material used in edible	Types of edible	Component/ product	Main advantages	References
Chitosan based film incorporating green tea extract	Edible film	Pork sausages	Extension of shelf life	Siripatrawan and Noipha (2012)
Nano zinc oxide to phenyllactic acid grafted chitosan	Edible film	Taiwan green jujube	Reduced weight loss and rotting rate	Li et al. (2018a, b)
Starch based hydrogel (potato and maize)	Edible polymers	Theophylline	Faster drug dissolution,	Szepes et al. (2008)
Kappa carrageenan based hydrogel beads	Edible polymers	Platelet derived growth factor	Performed as a proficient encapsulation means and shows an enhanced organized release profile	Santo et al. (2009)

2014). Natural polymers provide a particular taste and are easily recycled so that they could be applied as packaging materials alternative sources. In recent years edible polymers gained much importance due to safe nature, these polymers used as a replacement of synthetic plastic because it can be consumed with food products (Dhanapal et al. 2012). These polymers can be easily ingested by human beings and not affect human health as well; edible polymers mainly consist of polysaccharides, lipids, polypeptides, hydrocolloids, and polysaccharides, etc. The main reasons behind exploring the novel edible polymers are incorporating innovative, high-grade, safe, and appropriate foodstuffs in functional food industries (Kokoszka and Lenart 2007). Several edible polymers comprise proteins (wheat gluten, zein, gelatin, etc.), hydrocolloids (alginates, pectins, cellulose, carrageenans, starch, etc.), and lipids, within these edible polymers based on proteins are the most attractive ones (Jang et al. 2011; Álvarez et al. 2017). In a previous study, agar and kappa carrageenan-based hydrogels were developed using a natural cross-linker called genipin (Meena et al. 2009). Hydrogels that are formed by using edible polymers offer several valued attributes in comparison to other synthetic complements. Edible polymers also contribute to environmental contamination reduction, provide sustainability, advance recyclability, and improve its applicability together with as long as environmentally benign products. Hydrogels-based edible polymers applications contain several areas, including agricultural applications, food industry, biomedical field (drug delivery to tissue engineering), and additional attractive and safe products in the environmental fields, agricultural and pharmaceutical, etc. (Ali and Ahmed 2018).

19.3.2 Edible Films and Coatings

Packaging with an edible film made of bio-based polymer is an effective, simple, and low-cost method to avoid moisture loss; therefore, it reduces respiration rate and deterioration. Chitosan is interested exclusively in food wrapping purposes for coating and film production because of its intrinsic biodegradability, film making capability, and antimicrobial activity (Prashanth and Tharanathan 2007; Kumar et al. 2019). A study by Hu et al. (2016) reported that films made from quaternized chitosan combined with carboxymethyl cellulose (CMC) when applied on bananas enhanced shelf life. Furthermore, results concluded that high % containing quaternized chitosan-based films had shown many better-delayed effects for fruit decay. In another study, the chitosan-based film was developed using green tea extract as packing material for enhancing pork sausages shelf life. The findings proposed that the pork sausages samples covered with the films comprising an extract of green tea had shown lesser alteration in microbial growth, sensory characteristics, TBA value, texture, and color, in comparison to those that covered by chitosan and control film, respectively (Siripatrawan and Noipha 2012). Surface dehydration is preferred for preserving fresh fruit, so the water vapor permeability (WVP) property of films is one of the necessary properties (Souza et al. 2015). To enhance foodstuffs shelf life and quality, the coating is frequently used on fresh foods surfaces, for instance, meat, fish, vegetables, and fruits, etc. (Vargas et al. 2012). Coatings would be accomplished by immersing, dipping, spraying, or spreading food materials into chitosan-based complex solutions. Spatula or brush can be used for performing spread-coating, and it is an effective method to constrain the microorganism's growth, which increases shelf life and helps sustain food value (Dong et al. 2004; Pushkala et al. 2013; Yu et al. 2017). One of the recent studies by Martínez et al. (2018) reported that in cold storage, chitosan-based edible coating prolonged the strawberry fruit shelf life up to 15 days, comprising Thymus capitatus essential oil. Another study has described that chitosan-based coating on strawberries considerably decreased the flavonoid and anthocyanin contents, changes in the phenolic, ascorbic acid content, losses in titratable acidity, color loss, and water loss (Petriccione et al. 2015). Other evidence conveyed that edible coating based on olive oil, pectin, and cassava starch, improved the mango storage shelf life up to 12 days (Estrada Mesa et al. 2015). A previous study confirmed that coated strawberries using edible coatings based on shellac, beeswax, and tara gum dropped fruit physiological progressions (Pavón-Vargas and Valencia-Chamorro 2016). Edible coatings reduced the pH and acidity variations because of the effects that it is caused on the physiological processes of fruit. This type of performance was recorded in numerous edible coating containing fruits: hydroxypropyl methylcellulose based edible coating on blackberries (Villegas and Albarracín 2016), aloe vera mucilage based edible coating on blackberries (Ramírez et al. 2013), mango oxidized and native cassava starch-based edible coating on mango (Figueroa et al. 2013), and so on. Other study reported the properties of silk fibroin quintessential (hydrophobicity, conformability, polymorphism) to form a water protein suspension, which can accumulate during dip coating on the surface of the food through bananas and strawberries dip coating as evidence of principle, exposed that covering of micrometer-thin silk fibroin on the surfaces of the fruits, supports the supervision of postharvest structure of fruits. Therefore, coatings of silk fibroin improve the fruit's shelf life in room environments by dropping cell water evaporation as well as the rate of respiration (Marelli et al. 2016). The latest study estimated the consequence of an edible coating constructed by cassava starch, glacial acetic acid, stearic acid, glycerol, chitosan beeswax, and whey protein on the shelf life of fruit which stored at 4 °C. Comparison to fresh control fruit, sensory, microbiological, physical, chemical, and physical quality was estimated. Results have found that edible coating showed positive effects on the sensory (primarily in aroma, flavor, and texture) and physicochemical through diminishing physical progressions; however, alterations of color are mostly due to anthocyanin loss. Furthermore, chitosan decreased the yeast and mold count, while firmness was 81.4% higher and weight losses were 39.6% lower after storage of 10 days. The edible coating enhanced the beneficial life of the Andean blackberries by 100% (Rodríguez et al. 2020).

19.3.3 Edible Superhydrophobic Surfaces

Superhydrophobic edible can aid as novel food packing materials, which substitutes plastic because these kinds of materials can be eaten together with a food component. Furthermore, its alteration possibly will additionally deliver the packing with the utility of water storage, which could remove the coffee and tea secondary plastic packaging (Wang et al. 2020). The superhydrophobic edible materials work as self-regulating packaging that can exchange the plastic materials and offer innovative resolutions such as edible oxidation barriers, underwater storage, and rapid-slow release for fruit. Plant-derived components and edible materials such as polysaccharide and lipid can be selected to develop superhydrophobic ingredients. Cutin is an edible content, which tomato also contains; it is a type of hydrophobic lipid constituent (polymeric) of plant cuticle and could be functioning as an obstruction in contrast to the penetration of solutes, moisture and gas, consequently due to its low attraction to water as well as water vapors (Mérida et al. 1981; Järvinen et al. 2010; López-Casado et al. 2007; Wang et al. 2020). Due to superior surface assets, superhydrophobic materials are useful strategies for escape the adhesion of food particles and liquids (Cheng et al. 2013; Seo et al. 2013). Superhydrophobic are known as materials with a roll angle of less than 10° and a static water contact angle greater than 150° on their surface (Ueda and Levkin 2013). In recent times, superhydrophobic materials' potential applications have been identified based on their superior surface properties. For example, anticorrosion, oil-water separation, anti-bacterial adhesion, anti-scaling, drag reduction, anti-icing, and self-cleaning (Celik et al. 2020; Jiang et al. 2015; Hizal et al. 2017). Necessary food safety standards need to be met, so superhydrophobic materials use green production methods and should be non-toxic. Though, only a few studies have been existing in the literature on the use of such edible superhydrophobic surfaces. Previously, Bayer et al. (2011) developed a superhydrophobic surface using carnauba wax-alcohol emulsion. Another study by Wang et al. 2016 reported the development of edible superhydrophobic containers using sprayed beeswax nanoparticles on the surface of the container. A recent study by Li et al. (2018a, b), reported the formation of superhydrophobic food packaging coating using gelatin, arabic gum, and sprayed honeycomb wax. Another previous study stated the development of a superhydrophobic layer utilizing ethanol-containing spray-coating homogeneous wax suspensions (Zhao et al. 2018). All of these edible superhydrophobic surfaces demonstrated attractive scenarios for application in food packaging.

19.3.4 Edible Colloids

In different food products, food colloids are used, which provide mouth-feel, texture, and structure, such as mayonnaise, ice cream, jam, etc. Food colloid comprises hydrocolloid, which provides stabilizing, emulsifying, gelling, and thickening characteristics in food products (Milani and Golkar 2019). Food hydrocolloids are used as functional components in the processing of food to change shelf-life, flavor, texture, and microstructure, and these are also known as high molecular weight hydrophilic biopolymers (Milani and Maleki 2012). One of the chief purposes of the colloid comprising method is the organized bio-polymer connections control to produce definite nanoscale arrangements of emulsion droplet stabilization. The principal physicochemical methods to colloid constitute are precisely classified through three generic terms of layering, embedding, and clustering (McClements 2012). For example, shellac-based nanoscale colloidal particles had applied to cover silibinin that is acid-labile bioactive, to protect its stomach deprivation and accomplish an intestinal targeted delivery (Patel et al. 2011). A previous study reported that Patel et al. (2010) prepared the complex particles using mucoadhesive zein–curcumin to enhance curcumin solubility in gastro-intestinal tract (GIT), protect curcumin degradation under physiological situations, and during the same period via particles mucoadhesion improves the curcumin intake (absorption) via keeping the slowly liberating distribution scheme in close vicinity to the membrane of the cell for a lengthier period as well as improving its settle period in the upper GIT. Through the efficiency comparison of three diverse transport methods, that is, nanoemulsions, nanoparticles of lipid, and protein particles, Zou et al. (2016) reported that protein particles (zein) providing the utmost filling of curcumin and highest protection against GIT chemical instability. Current development made in the production of advanced materials, such as engineered and functional food colloids, promotes the designing of food and transforms the future. Besides, also resolve the majestic social challenges that are affecting environmental sustainability and human health. Numerous developing applications based on engineered and functional food colloids focused on food product development, which is sustainable, safer, tastier, healthier, and most essentially consistent with consumer beliefs (Patel 2020). Delivery systems based on colloidal materials comprise a wide range of micro–nanoparticulate schemes, which are made-up for controlling the release, protection, and encapsulation of active agents to diminish problems, for instance, chemical instability, undesirable flavor profiles, low bioactivity, poor solubility of active agents, and to produce innovative functional qualities, for example, on-demand delivery of actives at a particular place inside the GIT (Pan and Zhong 2016; McClements 2018). Due to sensitive properties and structures to alter in pH and temperature, the microgel constituent integrates into the food colloid formulations that deliver edible schemes retaining new stimulation responsive potentials (Richtering 2012). Therefore, recognizing fresh edible ingredients, discovering new methods of producing different colloid structures from distinguished but underutilized edible constituents, and ultimately using them in foodstuffs designing are some regions of quickly rising investigation (research).

19.4 Future Prospectus

So far, several treatments have been discovered that enhanced the vegetables and fruit postharvest life, includes heat and hypobaric treatments, osmotic treatments, modified atmosphere packaging, exposure to synthetic chemical fungicides, and cryopreservation (Han 2005). Consequently, the development of an edible coating signifies an alternate method to enhance the freshness of the crop by merging prolonged storage times with comfort of handling (Vargas et al. 2008; Janjarasskul and Krochta 2010). These are some of the properties that should be holding for becoming an edible coating material, such as safety, membrane forming capacity, antifungal and antibacterial activities, biodegradability, and biocompatibility. Up to the present time, lipids, resins, proteins, polysaccharides, and their combinations are the frequently used possibilities for coating preparations. Proteins and polysaccharides are documented to form films with worthy mechanical capabilities but show poor permeability, whereas resins and lipids form brittle films with better permeability (Marelli et al. 2016). These multifunctional edibles are currently used very frequently, and incoming time they would replace the other synthetic and toxic ingredients.

19.5 Conclusions

In the food industries packaging of food is the necessary process. On the other hand, liquid and food remains in food containers and holds to the walls are expected for containers. This will lead to the large wastage of the food that is predominantly disapproving for the foods products having a high value, such as fruit juice, honey, and even liquid medicine. Literature evidence revealed that by diverse edibles use; we can improve and keep the food attributes and also their shelf life. Research and development studies that are deliberated here are heightened and supported via the rising demand of consumers for safer as well as natural extracts having efficient potentials capable of responding to the requirement for active packaging constituents and growing concerns of the environment. Applicability of these food packaging edibles for industries still requires additional and much more in-depth studies; subsequently, numerous of them exhibited significant influence on the food product's organoleptic properties. Furthermore, it would also be essential to investigate detailed probable interactions between the packaged food and coatings/films.

References

Aidilla M, Thevinta RE (2017) Aloe vera edible coating retains the bioactive compounds of wax apples (Syzygium samarangense). Malays Appl Biol J 46(4):153–160

Ali A, Ahmed S (2018) Recent advances in edible polymer-based hydrogels as a sustainable alternative to conventional polymers. J Agric Food Chem 66(27):6940–6967

Álvarez K, Famá L, Gutiérrez TJ (2017) Physicochemical, antimicrobial and mechanical properties of thermoplastic materials based on biopolymers with application in the food industry. Adv Physicochem Propert Biopolym Part 1:358–400

Andrade RD, Skurtys O, Osorio FA (2012) Atomizing spray systems for application of edible coatings. Compr Rev Food Sci Food Saf 11(3):323–337

Andrade RM, Ferreira MS, Gonçalves ÉC (2016) Development and characterization of edible films based on fruit and vegetable residues. J Food Sci 81(2):E412–E418

Atarés L, Chiralt A (2016) Essential oils as additives in biodegradable films and coatings for active food packaging. Trends Food Sci Technol 48:51–62

Aydogdu A, Kirtil E, Sumnu G, Oztop MH, Aydogdu Y (2018) Utilization of lentil flour as a biopolymer source for the development of edible films. J Appl Polym Sci 135(23):46356

Bayer IS, Fragouli D, Martorana PJ, Martiradonna L, Cingolani R, Athanassiou A (2011) Solvent resistant superhydrophobic films from self-emulsifying carnauba wax–alcohol emulsions. Soft Matter 7(18):7939–7943

Bourtoom T (2008) Factors affecting the properties of edible film prepared from mung bean proteins. Int Food Res J 15(2):167–180

Celik N, Torun I, Ruzi M, Esidir A, Onses MS (2020) Fabrication of robust superhydrophobic surfaces by one-step spray coating: evaporation driven self-assembly of wax and nanoparticles into hierarchical structures. Chem Eng J 396:125230

Cheng Z, Du M, Lai H, Zhang N, Sun K (2013) From petal effect to lotus effect: a facile solution immersion process for the fabrication of super-hydrophobic surfaces with controlled adhesion. Nanoscale 5(7):2776–2783

Cox KD, Covernton GA, Davies HL, Dower JF, Juanes F, Dudas SE (2019) Human consumption of microplastics. Environ Sci Technol 53(12):7068–7074

de Aquino AB, Blank AF, de Aquino Santana LC (2015) Impact of edible chitosan–cassava starch coatings enriched with Lippia gracilis Schauer genotype mixtures on the shelf life of guavas (Psidium guajava L.) during storage at room temperature. Food Chem 171:108–116

de São José JF, de Andrade NJ, Ramos AM, Vanetti MC, Stringheta PC, Chaves JB (2014) Decontamination by ultrasound application in fresh fruits and vegetables. Food Control 45:36–50

del C Robles-Flores G, Abud-Archila M, Ventura-Canseco LM, Meza-Gordillo R, Grajales-Lagunes A, Ruiz-Cabrera MA, Gutiérrez-Miceli FA (2018) Development and evaluation of a film and edible coating obtained from the Cajanus cajan seed applied to fresh strawberry fruit. Food Bioprocess Technol 11(12):2172–2181

Devlieghere F, Vermeiren L, Debevere J (2004) New preservation technologies: possibilities and limitations. Int Dairy J 14(4):273–285

Dhanapal A, Sasikala P, Rajamani L, Kavitha V, Yazhini G, Banu MS (2012) Edible films from polysaccharides. Food Sci Qual Manag 3:9

Díaz O, Ferreiro T, Rodríguez-Otero JL, Cobos Á (2019) Characterization of chickpea (Cicer arietinum L.) flour films: effects of pH and plasticizer concentration. Int J Mol Sci 20(5):1246

Dong H, Cheng L, Tan J, Zheng K, Jiang Y (2004) Effects of chitosan coating on quality and shelf life of peeled litchi fruit. J Food Eng 64(3):355–358

Estrada Mesa EM, Padilla Reyes F, Márquez Cardozo CJ (2015) Efecto de recubrimientos protectores sobre la calidad del mango (Mangifera indica L.) en poscosecha. Rev UDCA 18 (1):181–188

Fabra MJ, López-Rubio A, Lagaron JM (2014) Biopolymers for food packaging applications. Smart Polym Appl:476–509

Farahmandfar R, Mohseni M, Asnaashari M (2017) Effects of quince seed, almond, and tragacanth gum coating on the banana slices properties during the process of hot air drying. Food Sci Nutr 5 (6):1057–1064

Figueroa JA, Salcedo JG, Narváez GJ (2013) Effect of edible coatings based on native and oxidized cassava starch on quality of mango (Tommy Atkins). Revista Temas Agrarios 18(2):94–105

Fortunati E (2016) Multifunctional films, blends, and nanocomposites based on chitosan: use in antimicrobial packaging. In: Antimicrobial food packaging. Academic Press, San Diego, CA, pp 467–477

Galus S, Arik Kibar EA, Gniewosz M, Kraśniewska K (2020) Novel materials in the preparation of edible films and coatings—A review. Coatings 10(7):674

Giosafatto CV, Al-Asmar A, D'Angelo A, Roviello V, Esposito M, Mariniello L (2018) Preparation and characterization of bioplastics from grass pea flour cast in the presence of microbial transglutaminase. Coatings 8(12):435

Han JH (ed) (2005) Innovations in food packaging. Elsevier

Harding KG, Dennis JS, Von Blottnitz H, Harrison ST (2007) Environmental analysis of plastic production processes: comparing petroleum-based polypropylene and polyethylene with biologically-based poly-β-hydroxybutyric acid using life cycle analysis. J Biotechnol 130 (1):57–66

Hizal F, Rungraeng N, Lee J, Jun S, Busscher HJ, van der Mei HC, Choi CH (2017) Nanoengineered superhydrophobic surfaces of aluminum with extremely low bacterial adhesivity. ACS Appl Mater Interfaces 9(13):12118–12129

Hu D, Wang H, Wang L (2016) Physical properties and antibacterial activity of quaternized chitosan/carboxymethyl cellulose blend films. LWT Food Sci Technol 65:398–405

Jang SA, Lim GO, Song KB (2011) Preparation and mechanical properties of edible rapeseed protein films. J Food Sci 76(2):C218–C223

Janjarasskul T, Krochta JM (2010) Edible packaging materials. Annu Rev Food Sci Technol 1:415–448

Järvinen R, Kaimainen M, Kallio H (2010) Cutin composition of selected northern berries and seeds. Food Chem 122(1):137–144

Jiang W, He J, Xiao F, Yuan S, Lu H, Liang B (2015) Preparation and antiscaling application of superhydrophobic anodized CuO nanowire surfaces. Ind Eng Chem Res 54(27):6874–6883

Kaushik N, Kaur BP, Rao PS, Mishra HN (2014) Effect of high pressure processing on color, biochemical and microbiological characteristics of mango pulp (Mangifera indica cv. Amrapali). Innovative Food Sci Emerg Technol 22:40–50

Kokoszka S, Lenart A (2007) Edible coatings-formation, characteristics and use-a review. Polish J Food Nutr Sci 57(4):399–404

Kraśniewska K, Gniewosz M (2012) Substances with antibacterial activity in edible films–a review. Polish J Food Nutr Sci 62(4):199–206

Kumar S, Ye F, Dobretsov S, Dutta J (2019) Chitosan nanocomposite coatings for food, paints, and water treatment applications. Appl Sci 9(12):2409

Kumar S, Mukherjee A, Dutta J (2020) Chitosan based nanocomposite films and coatings: emerging antimicrobial food packaging alternatives. Trends Food Sci Technol 97:196–209

Li R, Sun X, Zhu J, Wang D, Xu Y (2018a) Novel multifunctional and edible film based on phenyllactic acid grafted chitosan derivative and nano zinc oxide. Food Biophys 13(1):102–111
Li Y, Bi J, Wang S, Zhang T, Xu X, Wang H, Cheng S, Zhu BW, Tan M (2018b) Bio-inspired edible superhydrophobic interface for reducing residual liquid food. J Agric Food Chem 66 (9):2143–2150
López-Casado G, Matas AJ, Domínguez E, Cuartero J, Heredia A (2007) Biomechanics of isolated tomato (Solanum lycopersicum L.) fruit cuticles: the role of the cutin matrix and polysaccharides. J Exp Bot 58(14):3875–3883
Lu F, Ding Y, Ye X, Liu D (2010) Cinnamon and nisin in alginate–calcium coating maintain quality of fresh northern snakehead fish fillets. LWT Food Sci Technol 43(9):1331–1335
Mahfoudhi N, Hamdi S (2015) Use of almond gum and gum Arabic as novel edible coating to delay postharvest ripening and to maintain sweet cherry (P. runus avium) quality during storage. J Food Process Preserv 39(6):1499–1508
Maqbool M, Ali A, Alderson PG, Mohamed MT, Siddiqui Y, Zahid N (2011) Postharvest application of gum arabic and essential oils for controlling anthracnose and quality of banana and papaya during cold storage. Postharvest Biol Technol 62(1):71–76
Marelli B, Brenckle MA, Kaplan DL, Omenetto FG (2016) Silk fibroin as edible coating for perishable food preservation. Sci Rep 6(1):1–11
Martínez K, Ortiz M, Albis A, Gilma Gutiérrez Castañeda C, Valencia ME, Grande Tovar CD (2018) The effect of edible chitosan coatings incorporated with Thymus capitatus essential oil on the shelf-life of strawberry (Fragaria x ananassa) during cold storage. Biomolecules 8(4):155
McClements DJ (2012) Advances in fabrication of emulsions with enhanced functionality using structural design principles. Curr Opin Colloid Interface Sci 17(5):235–245
McClements DJ (2018) Recent developments in encapsulation and release of functional food ingredients: delivery by design. Curr Opin Food Sci 23:80–84
Meena R, Prasad K, Siddhanta AK (2009) Development of a stable hydrogel network based on agar–kappa-carrageenan blend cross-linked with genipin. Food Hydrocoll 23(2):497–509
Méndez-Vilas A (ed) (2013) Microbial pathogens and strategies for combating them: science, technology and education. Formatex Research Center, Badajoz
Mérida T, Schönherr J, Schmidt HW (1981) Fine structure of plant cuticles in relation to water permeability: the fine structure of the cuticle of Clivia miniata Reg. Leaves. Planta 152 (3):259–267
Milani JM, Golkar A (2019) Introductory chapter: some new aspects of colloidal systems in foods. In: Some new aspects of colloidal systems in foods. IntechOpen, Rijeka
Milani J, Maleki G (2012) Hydrocolloids in food industry. Food industrial processes–Methods and equipment 2:2–37
Muller J, González-Martínez C, Chiralt A (2017) Combination of poly (lactic) acid and starch for biodegradable food packaging. Materials 10(8):952
Oriani VB, Molina G, Chiumarelli M, Pastore GM, Hubinger MD (2014) Properties of cassava starch-based edible coating containing essential oils. J Food Sci 79(2):E189–E194
Pan K, Zhong Q (2016) Organic nanoparticles in foods: fabrication, characterization, and utilization. Annu Rev Food Sci Technol 7:245–266
Patel AR (2020) Functional and engineered colloids from edible materials for emerging applications in designing the food of the future. Adv Funct Mater 30(18):1806809
Patel A, Hu Y, Tiwari JK, Velikov KP (2010) Synthesis and characterisation of zein–curcumin colloidal particles. Soft Matter 6(24):6192–6199
Patel A, Heussen P, Hazekamp J, Velikov KP (2011) Stabilisation and controlled release of silibinin from pH responsive shellac colloidal particles. Soft Matter 7(18):8549–8555
Pavón-Vargas D, Valencia-Chamorro S (2016) Efecto de recubrimientos comestibles compuestos a base de goma tara en la calidad poscosecha de frutilla (Fragaria ananassa). Revista Iberoamericana de Tecnología Postcosecha 17(1):65–70

Pelissari FM, Andrade-Mahecha MM, do Amaral Sobral PJ, Menegalli FC (2013) Comparative study on the properties of flour and starch films of plantain bananas (Musa paradisiaca). Food Hydrocoll 30(2):681–690
Petriccione M, Mastrobuoni F, Pasquariello MS, Zampella L, Nobis E, Capriolo G, Scortichini M (2015) Effect of chitosan coating on the postharvest quality and antioxidant enzyme system response of strawberry fruit during cold storage. Foods 4(4):501–523
Prashanth KH, Tharanathan RN (2007) Chitin/chitosan: modifications and their unlimited application potential-an overview. Trends Food Sci Technol 18(3):117–131
Pushkala R, Raghuram PK, Srividya N (2013) Chitosan based powder coating technique to enhance phytochemicals and shelf life quality of radish shreds. Postharvest Biol Technol 86:402–408
Ramírez JD, Aristizabal ID, Restrepo JI (2013) Conservación de mora de castilla mediante la aplicación de un recubrimiento comestible de gel de mucílago de penca de sábila. Vitae 20 (3):172–183
Richtering W (2012) Responsive emulsions stabilized by stimuli-sensitive microgels: emulsions with special non-Pickering properties. Langmuir 28(50):17218–17229
Roberta MS, Mariana SL, Édira CB (2014) Functional capacity of flour obtained from residues of fruit and vegetables. Int Food Res J 21(4):1675
Rodríguez MC, Yépez CV, González JH, Ortega-Toro R (2020) Effect of a multifunctional edible coating based on cassava starch on the shelf life of Andean blackberry. Heliyon 6(5):e03974
Rodríguez-Rojas A, Ospina AA, Rodríguez-Vélez P, Arana-Florez R (2019) ¿ what is the new about food packaging material? A bibliometric review during 1996–2016. Trends Food Sci Technol 85:252–261
Santo VE, Frias AM, Carida M, Cancedda R, Gomes ME, Mano JF, Reis RL (2009) Carrageenan-based hydrogels for the controlled delivery of PDGF-BB in bone tissue engineering applications. Biomacromolecules 10(6):1392–1401
Seo J, Lee S, Han H, Jung HB, Hong J, Song G, Cho SM, Park C, Lee W, Lee T (2013) Gas-driven ultrafast reversible switching of super-hydrophobic adhesion on palladium-coated silicon nanowires. Adv Mater 25(30):4139–4144
Shit SC, Shah PM (2014) Edible polymers: challenges and opportunities. J Polym 2014
Silva KS, Garcia CC, Amado LR, Mauro MA (2015) Effects of edible coatings on convective drying and characteristics of the dried pineapple. Food Bioprocess Technol 8(7):1465–1475
Siripatrawan U, Noipha S (2012) Active film from chitosan incorporating green tea extract for shelf life extension of pork sausages. Food Hydrocoll 27(1):102–108
Souza M, Vaz AF, Silva H, Cerqueira MA, Vicente AA, Carneiro-da-Cunha MG (2015) Food Bioprocess Technol 8(11):2183–2191
Szepes A, Makai Z, Blümer C, Mäder K, Kása P Jr, Szabó-Révész P (2008) Characterization and drug delivery behaviour of starch-based hydrogels prepared via isostatic ultrahigh pressure. Carbohydr Polym 72(4):571–578
Tammineni N, Ünlü G, Min SC (2013) Development of antimicrobial potato peel waste-based edible films with oregano essential oil to inhibit L isteria monocytogenes on cold-smoked salmon. Int J Food Sci Technol 48(1):211–214
Tavassoli-Kafrani E, Shekarchizadeh H, Masoudpour-Behabadi M (2016) Development of edible films and coatings from alginates and carrageenans. Carbohydr Polym 137:360–374
Thomas AB, Nassur RD, Boas AC, Lima LC (2016) Cassava starch edible coating incorporated with propolis on bioactive compounds in strawberries. Ciência e Agrotecnologia 40(1):87–96
Ueda E, Levkin PA (2013) Emerging applications of superhydrophilic-superhydrophobic micropatterns. Adv Mater 25(9):1234–1247
Valero D, Serrano M (2010) Postharvest biology and technology for preserving fruit quality. CRC Press, New York
Vargas M, Pastor C, Chiralt A, McClements DJ, Gonzalez-Martinez C (2008) Recent advances in edible coatings for fresh and minimally processed fruits. Crit Rev Food Sci Nutr 48(6):496–511

Vargas M, Sánchez-González L, Cháfer M, Chiralt A, González-Martínez C (2012) Edible chitosan coatings for fresh and minimally processed foods. In: Emerging food packaging technologies. Woodhead Publishing, pp 66–95. https://www.elsevier.com/books/emerging-food-packagingtechnologies/9781845698096

Villegas C, Albarracín W (2016) Aplicación y efecto de un recubrimiento comestible sobre la vida útil de la mora de castilla (Rubus glaucus Benth). Vitae 23(3):202–209

Wang W, Lockwood K, Boyd LM, Davidson MD, Movafaghi S, Vahabi H, Khetani SR, Kota AK (2016) Superhydrophobic coatings with edible materials. ACS Appl Mater Interfaces 8 (29):18664–18668

Wang D, Huang J, Guo Z (2020) Tomato-lotus inspired edible superhydrophobic artificial lotus leaf. Chem Eng J 400:125883

Yu D, Jiang Q, Xu Y, Xia W (2017) The shelf life extension of refrigerated grass carp (Ctenopharyngodon idellus) fillets by chitosan coating combined with glycerol monolaurate. Int J Biol Macromol 101:448–454

Zhao X, Hu T, Zhang J (2018) Superhydrophobic coatings with high repellency to daily consumed liquid foods based on food grade waxes. J Colloid Interface Sci 515:255–263

Zheng K, Xiao S, Li W, Wang W, Chen H, Yang F, Qin C (2019) Chitosan-acorn starch-eugenol edible film: Physico-chemical, barrier, antimicrobial, antioxidant and structural properties. Int J Biol Macromol 135:344–352

Zou L, Zheng B, Zhang R, Zhang Z, Liu W, Liu C, Xiao H, McClements DJ (2016) Food-grade nanoparticles for encapsulation, protection and delivery of curcumin: comparison of lipid, protein, and phospholipid nanoparticles under simulated gastrointestinal conditions. RSC Adv 6(4):3126–3136

Part IV

Packaging Edibles and Their Regulatory Aspects

Packaging Edibles: New Challenges and Regulatory Aspects

20

Mamta Prajapati

Abstract

Packing of any food commodity is an intrinsic part, a resultant outcome of final finished good with enhanced shelf life and a perfect barrier to face the extrinsic factors of environment – physical, chemical and biological. The first and the foremost requirement of a packaging material is its quality of being food grade in nature. The material in use should be sound enough to bear mechanical, chemical or thermal stresses faced during the process of transportation. The best food grade packaging material used since times immemorial are different types of broad surfaced leaf, used in dried as well as fresh conditions. The restriction in using a leaf as a packaging material is – could not be sealed air tight to completely isolate it from the hassles of external environment. Development in packaging process leads to use of paper of various types from non-absorbing butter paper to sturdy kraft paper. Further progress gave new packaging material in food sector in the form of different types of plastic polymers which successfully covered the entire market. Later the negative outcomes of plastic polymers created burden on our environment because of its non-biodegradable nature and its entry into the ecosystem as well as living beings in the form of micro- and nanoparticles which created an alarm. Scientists and industry personnel started working on biodegradable packaging material. Biopolymers like polyester, starch-based polymers, starch-based polymers have replaced petrochemical-based plastic packaging material. Biopolymers, though compatible with the environment, need to be economical, act as characteristically strong barrier, heat insulator with good mechanical properties. Also, though biopolymers are compostable, still composting needs to be carried out religiously which also adds to greenhouse

M. Prajapati (✉)
Sector Expert (Food), National Consumer Helpline, Indian Institute of Public Administration, Delhi, India

A. Poonia, T. Dhewa (eds.), *Edible Food Packaging*,
https://doi.org/10.1007/978-981-16-2383-7_20

gases. So, a new alteration in packaging material innovated giving way to edible packaging material. With the idea of edible packaging material, the challenges associated still persists. The edible packaging material should be suitable to the type of food product and can be filled and sealed under any conditions. Pinholes and hairline cracks are a major issue in the development of edible packaging. The mechanical strength and brittleness of the material need to be checked for the development of standard packaging material of edible nature. Overall migration limit and specific migration limit though not a concern, but the new specification required is shelf life of packaging material so that the food product packed inside remains intact and fresh till end use.

Keywords

Packaging process · Biopolymers · Edible packaging material

20.1 Development of Packaging for Food Products

Packing of any food commodity is an intrinsic part, a resultant outcome of final finished good with enhanced shelf life and a perfect barrier to face the extrinsic factors of environment – physical, chemical and biological. The first and the foremost requirement of a packaging material is its quality of being food grade in nature. It should be capable to bear the force of mechanical, chemical or thermal nature faced during typical transportation. The first food grade packaging material used since times immemorial are different types of broad-surfaced leaf, used in dried as well as fresh conditions reaching out to paper packaging (non-absorbing butter paper to sturdy kraft paper and cardboard), metal tins and glass bottles, to plastic bags and foil-lined packets.

Clay pot, glass and wood packaging have an extremely long history in food packaging. All three being recyclable ruled the packaging material in use quiet a long till the development of metallic can, foil and laminates. The advent of plastic as packaging material was the turning point of the field. Plastics because of their characteristics of being hard, strong, malleable can be easily casted into sheets, shapes and structures of variable configurations. Economical, low-priced, light-weight and owing to their chemically resistant nature plastics are widely used in packaging. Further progress gave new packaging material in food sector in the form of different types of plastic polymers which successfully covered the entire market. Later the negative outcomes of plastic polymers created burden on our environment because of its non-biodegradable nature and its entry into the ecosystem as well living beings in the form of micro- and nanoparticles which created an alarm. The major disadvantage of petrochemical-derived packaging material like plastic, nylon and polyester is the burden on the ecological system. The findings of scientists have highlighted of shedding of small pieces of plastic called microplastic into the environment polluting soil, water and air. Being non-biodegradable and non-renewable, these materials represent a warning to the existence of flora and fauna, and mankind, by the generation of greenhouse gases and Chlorofluorocarbon, cause waste management issues around the world leading to create a hurdle in sustainable development. To tackle the problem of global warming, climate change and environmental contamination, the world is approaching towards green lifestyle, biodegradable and edible packaging being one of them. This has increasingly promoted the use of biopolymers including edible packaging material. Looking into the seriousness of the problem, scientists and industry personnel started working on biodegradable packaging material. Bio-based polymers like polyester, starch-based polymers have replaced petrochemical-based plastic packaging material. Biopolymers satisfy the ecological concerns, but they manifest some constrictions in terms of execution of properties like heat resistance, shielding and physical strength, associated with the costs. Also, though biopolymers are compostable, still composting needs to be carried out religiously which also adds to greenhouse gases. So, a new alteration in packaging material innovated giving way to edible packaging material.

20.2 History of Packaging Edibles

The use of edible films dates back somewhere between twelfth and fifteenth century. The development of edible films was based on its added attribute towards preservation and being economical in the absence of efficient packaging material. Chinese used wax coating to restrict moisture loss from citrus fruits in the twelfth century. Meat, fish fillets and poultry products being highly perishable in nature were coated and protected using edible films to restrict loss in colour, flavour and quality. The protective layer also reduces the chance of microbial contamination.

Yuba, an edible film composed of soy protein, was traditionally used by Japan in fifteenth century to preserve foods. Yuba is produced in the form of thin skin formed on the surface of soy milk when heated. It has a soft and smooth texture. Water vapour permeability, tensile strength and mechanical properties of edible film play an important role in its application as packaging material. Plasticizer, emulsifier, lipid and cross-linking compounds added to Yuba provide increased physical, mechanical and water-resistant properties. Addition of thickening agent - sodium carboxymethyl cellulose, lipid—bee wax, plasticizer—glycerol alone or in combination greatly induced the edible film attributes pertaining to water resistance, water vapour transmission and elongation (Kim et al. 2019). Ferulic, caffeic and gallic acid added to Yuba improved the colour and transparency of film along with reduction in moisture permeability and solubility (Insaward et al. 2015).

20.2.1 Textural Properties of the Yuba Film

The property of fat acting as oxygen and moisture barrier was a good means to enhance the shelf life of meat by coating. Larding, an age-old practice, had been successfully used to increase the shelf life of meat and its products by coating with lipid and fat. This greatly enhanced the shelf life of meat at refrigerated storage preserving the characteristic red colour of oxyhaemoglobin. Bee wax, plant wax (candelilla), paraffinic acid and acylated acylglycerol compounds made a perfect coating for refrigerated meat products by cutting off-flavours, reducing freezer burn and preserving colour (Bharti et al. 2020).

Early practice of edible packaging film and coatings was particularly for food preservation which currently can be successfully developed to deal with ongoing issue of sustainability environmental pollution.

20.3 Edible Coating and Films

Edible coating and films are two modes used in food preservation and packaging. Edible film is defined as thin and continuous standalone matrices composed of protein, lipid and polysaccharides, while coating on food is carried out by dipping or spraying which remains adhered to the product. Naturally occurring polymers like polysaccharides, lipids and protein are used to produce eco-friendly and sustainable edible films. The common components of edible films are polysaccharides, proteins and lipids either alone or in combination (Guilbert et al. 1996). Edible films have been used to segregate different food elements, to wrap foods and to encase foods held therein.

Coating and wrapping of foods extend the shelf life of food with edible packaging which can be consumed with the product encased. Edible films act as a tool for fortification. Controlled and selective permeability of edible film restricts loss of critical ingredients and maintains proper protection to food (Kerry and Milda 2009).

The various forms of edible packaging are film, sheet, coating and pouch. Packaging films of edible nature are independent structures applied to food both as external covering as well as in between food layers (Krochta and De 1997). Compared to edible films, coating is thin and inherently in contact with the food surface (Bharti et al. 2020). Freshness of fruits and vegetables can be retained with a waxy layer imparting shining and glossy finish and maintaining the acceptable traits like reduction in wilting and shriveling (El-Anany et al. 2009).

20.4 Traits Used in Packaging Material

Oxygen barrier and moisture barrier properties are extremely important especially for edible food packaging materials. Biodegradable packaging material (BPM) is expected to protect the food from the environment and maintain the food quality. Mechanical strength and barrier properties are important characteristics of BPM.

It is additionally critical to consider the chemical changes that can happen on the characteristics of the BPM during the shelf life of food. Moisture and oxygen are two of the fundamental permeants to be considered since they may move from the inward or outside environment through the polymer sheath, bringing about a persistent change in food quality and final shelf life of the product (Marsh and Bugusu 2007).

20.5 Functions of Packaging

The main function of food packaging is to land in the hands of end user in the most safe, economical manner without any hazards associated to food safety and leave no detrimental impact on the environment. Food packaging aids in protecting food products from outside influences and damage, contain the various shapes and state of food and illustrates ingredients present in pack with the required nutritional information (Coles 2003).

20.5.1 Contain

Food exhibits in numerous physical forms with a versatile nature. This inherent property of food is maintained by packaging. Packaged products increase the smooth handling of product. Food packaging contains the product and maintains the integrity of product during processing, transportation, marketing and dispensing of food. Containers, pouches, bags, paper packs and wrapping films allow the manufacturer to control the weight of the commodity, ensure quality control, maintain consistency and provide convenience during purchase.

20.5.2 Protect

The primary role of food packaging is to keep the food safe for consumption while maintaining the flavour, taste and texture—the essential characteristics of food. Packaging provides protection to the product encased against all types of risks associated, thus enhancing the food products' shelf life. Physical risks involved with food are any of the foreign particles (glass, wood, metal and objects associated with human), chemical risks are invisible (cleaning agents, insecticides, pesticides or other agricultural chemicals) and the biological risks involved are microorganisms (bacteria, fungi and moulds). Most critical aspect of packaging material is its own shelf life which should be comparable to the shelf life of food.

Protection is also required from external factors (light, heat, moisture). Light can degrade the photo-sensitive components of food (xanthophyll, carotenoid). Degradation of polyunsaturated fatty acid occurs on exposure to UV light catalysing the formation of peroxides. Heating accelerates the sequence of chemical reactions in the different components of food. The insulating capacity, determined by thermal conductivity of packaging material, retards the chemical changes in between molecules. Water vapour permeability of packaging material controls the transmission of moisture and is a significant controlling factor that determines the shelf life of food products. Unsaturated fat is susceptible to oxidative rancidity and is directly proportional to permeability of membrane.

The significance of packaging is its overall protection against external factors (dust and mechanical forces), gas permeability and biological factors (microorganisms) (Robertson 2005).

20.5.3 Labelling and Communication

Packaging becomes a mode of communication between manufacturer and consumer. Labelling requirement is mandatory by CODEX. The mandatory information provided by the package resolves the legal requirement and advertisement goals. The conventional facts given on pack are: common name of the food product, ingredients and composition of food including allergen, nutritional data, net weight of the contents, name and address of the manufacturers, maximum retail price (MRP) and country of origin.

20.5.4 Traceability

The label though designed to attract consumers also allows to track the food product throughout marketing chain. The Codex Alimentarius Commission defines traceability as "the ability to follow the movement of a food through specified stage(s) of production, processing and distribution" (Codex Alimentarius Commission 2004).

20.6 Components of Edible Packaging Material

20.6.1 Pectin

Pectin is an anionic polysaccharide composed of 1,4-D-galacturonic acids (Sato et al. 2008). Pectin present in ripe fruits is a white-coloured colloidal carbohydrate exhibiting amorphous properties. It finds wide use in jam and jelly, pharmaceutics and cosmetics manufacture. It possesses good thickening, solidifying and emulsifying properties which add to its potential in new possibilities (Valdés et al. 2015).

Being a surplus by-product of fruit and vegetable processing industry from pomace (apples, lime fruit, kiwi), peels (mango, pomegranate, papaya, dragon fruit, grape fruit), pulp from beet, albedo from citrus fruits (Canteri et al. 2011; Valdés et al. 2015), it could certainly be utilized for the making of edible film (Galus et al. 2012; Galus and Lenart 2013).

Edible food coating is derived from pectin with low degree of methyl esterification (Valdés et al. 2015). Pectin films successfully preserve flavour of the packed food product, are good barrier to oxygen and lipid and exhibit good mechanical strength. The hydrophilic nature of biopolymer allows easy transmission of water vapour. Pectin containing low methoxyl content produces firm gels in combination with calcium ions give a more sturdy film that inhibits moisture transfer (Ciolacu et al. 2014; Zhang et al. 2014).

The combined effect of pectin–alginate or alginate–whey protein chemistry on the characteristics of edible film is positive (Chakravartula et al. 2019). Thermal, mechanical and gas barrier properties turned out be increased with lower pectin and whey protein content and higher concentration of alginate in edible packaging film.

The properties of edible film also enhanced with pectin, whey protein and alginate in equal amounts. Lower pectin and whey protein content and higher concentration of alginate in film positively strengthened the thermal, mechanical and gas barrier behaviour. The same pattern of traits was observed with equal amount of pectin, whey protein and alginate in the proposed film. Whey protein incorporation into film has been reported to have decreased wettability for food wrapping film, thus providing the foldability required for wrapping (Chakravartula et al. 2019). Pectin extracted from papaya skin in combination with 25% cassava and 10% glycerol significantly improved the thickness, tensile strength, elasticity and water vapour transmission rates of the edible film (Rosida et al. 2018). Edible film composed of whey dangke and pectin in the ratio of 1:4 and 0.125% of beeswax and 2% of butter aroma produced good physical and barrier properties for a packaging material.

20.6.2 Cellulose

Cellulose, the cementing material in cell wall of plant cells, is utilized to produce cellulose films. Viscose is generated from the reaction of sulphite paper pulp with sulphur dioxide and carbon bisulfide. Extrusion of viscose into an acid salt bath forms cellulose hydrate. Cellulose hydrate is softened using glycerol, and the film so produced is dried on drum dryers. Cellulose could be the futuristic edible film for packaging due to its transparent and glossy profile, flat taste and odourless nature. Positive traits linked are toughness of cellulose-bound film, making it puncture resistant, and flexibility, which provides foldability to the material. Cellophane-like film is developed by incorporating zinc chloride and calcium ions into solubilized cellulose, converting the large molecules of cellulose into small fibrils with increased tensile strength. Biodegradability and abundance of cellulose in nature supports the idea of edible film in food packaging. Composite films have been created to integrate suitable barrier properties related to packaging using microcrystalline cellulose, protein isolates and cocoa powders in one formulation and carboxymethyl cellulose and fatty acid sucrose ester in other (Gontard and Guilber 1994).

20.6.3 Carageenan

Carrageenan stands out as one of the most accepted biopolymers within industrial standards. The initial usage of seaweeds as food dates back to 600 BC in China. The Irish used it as a component of milk puddings, whereas in the Eastern countries, they were mostly used in salads. During the 1930s, the United States patented carrageenan as a commercial food additive. Carrageenan was declared as GRAS in the United States in 1959 (FDA CFR title 21172.60).

Carrageenan are polysaccharides produced by red seaweeds *Rhodophyceae* (Jancikova et al. 2020). Carrageenans are composed of linear polysaccharides found in the cell wall and intercellular matrix of red seaweeds (Rhodophyta)

(Therkelsen 1993; Knutsen et al. 1994). Carrageenans are extensively used as thickeners, stabilizers and gelling agents in food industry. Isomers of carrageenan are differentiated on the basis of the number and position of the ester sulphate groups on the repetitive galactose units (Zablackis and Santos 1986). There are in principle three isomers of carrageenan depending upon the amount and location of the ester sulphate clusters that are present on the repeating glycosidic linkages (Mahadevappa et al. 2014).

Starch blended with carrageenan in different ratios using glycerol as plasticizer gives a blend having non-Newtonian pseudoplastic behaviour. The tensile strength and elongation at break increased with increasing carrageenan content in the blend for all films prepared. Significant increase in water vapour permeability of the blended films was reported with increase in carrageenan percent (Abdou and Sorour 2014).

Carrageenan is one such polysaccharide that can enhance the shelf life and maintain quality of food products. The native carrageenan films are known to produce hard, strong and flexible films which have applications in various fields. The films have properties such as aroma barrier, oil barrier, moisture barrier, barrier to mass transfer, oxygen barrier, acts as a carrier for food additives (antimicrobial, antioxidant, anti-browning, probiotics, minerals, etc.), product structural integrity enhancer, product appearance enhancer, food ingredients dispenser. For the production of composite edible film or coatings, carrageenan can be combined with other polysaccharides such as chitosan, pectin, alginate, etc. Carrageenan films have also been reinforced with food additives having antimicrobial, antioxidant, anti-browning and plasticizing properties, which help in the overall enhancement of chemical and mechanical properties of the film. Carrageenan has the raw potential to be an attractive biopolymer that can be used in food packaging applications and also minimize plastics (Bhat et al. 2020).

Composite carrageenan products have been used to improve various properties. Composite films of carrageenan and grapefruit seed extracts have shown good antimicrobial properties, UV barrier properties, high transparency and improvement of shelf life (Kanmani and Rhim 2010).

The impact of the addition of olive leaf extract on the mechanical properties and water vapour permeability of the biodegradable carrageenan films was positive, as it increased the elongation at break, decreased the tensile strength and decreased the water vapour. Olive leaf extract added film exhibited more red and yellow colour as compared to control. Carageenan–Olive leaf extract retarded the growth of aerobic mesophilic bacteria in lamb meat during storage. The results of the study suggested that olive leaf extract acts as a natural antimicrobial for use in biodegradable carrageenan films, for packaging food products (Martiny et al. 2020).

20.6.4 Corn Protein Films

The principal protein in corn is zein. Corn and maize processing industries manufacture edible oil and biofuel and leave behind a residue rich in corn protein, Zein, the corn protein is a byproduct of bioethanol and oil industries. The protein is characterized by superb film-forming ability, well-defined solubility in ethanol and easily utilized with many naturally active substances (Arcan et al. 2017).

Zein is composed of prolamin protein soluble in 70–80% ethanol (Parris and Dickey 2001) characterized as hydrophobic and thermoplastic material owned to its high content of non-polar amino acids (Shukla and Cheryan 2001). Hydrophobic property of zein attributes to its high shielding characteristics as compared to other proteins. The magnificent film-forming properties can consummately be used for creation of biodegradable films. The science associated with the development of film are the hydrophobic, hydrogen and restricted disulphide connections between zein chains. The brittleness of zein film is lowered by the addition of plasticizer. Guilbert (1986) has reported zein films to possess relatively good moisture-resistant properties as compared to other edible films. Zein coating forms low permeable layer on fruits reducing the exchange of oxygen and carbon dioxide. This maintains the firm texture and delay in any colour change of fresh fruits (Thawien 2012).

Addition of monolaurin (ML) and *Zataria multiflora* Boiss. essential oil (ZEO) to zein film improved the antioxidant and antimicrobial property of the casted film. ML and ZEO containing films significantly inhibited the growth of *L. monocytogenes* and *E. coli* in minced meat stored at 4 °C (Moradi et al. 2016). The composite films of zein and chitosan presented better barrier and mechanical properties than single ingredient films (García et al. 2014).

20.6.5 Gluten Protein Films

Gluten is a globular, water-insoluble protein present in wheat, rye and barley grains. Gluten bears the property of cohesiveness and elasticity, thus facilitating film formation. Gliadin and glutenin are the constituents of gluten. Gliadins are low-molecular-weight, monomeric proteins with few intra-chain disulphide bonds while glutenins are high-molecular-weight, polymeric proteins containing inter-chain disulphide bonds. Films formulated out of glutenin were mechanically stronger and had more noteworthy boundary properties than films from gliadins or entire gluten. Gliadin films introduced better optical properties; however, they were not moisture safe. The new disulphide bonds formed during heating the solubilized wheat gluten along with hydrogen and hydrophobic bonds (Gennadios and Weller 1990) give the integrity to the film.

Plasticizer like glycerine impart flexibility to the wheat gluten film though strength, flexibility and moisture hindrance properties are contrarily affected. The strength and transparency of wheat gluten film is directly proportional to purity of the original raw material. Gluten is a good oxygen barrier but permeable to moisture. This can be overcome with a good cross-linking agent—glutaraldehyde or through heat curing.

20.6.6 Soy Protein Films

Edible oil from soybean leaves behind a nourishing versatile residue which is easily incorporated into a variety of food items. Being abundant and inexpensive, it is applicable for making edible packaging film. It is composed of 38–44% protein, the main constituent of the simple amino acid—globulin. The presence of polar and non-polar side chain in soy protein magnifies intra- and intermolecular interactions in the form of hydrogen bonding, dipole–dipole, charge–charge and other hydrophobic interactions. Above interactions result in durable film of comparable tensile strength (Zhang et al. 2001). New disulphide bonds are formed in combination with hydrogen bonds and hydrophobic bonds by splitting of disulphide bond, opening up of sulfhydryl groups and hydrophobic groups present in original raw protein which are formulated into a protein-based film. Totally new intermolecular interactions form the matrix of film upon final drying process. Addition of plasticizer and hydrophilic nature imparts poor moisture barrier traits which can be enhanced by integrating hydrophobic lipids into film-forming solutions. Alternatively, protein network may be modified through cross-linking process through any of the treatments—chemical, enzymatic or physical—to improve the attributes of soy protein film.

20.6.7 Casein Films

The principal proteins present in milk are casein and whey protein. Casein component is soluble and can be modified into films with stability against denaturation for a wide array of parameters like temperature, pH and salt concentrations. The aqueous nature of casein gives formation of durable films because of the molecular coiled structure which allows intermolecular hydrogen bonding, hydrophobic and electrostatic bonds. A change in pH modifies their water-soluble structure to water-insoluble type.

Though durable with good mechanical strength and low oxygen permeability, casein films are moisture sensitive and less flexible. The desirable properties can be achieved by formulating it with minerals and other biopolymers like pectin.

20.6.8 Whey Protein

Whey proteins are used singly as well as in combination with other ingredients to develop coatings, laminated film made out of heterogeneous substances and pouches produced to carry food. Whey proteins are classified into whey protein concentrates (WPC) and whey protein isolates (WPI) based on the protein content. Whey proteins have all the basic traits to be used as a packaging material showing good shielding for oxygen, volatile content at low to intermediate humidity and form a durable film with comparable mechanical strength.

Whey protein concentrate (WPC) constituted of 25–80% protein is utilized for manufacture of protein packaging films (DeWit 2001). Whey proteins, α-lactalbumin (α-LA) and β-lactoglobulin (β-LG) interact into a matrix; transglutaminase acting as a crosslinking agent are availed for to produce films. Mixed films composed from α-LA and β-LG were found more permeable to moisture as compared to pure protein films (Mahmoud and Savello 1993).

Whey protein isolate (WPI), which has a protein content of about 90% (DeWit 2001), forms an edible film which is transparent, heat-sealable, good resistance for oxygen transmission and comparable tensile strength at low to medium relative humidity (RH) conditions (McHugh and Krochta 1994). The carrier molecules in the form of antioxidants and antimicrobial can be easily incorporated into WPI film characterized by flat taste and no odour (Han and Krochta 2007).

20.6.9 Collagen

Collagen is widespread structural protein found in animal tissue. It is fibrous protein used as a precursor to gelatin. Gelatin is a USFDA-approved GRAS additive. The fibrous and structural traits which provide gel strength allow its applicability in film production. Gelatin can be modified to form a stable colloid, emulsion or foam exhibiting a perfect gelling behaviour. However, the highly hygroscopic nature of gelatin and limited thermal stability are the shortcomings of its use in packaging. It is reported that minor change in helical structure occurs with application of heat to collagen, but it does not modify the strong covalent bonds in the helixes to produce gelatin. The combination of the use of the extract from collagen with glycerol as plasticizer is considered to improve the properties of edible film (Said et al. 2016).

20.7 Utilization of Edible Films

Edible packaging film should be used in place of primary packaging in conjunction to suitable ecological secondary packaging to restrict any direct contamination to food in the packet. Basically, edible films are best employed in wrapping the freshly cut fruits and vegetables to avoid enzymatic browning and oxidation. This also secures the nutrition of fresh foods. Daily food left over at home, sandwich, vegetable salad and peeled fruits could be safeguarded from the external environment by applying edible films on the open surface. Edible films furnish a good option to wrap any food product before storing in refrigerator.

Corn zein is already in practice to shield and extend the shelf life of nuts and fruits. Oxidation of fatty acids in nuts is retarded using zein-based films. Fruits are screened with edible films composed of casein and whey protein-based film to retain freshness and crunchy texture due to the excellent moisture barrier trait.

Frozen fish loses its original taste and flavour because of lipid oxidation. Protein-based films support the shelf life of frozen fish by blocking the moisture in package. Isolated whey protein and acetylated monoglyceride-composed film delayed oxidation of lipid in the substrate and lessened moisture depletion rate in frozen king salmon (Stuchell and Krochta 1995). The organoleptic characteristics of cooked turkey was stabilized with corn zein dipping integrated with an antioxidant (butylated hydroxyanisole) and emulsifier (Herald et al. 1996).

Bread, the daily food of common man, exhibited increased shelf life with sodium caseinate wrapping over it. Compression test data revealed lower values for bread taken into observation with onefold or twofold sodium caseinate film. Hardening in bread was also reduced with sodium caseinate wrapping. The value for hardness though comparable to synthetic polyvinyl chloride film wrapping of bread but was found not effective for long storage period (Schou et al. 2005).

20.8 Advantage of Edible Films

The superiority of edible films lies in the fact that it can be consumed with packed product leaving no residue in the form bioburden to the environment (Janjarasskul and Krochta 2010). It is a form of green and clean packaging contributing to the reduction of environmental pollution. Resources utilized in edible film production are agricultural products, thus edible and renewable and marked as biocompatible material (Bourtoom 2008). The principal function of edible coatings is to maintain and conserve the high-quality characteristics of a food incorporated (Longares et al. 2004; Wan et al. 2005). Edible films protect the food, create an invisible hurdle for transmission of moisture, gas and lipid, in turn extending the life of the product (Jang et al. 2011).

One significant benefit of utilizing edible films and coatings is that some dynamic substances can be fused into the intercellular grid and utilized with the food, consequently adding to food safety and wholesomeness of food. The attributes of films and coating can be improved remarkably by integrating food additive in the form of colourants, flavouring agents, antioxidants and antimicrobials (Dhall 2012). The external coating/films applied to the food product strengthen the moisture and gas barrier traits, retards biological action and maintains colour and texture, enhancing the overall acceptability and shelf life of the product (Kumar and Sethi 2018).

Specific constituents added in the form of favour, colour, sweetening agent, etc. augment sensory and organoleptic properties (Bourtoom 2008). Antimicrobials, antioxidants, nutrients, colour, herbs, and spices, added to edible film and coating, act as good conveyors to add on to the properties of packaging films (Bharti et al. 2020). Multi-level packaging can be taken over by single-layer packaging leading to remarkable reduction of utilized resources and complexity of the process, with better recyclability (Khwaldia et al. 2010) without compromising protective functions (Krochta 2002).

The basic principle behind conserving the quality and shelf life of produce using edible packaging is the development of modified atmosphere within the fruit with each entity in function. The application of edible coatings efficiently decreases the rate of respiration retarding the process of ripening of fruits. The tropical fruit banana reported to exhibit retarded ripening action on application of sucrose esters of fatty acids (SPE), sodium carboxymethyl cellulose and monodiglycerides of fatty acids coating (Banks 1984) and apples (Santerre et al. 1989; Drake et al. 1987). Rate of ripening was effectively reduced on implementation of SPE coating on pears after cold storage (Van Zyl et al. 1987). The delaying effect of SPE on ripening was also observed by Meheriuk and Lau (1988).

20.9 Commercially Available Edible Coating

Agricoat-Natureseal Ltd. (Berkshire, UK), a subsidiary of Mantrose-Haeuser Co., Inc., is an age-old manufacturer of natural, edible coatings and specialty films for a variety of artistic, practical and workable solutions for fruits, nuts, confectionary, baked goods and other food products. Coatings, especially in fruits and vegetables, are used as a tool to enhance post-harvest shelf life.

Freshseel is one of the state-of-the-art, specialty coating from the post-harvest protection of melons. Semperfresh, made out of ester is thin and taste-neutral coating utilized in post-harvest care of fruits and vegetables during the supply chain movement including the steps of packing, storage, shipping, and marketing.

The most recent version is Semperfresh® 1.9, known as a latest tool in coating of fruits and vegetables for extension of shelf life, conserving quality and locking the freshness of the produce. Advancing with the new characteristics, Semperfresh is easy to use and bears fast drying properties making it suitable for coating stone fruit, pineapple, melon, avocadoes, citrus fruit, pears, apples, sweet cherries and other tropical fruits.

Semperfresh 1.9 functions by making a modified atmosphere for each entity coated with invisible, edible and biodegradable protective film. This reduces the respiration rate enhancing the shelf life of the fresh produce. Basic advantage of the film is its biodegradable nature in contrast to plastic packaging which creates a burden on the environment. Semperfresh 1.9 prevents common disorders such as chilling injury, skin and rind pitting when applied after hydrocooling of the fruit and vegetable entities.

The other brand available for edible coating is NatureSeal manufactured out of blend of vitamins and minerals applied to ready-to-eat fresh produce functions in the same manner., is a blend of vitamins and minerals, easy to apply and there is no requirement for packaging with modified atmosphere. NatureSeal is also easy to apply, hinders enzymatic browning and preserves the colour, texture, flavour of fresh cut ready to eat fruits. The crisp texture of sliced fruits could be enjoyed with NatureSeal. It becomes easy to work in the supply chain with a satisfactory shelf life thus increasing final production and processing efficiency.

All the constituents used in edible coating are approved by the European Union and the Food and Drug Administration (FDA) of the United States for the external treatment of produce (Post harvest 2021).

Pears in turkey are coated with Semperfresh, and Johnfresh exhibited good colour, firmness, acidity and soluble solid retention causing a delay in ripening, thereby extending shelf life. No notable variation was found in fruits coated with Johnfresh and Semperfresh (Gulum and Levent 1994).

Resin extracted from conifers, Longleaf (*Pinus palustris*) and Slash pine (*Pinus elliottii*), was developed into a commercially available biodegradable coating named Wood Rosin. Solvent extraction and fractional refining gave a good yield for the compound (Beglinger 1958). Commercially it is produced at Pinova Solutions Inc., New Brunswick, Georgia (Merck 2013). It is a permitted organic and biodegradable ingredient for conventional citrus fruit waxes according to FDA regulations, 21 CFR 172.210(b)(2), and also incorporated as indirect food additive for making food contact material used in packing, processing, treating, packaging, transporting or holding food under 21 CFR 178.3780.

Flo Zein Products is commercially producing zein protein, extracted from corn, for being utilized in edible coating and films since 1976. UDFDA had recognized zein as GRAS ingredient for food processing sector under 21 CFR 184.1984. Zein is characterized as water-insoluble odourless, tasteless, clear, hard and almost invisible protein employed in the manufacture of edible coating (FloZein Products 1976).

Edible film produced from zein acts as a good moisture barrier thus finding its use in coating and encapsulation. It can combat microbial attack exhibiting outstretched shelf life for fresh fruits and vegetables, whole and shelled nuts. The protein macromolecule can be easily casted/sprayed into film meant for making receptacle for holding food.

Zein is a hydrophobic protein exhibiting easy breakdown naturally with the action of biological agents. The macromolecule has been integrated with different ingredients to develop better coating properties. Zein is a real boon to develop antimicrobial wrap, bioplastics, coating paper, grease resistant food boxes and biodegradable packaging.

Application of zein involves considerable extension of shelf life of nuts and dried fruit. It is an excellent barrier to restrict oxygen permeability, moisture loss and microbial activity. Pome fruit—apple and pear, citrus fruits and tropical fruits coated with zein are protected from microbes, show delay in ripening and prevent moisture loss. Frozen fruits and vegetables are shielded towards freezer burn with zein.

Plum (*Prunus salicina* L.) fruits of cv. Santa Rosa shelf life was enhanced after treatment with edible coating of SemperfreshTM (1:3), vegetable wax (1:5) and lac based (2:3) on the farm as well as after transportation to the laboratory (off-farm).

Analytical determinations were made after 3, 6, 9, 12 and 15 days at 20 ± 2 °C. All surface coatings, especially lac-based wax, were effective in inhibiting loss of moisture, ascorbic acid, total antioxidant activity and total phenols content. Lac based and SemperfreshTM displayed better efficacy in maintaining firmness followed by vegetable wax. At the end of storage period, lac-based coated fruits showed higher fruit firmness (8.87 and 8.53 N) than control (6.50 and 6.30 N) in both on-farm and off-farm treatments, respectively, followed by SemperfreshTM coated fruits (7.84 and 7.79 N). Maximum loss in titratable acidity (52%) was observed in control fruits, whereas minimum loss was observed in on-farm treated lac based and SemperfreshTM-coated fruits. The delay of the ripening process was also related to lower anthocyanin accumulation and least colour changes. After

15 days of storage, lac-based coated fruits showed ~13% lower anthocyanin content. The maximum total antioxidant activity and ascorbic acid content at end of storage were recorded in on-farm-treated fruits with lac-based coating (17.49 μmole Trolox/g and 3.92 mg/100 g, respectively), followed by SemperfreshTM and vegetable wax. Overall, results suggest that the surface treatments with lac-based and SemperfreshTM coatings could effectively maintain the acceptability of plum fruits (Kumar et al. 2016).

20.10 Regulations for Packaging (India)

The amendment made in the Packaging and Labelling Regulation, 2011 enacted the new Packaging Regulation, 2018 by Food Safety and Standards of India. Food grade packaging materrial is composed of safe and suitable substances as per intended use and does not produce and allow any unwanted alteration in the composition of the food or sensory traits. The regulations permit the use of cardboard, paper, glass, metal and plastic as principal constituents of food packaging. The differentiation of primary and secondary food packaging is based on their contact with food. A primary food packaging is in direct contact with food while secondary one only covers the primary food packaging, thus separated by a layer from original food packed. The regulation restricts the use of newspaper as primary packaging material due to the presence of harmful chemicals carcinogenic in nature in the printed ink. Printing inks shall hold to IS: 15495 standards as a requirement of food safety. It is also obligatory to prevent any direct exposure of food to ink used on the package. Schedule I of Packaging Regulation, 2018 mentions the specification of paper material meant to be utilized in packaging of food. Food Safety and Standards (Packaging) Regulation 2018 states the packaging paper to have the desirable uniform thickness, devoid of any negative traits—specks, grease marks, cuts, pinholes and other blemishes (FSSAI 2018). Table 20.1 shows the list of Indian Standards for Paper as Food Packaging Material.

Primary packaging plays an important role in maintaining the integrity of food packed. It all depends on the capability of primary packaging, on how it encounters the physical, chemical and thermal forces involved during the process of transportation. Packaging edible are still on the way towards development in India.

Table 20.1 List of Indian standards for paper as food packaging material

Type of paper in food packaging	Indian standard
Grease-proof paper	IS 6622
Vegetable parchment	IS 7161
General purpose packing or wrapping paper	IS 6615
Paraffin wax used for coating the paper	Type I of IS 4654

20.11 Regulation for Edible Coating

All edible films and coatings to be used in accordance with good manufacturing practices should have the approval of the Food and Drug Administration (FDA) and be listed in the category of Generally Recommended as Safe (GRAS) items.

Edible coating and film belong to the category of indirect additive described in GRAS list of food additives. Indirect additives are defined as substances used during processing, packaging, holding and transporting of food. This category of additives has no functional impact on food but may have an indirect effect in terms of becoming a component of the product or may affect the characteristic of any food including any substance intended for use in coating, packing or holding the food. Packaging edibles correctly fall into the definition of indirect food additives.

20.11.1 Regulation for Food Contact Substances

US Food and Drug Administration controls the security of substances added to food, keeps up instructive data, information bases and listings identified with food allergens, ingredients, food additives, colour additives and GRAS substances.

"GRAS" is an abbreviation for the expression Generally Recognized as Safe. Under segments 201(s) and 409 of the Federal Food, Drug, and Cosmetic Act (the Act), any substance that is purposefully added to food is a food additive, that is liable to premarket audit and endorsement by FDA, except if the substance is by and large perceived, amongst qualified specialists, as having been satisfactorily demonstrated to be protected under the states of its proposed use, or except if the utilization of the substance is generally excepted from the term of a food additive.

Under sections 201(s) and 409 of the Act, and FDA's implementing regulations in 21 CFR 170.3 and 21 CFR 170.30, the use of a food substance may be GRAS

1. Through the guidance based on technically scientific procedures or,
2. The substance in use as common food before 1958, under 21 CFR 170.30(b),
3. General acknowledgment of safety through technically scientific procedures needs substantial quantity and quality of scientific proof for acceptance as a food additive.

Unpublished scientific information along with published and accepted scientific data, information or techniques becomes the base for general perception of safety of food additives in use.

Substantial history of a substance regarding its utilization in common food by mass gives it a recognition as USFDA, Generally recognized as Safe (GRAS) as mentioned in 21 CFR 170.30(c) and 170.3(f) (USFDA 2019).

Food, Drug & Cosmetic Act, 1958 controls the specifications and required parameters for Food Contact Substance (FCS) in United States with Food & Drug Administration as the regulatory authority. Food Additives Amendment was implemented to the Federal Food, Drug, and Cosmetic Act (FFDCA) in 1958.

This guideline permitted the US Food and Drug Administration (FDA) with the position to guarantee the safety of the food supply by controlling food additives either straightforwardly or implicated as secondary ingredients. The law binds to acquire permission for presenting any food additive during processing. FDA had issued the lists of sanctioned food contact substances in the Code of Federal Regulations (CFR) Section 21, Parts 170–199. The catalogue has indexed some Generally Recognized as Safe (GRAS) chemical ingredients used as food additive to be a part of petition process.

Threshold of Regulation (TOR) was executed by FDA for food contact materials with limited migration in 1990s to shorten the time period required for reviewing the petition. Time period of about 18 months was provided on an average for petition to be filed for TOR. The TOR review period ranged somewhere between one and a half year and the consent required quite similar to petitions for the food additive. Still the stakeholders found GRAS self-determination method more suitable as comparison to petitions for food additive, GRAS and TOR due to its speedy process.

The Food and Drug Administration Modernization Act (FDAMA) was enacted in 1997 with an amendment in FFDCA. The amendment added the clause for submission of notifications for food contact substances as an alternative of petitions and threshold of regulation submissions. In such a situation, FDA works in a step-by-step manner—first it evaluates the notice, studies the basic requirements for a GRAS determination and or FDA puts up the matter to question whether the substance could be used as GRAS ingredient. FDA responds to the notifier by letter.

Notifications for new food contact substances came into force on January, 2000. It allowed the producer or distributor of a food contact substance (FCS) to voluntarily submit a food contact notification (FCN) to FDA revealing the identity, use and the supporting information of the FCS to take a decision regarding safety of FCS for the intended use.

GRAS affirmation petitions sponsored by industry in between 1972 and 1990 are listed in 21 CFR Part 184 (USFDA, FDA's Approach to the GRAS Provision: A History of Processes) (USFDA 2018).

The GRAS final rule 81 FR 54960 enacted on August 17, 2016, formalized a notification procedure and established the regulations in Subpart E of part 170 (USFDA, About the GRAS Notification Program) (USFDA 2018).

Food, Drug & Cosmetic Act defines food additive as any substance purposefully used directly or indirectly and may become a constituent or may affect the properties of food with due adherence of its impact in either production, manufacturing, packing, processing, preparing, treating, packaging, transporting or holding food, and including any source of radiation intended for any such use.

Food Contact Substance (FCS), as incorporated in section 409 of the Food, Drug & Cosmetic Act, defines it as any substance being in use as an integral part of materials utilized in manufacturing, packing, packaging, transporting or holding food and is not known to have any technical effect in food involved.

Any of the substances utilized during the array of manufacturing process can be broadly classified as Food Contact Substance (FCS), Food Contact Material (FCM) and Food Contact Article (FCA) (USFDA 2018).

1. Food Contact Substance (FCS) is generally composed of single ingredient, may be a polymer in nature or an antioxidant component of a polymer and chemically pure with an unambiguous composition.
2. Food Contact Material (FCM) is made with the FCS and other substances may exhibit a variable composition.
3. Food Contact Article is the final holder of finished product in the form of film, bottle, tray or anything made from FCM (USFDA, Food Ingredient & Packaging Terms).

20.12 The Future of Packaging

Proteins are good moisture and oxygen, inhibit lipid migration and enhance mechanical properties of the edible film in use for packaging. They provide better physical covering and also a good alternative to synthetic commercial packaging material. Proteins go about as a durable, underlying lattice in multicomponent frameworks, yielding films and coatings with great mechanical properties. Lipids also act as good moisture barrier with limited low barrier properties for gas, lipid and flavour. By consolidating proteins and lipids in emulsion or bilayer (a film comprising of two sub-atomic layers), the positive credits of both can be joined and the negatives limited.

Starch-based films are non-toxic and has a great potential for edible packaging alternative. Starch is highly hydrophilic in nature. This characteristic makes it soluble in water leaving with low water vapour permeability. Based on the moisture percentage in the film, it exhibits low resistant and tensile strength for a packaging material. Polylactic acid obtained from agricultural products is transparent and exhibits water vapour permeability comparable to synthetic plastic material. It is brittle but still contains good tensile strength (Muller et al. 2017).

Laminated films provide good benefits than single, emulsion-based biopolymer films because of their characteristic of improved hindrance properties against required external components. The formation of laminated layers can possibly conquer these inadequacies by designing edible/biodegradable films with numerous functional layers. The composite properties of starch, which is highly hydrophilic, low water vapour permeable (WVP) and shows low barrier and tensile strength, and poly lactic acid, exhibiting good WVP and tensile strength but brittle in nature, can be utilized to form good laminates.

Edible/biodegradable films as food packaging materials has a brilliant future as it can offer a characteristic natural protection to food sources as per explicit packaging requirements. Notwithstanding the benefits of edible/biodegradable materials which have been introduced by researchers, various obstacles in their advancement potential must be survived, like expense viability, improved water vapour permeability barrier and innovative application strategies. Complete and hundred percent commercial application is still some way off; however, the capability edible/biodegradable films have been figured out (Kerry and Wang 2018).

Current consumer demand and requirement have introduced the most innovative mode of packaging "active packaging." The term can be defined as a packaging material which plays an additional role other than protection from the external environment. In "active packaging," the shelf life of the product might be increased due to antimicrobial property, oxygen or moisture scavenging trait of the edible packaging film. Sensory characteristics of food could also be enhanced with the implementation of "active packaging." Thus, "active packaging" improves quality, shelf life and microbiological safety of the food product involved.

20.13 Conclusions

The commercially produced edible packaging includes alginates, starch-based, carrageenan, seed gums, pectin and proteins, amongst others. Protein-based films have diverse benefits over synthetic film manufactured out of plastic as a base ingredient, the first being its edible nature. Proteinaceous films have a remarkable gas barrier property in contrast to those composed of lipids and polysaccharides. The novel structure of proteins supports superior mechanical properties of protein-based edible films to those of polysaccharide and fat-based films. Scientifically, the characteristics of edible films composed out of protein depend on the class of proteins, macromolecule interaction, type of process involved and the additives incorporated. On the contrary, the negative aspect associated with protein-based edible films are inferior moisture inhibition and physical strength as compared to synthetic polymers. These two properties restrict their usage in food packaging. To improve the water vapour and mechanical resistance of protein-based films, chemical and enzymatic treatments could be applied, or aggregated with some hydrophobic element or some artificial polymers or any physical method applicable. The modification methods applied would enhance the properties of protein-based edible packaging material. The chemical and enzyme modifications efficiently boosted the mechanical and water vapour barrier properties, but more focus is needed in addition of hydrophobic material. Pertaining to the impressive gas barrier properties of protein-based edible film as compared to lipid and polysaccharide based edible films, protein can provide an alternative for synthetic films in packaging. Studies conducted for soy protein-based film revealed oxygen permeability quite comparable to low-density polyethylene, methylcellulose, starch and pectin (Cuq et al. 1998). The potency of physical strength is greatly induced in protein-based films as compared to lipid and polysaccharides. The novel structure of protein macro-molecules with high intermolecular binding potential grants them comprehensive functional traits as packaging material. The practical application of protein-based films in amalgamation of hydrophobic material is greatly induced in terms of mechanical and barrier traits, characteristic of packaging films. Protein-based films exhibited better mechanical and barrier traits when treated with irradiation. Thus, by enhancing the essential packaging traits, good gas barrier, limiting the growth of microbes and preventing oxidation of lipid proteins, could act as good ingredient for edible packaging. Edible protein-based edible films in combination with hydrophobic materials and synthetic polymer can result in better functionality than films produced with only proteins, especially with respect to their mechanical and barrier properties. Irradiation was found to be an effective method for the improvement of both barrier and mechanical properties of protein-based edible films. Protein-based edible films show impressive gas barriers; hence, protein films would be a suitable food packaging material for preventing the growth of aerobic microbial and lipid oxidation in lipid enriched foods. The major attribute of edible/biodegradable film is its considerable resistance to oxygen permeability. The inconvenience associated is regarding water vapour permeability of edible/biodegradable films which in general is higher than that of synthetic plastic films (Kerry and Wang 2018).

Edible films provide an encouraging technique in the coming future to enhance the food quality and preservation during transportation, storage and selling point. Edible film provides numberless myriad of benefits over the conventional petroleum-based packaging materials. The intrinsic parameter of films being edible and biodegradable are the best advantages adhered to these innovative items. Supplementation of active ingredients into these consumable film/coating boosts food safety in parallel to their eco-friendly nature. Edible films that are all more precisely adaptable to mechanical strength and stable to the environment would withstand actual processing and end-use conditions and allow numerous workable applications (Shendurse et al. 2018). Starch is a natural polymer, extractable from agricultural fractions. It is most widely available as storage polysaccharide in the staples. Abundant in nature, starch is a suitable raw material for edible packaging film. Starch is economically feasible as compared to its synthetic petroleum counterpart. It is a challenge to develop edible starch-based film owing to its low mechanical strength in terms of flexibility, elongation percentage and tensile strength. Brittleness and permeability to gases of starch-based films could be overcome with the incorporation of plasticizer. Synthetic packing fulfils the function by providing good protection, but the non-biodegradable material left at the end has created the new space for edible packaging. Biopolymers in use as raw material for the edible packaging are fully utilized at the end of the process.

References

Abdou ES, Sorour MA (2014) Preparation and characterization of starch/carrageenan edible films. Int Food Res J 21(1):189–193

About the GRAS Notification Program (2018) USFDA. https://www.fda.gov/food/generally-recognized-safegras/about-gras-notification-program#f1. Content current as of: 01/04/2018

Arcan I, Boyacı D, Yemenicioğlu A (2017) The use of Zein and its edible films for the development of food packaging materials. In: Reference module in food science. Elsevier, Netherlands. https://doi.org/10.1016/B978-0-08-100596-5.21126-8

Banks NH (1984) Some effects of TAL prolong coating on ripening of bananas. J Exp Bot 35:127–137

Beglinger E (1958) Distillation of resinous wood. USDA Forest Services, Madison, WI, pp 1–7. Wood Rosin

Bharti SK, Pathak V, Alam T (2020) Materiality of edible film packaging in muscle foods: a worthwhile conception. J Package Technol Res 4:117–132. https://doi.org/10.1007/s41783-020-00087-9
Bhat KM, Jyothsana R, Sharma A, Rao NN (2020) Carrageenan-based edible biodegradable food packaging: a review. Int J Food Sci Nutr 5(4):69–75
Bourtoom T (2008) Edible films and coatings: characteristics and properties. Int Food Res J 15(3):237–248
Canteri M, Moreno L, Wosiacki G, Scheer A (2011) Pectin: from raw material to the final product. Polímeros 22:149–157
Chakravartula SSN, Soccio M, Lotti N, Balestra F, Dalla Rosa M, Siracusa V (2019) Characterization of composite edible films based on pectin/alginate/whey protein concentrate. Materials 12(15):2454. https://doi.org/10.3390/ma12152454
Ciolacu L, Nicolau AI, Hoorfar J (2014) Global safety of fresh produce. A handbook of best practice, innovative commercial solutions and case studies. Woodhead Publishing Ltd., Sawston
Codex Alimentarius Commission (2004) Report of the 20th session of the Codex committee on general principles. Joint FAO/WHO food standards programme. 2004 May 2–7. Paris, France, p 44
Coles R (2003) Introduction. In: Coles R, McDowell D, Kirwan MJ (eds) Food packaging technology. Blackwell Publishing Ltd, CRC Press LLC, London, pp 1–31
Cuq B, Gontard N, Guilbert S (1998) Proteins as agricultural polymers for packaging production. Cereal Chem J 75(1):1–9. https://doi.org/10.1094/CCHEM.1998.75.1.1
DeWit JN (2001) Lecturer's handbook on whey and whey products, 3rd edn. European Whey Products Association, Belgium
Dhall RK (2012) Advances in edible coatings for fresh fruits and vegetables: a review. Crit Rev Food Sci Nutr 53:435–450
Drake SR, Fellman SK, Nelson JW (1987) Postharvest use of sucrose polyesters for extending the shelf-life of stored "Golden delicious" apples. J Food Sci 52:1283–1285
El-Anany AM, Hassan GFA, Ali FMR (2009) Effect of edible coatings on the shelf life and quality of 'Anna' apple (Malus domestica Borkh.) during cold storage. J Food Technol 7:5–11
FloZein Products (1976) A Flo Chemical Corp Company. https://www.zeinproducts.com/
Food Ingredient & Packaging Terms (2018) USFDA. https://www.fda.gov/food/food-ingredients-packaging/food-ingredient-packaging-terms. Content current as of: 01/04/2018
Food Safety and Standards (Packaging) Regulation (2018) FSSAI. https://www.fssai.gov.in/upload/uploadfiles/files/Gazette_Notification_Packaging_03_01_2019.pdf
Galus S, Lenart A (2013) Development and characterization of composite edible films based on sodium alginate and pectin. J Food Eng 115:459–465. https://doi.org/10.1016/j.jfoodeng.2012.03.006
Galus S, Turska A, Lenart A (2012) Sorption and wetting properties of pectin edible films. Czech J Food Sci 30:446–455. https://doi.org/10.17221/444/2011-CJFS
García EA, Feria JS, Villalobos JRR, González FR, Pérez AJ, Huicochea EF (2014) Properties of edible films based on oxidized starch and Zein. Int J Polym Sci. https://doi.org/10.1155/2014/292404
Generally Recognized as Safe (GRAS) (2019) USFDA. https://www.fda.gov/food/food-ingredients-packaging/generally-recognized-safe-gras. Content current as of: 09/06/2019
Gennadios A, Weller CL (1990) Edible films and coating from wheat and corn proteins. Food Technol 44(10):63–69
Guilbert S (1986) Technology and application of edible protective films. In: Mathlouthi M (ed) Food packaging and preservation: theory and practice. Elsevier Applied Science Publishing Co., London, pp 371–399
Guilbert S, Gontard N, Gorris LGM (1996) Prolongation of the shelf life of perishable food products using biodegradable films and coatings. Lebensm-Wiss Technol 29:10–17
Gulum S, Levent B (1994) Effects of Semperfresh and Johnfresh fruit coatings on poststorage quality of "Ankara" pears. J Food Process Preserv 18:189–199

Han J, Krochta J (2007) Physical properties of whey protein coating solutions and films containing antioxidants. J Food Sci 72:308–314. https://doi.org/10.1111/j.1750-3841.2007.00358.x

Herald TJ, Hachmeister KA, Huang S, Bowers JR (1996) Corn Zein packaging materials for cooked Turkey. J Food Sci 61(2):415–418. https://doi.org/10.1111/j.1365-2621.1996.tb14206.x

Insaward A, Duangmal K, Mahawanich T (2015) Mechanical, optical, and barrier properties of soy protein film as affected by phenolic acid addition. J Agric Food Chem 63(43):9421–9426. https://doi.org/10.1021/jf504016m

Jancikova S, Dordevic D, Jamroz E, Behalova H, Tremlova B (2020) Chemical and physical characteristics of edible films, based on κ- and ι-Carrageenans with the addition of Lapacho tea extract. Foods 9(3):357. https://doi.org/10.3390/foods9030357

Jang SA, Shin YJ, Song KB (2011) Effects of rapeseed protein–gelatin film containing grapefruit seed extract on 'Maehyang' strawberry quality. Int J Food Sci Technol 46:620–625

Janjarasskul T, Krochta JM (2010) Edible packaging materials. Annu Rev Food Sci Technol 1:415–448

Kanmani P, Rhim JW (2010) Development and characterization of carrageenan/grapefruit seed extract composite films for active packaging. Int J Biol Macromol 68:258–266

Kerry H, Milda E (2009) Edible films and coatings for food applications. https://doi.org/10.1007/978-0-387-92824-1

Kerry JP, Wang L (2018) Edible/biodegradable packaging for food. https://www.newfoodmagazine.com/article/215/edible-biodegradable-packaging-for-food. Accessed 6 Feb 2021

Khwaldia K, Arab TE, Desobry S (2010) Biopolymer coatings on paper packaging materials. Compr Rev Food Sci Food Saf 9:82–91

Kim N, Seo E, Kim Y (2019) Physical, mechanical and water barrier properties of yuba films incorporated with various types of additives. J Sci Food Agric 99:2808–2817. https://doi.org/10.1002/jsfa.9490

Knutsen SH, Mylabodski DE, Larsen B, Usov AI (1994) A modified system of nomenclature for red algal galactans. Bot Mar 37:163–169

Krochta JM (2002) Protein as raw materials for films and coatings: definitions, current status, and opportunities. In: Gennadios A (ed) Protein-based films and coatings. CRC Press, New York, p 40

Krochta JM, De MJC (1997) Edible and biodegradable polymer films: challenges and opportunities. Food Technol 51(2):61–73

Kumar P, Sethi S (2018) Edible coating for fresh fruit: a review. Int J Curr Microbiol App Sci 7(5):2619–2626. https://doi.org/10.20546/ijcmas.2018.705.303

Kumar P, Sethi S, Sharma RR, Varghese E (2016) Effect of edible coatings on eating and functional quality of Japanese plum cv. Santa Rosa. Indian J Horti 73(3):416–422. https://doi.org/10.5958/0974-0112.2016.00088.8

Longares A, Monahan ED, O'Riordan ED, O'Sullivan M (2004) Physical properties and sensory evaluation of WPI films of varying thickness. Lebensm Wiss Technol 37:545–550

Mahadevappa YK, Arjumand AK, Ravindra R (2014) Polymer synthesis and processing. Natural and Synthetic Biomedical Polymers. pp. 1–31

Mahmoud R, Savello P (1993) Solubility and hydrolyzability of films produced by transglutaminase catalytic crosslinking of whey protein. J Dairy Sci 76(1):29–35

Marsh K, Bugusu B (2007) Food packaging—roles, materials, and environmental issues. J Food Sci 72(3):39–55. https://doi.org/10.1111/j.1750-3841.2007.00301.x

Martiny TR, Pacheco BS, Pereira CMP (2020) A novel biodegradable film based on κ-carrageenan activated with olive leaves extract. Food Sci Nutr 8:3147–3156. https://doi.org/10.1002/fsn3.1554

McHugh AJF, Krochta JM (1994) Plasticized whey protein edible films: water vapor permeability properties. J Food Sci 59:416–419. https://doi.org/10.1111/j.1365-2621.1994.tb06980.x

Meheriuk M, Lau OL (1988) Effect of two polymeric coatings on fruit quality of "Bartlett" and "d'Anjou" pears. J Am Soc Hort Sci 113:222–226

Merck B (2013) Interview by Lindsay Frenandez-Salvador. Wood Rosin (September 5, 2013). Wood Rosin

Moradi M, Tajik H, Ri R, Seyed M, Mahmoudian A (2016) Antioxidant and antimicrobial effects of zein edible film impregnated with Zataria multiflora Boiss. essential oil and monolaurin. Lebensm Wiss Technol 72:37–43. https://doi.org/10.1016/j.lwt.2016.04.026

Muller J, González-Martínez C, Chiralt A (2017) Combination of poly(lactic) acid and starch for biodegradable food packaging. Materials 10(8):952. https://doi.org/10.3390/ma10080952

Parris N, Dickey LC (2001) Extraction and solubility characteristics of zein proteins from dry-milled corn. J Agric Food Chem 49(8):3757–3760. https://doi.org/10.1021/jf0011790

Post Harvest (2021) Product Catalog Agricoat Natureseal Ltd. https://www.postharvest.biz/en/company/agricoat-natureseal-ltd/_id:49906,seccion:productcatalog,producto:10816/. Accessed 20 Feb 2021

Robertson, GL (2005) An introduction to food packaging. University of Queensland. https://iufost.org/wp-content/uploads/disteducation/AN_INTRODUCTION_TO_FOOD_PACKAGING_07.02.09.pdf

Rosida R, Sudaryati, Yahya A (2018) Edible film from the pectin of papaya skin (The Study of Cassava Starch and Glycerol Addition). J Phy Conf Ser 953:012248. https://doi.org/10.1088/1742-6596/953/1/012248

Said MI, Erwanto Y, Abustam E (2016) Properties of edible film produced using combination of collagen extracts of Bligon goatskin with glycerol. Am J Anim Vet Sci 11(4):151–159. https://doi.org/10.3844/ajavsp.2016.151.159

Santerre CR, Leach TF, Cash JN (1989) The influence of the sucrose polyester, Semperfresh, on the storage of "McIntosh" and "Golden delicious" apples. J Food Process Preserv 13:293–305

Sato ACK, Oliveira PR, Cunha RL (2008) Rheology of mixed pectin solutions. Food Biophys 3: 100–109

Schou M, Longares A, Montesinos-Herrero C, Monahan FJ, O'Riordan D, O'Sullivan M (2005) Properties of edible sodium caseinate films and their application as food. LWT Food Sci Technol 38(6):605–610. https://doi.org/10.1016/j.lwt.2004.08.009

Shendurse AM, Gopikrishna G, Patel AC (2018) Milk protein based edible films and coatings–preparation, properties and food applications. J Nutr Health Food Eng 8(2):219–226. https://doi.org/10.15406/jnhfe.2018.08.00273

Shukla R, Cheryan M (2001) Zein: the industrial protein from corn. Ind Crop Prod 13(3):171–192. https://doi.org/10.1016/S0926-6690(00)00064-9

Stuchell YM, Krochta JM (1995) Edible coatings on frozen king Salmon: effect of whey protein isolate and acetylated monoglycerides on moisture loss and lipid oxidation. J Food Sci 60(1):28–31. https://doi.org/10.1111/j.1365-2621.1995.tb05599.x

Thawien W (2012) Protein-based edible films: characteristics and improvement of properties. In: Eissa AA (ed) Structure and function of food engineering. IntechOpen, London. https://doi.org/10.5772/48167

Therkelsen GH (1993) Carrageenan. In: Whistler RL, BeMiller JN (eds) Industrial gums, 3rd edn. Academic Press, San Diego, CA, pp 145–180

Valdés A, Burgos N, Jiménez A, Garrigós M (2015) Natural pectin polysaccharides as edible coatings. Coatings 5(4):865–886. https://doi.org/10.3390/coatings5040865

Van Zyl HJ, Torman H, Von Mollendorff LJ (1987) Effect of wax treatment on fruit quality of pears, apples and nectarines after cold storage. DFGA:169–172

Wan V, Ch H, Kim MS, Lee SY (2005) Water vapor permeability and mechanical properties of soy protein isolate edible films composed of different plasticizer combinations. J Food Sci 70:387–391
Zablackis E, Santos GA (1986) The carrageenan of *Catenella nipae* Zanard., a. Marine Red Alga 29(4):319–322. https://doi.org/10.1515/botm.1986.29.4.319
Zhang J, Mungara P, Jane JL (2001) Mechanical and thermal properties of extruded soy protein sheets. Polymer 42(6):2569–2578. https://doi.org/10.1016/S0032-3861(00)00624-8
Zhang Y, Rempel C, McLaren D (2014) Edible coating and film materials: carbohydrates. In: Innovations in food packaging, 2nd edn. Elsevier, Amsterdam, pp 305–323

Edible and Oral Thin Films: Formulation, Properties, Functions, and Application in Food Packaging and Pharmaceutical Industry

21

Himani, Mohit Singla, Pramod K. Prabhakar, Arun Sharma, and Murlidhar Meghwal

Abstract

Edible and oral films are the directly consumable materials in the form of food packaging or a drug dosage. The edible oral films are developed using edible biopolymers and food grade additives. The packaging application of film provides good barrier properties to the food and thus enhances the overall quality and safety of a food product. In pharmaceutical application, these films proved as an effective form of drug administration as they disintegrate and dissolve instantly when come in contact with saliva. Thus, these films can provide antioxidant and antimicrobial properties, extend shelf life of a product, and can provide better delivery of active ingredients. These advantages of the edible oral film have paved its growth in both the food and pharmaceutical industries. This chapter will discuss the components of the film and their selection, method of the formulation, film properties and effect of different components on film, and finally the application of these films in the food and pharmaceutical industries.

Himani
Department of Food Business Management and Entrepreneurship Development, National Institute of Food Technology Entrepreneurship & Management, Kundli, HR, India

M. Singla
Department of Food Engineering and Technology, Tezpur University, Assam, India

P. K. Prabhakar (✉) · M. Meghwal
Department of Food Science and Technology, National Institute of Food Technology Entrepreneurship & Management, Kundli, HR, India
e-mail: pramodkp@niftem.ac.in

A. Sharma
Department of Food Engineering, National Institute of Food Technology Entrepreneurship & Management, Kundli, India

A. Poonia, T. Dhewa (eds.), *Edible Food Packaging*,
https://doi.org/10.1007/978-981-16-2383-7_21

Keywords

Edible biopolymers · Oral thin films · Food packaging · Active ingredients · Shelf life · Food and pharmaceutical industries

21.1 Introduction

Edible and oral films are the thin and flexible layer composed of a single polymer or a combination of polymers formed with or without using plasticizers (Borges 2015). These films are water soluble in nature thus containing polymers which are hydrophilic in nature (Kalyan and Bansal 2012; Mahboob et al. 2016). The film can be composed of a single biopolymer or a mixture of biopolymers. The biopolymers used for film can belong to a group of proteins, polysaccharides (carbohydrates and gums), and lipids (Han 2014). Among these groups, some of the commonly used polymers for the development of edible oral films are hydroxyethyl cellulose, chitosan, polyvinyl alcohol (PVA), pullulan, gelatin, and maltodextrin. The plasticizers and other additives are added to biopolymers which modify the physical or functional properties of the film. The film-forming mechanism of biopolymers involves intermolecular forces such as covalent bonding (e.g., disulfide bonds and crosslinking) and electrostatic, hydrophobic, or ionic interactions (Han 2014). The film development can be done with methods including solvent casting, semisolid casting, hot melt extrusion, solid dispersion extrusion, and rolling. The edible and oral films can be used as a food packaging material and a drug dosage form, respectively.

The edible film as a packaging application provides barrier properties to the food product against physical, chemical, and biological properties thus enhancing overall quality and safety of food products (Cerqueira 2018; Trinetta 2016). The selection of material depends on the type of the food and the storage conditions, that is, temperature and relative humidity (Cerqueira 2018). The edible packaging has advantages over other packaging materials in being edible and biodegradable and thus leads to less waste generation. Further, it can add additional functional properties like antioxidant and antimicrobial properties to the film.

The oral film emerged in the late 1970s as an alternative to conventional drug dosage forms like tablets, capsules, and syrups (Reddy et al. 2018; Keshari et al. 2014; Bhyan et al. 2011). These films are proved to have advantages over conventional drug systems as they can be administered without water and disintegrate and dissolve rapidly when comes in contact with saliva. This enhanced efficacy results due to larger surface area thus providing better drug delivery to the targeted site (Jain et al. 2018; Karki et al. 2016; Mahboob et al. 2016; Saini et al. 2012). This is especially an effective way of drug delivery in patients having swallowing problems and in some specific patient categories including geriatric, pediatric, and dysphasic patients (Kathpalia and Gupte 2013; Keshari et al. 2014; Nagar et al. 2011). Oral thin film also helps in eliminating bad breath and can provide a better alternative to chewing gums or is not socially acceptable (Karki et al. 2016; Mahesh et al. 2010; Arya et al. 2010).

The edible and oral films thus have great potential to serve as a food packaging and drug dosage alternative in respective sectors. This chapter will discuss the components of the film and their selection, method of the formulation, film properties and effect of different components on film, and finally the application of these films in the food and pharmaceutical industries.

21.2 Components of Edible and Oral Films

21.2.1 Film Forming Polymer

The film polymer is the integral component of edible and oral films (Jain et al. 2018), which act as a base material for the film formation. A wide variety of biopolymers, alone or in combination, can be used for formulating the film. The polymer concentration can range between 40% and 45% w/w of total film weight; however, a higher polymeric concentration of 65% w/w of film weight can be used to obtain the desired film properties (Jain et al. 2018; Kalyan and Bansal 2012; Mahboob et al. 2016). A single polymer or a combination of polymers is used for the film formulation. The used polymers should be hydrophilic in nature, so as to provide water solubility to the film. In addition, the film-forming biopolymers should be nontoxic, tasteless, and with good shelf life, spreadability, and wetting property (Bhyan et al. 2011; Kalyan and Bansal 2012; Saini et al. 2012). The polymers that can be effectively used in development of edible and oral films are hydroxy methylcellulose, chitosan, polyvinyl alcohol (PVA), pullulan, gelatin, and maltodextrin.

Since diverse varieties of polymers are available for film formation, the film can be imparted with specific properties. For example, pullulan is commonly used for film formation to impart high solubility, mechanical strength, and stability over a wide temperature range. The high molecular weight gelatin is usually added to film for making it more appealing and glossier. The film-forming polymers which are cellulose derivatives like methyl cellulose, carboxymethyl cellulose being hydrophilic in nature, result in a film with low water vapor barrier and thus help in water retention (Karki et al. 2016). Thus, the film strength is largely determined by the type and amount of polymer used for film formation (Kathpalia and Gupte 2013). As polymer and active ingredient are the primary film components, their proportion with respect to each other is determined by the following two factors (Kathpalia and Gupte 2013):

- Minimum % w/w polymeric concentration is required for formation of film matrix which incorporates the active substance and other components with desired mechanical strength and viscoelasticity.
- % w/v polymeric concentration in film formulation is mainly determined by the required viscosity. The film solution should have enough viscosity to prevent settling of the suspended solids and can form a film with smooth spreadability.

21.2.2 Active Substance

It is the principle active ingredient which gives the film its desired properties. The active substance should be soluble and stable in saliva. The ingredient can be added in milled, micronized, or nanocrystal form, depending on the desired release profile (Mahboob et al. 2016). The active ingredient like clove or fruit peel extract can be added in edible film in order to provide antioxidant and antimicrobial properties to the film (Himani et al. 2020; Kumar et al. 2019). In a similar manner, the oral film can be incorporated with various classes of drugs, like anti-diarrheal, anti-depressants, anti-emetic, etc. (Muhammad et al. 2016). The active substance is added in a concentration of 5–30% as it is difficult to add a higher amount in oral film due to the size of the dosage (Mahboob et al. 2016; Kalyan and Bansal 2012).

21.2.3 Plasticizers

These are added to the film formulation to impart good strength, flexibility, and gloss, reduce the brittleness and glass transition temperature, and thus enhance the thermal stability of the added polymers (Saini et al. 2012; Bhyan et al. 2011). The plasticizer should have compatibility with the polymer. Further, it should have good compatibility with the solvent and method of formulation (Keshari et al. 2014; Bhyan et al. 2011). These are added to the film formulation in a concentration of 0–20% w/w of polymers dry weight (Kalyan and Bansal 2012; Saini et al. 2012; Bhyan et al. 2011). The plasticizers commonly used for film formulation are glycerol, PEG, dibutyl phthalate, and PEG.

The process of plasticization can happen with two mechanisms which are internal plasticization and external plasticization. The internal plasticization happens due to chemical interaction between molecules of the polymers, while in external plasticization a physical plasticizer is added separately. Since there is no chemical interaction between the polymers in external plasticization, it is the preferred plasticization mechanism (Kathpalia and Gupte 2013).

21.2.4 Surfactants

Surfactants added to the film enhances the solubility and wetting property of film, thereby leading to fast release of active substance within seconds. The common type of surfactants used in film formation are sodium lauryl sulfate, tweens, benzalkonium chloride, and benzethonium chloride (Muhammad et al. 2016; Keshari et al. 2014; Kalyan and Bansal 2012; Panda et al. 2012).

21.2.5 Saliva Stimulating Agent

They aid in rapid disintegration of the oral thin film by increasing the saliva production. The surface-active agents can be added with a concentration level of 2–6% w/w of film's weight, alone or in combination. Saliva-stimulating agents such as succinic acid, lauric acid, ascorbic acid, tartaric acid, citric acid, and malic acid are used in the film (Mahboob et al. 2016; Kalyan and Bansal 2012; Saini et al. 2012; Bhyan et al. 2011).

21.2.6 Sweeteners

These are added to the films as a taste masking agent to eliminate astringent taste of active substance. Both natural and artificial sweeteners are used for this purpose (Mahboob et al. 2016; Keshari et al. 2014; Saini et al. 2012). Some of the natural sweeteners such as mannitol and isomalt are used for providing good mouthfeel and oral cooling sensation. The reports indicated that the use of artificial sweeteners is more preferred by the diabetic patients and people on diet (Bhyan et al. 2011). The sweeteners are added in the concentration of 3–6%, both alone and in combination (Kalyan and Bansal 2012; Saini et al. 2012; Bhyan et al. 2011). The commonly used sweeteners are glucose, galactose, aspartame, acesulfame K, sucralose, mannose, and neotame.

21.2.7 Flavors

Flavors are added to the film to mask the taste and odor of the oral films or in other words enhance the appeal of the film. Identification of the film formulation depends on the initial flavor quality perceived and the after taste of the film which normally stays for a minimum of 10 min. The quantity of flavor required to be added is determined by the type of flavor and its strength; however, a 10% w/w of flavor is preferably added (Mahboob et al. 2016; Kalyan and Bansal 2012; Saini et al. 2012; Bhyan et al. 2011).

21.2.8 Coloring Agent

They impart color to the film and help in the enhancement of appeal of the film (Mahboob et al. 2016; Bhyan et al. 2011). FD&C approved colors with the concentration of 1% (Mahboob et al. 2016; Kalyan and Bansal 2012).

21.3 Selection of Polymer

The polymer properties define the physical and mechanical properties of the edible and oral film, which constitute 20–75% (w/w) of total dry weight of the film (Dahiya et al. 2009). The polymer provides strength to the film (Reddy et al. 2018). Thus, this makes the selection of polymer the most crucial factor to successfully develop the film (Dahiya et al. 2009). Thus, the selected polymers should be hydrophilic in nature, rapidly disintegrate, have good mouthfeel, and have suitable mechanical properties. Additionally, it should have good physicochemical and permeability properties with composition satisfying the required film properties (Borges 2015; Dahiya et al. 2009). The film should have mechanical strength with adequate elongation and elasticity in order to sustain the internal and external stresses during storage and when exposed to environmental conditions. The properties of film are largely determined by polymer chemistry, solvent effects, and additives like plasticizer, surfactants, etc. As already mentioned, a single polymer or combination of polymers can be used for film development in order to optimize the properties. The types of polymers that can be used for film are cellulose derivatives, synthetic polymers, natural gums, starch derivatives, polysaccharides, peptides etc. (Reddy et al. 2018; Borges 2015; Dahiya et al. 2009). The polymer selection can be done based on properties exhibited by polymer in the film formation, as listed in Table 21.1, and can help in selection of a desired polymer or a blend of polymer. The selected polymer should have the following properties (Kathpalia and Gupte 2013):

1. Should not be toxic.
2. Do not cause irritation.
3. Should have a bland taste.
4. Should have a good mouthfeel.
5. Should have long stability.
6. Do not modify properties of active ingredient or other components in the film formulation.
7. Should not be expensive.
8. Should exhibit good wetting and spreading properties.
9. Exhibit less disintegration time of the film.
10. It should exhibit optimal peel strength and tensile strength.

21.4 Selection of Active Ingredient

It is the principal compound that helps to incorporate desired property in film. The selection of active ingredients depends on the properties of the substance like taste, allergenicity, and its possible effects like discoloration and unpalatability. The active substance should be water soluble and should have intraoral absorption. When a drug is incorporated in an oral film, it should have a better release and dissolve on saliva instantly. The active substance can preferably be added in milled form

Table 21.1 Properties of the polymers

S. no	Polymer	Properties	References
1.	Hydroxypropyl methylcellulose (HPMC)	• Molecular weight (mw): 10,000–1,500,000 • It is a white cream powder with no taste and odor • It has solubility in cold water • It is nonionic in nature with moderate muco-adhesiveness • The polymer solution is stable between pH 3 and 11 • Film formation at 2–20% concentration • Effective in controlled/delayed release of active substance	Karki et al. (2016), Borges (2015)
2.	Carboxymethyl cellulose (CMC)	• Mw: 90,000–700,000 • It is a white, odorless powder • Easily dispersible in water and result in a clear or colloidal solution • It has high swelling index with good bio-adhesiveness • It is highly compatible with starch forming single-phase polymeric film which enhances the mechanical as well as barrier properties • The film-forming property can be improved by enzymatic modification	Karki et al. (2016), Borges (2015)
3.	Hydroxypropyl cellulose (HPC)	• Mw: 50,000–1,250,000 • It is white to slightly yellow in color, inert in nature with no odor and taste • It has solubility in cold and hot polar organic solvents such as absolute ethanol, methanol, isopropyl alcohol, and propylene glycol • It has moderate muco-adhesiveness • It can be used with starch to replace synthetic polymers or HPMC for improved solubility • Good film-forming ability and 5% (w/w) is generally for film coating	Karki et al. (2016), Borges (2015)
4.	Poly (vinyl pyrrolidone) (PVP)	• It has high range of solubility • Nonionic in nature • It has high solubility • It is used as a co-adjuvant to increase muco-adhesiveness • The film-forming property can be enhanced with blending with PVA and HPMC • PVP blending with ethyl cellulose and HPC can produce film with enhanced flexibility, softness, and tougher properties • PVP-alginate blends in different ratios can be used for designing controlled drug release	Karki et al. (2016), Borges (2015)

(continued)

Table 21.1 (continued)

S. no	Polymer	Properties	References
5.	Poly (vinyl alcohol) (PVA)	• Mw: 20,000–200,000 • It is white to cream in color and granular in nature • It is a hydrophilic synthetic polymer • Nonionic polymer • Moderate muco-adhesiveness • Forms flexible film • It has higher value of elongation at break	Karki et al. (2016), Borges (2015)
6.	Poly (ethylene oxide) (PEO)	• It is nonionic in nature • It has high molecular weight and has high muco-adhesiveness • Tear resistance, dissolution rate of film, and adhesiveness can be optimized by blending low mw and high mw PEO with cellulose • It has good mouthfeel and produces high viscosity gel with no stickiness	Karki et al. (2016), Borges (2015)
7.	Pullulan	• Mw: 8000–2,000,000 • It is white powder with no odor and taste • It has solubility in both hot and cold water • It has moisture more than 6% w/w • It is biodegradable in nature and nontoxic and good mechanical properties • It can be blended with sodium alginate and/or CMC to synergistically enhance the film properties • It can be added with HPMC for improving thermal and mechanical properties • A 5–25% (w/w) of pullulan results in flexible film • The film has stability with less oxygen permeability	Himani et al. (2020), Karki et al. (2016), Borges (2015)
8.	Pectin	• Mw: 30,000–100,000 • It is a yellowish white colored powder with mucilaginous taste and no odor • It has solubility in water and is not soluble in most of the organic solvents • It has strong muco-adhesiveness • Does not produce fast-dissolving film, but pectin can be modified for improving dissolution rates • Forms good film at low temperature	Karki et al. (2016), Borges (2015)
9.	Chitosan	• It is white or creamy-colored powder or flake with no odor • It is biodegradable in nature and biocompatible • It has solubility in water sparingly and with no solubility in ethanol and other organic solvents • It is also insoluble in neutral or alkali	Himani et al. (2020), Karki et al. (2016), Borges (2015)

(continued)

Table 21.1 (continued)

S. no	Polymer	Properties	References
		solutions at pH above 6.5 • It has excellent film-forming property with good mechanical properties	
10.	Sodium alginate	• It is a white or buff powder, with no odor and taste • It is not soluble in organic solvents and acid where the pH of the formed solution is below 3 • It is anionic in nature with high muco-adhesiveness • It is safe and nonallergic with biodegradable nature • It rapidly swells and dissolves in water • It has excellent gel and film-forming ability • It has compatibility with most water-soluble thickeners and resins	Karki et al. (2016), Borges (2015)
11.	Carrageenan	• It is anionic in nature • It has three structural types: Iota, kappa, and lambda. These differ in solubility and rheology • The sodium form of these types has solubility in cold as well as hot water • The solution has maximum solubility between pH 6 and 10 • It has moderate muco-adhesiveness	Karki et al. (2016), Borges (2015)
12.	Gelatin	• Mw: 15,000–250,000 • It is light amber to faintly yellow in color • It has solubility in glycerin, acid, alkali, and hot water • It has a moisture content 9–11% (w/w) • It has excellent good forming properties • Commonly used for developing sterile ophthalmic film	Karki et al. (2016), Borges (2015)

(powder form). Since incorporation of higher amounts in OTFs is difficult due to size of the dosage, only 5–30% of active compounds are added (Mahboob et al. 2016; Kalyan and Bansal 2012). The active ingredient should have the following properties (Jain et al. 2018; Bala and Sharma 2018; Kathpalia and Gupte 2013):

- Low dosage level.
- Should be palatable.
- Small molecular weight.
- Should have the ability to permeate the oral mucosa.
- Should be soluble and stable in saliva.

21.5 Method of Manufacture of Edible Oral Film

21.5.1 Solvent Casting Method

The solvent casting method is the most preferred method of film development as it is the simple and low-cost method of manufacture. In this method, hydrophilic polymers are dissolved for formation of a homogenous solution. The active substance and other components are mixed with a soluble solvent. Both solutions are mixed and homogenized to obtain a uniform solution. The entrapped air bubbles in the film formulation are then allowed to release naturally or by applying vacuum, which is called deaeration (Karki et al. 2016; Panda et al. 2012; Saini et al. 2012; Naik et al. 2014). The film-forming solution is then casted on the casting mold and dried by various drying instruments such as oven, microwave, tray oven, and vacuum between 40 and 50 ° C (Saini et al. 2012; Kalyan and Bansal 2012). After drying, the film is removed from mold and cut into desired shape and size. The drying rate, the thickness, morphology, and content uniformity depend on the rheological properties of the polymer solution. This method is beneficial in terms of producing film with improved physical properties and an easy and cheap processing method (Karki et al. 2016).

21.5.2 Hot Melt Extrusion (HME)

The HME method is an alternative to the solvent casting method for film formation used especially when no organic solvent is needed. The method is commonly used for developing granules, tablets, pellets, and film formation. The method involves casting a film by forming a blend of polymers, drug substance, and other components by melting all the components using heat. The mixture is then homogenized using the extruder screw until the components are melted and mixed. The melted film material is forced down a flat extrusion die (orifice) that presses the extrudate into a homogenous film. The film can be modified for its thickness and strength by the elongation rollers when the material is still hot and malleable. The film is then left for cooling and then cut to the desired shape (Suhag et al. 2020; Karki et al. 2016; Naik et al. 2014). The following are the main steps of HME process:

- Components are fed via a hopper to the extruder.
- Mix, grind, and knead.
- The molten and blended mass is allowed to flow through the die.
- The mass is extruded through the die and further downstream processing.

21.5.3 Semisolid Casting Method

The semisolid casting is preferred when film formulation is composed of an acid insoluble polymer. In this method, a solution of hydrophilic polymers is prepared.

This solution is then mixed with a solution containing acid insoluble polymers like cellulose acetate butyrate, cellulose acetate phthalate, etc. Further, the plasticizer is added to the resulting solution to obtain a gel mass, which is casted using heat-controlled drums to form film, which should be 0.015–0.05 inches thick. The acid insoluble polymer and film-forming biopolymer should have a ratio of 1:4 in the film solution (Naik et al. 2014; Saini et al. 2012).

21.5.4 Rolling Method

In this method, a pre-mix is prepared and added with an active substance and then developed into a film. The batch of a pre-mix is prepared using biopolymer, a polar solvent, and other additives leaving the active substance, which is mixed in a master feed tank. A predetermined quantity of pre-mix batch is passed via the first metering pump and control valve. The drug is added to the resulting solution and mixed sufficiently into a homogenized matrix. A fixed quantity of matrix is passed via a second metering pump into the pan. The film thickness is determined using the metering roller. Finally, the film is developed onto the substrate and moved via the support roller. The wet film is allowed to dry with the help of controlled bottom drying. The water or a mixture of water and alcohol is used as a solvent (Mahboob et al. 2016).

21.6 Evaluation Test Method for Edible and Oral Film

The film can be evaluated for the following properties:

- Mechanical properties.
  - Tensile strength.
  - Percent elongation.
  - Young's modulus.
  - Tear resistance.
  - Folding endurance.
- Other physical properties.
  - Thickness.
  - Weight.
  - Moisture.
  - Contact angle.
  - Transparency.
- Chemical properties.
  - Surface pH.
  - Disintegration time.
  - In vitro dissolution.
- Morphology study.
- Thermal properties.

- Swelling test.
- Solubility.
- Assay/content uniformity.
- Organoleptic properties.

21.6.1 Mechanical Properties

21.6.1.1 Tensile Strength

Tensile strength of a film is the maximum strength which a film can withstand before it breaks. It can be determined by dividing the applied load at break (N) with the film's cross-section area (mm^2) as mentioned below:

$$\text{Tensile strength}\left(\text{N/mm}^2\right) = \frac{\text{Breakingforce (N)}}{\text{Cross section area of the film strip}\left(\text{mm}^2\right)} \tag{21.1}$$

21.6.1.2 Percent Elongation

Elongation is a type of deformation that a film undergoes when stress is applied. The percent elongation is determined by applying stress, which causes a film strip to stretch and this is known as strain. The film elongation can be increased by increasing plasticizer content (Mahboob et al. 2016). The film sample was pulled using a pulley system. Weight is gradually added to increase the pulling force until the film breaks. The distance traveled by pointer on the graph before the film break determines the elongation. The percent elongation can be calculated using the following formula:

$$\text{Percent elongation} = \frac{L_1 - L_0}{L_0} \times 100 \tag{21.2}$$

Where L_1 was the final length and L_0 was initial length.

21.6.1.3 Young's Modulus

Young's modulus determines the stiffness of film. It can also be determined using the measurement method of tensile strength. It can be calculated using a ratio of applied stress and strain in the elastic deformation (Mahboob et al. 2016; Naik et al. 2014). The following formula is used for calculating Young's modulus:

$$\text{Young's modulus} = \frac{\text{Slope} \times 100}{\text{Film thickness} \times \text{cross head speed}} \tag{21.3}$$

21.6.1.4 Tear Resistance

Tear resistance of film also known as sheeting determines film's resistance to rupture. It is measured by employing a very low loading rate at 51 mm/min, which starts the tearing process. The maximum force recorded for tearing the film specimen is known as tear resistance, which is measured in Newton (Mahboob et al. 2016; Naik et al. 2014).

21.6.1.5 Folding Endurance

Folding endurance is crucial to determine the flexibility of film. The film with high folding endurance can be administered without breakage (Karki et al. 2016). It determines the tensile properties of the film (Bala and Sharma 2018). It is determined using a film of uniform thickness and cross-sectional area and it is folded until it breaks. The number of times a film can be folded till it breaks is taken as folding endurance. Folding endurance of film typically lies between 100 and 150 (Bala and Sharma 2018; Patil and Shrivastava 2014; Lakshmi et al. 2011).

21.6.2 Other Physical Properties

21.6.2.1 Thickness

Thickness is the important parameter as it defines the content uniformity of the film (Bhyan et al. 2011). The thickness of each sample is measured from six different points in mm diameter with the help of a digital micrometer. The mean value of oral film thickness is further used for calculating transparency of the film (Bala and Sharma 2018; Patil and Shrivastava 2014).

21.6.2.2 Weight Variation

Weight helps to determine the content uniformity of the film sample (Bhyan et al. 2011). An individual film strip of 2 cm × 2 cm is taken and weighed using an electronic balance. The value of average weight is calculated, which is subtracted from the individual weight (Bala and Sharma 2018; Karki et al. 2016). If the weight varies largely, this would indicate the inefficiency of the method used and indicate the nonuniform drug contents (Naik et al. 2014).

21.6.2.3 Moisture

Moisture content is determined to define the mechanical strength and friability of the oral thin film (Karki et al. 2016). The moisture content (MC) for each film sample is determined by taking a film sample of size 3 × 2 cm and drying at a 100° C temperature in a hot air oven for 12 h (Karki et al. 2016; Kanmani and Lim 2013). The weight difference before and post drying process with respect to initial weight is taken as moisture content. The moisture content is expressed as a percentage based on the initial film weight. The experiment should be conducted in triplicates. Following formula was applied to determine the moisture content (Panda et al. 2012):

$$\text{Moisure content } (\%) = \frac{\text{Initial weight} - \text{final weight}}{\text{Initial weight}} \times 100 \tag{21.4}$$

21.6.2.4 Contact Angle

Contact angle can be determined using the Goniometer at a room temperature. This is measured by putting a drop of distilled water over the dry film surface. The image of the water droplet is recorded using the digital camera within 10 s of deposition. The process is repeated for both sides and average value is taken as contact angle (Mahboob et al. 2016; Naik et al. 2014).

21.6.2.5 Transparency

Transparency is an important property which is directly related to product appearance and hence consumer acceptability (Rawdkuen 2018). It can be measured with the help of a UV-Vis spectrophotometer. A rectangular shape film sample was taken and placed in the test cell of the UV spectrophotometer and air is taken as a reference. The transparency of the film is determined at 600 nm. It can be determined using following formula (Kim et al. 2015; Naik et al. 2014; Bala et al. 2013):

$$\text{Transparency } (\%) = \text{Abs } 600/L \tag{21.5}$$

Abs 600 = absorbance value of the sample at 600 nm wavelength, L = thickness of the film (mm).

21.6.3 Chemical Properties

21.6.3.1 Surface pH Test

The pH of the film surface is evaluated by putting the sample on the surface of 1.5% w/v agar gel and then putting the pH paper of range 1–11 on the film surface. The change in the color of pH paper is observed and recorded (Pawar et al. 2019).

21.6.3.2 In Vitro Disintegration Time

In vitro disintegration time demonstrates the disintegration and dissolution properties of the film. This is performed by simulating the disintegration condition of saliva (Prajapati et al. 2018). It is evaluated using a phosphate buffer which can be prepared using Na_2HPO_4 (2.38 g), KH_2PO_4 (0.19 g), and NaCl (8.00 g) per liter of distilled water. The pH of the buffer can be adjusted to 6.8 using phosphoric acid (Mashru et al. 2005). The film sample (2×2 cm^2) is placed in a petri plate containing 25 mL of the prepared phosphate buffer (pH 6.8) and stirred (Bala and Sharma 2018; Hussain et al. 2018). The time when the film starts to break and dissolve is noted. The test is performed in triplicates and an average value is taken.

21.6.3.3 In Vitro Dissolution Test

The dissolution test for the film is performed with the standard basket or paddle apparatus. The selection of dissolution medium depends on the sink conditions and highest dose of the active ingredient. The test is difficult to conduct due to the film tendency to float over the dissolution medium. The problem mostly occurs in case of paddle apparatus, thus the basket apparatus is mostly preferred. The dissolution media consists of 300 mL phosphate buffer (pH 6.8) and 900 ml of 0.1 N HCl. The temperature is maintained at 37 ± 0.5° C and rotation speed is adjusted at 50 rpm. The active substance dissolved is collected at predetermined time intervals and further analyzed with UV-spectrophotometer (Pawar et al. 2019; Naik et al. 2014).

21.6.4 Morphology Study

The film morphology should be homogenous and continuous to ensure uniformity in terms of drug distribution throughout the film matrix. During the drying process, self-aggregation may happen due to the intermolecular and convective forces which results in wrinkling of the film surface. In addition, the rough film surface may be formed due to interaction of active ingredient/drug with used polymers and due to crystallinity of the active ingredient. Thus, the assessment of surface morphology and the texture is critical for ensuring uniformity in distribution of active ingredients without any interaction with film polymers. The film surface properties that include texture, thickness, and drug distribution (aggregated or scattered) can be evaluated with light microscopy, scanning electron microscopy (SEM), transmission electron microscopy (TEM), etc. Among these, SEM is thought to be the most reliable method for examination of the film surface. SEM is conducted by mounting the films on stubs and gold coated sputter in inert surroundings, and further at a suitable magnification the photographs are taken (Mahboob et al. 2016; Karki et al. 2016; Nair et al. 2020).

21.6.5 Thermal Properties

The differential scanning calorimetry (DSC) is used for evaluating the film's thermal stability under nitrogen atmosphere conditioning. This determines the glass transition temperature (Tg) by weighing 5–8 mg sample by setting 10 mL/min flow capacity at 20–300° C with 10 ° C/min heating rate. The DSC instrument can be calibrated using the Indium (melting point −156.6° C, ΔH = 28.5 J/g). T_g is estimated with the first derivative of the thermograms. Tg indicates the baseline of DSC plot in the glass transition and is the midpoint of the steps (Kumar et al. 2014).

21.6.6 Swelling Property

Swelling property determines the control release of active constituents of the film. The swelling test is conducted in simulated saliva solution (Karki et al. 2016). It is evaluated by weighing the film specimen (2×2 cm^2) and placing it on a pre-weighed stainless steel wire mesh. The mesh with the film sample is then placed in a vessel with 50 mL simulated saliva solution. The change in the weight is recorded at regular time intervals (0, 10, 20, 30, 40, 50, 60, 70, 80, 90, 100, 110, 120 min) until constant weight is achieved. The following formula is used to determine degree of swelling (Tejada et al. 2017; Reyad-ul-ferdous et al. 2015; Patil and Shrivastava 2014; Panda et al. 2012):

$$\mathrm{SI} = \frac{W_t - W_o}{W_o} \times 100 \tag{21.6}$$

Where SI = swelling index, W_t = film weight at time 't,' W_o = film weight at the time = 0.

21.6.7 Solubility

The film solubility is determined by taking the film specimen (3 × 2 cm^2) and weighing the initial water content using weighing balance. The film is kept for drying in an oven at 100° C for 24 h. The film sample is placed in 30 mL distilled water and kept for incubation at a temperature of 25 °C for 24 h. After that, the leftover of the film is removed from the water and dried in an oven at 100° C for 24 h. After drying, the final weight is taken and the water solubility of the film is computed as follows (Toth and Halasz 2019; Kanmani and Lim 2013):

$$\text{Water solubility } (\%) = \frac{W_o - W_f}{W_o} \times 100 \tag{21.7}$$

Where W_O = Initial weight of the dried film, W_f = Final weight of the dried film. The film solubility (%) is measured in triplicates.

21.6.8 Assay/Content Uniformity

This can be evaluated using any standard assay method defined for the particular active ingredient. The content uniformity is evaluated by calculating the active ingredient in the individual film. The content uniformity is limited to 85–115% (Pawar et al. 2019; Naik et al. 2014).

21.6.9 Organoleptic Properties

The organoleptic properties in terms of sweetness and flavor determine the acceptability of product. The organoleptic properties are determined by controlled human taste panels. This is determined by in vitro methods which utilize taste sensors, especially designed apparatus, and further in case of oral thin film, drug release is determined by modified pharmacopeial methods (Mahboob et al. 2016). The differentiation between levels of sweetness is determined by performing electronic tongue measurements (Anand et al. 2007).

21.7 Properties of the Edible and Oral Films

21.7.1 Physical Properties

The physical parameters of edible and oral films are largely defined by the properties of the key ingredients. Additionally, chemical structure and solubility of film additives are crucial for the multicomponent formulations and thus impact the physical properties of biopolymer film. Further, the compatibility between the film components plays a crucial role for forming composite structure, and this eventually affects the physical properties of edible and oral films. The physical properties are important as they help in determining uniformity of the film and help to maintain aesthetic appeal of the final formulation (Naik et al. 2014). The most commonly determined physical properties are water vapor permeability and mechanical resistance. Relative humidity and temperature are the most important parameters with regard to water vapor permeability analysis (Kraśniewska et al. 2020).

21.7.2 Mechanical Properties

The edible and oral films should have sufficient mechanical properties, as weak flexibility or strength can result in film rupture or cracking during processing, handling, storage, or application (Otoni et al. 2014; Sothornvit and Rodsamran 2008). The mechanical characteristics of edible films depend strongly on the nature of the material and final composition of the film (Sothornvit and Pitak 2007). The mechanical properties of the film can be modified by adding binding agents, fillers, crosslinkers, and/or plasticizers to the film formulation (Otoni et al. 2014). The most usual method to improve mechanical properties of the film is addition of plasticizer in the film formulation which reduces the biopolymer chain-to-chain interaction which thus improves film flexibility and stretchability. However, it also increases the film permeability (Sothornvit and Pitak 2007). The mechanical properties like tear resistance, folding endurance, and elongation at break are the crucial parameters in the manufacturing of the edible and oral film. These properties support the correlation of the film mechanical properties with the compositions and chemical structures (Otoni et al. 2014).

21.7.3 Functional Properties

The functional properties of the film are largely determined by the source of the biopolymer which thus determine characteristics of biopolymer films. The intermolecular bonding formed between the biopolymers during the film formation has a primary role in determining the functional properties of the films (Phanthe et al. 2009). The blending of a variety of polymers is one of the effective methods for improving the film's functional properties. Water resistance is one of the important functional parameters of the edible and oral film. Water resistance is determined by evaluating the water solubility, water content, and water vapor permeability (Jamroz et al. 2019). The functional properties of film are highly important for application of film in food packaging (Galus et al. 2020). In the packaging application, the functional properties of the films are determined by its barrier properties against solute and gas and which thus maintain the food quality and prolong the shelf life (Sothornvit and Pitak 2007). For its application as a dosage form, the film is required to have functional properties which include muco-adhesion, hydration and swelling on contacting saliva, drug release from the swollen gel, and eventually permeation through the film membrane (Khan et al. 2016).

21.7.4 Thermal Properties

The mobility of the polymeric chain which defines the physical characteristic of the film is a function of atomic movement, which increases with rise in temperature. Thus, the thermal property of the polymer is crucial while forming film. However, the films have slightly different thermal properties than their components (Umaraw and Verma 2015). The thermal properties of film differ from its component properties due to the inner film structure (Gould et al. 2017). The thermal properties of the film can be largely affected by the degree of crystallinity and the morphology of the materials (Ortiz et al. 2009). It has been determined by studies that films with crystalline structure undergo thermal degradation at higher temperature than the amorphous structure and thus exhibit higher thermal stability than the amorphous structure (Othman et al. 2011). At industrial level for packaging application, the thermal properties of film are highly important as high temperature may be applied to reduce processing time, thus it should be known in advance that the materials degradation temperature will not be reached (Lorevice et al. 2012). The thermal properties of film can be analyzed using a differential scanning calorimeter (DSC) (Himani et al. 2020; Kanth et al. 2019). The determination of thermal properties of a film using DSC is used as a drug dosage form to evaluate combination of drug-polymer and polymer-polymer blend (Steven 2010).

21.8 Advantages of Edible Oral Film

21.8.1 Advantages as a Drug Dosage Form

The common terminology used for these films in pharmaceuticals is "Oral Thin Film (OTF)." The OTFs are usually preferred over the conventional tablets due to the ease in administration without requiring water. This is attributed due to their faster disintegration and dissolution in the oral mucosa due to their large surface area (Mahboob et al. 2016; Saini et al. 2012). Further, they are flexible and not brittle which makes their transportation easy as compared to tablets. They are especially beneficial for patients including geriatric, pediatric, and mentally ill patients and those with swallowing problems known as dysphasic patients (Jain et al. 2018; Karki et al. 2016). As these dissolve instantly, OTFs reduce the fear of choking (Karki et al. 2016; Saini et al. 2012). The film releases the active ingredient rapidly and thus is effective in inducing desired effect. OTF leaves least or no residues on the mouth and has a pleasant mouth feel, thus making it more acceptable over tablets or other dosage form (Jain et al. 2018; Karki et al. 2016; Saini et al. 2012).

21.8.2 Advantages in a Food Industry

The film in a food industry is known by the term "edible film," which is commonly used for its packaging application. As a packaging application, it provides advantages over convention coating or films such as better spreadability, diffusivity, and solubility (Ulusoy et al. 2018). They further improve properties such as flavor and color, reduce moisture and weight loss, and prevent oxidative rancidity for a food with high fat content. Further, they can also prove to be a carrier of food additives such as antimicrobial, antioxidative, and antibacterial agents to the food products (Kamal 2019). With respect to environmental concern which arises due to other packaging material, edible film reduces the disposal problem as they are biodegradable in nature (Kamal 2019; Farias et al. 2012). Recently, these films have also been used as a mouth freshener also, with products like Listerine and Ziminta mouth freshener strips. The film as a mouth freshener provides advantages of providing instant breath freshening. Further, they are also preferred as consumption of chewing gums is not accepted socially as it requires a lot of sucking and choking (Himani et al. 2020; Xu et al. 2002). Thus, these films have numerous advantages and applications in a food industry.

21.9 Conclusions

Edible and oral films are one of the novel approaches in the field of food and pharmaceutical sectors. In food packaging, these are a promising alternative to the conventional packaging materials due to their edibility and their potential to increase the shelf life of the products. While as a drug dosage form, the film provides greater patient acceptance and compliance without the associated risk of choking with greater efficacy in comparison to the conventional dosage form. The film has added advantages of carrying additional food additives like antimicrobial, antibacterial, and antioxidant agents. The major advantage of edible and oral films is their solubility and functionality. The film efficiency and functionality are largely determined by the intrinsic characteristics of the film-forming materials, including biopolymers, plasticizers, and other additives. The basic characterization of the film is the evaluation of mechanical properties and disintegration properties. The edible and oral films have several advantages; however, it has a lesser industry application. The film faces several challenges during its formulation and development, which should be overcome for optimization of overall formulation. The future of the film technology is very promising with new technologies evolving rapidly for preparation of edible and oral film.

References

Anand V, Kataria M, Kukkar V, Saharan V, Choudhury PK (2007) The latest trends in the taste assessment of pharmaceuticals. Drug Discov Today 12(5):257–265

Arya A, Chandra A, Sharma V, Pathak K (2010) Fast dissolving oral films: an innovative drug delivery system and dosage form. Int J ChemTech Res 2(1):576–583

Bala R, Sharma S (2018) Formulation optimization and evaluation of fast dissolving film of aprepitant by using design of experiment. Bull Fac Pharmacy 56:159–168

Bala R, Khanna S, Pawar P, Arora S (2013) Orally dissolving strips: a new approach to oral drug delivery system. Int J Pharm Investig 3(2):67

Bhyan B, Jangra S, Kaur M, Singh H (2011) Orally fast dissolving films: innovations in formulation and technology. Int J Pharm Sci Rev Res 9(2):50–57

Borges, AFS (2015) Development of novel pharmaceutical forms for oral administration of bioactive agents, doctorate dissertation, University of Coimbra Ch. 9, 45

Cerqueira AMPR (2018) Edible packaging, Encyclopedia of food chemistry, 173–176. Elsevier Inc, pp 1–4

Dahiya M, Saha S, Shahiwala AF (2009) A review on mouth dissolving films. Curr Drug Deliv 6:469–476

Farias MG, Fakhouri FM, de Carvalho CWP, Ascheri JLR (2012) Physicochemical characterization of edible starch films with Barbados cherry (*Malphigia emarginata* DC). Quim Nova 35 (3):46–552

Galus S, Kibar EAA, Gniewosz M, Krasniewska K (2020) Novel materials in the preparation of edible films and coatings—a review. Coatings 10(7):674

Gould RD, Kasap S, Ray AK (2017) Thin films. In: Springer handbook of electronic and photonic materials, Springer Handbooks. Springer, Cham. https://doi.org/10.1007/978-3-319-48933-9_28

Han JH (2014) Edible films and coatings: a review. Innovations in food packaging (second edition), Ch.9, Academic Press, Cambridge, MA, pp 213–255

Himani KN, Prabhakar PK, Pant V (2020) Physical, mechanical, functional, and thermal characterization of chitosan: maltodextrin blends edible Oral film incorporated with aqueous clove extract. Starch 1900220:1–12

Hussain A, Latif S, Abbas N, Irfan M, Arshad MS, Bukhari NI (2018) Hydroxypropyl cellulose-based orally disintegrating films of promethazine HCl for the treatment of motion sickness. Trop J Pharm Res 17(6):991–996

Jain A, Ahirwar HC, Tayall S, Mohanty PK (2018) Fast dissolving Oral films: a tabular update. J Drug Deliv Ther 8(4):10–19

Jamroz E, Kulawik P, Kopel P (2019) The effect of nanofillers on the functional properties of biopolymer-based films: a review. Polymers 11(675):1–42

Kalyan S, Bansal M (2012) Recent trends in the development of oral dissolving film. Int J PharmTech Res 4(2):725–733

Kamal I (2019) Edible films and coatings: classification, preparation, functionality and applications-a review. Arch Org Inorg Chem Sci 4(2):501–510

Kanmani P, Lim ST (2013) Development and characterization of novel probiotic-residing pullulan/starch edible films. Food Chem 141:1041–1049

Kanth NP, Prasad G, Kumar BV (2019) Oral dissolving films of chlorpheniramine maleate. Int J Pharm Sci Res 5(5):1859–1873

Karki S, Kim H, Na SJ, Shin D, Jo K, Lee J (2016) Thin films as an emerging platform for drug delivery. Asian J Pharm Sci:559–574

Kathpalia H, Gupte A (2013) An introduction to fast dissolving oral thin film drug delivery systems: a review. Curr Drug Deliv 10(6):667–684

Keshari A, Kumar Sharma P, Parvez N (2014) Fast dissolving oral film: a novel and innovative drug delivery system. Int J Pharma Sci Res 5:92–95

Khan S, Trivedi V, Boateng J (2016) Functional physico-chemical, ex vivo permeation and cell viability characterization of omeprazole loaded buccal films for paediatric drug delivery. Int J Pharm Int J Pharm 500:217–226

Kim SRB, Choi YG, Kim JK, Lim ST (2015) Improvement of water solubility and humidity stability of tapioca starch film by incorporating various gums. LWT Food Sci Technol 64:475–482

Kraśniewska K, Galus S, Gniewosz M (2020) Biopolymers-based materials containing silver nanoparticles as active packaging for food applications–a review. Int J Mol Sci 21(3):698

Kumar GP, Phani AR, Prasad RGSV, Sanganal JS, Manali N, Gupta R et al (2014) Polyvinylpyrrolidone oral films of enrofloxacin: film characterization and drug release. Int J Pharm 471(1–2):146–152

Kumar N, Neeraj, Singh R (2019) Preparation and characterization of chitosan-pullulan blended edible films enrich with pomegranate peel extract. React Funct Polym 144:104350. https://doi.org/10.1016/j.reactfunctpolym.2019.104350

Lakshmi PK, Sreekanth J, Sridharan A (2011) Formulation development of fast releasing oral thin films of levocetrizine dihydrochloride with Eudragit® Epo and optimization through Taguchi orthogonal experimental design. Asian J Pharm 5:84–92

Lorevice MV, Moura MRD, Aouada FA, Mattoso LH (2012) Development of novel guava puree films containing chitosan nanoparticles. J Nanosci Nanotechnol 12(3):2711–2717

Mahboob MB, Riaz T, Jamshaid M, Bashir I, Zulfiqar S (2016) Oral films: a comprehensive review. Int Curr Pharm J 5(12):111–117

Mahesh A, Shastri N, Sadanandam M (2010) Development of taste masked fast disintegrating films of Levocetirizine Dihydrochloride for Oral use. Curr Drug Deliv 7:21–27

Mashru R, Sutariya V, Sankalia M, Parikh P (2005) Development and evaluation of fast-dissolving film of salbutamol sulphate. Drug Dev Ind Pharm 31(1):25–34

Muhammad I, Rabel S, Bukhtar Q, Qadir MI, Jabeen F, Khan A (2016) Orally disintegrating films: a modern expansion in drug delivery system. Saudi Pharm J 24:537–546

Nagar P, Singh K, Chauhan I, Verma M, Yasir M, Khan A, Gupta N (2011) Orally disintegrating tablets: formulation, preparation techniques and evaluation. J Appl Pharm Sci 1(4):35–45

Naik TS, Khale A, Kanekar H (2014) Evaluation of mouth dissolving films: physical and chemical methods. Int J Pharm Phytopharmacol Res 4(1):62–65

Nair AB, Al-Dhubiab BE, Shah J, Jacob S, Saraiya V, Attimarad M, SreeHasha N, Akrawi SH, Shehata TM (2020) Mucoadhesive buccal film of almotriptan improved therapeutic delivery in rabbit model. Saudi Pharm J 28(2):201–209

Ortiz AV, Nogueira BR, Oliveira VM, Moura EAB (2009) Mechanical and thermal properties of commercial multilayer pet/PP film irradiated with electron-beam. International Nuclear Atlantic Conference, 978-85-99141-03-8

Othman N, Azahari NA, Ismail H (2011) Thermal properties of polyvinyl alcohol (PVOH)/corn starch blend film. Malays Polym J 6(6):147–154

Otoni CG, de Moura MR, Aouada FA, Camilloto GP, Cruz RS, Lorevice MV, Mattoso LH (2014) Antimicrobial and physical-mechanical properties of pectin/papaya puree/cinnamaldehyde nano-emulsion edible composite films. Food Hydrocoll 41:188–194

Panda PB, Dey NS, Rao MEB (2012) Development of innovative orally fast disintegrating film dosage form: a review. Int J Pharm Sci Nanotechnol 5(2):1666–1674

Patil P, Shrivastava SK (2014) Fast dissolving oral films: an innovative drug delivery system. Int J Sci Res 2(4):2088–2093

Pawar R, Sharma R, Sharma P, Darwhekar GN (2019) A review on mouth dissolving film. J Drug Deliv Therap 9(6):206–210

Phanthe D, Debeaufort F, Voilley A, Luu D (2009) Biopolymer interactions affect the functional properties of edible films based on agar, cassava starch and arabinoxylan blends. J Food Eng 90:548–558

Prajapati VD, Chaudhari AM, Gandhi AK, Maheriya P (2018) Pullulan based oral thin film formulation of zolmitriptan: development and optimization using factorial design. Int J Biol Macromol 107:2075–2085

Rawdkuen S (2018) Edible films incorporated with active compounds: their properties and application. In Isil Var. Active antimicrobial food packaging. pp 71–91

Reddy TUK, Reddy KSK, Manogna K, Thyagaraju PK (2018) A detailed review on fast dissolving Oral films. Indo Am J Pharm Res 8(6):1315–1326

Reyad-ul-ferdous M, Shahjahan DMS, Tanvir MMI, Sumi S, Alam MN, Chowdury MIA, Mazumdar MMU, Chowdhury S, Alam SMAB (2015) Effective development and evaluation of oral thin film of etoricoxib. World J Pharm Res 4(9):257–272

Saini P, Kumar A, Sharma P, Visht S (2012) Fast disintegrating oral films: a recent trend of drug delivery. Int J Drug Dev Res 4(4):80–94

Sothornvit R, Pitak N (2007) Oxygen permeability and mechanical properties of banana films. Food Res Int 40(3):365–370

Sothornvit R, Rodsamran P (2008) Effect of a mango film on quality of whole and minimally processed mangoes. Postharvest Biol Technol 47(3):407–415

Steven PS (2010) Thermal analysis – a review of techniques and applications in the pharmaceutical sciences. Am Pharm Rev 13(2)

Suhag R, Kumar N, Petkoska AT, Upadhyay A (2020) Film formation and deposition methods of edible coating on food products: a review. Food Res Int 136:10958

Tejada G, Barrera MG, Piccirilli GN, Sortino M, Frattini A, Salomon CJ, Lamas MC, Leonardi D (2017) Development and evaluation of buccal films based on chitosan for the potential treatment of oral candidiasis. American Association of Pharmaceutical Scientists. AAPS PharmSciTech 18(4):936–946

Toth A, Halasz K (2019) Characterization of edible bio-composite films directly prepared from psyllium seed husk and husk flour. Food Packag Shelf Life 20:100299. https://doi.org/10.1016/j.fpsl.2019.01.003

Trinetta V (2016) Edible packaging. Elsevier Inc., Amsterdam, pp 1–4

Ulusoy BH, Yildirim FK, Hecer C (2018) Edible films and coatings: a good idea from past to future technology. J Food Technol Res 5(1):28–33

Umaraw P, Verma AK (2015) Comprehensive review on application of edible film on meat and meat products: an eco-friendly approach. Crit Rev Food Sci Nutr 57(6):1549–7852

Xu G, Recci M, Blackwell BL, Robinson RS, Viscio D, Curtis JP (2002) Breath freshening film (Patent No.: US 6,419,903 B1)

Consumer Acceptance to Commercial Applications of Packaging Edibles

22

Sandeep Kaur, Anila Kumari, and Vandana Sharma

Abstract

The advent of novel green technologies revolutionizing the food industry has marked the beginning of a new era in the food sector with an aim to enhance the safety, nutritional value and sustainability of food. One of these technologies is the bio-based or edible packaging arising from concerted research focused on studying alternatives to regular plastic packaging in an attempt to reduce the CO_2 footprint. Although edible packaging enjoys the reputation of being biodegradable, eco-friendly and sustainable, its success in the market is finally dictated by consumer's behaviour and attitude towards adopting the newly introduced packaging technology. Many consumers exhibit the pre-conceived notion that technology may interfere with the state of naturalness, taste, nutrition and health aspects of their food item leaving them in a state of confusion enhancing their food neophobia. This along with limited knowledge about food processing and incorrect perceptions of the negative environmental impact of using plastic-based conventional packaging technique represent major challenges for the success of new packaging-based ideas in the food industry. The present chapter, thus, aims to discuss the factors dictating consumer perceptions and their behaviour. Further, it focusses on delineating the strategies to enhance the acceptability and adoption rates of edible packaging for their market stay and survival. This will enable to understand the barriers faced and the solution factors which must be stressed by the innovators and the launching platform while planning to introduce such novel packaging edibles in the market for longer sustenance and success.

S. Kaur · A. Kumari · V. Sharma (✉)
Department of Food Science, Mehr Chand Mahajan DAV College for Women, Chandigarh, India

A. Poonia, T. Dhewa (eds.), *Edible Food Packaging*,
https://doi.org/10.1007/978-981-16-2383-7_22

Keywords

Edible packaging · Consumer acceptance · Sustainable · Bio-based packaging · Biodegradable

22.1 Introduction

Food packaging refers to the process of enclosing the food with the primary aim of protecting the food from contamination, damage, or decay during the process of delivery or storage; thus, preserving and maintaining the food's shelf life. There has been a wide array of materials currently being used in food packaging ranging from plastic, metal, ceramics, paper, glass and wood or combination of any of these (Pongracz 2007). However, packaging materials used worldwide accounts for a major portion of waste generated that contribute to the growing environmental issues as resultant problems with waste management have been the subject of environmental concern and research (Pongracz 1998; Davis and Song 2006). It has been estimated that food packaging accounts for almost two-thirds of total packaging waste by volume (Hunt et al. 1990). As per 2018 study, containers and food packaging make up a major portion of municipal solid waste (MSW) that amounts to a significant 82.2 million tons of waste generation which is 28.1 percent of total generation (EPA, Facts and Figures about Materials, Waste and Recycling 2018). A total of 170 kg of waste generated from food packaging was produced in Europe, in the year 2017, according to the data provided by Eurostat statistics. In India itself, 26,000 tonnes of plastic waste is generated every year and food is a major generous contributor to this. In total, the global packaging waste accounts for a large portion of approximately 30–35% of the total municipal solid waste (MSW) and almost 60% of all packaging is for food products (Nemat et al. 2020; Tencati et al. 2016).

The waste generated due to plastic-based food packaging has become a major disposal and environmental issue (Coltelli et al. 2018). Although plastics pose a major environmental threat, plastic has dominated the food packaging market and an estimated 141 million tonnes per-year, is used in various types of food packaging materials (US Environmental Protection Agency. 2014; Thompson Richard et al. 2009). The main reason being that plastic-based food packaging represents a fit-for-purpose, cost-effective and durable option allowing long distance travelling of food items and increasing their shelf life; thus, ensuring that food reaches the consumer and does not go waste (Verghese et al. 2015). Moreover, owing to its mechanical strength, degree of rigidity, heat and chemical resistance properties, heat seal ability, lower energy consumption, ease of melding and high flexibility, plastic dominated the packaging industry in the twentieth century (World Economic Situation and Prospects. 2012). But the sad part, threatening public health and environment worldwide is the huge amount of long-standing non-renewable and non-biodegradables ending in landfills with a poor recycling rate (less than 5%) (Weller et al. 1998; De Ore 1933; Helmut Kaiser Consultancy 2016). Therefore, this has led to the existence of a paradoxical condition between industry and government with impacts and benefits at the two extremes. This calls for use and adoption of novel and eco-friendly solutions for food packaging of equal strength and potential so as to mitigate the overall use of plastic-based packaging.

22.2 Emerging Sustainable Strategies of Food Packaging

The current trend in innovations and developments in new food packaging technologies focuses on creating sustainable packaging that is biodegradable and environmental friendly. A sustainable packaging should fulfil the following criteria (Robertson 2009; Pavlath and Orts 2009):

1. Should be made up of biodegradable, non-toxic and non-allergic components.
2. Should have mechanical strength to avoid physical damage during handling, processing and transportation of food.
3. Packaging material used should be derived from non-renewable energy sources.
4. In order to sustain organoleptic properties of the food, the packaging material should provide a barrier to moisture, gases and solutes during storage.
5. To maintain food safety and quality, it should protect the food from microbial contamination, proliferation of spoilage and pathogenic microbes as well as pest infestation.
6. Packaging material should be renewable and recyclable.
7. Last but not least, the packaging should be functional and economically viable.

Today consumers are presented with an increasing number of sustainable, environmentally friendly, packaging materials that include use of polylactide acid (PLA) plastics, sugar cane pulp based and starch-based edible films and so on (Nitaigour 2014; Ahvenainen and Hurme 1997).

Use of biopolymers to develop edible packaging has generated considerable interest in recent years (Chiellini et al. 2004). Edible packaging are thin coatings/ layers of biodegradable material that either coat the food product or can be placed on or between food components (Hassan et al. 2018). A liquid/solid form of biopolymer can be used depending on their specific application and type of food product involved. The edible packaging has dominated the specific market of sustainable food packaging by reducing the negative burden of the conventional non-biodegradable food packaging waste (Avramescu et al. 2020). Using renewable resources, edible packaging could improve the mechanical properties of the food; avoid loss of moisture and components that stabilize aroma, flavour, nutritional and sensorial characteristics of the food necessary for consumer acceptance, particularly during storage; limit the load of spoilage and pathogenic microorganism on the surface of foods and provide antimicrobial or antioxidant capabilities to the product (Reichert et al. 2020). Edible packaging aims to address environmental issues and is a eco-friendly alternative and this factor enhances overall customer acceptance for this new generation of edible packaging. (De Ore 1933; Fernqvist et al. 2015; Lindh et al. 2016; Magnier et al. 2016; Prakash and Pathak 2017).

22.3 Understanding Consumer Acceptance and Behaviour

Packaging represents a major motivation for food consumption (Chandon and Wansink 2010); and besides playing its conventional role, it is also an important marketing tool dictating consumer acceptability and sale (Sehrawat and Kundu 2007; Shekhar and Raveendran 2013). Global consumers have been equally aware and concerned regarding the negative impacts of conventional packaging waste (Rokka and Uusitalo 2008; Steenis et al. 2017; Steenis et al. 2018). Health and environment concern are a major factor that positively affects the consumer's decision to purchase a product packed in sustainable packaging (Koenig-Lewis et al. 2014). Another US-based Deloitte study-related finding also emphasized on the increasing consumer demand for environment-friendly products as they become more aware of consequences of environmental issues pertaining to packaging waste. This has compelled the multinationals to develop green products adopting green practices so as to achieve a competitive advantage in terms of profitability and market shares (Menguc and Ozanne 2005). Despite the increasing interest, however, not all innovative food technologies get acceptance and market success (Siegrist 2008). One biggest challenge to promote these novel packaging strategies is to understand its level of acceptability and willingness by the end users, that is, our consumers (Kuznesof 2010). Therefore, acquiring data on how the consumers react to the novel packaging method and their opinion and viewpoint definitely affects the overall success and sustenance of this new era of edible packaging materials.

There are multiple reasons why consumers, despite the favourable attitude, fail to choose eco-friendly sustainable packaging. A major factor that makes a food packaging unacceptable to the consumers includes the surrounded fear, doubts and related misconceptions about the technology that has been adopted (Behrens et al. 2009). The results obtained from a focus group study revealed that consumers were more likely interested in knowing the nature of the coating on food products before deciding on its purchase (Wan et al. 2007). They expressed their preference for coatings made up of natural ingredients as food safety was the major concern among the participants of the focus group study. They also made a choice on coatings that could be easily removed before consumption of the coated food product. Best (1991) also reported that consumers do not trust the technologies applied to food products which are invisible to them (Best 1991). Another focus group study revealed consumer's misconceptions regarding the benefits of edible food packaging. Consumers who are frequent grocery buyers pay less attention to the coated food products with extended shelf life which have been lying on the shelves for prolonged periods. The consumers perceive that the extended shelf life of the product is more beneficial to the manufactures (Wan et al. 2007). Use of gene technology in food production is also one such technique that is surrounded with high consumer scepticism (Ronteltap et al. 2007). Grunert et al. (2003) figured out that consumer's disquiet and distrust in GM foods are due to their negative attitude against genetic modifications applied to natural food products. A survey-based study done in Turkey revealed that 80% of consumers were having their doubts regarding safety of irradiated foods (Gunes and Tekin 2006), and in another survey-based study, 46% of consumers believed that food irradiation makes it radioactive (Junqueira-Goncalves et al. 2011). Therefore, the high level of unawareness and lack of education among the consumers towards: (1) the benefits of these new approaches such as edible packaging strategies, (2) the need to use them now and c) the related risks of the continued use of traditional packaging on health and environment are impeding factors that require to be focused.

Second factor that dictates the acceptability is the price that consumers have to pay. With food quality and nutrition being a key deciding criterion, the judgments are also driven by the price of the packaged food product; whether the price is appropriate to pay and whether the product is worth the price. Moreover, price is a long-term dictating parameter and it will affect the long-term sustainability of such packaged food items in the market (Brunso et al. 2002). Besides this, consumers also have a perceived threat of uncontained bioactivity associated with use and adoption of bio-based technological application and the inability of consumers to differentiate between more and comparatively less eco-friendly package alternatives that may result in rejection (Bech-Larsen 1996; Thøgersen 1996). Moreover, the widespread 'green washing'-related false claims have created more confusion on various aspects of package sustainability. Improper labelling and lack of disclosure of complete information on the food label is another impending factor governing end-user decision and, hence, related sale. In case of lack of environmental information on labels, many consumers may fail to understand the connection between their buying decision and environment-related impacts and benefits. Laroche et al. (2001) showed that today's consumers are more aware of environmental threats and do prefer checking the packages if it is made of recycled material or not. According to Rashid (2009), giving adequate information through putting eco-labels has a major influence on the consumers at the point of purchase. As per recent research, today's consumers prefer clear claims on bio-based packaging written as 'organic,' biodegradable or compostable, etc. The more concrete the sustainability benefits, the better the packaging is perceived, as product claims is a solid factor for decisions made during in-store purchase (Wageningen Economic Research). Another aspect to the labelling part is the role of government approval agencies and the seals allowed on the packaged material. This has a major influence on the purchase decision before the consumer pays the price and takes it to home (Aday and Yener 2014). Also worth mentioning is the overall appearance, colour, design and shape of the packaging final product. This does have an influence on the decision of the consumer population especially the younger segment. Becker et al. (2011) reported a study on consumer perception in which he studied yogurt containers of different curvatures and colours. The team found that the designs that were bolder did influence the taste expectations and acceptance by consumers. Similar study on shape and colour of packaging used for milk desserts also influenced consumer expectations (Ares and Deliza 2010). Therefore, due focus should also be given on the overall shape, presentation and appearance of the novel packaging used while preserving its sensory and nutritional values (Marshall et al. 2006). Social norms set for behaving towards sustainable development so as to mitigate threat to environments will also help to motivate consumer's behaviour and attitude to think in positive manner (Thogersen et al. 1999). For example, if in a given residential area/society, majority of consumers avoid buying and using non-recyclable bottles or plastic-based packaging, it will influence the rest of the consumer population and they will more likely adopt that kind of behaviour. In addition to these, consumer acceptance is also influenced by additional parameters that go hand in hand, such as psychological, sociological, religious and environmental factors acting as bridges between perceived benefits and risks presented by the novel food technology used.

22.4 Enhancing Consumer Acceptance: Critical for Success of Edible Packaging

After having discussed the major factors dictating consumer willingness to accept the novel packaging concepts, this section focuses on the possible strategies required to be adopted for improving consumer acceptance rates. Consumers are concerned and prefer more about the safety of the food intake on their health rather than launching an innovative technology. Therefore, simply by writing that this is an innovative packaging methodology on the food label will not prove to be useful to enhance its adoption and many customers will walk away owing to the uncertainty and complexity of the innovation involved. This calls for working at many folds.

22.4.1 Marketing and Branding

Firstly, the manufacture or the person(s) involved in launching the edible packaging should have a concrete plan of marketing and publishing its innovative product packaging so as to reach consumers well before the final launch and throughout the initial period after launch. In a questionnaire-based study, when people were questioned about what is required for enhancing the overall acceptability of a novel packaging technology for wider adoption, the participants mostly answered that dispersion of maximum education on the new technology by various ways such as through commercials, media participation and publicity under trusted brands seem the most effective ways (Chen et al. 2013). This action influences consumer's confidence and reduces the risk perceptions allowing them to see and decide on the difference between innovative packaging and conventional packaging. Formation of focus group-based study on consumer behaviour and attitude about edible packaging might also help. Focus group-based studies are planned when the novel food product is still in its early stages of development and before the final launch to obtain date on market research. This helps to have a bird's view; thus, allowing to probe into consumer reaction to the new innovation well in advance to make necessary changes if required (McQuarrie and McIntyre 1986; Sheth et al. 1999;

Langford and MCDonagh 2003). For example, a focus group-based study was conducted to determine consumer attitudes towards irradiated poultry (Hashim et al. 1996) and the results were found to be useful later for the acceptance of such techniques. Hence, the focus group acts as a bridge between laboratory/innovating platform and actual reaction of the end users, enabling prior homework to be done before final launch (McNeill et al. 2000). Such studies also enable an easy platform for constructing useful questionnaires for future subjective and quantitative analysis.

22.4.2 Educating the Consumers

Educating the consumers on various aspects of use of bio-based packaging and related aspects of sustainability represents a long-term solution. Research data from past studies indicate that concept of sustainable packaging is not well understood and not well communicated to the end consumers (Lindh et al. 2016; Boz et al. 2020). In a similar study by Magnier and Schoormans (2015), it was reported that the perception of consumer also is dictated by their level of environmental awareness as consumers with poor environmental awareness and knowledge were not equally sensitive to the claims. Consumers have a pre-conceived notion of what makes a package sustainable (e.g., recycling) while not looking into its social and health impacts. In a recent study by Sijtsema et al. (2016), which was an exploratory study done in five European countries, it was found that the majority of the population was actually unfamiliar with the simple term 'Bio-based,' highlighting the overall complexity and non-familiar aspects in relation to this concept. Hence, percolation of complete information about the novel technology, its efficacy, its safety profile and benefits associated over traditional technology to the end users should be mandatory by the approving authorities and small pamphlets can be distributed or given along with the product by the retailers and common shopkeepers to the public. A government seal of approval enables to mitigate many threats and make them more secure as perceived by customers, especially safety aspects. Many sections of consumers show worry thinking that innovative packages might mislead them due to excessive false claims and green washing and, therefore, sales should be handled under trusted brands only as big brands in the food industry have high consumer trust and confidence with a positive effect on developing consumer relationship (Aday and Yener 2015).

22.4.3 Posting all Relevant Information on Labels

Third strategy is posting complete and updated information clearly and in a simplified form on the food labels along with government agency seals. This is required to achieve success in the marketplace (Chen et al. 2013). Clear labelling of claims about health has a major positive effect on consumers' attitude and buying behaviours (Roe et al. 1999). Williams (2005) showed that food labels including authentic health claims help to give more clarity to the consumer and such food products are considered healthier. Therefore, the more concrete, the better but the retailers and manufactures need to be clear yet careful about such claims. As per food norms in the UK, it is mandatory to inform about the ill effects of carbon foot printing while informing the consumer about the decarbonizing aspect in food use and its related benefits (Boz et al. 2020).

22.4.4 Regulatory Aspects

Another point of utmost importance is that the components of edible packaging must meet all required regulatory aspects regarding food products (Guilbert and Gontard 1995). This is essential for building trust and confidence in consumers from a safety point of view. All the materials that are used in formation of edible packaging must fall into the category of generally recognized as safe (GRAS) as sanctioned US Food and Drug Administration (FDA) Code of Federal Regulations and this can be clearly indicated on the outer label. This will add a lot of weight for the product item and help to secure consumer's concerns about safety. Also, use of any animal-based products (for vegetarians) as well as presence of component(s) that may act as a common allergens (soy, milk, eggs, nuts and fish oil) should be clearly mentioned on food label for the population prone to allergies as directed by Food Allergen Labeling and Consumer Protection Act, 2004. As the consumers are becoming more aware of labels depicting what they are about to eat, updated and all essential information would help to build trust in the consumer chain, promoting sustenance in the market.

22.4.5 Preserving the Taste, Appearance, Sensory and Nutritional Aspects

There has been an emerging interest in studying the hedonic aspects of consumer behaviour, pertaining to consumer taste, appearance, sensory parameters and packaging design that dictates consumer judgment and decision-making (Hoyer and Stokburger-Sauer 2012). As the conventional food item, the new packaging must have neutral sensory properties and should be entirely compatible so as to go undetected during consumption. Secondly, in edible packaging, how well the packaging is integrated with its principal food product is also important as study revealed that packaging materials which are easily separable (such as edible cups) are more likely to face greater hindrances to adoption. The innovators of such edible packaging food products should carry out detailed research devoted to minimize any change or lowering of the sensory and organoleptic properties, taste and physical appearance as compared to the already used products in market. Any changes in these parameters might be less entertaining, especially by the younger segment which may outweigh taste and appeal over health and eco-friendliness of the packaged component.

22.4.6 Reducing Price or Offering Minimal Start Price

With price as a major factor, all efforts must be put in by the innovator of the new technology to reduce the production cost and the initial launch price as well in order to promote the commercialization of new packaging. As taste and price are the primary determining factors of purchase, sustainable packaging needs to be comparable in price with the already used technologies (Van Birgelen et al. 2009). Since most of the edible packaging has still to see the light, working on reducing the costs associated with the scaling up process is very much essential (Han and Gennadios 2005). Post-production, the manufacturers can set a minimal launch price similar to the conventional technology and gradually increase the price later as the market demand is met. Also, consumers who are sceptical about paying the price for such innovative packages can be distributed free test samples and offered attractive schemes so as to promote and publicize the novel packaging to see the difference between innovative packaging and regular packaging (Inman et al. 2009).

22.5 Conclusions

Edible packaging made from human consumable ingredients is one of the emerging sustainable strategies with the potential to reduce and/or replace the already used plastic-based conventional, packaging materials. Knowing how consumers accept bio-based packaging and how consumer opinion reflects on the brand is a biggest challenge towards the success of these new generations of edible packaging materials. Various factors that dictate the consumer's acceptability are high levels of unawareness about the benefits of these edible packaging strategies, price the consumer has to pay, lack of disclosure of updated information on the ingredients/composition, innovation adopted, content and instructions for safe use. Furthermore, intrinsic factors such as colour, design and shape of the packaging final product also influence the decision of the consumer population, especially the younger segment. Consumer acceptance is in fact influenced by a complex mix of multiple factors. Awareness through commercials, media participation and publicity under trusted brands is one of the most effective ways for edible packaging to meet and pass the consumer acceptance standards. Percolation of the complete information about the novel technology, its efficacy, its safety profile and benefits associated over traditional technology to the end users is the key to win the consumer's trust (as depicted diagrammatically in Fig. 22.1). Edible packaging which enforces correct labelling and clear information regarding the material of edible packaging and processing technologies used develops consumer relationship and trust. It should meet the regulatory guidelines which guarantees the safety of the consumers. In addition, obstacles such as cost-effectiveness and retention of nutrition, sensory and organoleptic properties of the food need to be addressed in order to attain commercial success of edible packaging.

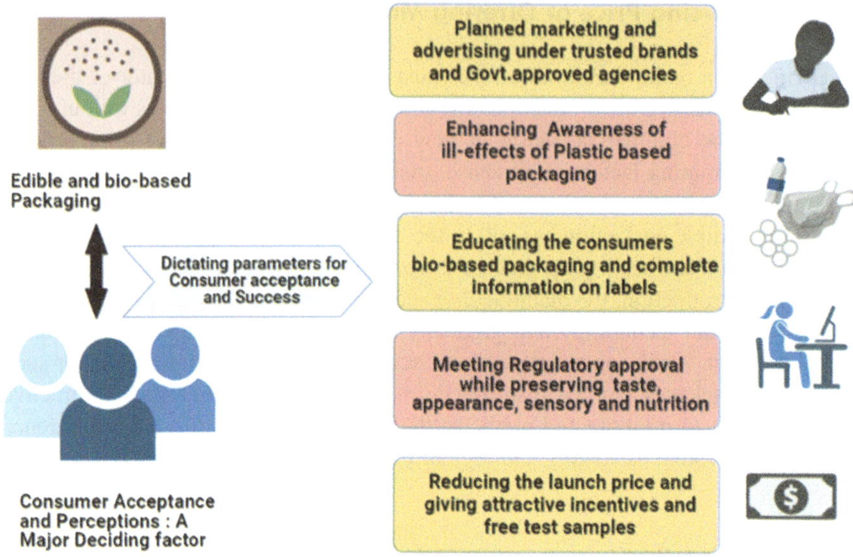

Fig. 22.1 Diagrammatic representation of deciding factors to enhance consumer acceptability and success of packaging edibles

Funding This work received no specific grant from any funding agency.

Conflicts of Interest The authors declare no conflict of interest.

References

Aday MS, Yener U (2014) Understanding the buying behavior of young consumers regarding packaging attributes and labels. Int J Consum Stud 38(4):285–393

Aday MS, Yener U (2015) Assessing consumers' adoption of active and intelligent packaging. Br Food J 117(1):157–177

Ahvenainen R, Hurme E (1997) Active and smart packaging for meeting consumer demands for quality and safety. Food Addit Contam 14(6–7):753–763. https://doi.org/10.1080/02652039709374586

Ares G, Deliza R (2010) Studying the influence of package shape and colour on consumer expectations of milk desserts using word association and conjoint analysis. Food Qual Prefer 21(8):930–937. https://doi.org/10.1016/j.foodqual.2010.03.006

Avramescu SM, Butean C, Popa CV, Ortan A, Moraru I, Temocico G (2020) Edible and functionalized films/coatings—performances and perspectives. Coatings 10:687

Bech-Larsen T (1996) Danish consumers' attitudes to the functional and environmental characteristics of food packaging. J Consum Policy 19(3):339–363

Becker L, van Rompay T, Schifferstein R, Galetzka M (2011) Touch package, strong taste: the influence of packaging design on taste impressions and product evaluations. Food Qual Prefer 22:17–23

Behrens JH, Barcellos MN, Frewer LJ, Nunes TP, Landgraf M (2009) Brazilian consumer views on food irradiation. Innov Food Sci Emerg Technol 10:383–398

Best D (1991) Designing new products from a market perspective. In: Graf E, Saguy IS (eds) Food prod develop. AVI Book, New York, NY, pp 1–28

Boz Z, Korhonen V, Koelsch Sand C (2020) Consumer considerations for the implementation of sustainable packaging: a review. Sustainability 12(6):2192. https://doi.org/10.3390/su12062192

Brunso K, Fjord TA, Grunert KG (2002) Consumers food choice and quality perception. MAPP working paper no 77. The Aarhus school of business, MAPP Centre, Aarhus

Chandon P, Wansink B (2010) Is food marketing making us fat? A multidisciplinary review. Found Trends Mark 5(3):113–196

Chen Q, Anders S, An H (2013) Measuring consumer resistance to a new food technology: a choice experiment in meat packaging. Food Qual Prefer 28(2):419–428. https://doi.org/10.1016/j.foodqual.2012.10.008

Chiellini E, Cinelli P, Chiellini F, Imam SH (2004) Environmentally degradable bio-based polymeric blends and composites. Macromol Biosci 4:218–231

Coltelli MB, Gigante V, Cinelli P, Lazzeri A (2018) Flexible food packaging using polymers from biomass. In: Morganti P (ed) Bionanotechnology to save the environment; plant and fishery's biomass as alternative to petrol. MDPI, Basel, pp 272–296

Davis G, Song JH (2006) Biodegradable packaging based on raw materials from crops and their impact on waste management, industrial crops and products. Int J 23(2):147–161

De Ore MR (1933) Process of treating fruit for the market. United States patents office US1900295 A. United States: FMC Corporation

Fernqvist F, Olsson A, Spendrup S (2015) What's in it for me? Food packaging and consumer responses, a focus group study. Br Food J 117(3):1122–1135

Grunert KG, Bredahl L, Scholderer J (2003) Four questions on European consumers' attitudes toward the use of genetic modification in food production. Innov Food Sci Emerg Technol 4: 435–445

Guilbert S, Gontard N (1995) Edible and biodegradable food packaging. In: Ackermann P, Jagerstad M, Ohlsson T (eds) Foods and packaging materials chemical interactions. The Royal Society of Chemistry, Cambridge, pp 159–168

Gunes G, Tekin MD (2006) Consumer awareness and acceptance of irradiated foods: results of a survey conducted on Turkish consumers. Food Sci Technol 39:443–444

Han JH, Gennadios A (2005) Edible films and coatings: overviews. Innovations Food Packag:239–262. https://doi.org/10.1016/B978-012311632-1/50047-4

Hashim IB, Resurreccion AA, McWatters KH (1996) Consumer attitudes toward irradiated poultry. Food Technol (Chicago) 50(3):77–80

Hassan B, Chatha S, Hussain AI, Zia KM, Akhtar N (2018) Recent advances on polysaccharides, lipids and protein based edible films and coatings: a review. Int J Biol Macromol 109:1095–1107

Helmut Kaiser Consultancy (HKC) (2016) Bioplastics Market Worldwide 2014–2015–2020-2025 and historical data 2011-12-13. http://www.hkc22.com/bioplastics.html

Hoyer WD, Stokburger-Sauer NE (2012) The role of aesthetic taste in consumer behavior. Acad Mark Sci Rev 40(1):167–180. https://doi.org/10.1007/s11747-011-0269-y

Hunt RG, Sellers VR, Franklin WE, Nelson JM, Rathje WL, Hughes WW et al (1990) Estimates of the volume of MSW and selected components in trash cans and landfills. Report prepared by the garbage project and franklins associates. Ltd. for the Council for Solid Waste Solutions, Tucson, AZ

Inman JJ, Winer RS, Ferraro R (2009) The interplay among category characteristics, customer characteristics, and customer activities on in-store decision making. J Mark 73(5):19–29. https://doi.org/10.1509/jmkg.73.5.19

Junqueira-Goncalves MP, Galotto MJ, Valenzuela X, Dinten CM, Aguirre P, Miltz J (2011) Perception and view of consumers on food irradiation and the Radura symbol. Radiat Phys Chem 80:119–122

Koenig-Lewis N, Palmer A, Dermody J, Urbye A (2014) Consumers' evaluations of ecological packaging e rational and emotional approaches. J Environ Psychol 37:94–105

Kuznesof SAB (2010) Consumer acceptance of novel foods: a grounded theory study. PhD thesis submitted to and available from Newcastle University
Langford J, McDonagh D (2003) Introduction. In: Langford J, McDonagh D (eds) Focus groups. Taylor & Francis, New York, NY, pp 1–20
Laroche M, Bergeron J, Barbaro-Forleo G (2001) Targeting consumers who are willing to pay more for environmentally friendly products. J Cons Mark 18(6):503–520
Lindh H, Olsson A, Williams H (2016) Consumer perceptions of food packaging: contributing to or counteracting environmentally sustainable development? Packag Technol Sci 29(1):3–23. https://doi.org/10.1002/pts.2184
Magnier L, Schoormans J (2015) Consumer reactions to sustainable packaging: the interplay of visual appearance, verbal claim and environmental concern. J Environ Psychol 44:53–62. https://doi.org/10.1016/j.jenvp.2015.09.005
Magnier L, Schoormans J, Mugge R (2016) Judging a product by its cover: packaging sustainability and perceptions of quality in food products. Food Qual Prefer 53:132–142
Marshall D, Stuart M, Bell R (2006) Examining the relationship between product package colour and product selection in preschoolers. Food Qual Prefer 17:615–621
McNeill KL, Sanders TH, Civille GV (2000) Using focus groups to develop a quantitative consumer questionnaire for peanut butter. J Sens Stud 15(2):163–178. https://doi.org/10.1111/j.1745-459X.2000.tb00263.x
McQuarrie EF, McIntyre SH (1986) Focus groups and the development of new products by technologically driven companies: some guidelines. J Prod Innov 3(1):40–47. https://doi.org/10.1016/0737-6782(86)90042-1
Menguc B, Ozanne LK (2005) Challenges of the 'green imperative': a natural resource based approach to the environmental orientation-business performance relationship. J Bus Res 58: 430–438
Nemat B, Razzaghi M, Bolton K, Rousta K (2020) The potential of food packaging attributes to influence consumers' decisions to Sort waste. Sustainability 12(6):2234
Nitaigour PM (2014) Advances in packaging methods, processes and systems. Challenges 5:374–389
Pavlath AE, Orts W (2009) Edible films and coatings: why, what, and how? In: Embuscado ME, Huber KC (eds) Edible films and coatings for food applications. Springer, New York, pp 1–23
Pongracz, E (1998) The Environmental Effects of Packaging. Licentiate thesis. Tampere University of Technology, Department of Environmental Technology, Institute of Environmental Engineering and Biotechnology, Tampere, Finland
Pongracz E (2007) The environmental impacts of packaging. Environ Conscious Mater Chem Process 2:237
Prakash G, Pathak P (2017) Intention to buy eco-friendly packaged products among young consumers of India: a study on developing nation. J Clean Prod 141:385–393
Rashid NA (2009) Awareness of eco-label in Malaysia's green marketing initiative. Int J Bus Manag 4(8):10
Reichert CL, Bugnicourt E, Coltelli MB, Cinelli P, Lazzeri A, Canesi I et al (2020) Bio-based packaging: materials, modifications industrial applications and sustainability. Polymers 12:1558
Robertson GL (2009) Sustainable food packaging. In: Waldron KW (ed) Handbook of waste management and co-product recovery in food processing, vol 2. Woodhead Publishing, Cambridge, pp 221–254
Roe B, Levy AS, Derby BM (1999) The impact of health claims on consumer search and product evaluation outcomes: results from FDA experimental data. J Public Policy Mark 18(1):89–105. https://doi.org/10.1177/074391569901800110
Rokka J, Uusitalo L (2008) Preference for green packaging in consumer product choices–do consumer's care? Int J Consum Stud 32(5):516–525
Ronteltap A, van Trijp JCM, Renes RJ, Frewer LJ (2007) Consumer acceptance of technology-based food innovations: lessons for the future of nutrigenomics. Appetite 49:1–17

Sehrawat M, Kundu SC (2007) Buying behavior of rural and urban consumers in India: the impact of packaging. Int J Consum Stud 31(6):630–638
Shekhar SK, Raveendran PT (2013) Role of packaging cues on consumer buying behavior. Int J Eng Manag Sci 4(1):61–69
Sheth JN, Mittal B, Newman BI (1999) Customer behavior: consumer behavior & beyond. Dryden Press, Fort Worth, TX
Siegrist M (2008) Factors influencing public acceptance of innovative food technologies and products. Trends Food Sci Technol 19:603–608
Sijtsema SJ, Onwezen MC, Reinders MJ, Dagevos H, Partanen A, Meeusen M (2016) Consumer perception of bio-based products-An exploratory study in five European countries. NJAS Wageningen J Life Sci 77:61–69. https://doi.org/10.1016/j.njas.2016.03.007
Steenis ND, van Herpen E, van der Lans IA, Ligthart TN, van Trijp HC (2017) Consumer response to packaging design: the role of packaging materials and graphics in sustainability perceptions and product evaluations. J Clean Prod 162:286–298
Steenis ND, van der Lans IA, van Herpen E, van Trijp HCM (2018) Effects of sustainable design strategies on consumer preferences for redesigned packaging. J Clean Prod 205:854–865
Tencati A, Pogutz S, Moda B, Brambilla M, Cacia C (2016) Prevention policies addressing packaging and packaging waste: some emerging trends. Waste Manag 56:35–45
Thøgersen J (1996) Recycling and morality: a critical review of the literature. Environ Behav 28(4):536–558
Thogersen VB, Jorgensen PE, Sorensen BS, Bross P, Orntoft T, Wolf H, Nexo E (1999) Expression of transforming growth factor alpha and epidermal growth factor receptor in human bladder cancer. Scand J Clin Lab Invest 59(4):267–277. https://doi.org/10.1080/00365519950185634
Thompson RC, Moore CJ, vom Saal FS, Swan SH (2009) Plastics, the environment and human health: current consensus and future trends. Philos Trans R Soc Lond Ser B Biol Sci 364 (1526):2153–2166. https://doi.org/10.1098/rstb.2009.0053
US Environmental Protection Agency (2014) Reducing wasted food & packaging: a guide for food services and restaurants. EPA. https://www.epa.gov/sites/production/files/2015-08/documents/reducing_wasted_food_pkg_tool.pdf
Van Birgelen M, Semeijn J, Keicher M (2009) Packaging and pro-environmental consumption behavior: investigating purchase and disposal decisions for beverages. Environ Behav 41(1):125–146. https://doi.org/10.1177/0013916507311140
Verghese H, Lewis S, Lockrey HW (2015) Packaging's role in minimizing food loss and waste across the supply chain. Packag Technol Sci 28:603–620
Wan VCH, Lee CM, Lee SY (2007) Understanding consumer attitudes on edible films and coatings: focus group findings. J Sens Stud 22:353–366
Weller CL, Gennadios A, Saraiva RA (1998) Edible bilayer films from zein and grain sorghum wax or carnauba wax. LWT Food Sci Technol 31(3):279–285
Williams P (2005) Consumer understanding and use of health claims for foods. Nut Rev 63(7):256–264. https://doi.org/10.1111/j.1753-4887.2005.tb00382.xy
World Economic Situation and Prospects (2012) Sales no. E.12.Ii.C.2, United Nations Publication. 2012

Part V

Innovations in Packaging Edibles in Other Sectors

Recent Trends in Packaging

23

Sujata Pandit Sharma

Abstract

Packaging and packing materials used in food packaging are on boom nowadays. Packaging is an efficient tool to enhance shelf-life of food materials. Recent technologies with a combination of different packaging materials help food to be acceptable on global level. Efficient packaging acts as an aid with food safety. The use of bioplastics, composite containers, aerosol containers, PET, Tetra packing, edible films, nanofilms, and biofilms is in fashion. The particular characteristics or properties of food packaging materials are the part of research and development aspects of packaging. Various combinations of polymers provide very good barrier properties and permeability characteristics for lower gas transmission rate, water vapor transmission rate which helps in preservations of odor and flavor. A good packaging material along with good packaging techniques is the only solution for food trade. Packaging material with biodegradables in nature is in trend. Replacement of non-biodegradable material and minimization of the use of packaging are environment needs. Use of active enzymes, plasticizers, additives, and preservatives in packaging materials helps in many food products to control their metabolic activities during preservation without adding extra additives as ingredients in metrics. Of course, packaging creates waste but its functions of containment, convenience, transportation, marketing ease out many other problems related to post-harvest management, food security, and food preservation. New innovative and recent trends of packaging material and packaging techniques help stake holders and consumers to eliminate all types of hazards and to sustain the environment.

S. P. Sharma (✉)
Department of Life-Sciences, Sharda University Greater Noida, Greater Noida, India
e-mail: Sujata.sharma@sharda.ac.in

A. Poonia, T. Dhewa (eds.), *Edible Food Packaging*,
https://doi.org/10.1007/978-981-16-2383-7_23

Keywords

Biodegradable packaging · Techniques · Biofilms · Nanofilms · Regulations

23.1 Introduction

This chapter shall set out a context for the consideration of recent trends and types of packaging technology available. The content highlights the food safety, products protection role during retail of packed products in the marketing process through innovative packaging materials, packaging design, packaging technology, and new innovative material developments. Packaging and packing materials used in food packaging are on boom nowadays. Packaging is an efficient tool to enhance shelf-life of food materials. Recent technologies with a combination of different packaging materials help food to be acceptable on global level. Efficient packaging acts as an aid with food safety. The use of bioplastics, composite containers, aerosol containers, PET, Tetra packing, edible films, nanofilms, and biofilms is in fashion. The particular characteristics or properties of food packaging materials are the part of research and development aspects of packaging. Various combinations of polymers provide very good barrier properties and permeability characteristics for lower (GTR) gas transmission rate, water vapor transmission rate (WVTR) which helps in preservations of odor and flavor as shown in Fig. 23.1.

Trend in new packaging develops business drive to reduce supply chain costs. Innovative packaging materials are balanced against the basic packaging techniques. Innovative packaging fulfills the requirements to the business objective of saving

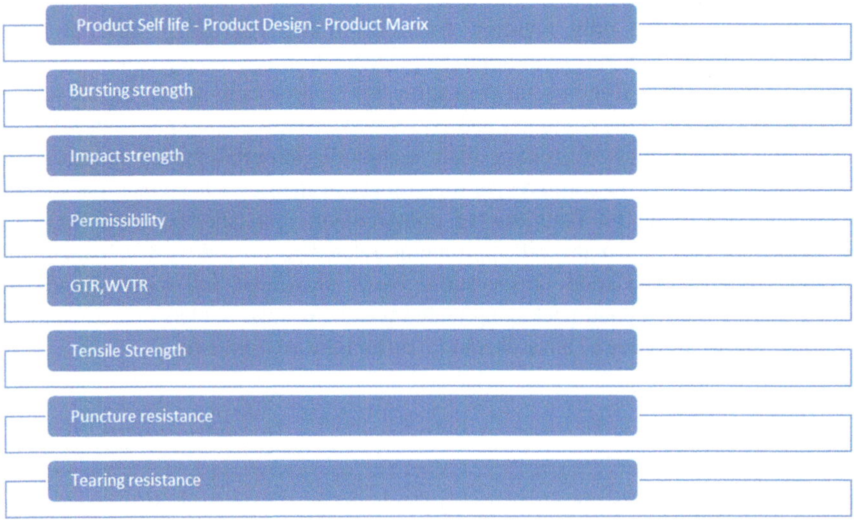

Fig. 23.1 Considerable factors in packaging material selection: a) product design, matrix, and its self-life in selection of packaging material

brand image through value-added pack design. New trends in packaging include design inputs and technological functions which communicate difference, esthetically pleasing, reduced energy consumption, functionally environmentally conscious. Food packaging is always a challenge to ensure low-cost packing performance to satisfy the needs and choices of users, food safety of the food product as well as a challenge to decrease the amount of packaging material in products for reduction waste to be disposed of. At the same time, the environmental impact of the products has been reduced. The need for global trading arises from packaging materials.

Developments in packaging materials and packaging technologies required impact in the packaging field and its influences on quality, production, engineering, and marketing, purchase, labelling issues, finance, supply chain demands, and environmental needs.

23.1.1 Historical View in New Packaging Materials with Several Technologies

Studies on packaging materials revealed that in the recent decade of 1939, first ethylene was polymerized, and later, polyethene (PE) was a product launched by ICI (Imperial Chemical Industries Ltd) in association with Du Pont. During the Second World War, many changes in food process industries have been noticed like the use of aerosol containers which became a grand success in the market for the distribution of food products in pasteurized processed products like cheese, milk, and dessert toppings. Polyvinylidene chloride (PVdC) was used in 1946 as a moisture barrier resin in packaging material.

In 1956, aluminum trays and aluminum cans and squeezable plastic bottles were used in the market for zip-molded lemon-shaped plastic packs of lemon juice. In heat-treated processed food, retort Pouch was developed with extended shelf-life of products used in military food products.

In 1960s commercially, the pouch of polyethene was most used in food products. Drawer Wall Iron (DWI) developed for carbonated drinks. Side seam welded developed in tin-plated metal cans; bottle neck shrink sleeve has been developed as precursor in aluminum roll-on pilfer-proof (ROPP) cap (Robertson 2009).

In 1970s as changes started in food Industry immediately after conversion of fresh produce to process products and easy-to -open with tin-plated ring was developed in canned drinks. Tetra Packing is launched as Tetra Brick Aseptic (TBA) packaging for (UHT) ultra-heat-treated milk. Nowadays, this Tetra Brick Aseptic carton has become the world's popular pack form for liquid foods treated under high pressure and temperature. Functional as in hurdle technology, these tetrahedral milk cartons were constructed in the form of low-density polyethylene extrusion coated paperboard.

In 1970s, bar code systems were introduced in retail packaging. Several methods were introduced to make food packaging modified into temperature-resistant boil-in-the-bag for ease to use. In 1970s, food technologies such as modified atmospheric packaging (MAP) were introduced for frozen meats, frozen foods with microwave-able heat-resistant rigid plastic containers. Sustainable packaging systems in the range of aseptic form fills and seal (FFS) developed.

In 1980s, PET bottles were used for colas and other carbonated drinks. PET bottles consist of plastics with oxygen barrier co-extruded plastic PET materials for bottling and plastic containers for retort packaging. These types of materials were applicable for microwaved heating. PET-coated dual ovenable paperboard materials were started to use in ready to use food categories.

In 1988, transformations in cans were introduced by Japan in beer. Japan launched ring pull with removal of the lid system. This innovation turned the pack into an easy to carry packet. Mandatory techniques for digital printing on cartons and on labels were introduced in 1990 in food packaging and adopted by several brands of drink industries in plastic materials. Processed food started to be treated for its product quality; food safety and convenience as majorly in packaging innovation.

In the twentieth century, this introduced ease-to-open packaging with high-quality food to consumer's respective busy lifestyles. New technologies like gas fluxing, extrusion, carbonation, vacuum packaging adoption by the perishable, non-perishable product in global trade were in trend. Gas barrier plastic materials utilized in aseptic FFS (form fill and seal) plastic containers were in fashion for soups and emulsions. Biodegradable plastic in retail tray packs for meat and dairy products with modified atmosphere technology and retort packaging technology.

Innovations in packaging closely related to technological improvements in products, materials, and its adoption. There is raised demand for packaging materials for the use of processed products, finished products, during transportation, storage in new retail markets. Materials have been made for the use of electronic domestic storage and cooking appliances such as refrigerators, freezers, and microwave ovens. Microwave oven materials are always demanded in convenience packaging for a wider range of foods. Compatible materials required for changed cultural needs, social trends, consumer's lifestyles, and climate changes under economic approachability.

23.2 Basis of Packaging Materials Development

Packaging places have an important role in the global market. Market's rapid growth for primarily processed packaged foods demanded in cooked and processed food catering service has diversified the range of various packaging materials.

Recent developments in packaging material depend on innovative designs and materials as per need of product and technological aspects of product packaging like the expansion of packing products like emulsions, ketchup in glass bottles exchanged in extruded layered plastic bottles in corporations with oxygen barriers for extended product life. Accidental inventions to fulfill the demand of military food products played key roles in packaging materials' development. These include canned products as well as composite containers (paper-based containers) in the shortage of tinplate, steel cans in marketing of various products.

Packaging in food supply is an area where it is rapidly changing because of the development of globalization of the packed product market with its suitable diversification of consumer demand. In new development of packaging materials distribution, legal and technological requirements also play an important role.

External influences on packaging as convenient product may be summarized as follows:

- Functionality and Technical Requirements.
- Label Requirements.
- Consumer Requirements
- Environment Requirements.
- Economic Requirements.

23.2.1 Functionality and Technical Requirements

Changing consumer lifestyles causes rapid growth of large retail markets and industries. A high competition in retail, trade, marketing, and expertise was involved in the requirements for the success of the packaging materials invention. This refers to the quality and technological functions of packaging materials. Technical product requirements for the material selection planning required for packaging to ensure package functionality and product protection throughout the shelf-life of the package during distribution, storage until product consumption.

Technological requirements always add value in packaging and product features, such as aesthetics, taste and comfort, functional and environmental performance. Other requirements for package design innovations are to develop different product, product function, quality, and brand integrity. In accordance with marketing strategy, this approach satisfies the future demands at an acceptable profit. Package compatibility with the current packet range and production system required to carry forward functionality and technical requirements of packaging. The operational, financial requirements of regulations for food hygiene, label requirements, specifications like weights and mass are necessary. The adoption of advanced distribution packaging systems has been encouraged in commercial food supply chains. Marketing research and consumer research professional agencies need to be employed according to the requirements and technical research to identify opportunities and minimize the financial costs and risks involved in the development, production, and marketing for new product packaging.

23.2.2 Label Requirements

Trade barriers, WTO (World Trade Organization) agreements between countries have helped to procure products from around the world. This affects enhanced competition and reduced the cost. Competition increased the rationalization in industrial structures. In packaging, with increase in automation, adoption of new

Table 23.1 Objectives of new innovative materials development

S.no	Concerns related to food packaging in industries
1	Increased trends for more layers packaging increase the packaging waste
2	Cost of disposal
3	Pollution associated with type of materials
4	Facilitation of conventional material
5	Perceptions related to over-packaging
6	Eligibility and integrity information of labels

materials with various pack sizes reduces unit cost. An internal part of packing in the distribution system plays an important role by preventing excess packaging layers to goods for reduction waste. Packaging materials are important factors in International trade. This needs appreciation for promotion of positive contributions in packaging for quality of life. Legibility of labels for the visually impaired or consumers with specific needs are important for product intended use. Various laws and regulations, codes of practice and guidelines govern food packaging. In legal requirements, packaging may include for the following:

1. Prevention of food spoilage.
2. Protection of nutrients.
3. Promotion to reduce solid waste disposal.
4. Cost-effectiveness.
5. Eliminates the risk of hazards (physical, chemical, and biological).
6. Aesthetically accepted and overall acceptability.
7. Appropriate labelling as per FSS Act 2006 or Codex guidelines to communicate important information about the food.
8. Functional convenience in use or preparation.
9. Promotes the development of modern retail.
10. After disposal biowaste capability of material.

In food industry, current public concerns related to packaging are in Table 23.1.

23.2.3 Consumer Requirements

Packaging innovation contributed to convenience in food choices for consumer demand. There are various attributes obtained through modern packaging. Some of these are easy access and opening, disposal and handling, visibility of the product, capacity for resealing, capacity for microwave, prolonged shelf-life, etc. Packaging products have more importance to companies. On the basis of end user's needs, desires, industries opened up new distribution systems for better presentation quality with decreasing costs, increasing margins. Improved product improves brand differentiation and services. Distribution system carried food from farm to folk. Food supply chain permits the reuse of packed material.

Packaging helps in preservation of the world's resources by protecting raw food spoilage, product spoilage, and waste. Modern distribution packaging systems enable customers, when and where they want, to buy food.

23.2.4 Economic Requirements

Mass production of packaged food has evolved in the economic growth of food industries. In production, processing and packaging play a key role in mass processed food products. The economies of scale-up of production involved industrial competition. This made more products affordable. Rapid growth in the global population increased demand for packaged food and increased economic requirements.

Countries grow in rapid urbanization as well as growth of industries. This needs scientific support and awareness of distribution constraints in marketing while designing a total product concept. Essential is to consider product production economically as per its marketability, with nominal cost factors, for their physical distribution practicability. Intend use of the product depends on the type of population.

23.2.5 Environment Requirements

Environmental requirements change the market conditions. Use, reuse and recycling before final disposal of non-biodegradable materials are manufacturer's responsibility related to the environment. Packaging means the safety and cost-effectiveness in delivering products to the consumer as per organization's strategy. Finished products require strategic and tactically importance in marketing functions. Brands compete in the market for their distinctive or innovative packaging materials. New innovative materials are usually environment-friendly and biodegradable in nature. A good packaging material along with good packaging techniques is the only solution for food trade. Packaging material with biodegradables in nature is in trend. Replacement of non-biodegradable and minimization of the use of packaging are environment needs. Use of active enzymes, plasticizers, additives, and preservatives in packaging materials helps in many food products to control their metabolic activities during preservation without adding extra additives as ingredients in metrics.

23.3 Basis of Innovation in Packaging

Packaging material plays a role in protection and preservation of food. Food waste can be minimized with this approach. Success in packaging materials and food products reflects the daily consumption of packs that the population consumes safely. Minimizing food waste throughout the supply chain saves cost. Innovative packaging is in demand. As per the World Health Organization data, significant food waste occurs in countries due to insufficient means of preservation, scientific storage, and retailing, between 30% and 50%. Food waste in developed nations with modern processing, packaging, and distribution systems is only 2–3%. And in the case of packaged food, less than 1%, compared with unpackaged food. According to the Industry Council for Packaging and the Environment (INCPEN), food wastage represents greater financial loss. Food distribution systems described a coordinated system for the preparation of goods for transport, delivery, storage, retail, and final use. The technical-commercial function was aimed at optimizing distribution costs. Recent developed packaging materials also provide the following functions of packaging as specifically stated:

1. Non-toxic in nature.
2. Non-corrosive.
3. Biodegradable nature material.
4. Good barrier properties for containment.
5. Functionally for preservation.
6. Mechanical properties for protection during distribution.
7. Labelling and information about product to be packed.
8. Easy to use and throw.
9. Aesthetically acceptable.
10. Brand promoter.
11. Marketing tool.
12. Cost-effective.

Generally, new product developments are successfully implemented as a total concept of packaging. Package design and distribution are considerable factors in marketable products. New products have always been manufactured using packaging materials, shape and design to fulfill its market requirements. Sometimes product failure tends to happen in the marketplace due to inadequate packaging and because of the product cost. Optimization in packaging is the main concern of the packaging function. It involves a detailed assessment of the packaging system approach, its evaluation, and its functionality. The total packaging system includes a number of different components, including the use of materials, efficiency of machinery and production line, distribution storage, management, and workforce. The total product concept is where the package and product are considered as together. Understanding of a product's characteristics is necessary as its intrinsic mechanism. Intrinsic factors indicate the product deterioration factor, distribution fragility, and possible leaching of the product with packaging materials. Compatibility is an essential factor in the design of suitable packaging for development. The product has a chemical, physical, and biological impact mainly in three environments. Products are basically in solid, liquid, granules, powders, emulsions, pastes, etc. Biochemical nature ingredients with chemical composition, nutritional value, volatile, perishable, odorous, etc. Physically dimensions are size, shape volume, weight & density method of fill, dispense, accuracy, legal obligation, etc.

Other considerable factors in new packaging materials, during designing of packaging materials for storage and distribution, are convenience, display, ease of handling, traceability of product include. Packaging is required to identify the optimum design of a packaging system:

1. *Primary packaging*—Where the packaging material is directly in contact with the food.
2. *Secondary packaging*—Where it contains and collates primary packs units.
3. *Tertiary packaging*—Where multi-pack contains primary and secondary units.

Design and development of packaging requires selection of the most suitable materials, which are machinery safe, environmentally friendly, cost-effective. A printed and coated oriented polypropylene (OPP) film replaces the aluminum foil wrap and printed paper label band. Good gas barriers, moisture barrier properties, and low-cost package materials save this advancement in packaging materials now applied instead of the two previously used in one wrapping operation. Production speeds were observed much higher in high tensile properties of OPP (oriented and biaxially oriented polypropylene).

Nowadays packaging operations in food manufacturing firms are automatic or semi-automatic. Automatic operations require packaging materials that can operate in unit operations efficiently and effectively. Needs for packaging are specific dimensions, type and formats, and strength. The characteristics of the material need an account as per the requirements of packing and food processing operations. Materials should have properties such as non-corrosive, non-toxic, as well as for each type of material, the tensile strength, and stiffness. Sometimes new designs and materials may require minor or major machine modification. Design modifications in primary packs always affect secondary packs. This effect retailing resulted in acceptability of product. It is important in designing and innovation of new materials should bring and informed at all stages of project team implementation. For the complete process of operation's filling and sealing techniques, the micro-electronic systems, packaging machinery has introduced into a wide range of equipment and integrated systems. Ideally, the product is made to meet the expectations of customers. Packaging is tool to attract consumer. Information available on the product communicates its desirability, acceptability, and informs the customer. Now a single matrix of foods is available in a vast range of packed products as well as in pack combinations. For example, raw packed milk, pasteurized chilled milk, ultra-heated (UHT) sterilized and dried powders of the dairy in package combinations for consumer perception. Quality attributes and sensory characteristics of food, such as flavor, taste, and smell, affect human sensory perception.

Packaging protects quality characteristics such as taste, color, texture, and nutritional content during processing with extended storage capability of product. Product quality depends not only on the quality of raw ingredients, processing adds, and packaging methods but also on the conditions of distribution and storage during the expected shelf-life of good packed materials. Significant improvements in food quality and a dramatic increase in the choice of packaged food have resulted in increased competition in the producers' and suppliers' market for quality audits of suppliers. The extension of shelf-life is a marketing tool for promoting the concept of freshness. Codex guidelines on lettering font and label, critical views drawn on the volume of packaging rather than the weight of packaging. The change in packaging increases the trend toward demand for prepackaged foods. Environmental compatibility is one of the marketing tactics used by retailers and manufacturers. Supply chain is responsible for explaining the functions and advantages of the packaging material itself.

In retail logistics, packaging plays multiple roles. The evolution of modern packaging materials such as nanomaterials, edible films for fast-moving consumer products, requires multiple food retail markets. This is important for the awareness of packaging suppliers regarding market demand and responses to market changes.

Packaging plays multiple roles in retail logistics. Multiple food retail market needs the evolution of modern packaging materials like nanomaterials, edible films for fast-moving consumer goods. This creates importance for packaging supplier's awareness regarding market demand and responses to market changes. In response to structural changes in retailing, producers may have to modify their distribution and packaging operations. Fast-moving consumer goods packaging (F.M.C.G) could be referred to as part of the marketing mix for food retail and affects all other marketing variables, that is, product, price, promotion, and location. It also relies upon demand forecasting. (Robertson 2009).

Electronic data interchange (EDI) is helpful in FIFO (first in first out) for selling well before the date of expiry. Bar codes allow the industry to identify retail product units by means of a unique reference number, with the electronic point of sale (EPoS) system at the retail checkout as its main application. The use of the barcode for packaging identification has enabled efficient management of distribution and stock control. GHP is another problem which concerns the distributor and store manager (Good Hygiene Practices).

23.4 Packaging Innovation

Innovation has been defined as solution-based inventions and its applications. New inventions cover all efforts for creating new ideas, concepts, devices, or processes and getting them to be implemented. An invention in process includes all aspects of commercial development of product from application to the inventions and specific objectives. Minimizing the problems caused by design consequences helps in improving the quality of the final product. Computer-aided design (CAD) and rapid prototyping facilities for design are helpful to trace products. Awareness about customer needs plays a powerful role in invention and innovation, leading to what is known as "market pull" in contrast to "technological push," which is less likely to be successful. Mostly, innovation is all about small changes that build on inherent flexibility in existing products or systems. Consumers always want innovation and value novelty, so the packaging industry must continue to innovate or risk stagnation.

An interesting way to view innovations is provided by the Gartner hype cycle (Morris 2011). It characterized the rapid progression of innovation from over-enthusiasm through a period of disillusionment to an ultimate understanding of the importance of technology and the role of packaging in a market (see Fig. 23.1). An eventual understanding of relevant technology in packaging and its role in a market or domain emerges, driven primarily through performance gains and adoption growth and release of second- and third-generation products (Fenn and Raskino 2008). Despite these criticisms, it has remained a popular and useful way for companies to evaluate innovations. In the area of food packaging, smart packaging is still subject to positive hype, together with bio-based polymers such as bio-PET and bio-HDPE. Antimicrobial packaging is also at this early stage but is unlikely to ever reach the plateau of productivity. Bio-based polymers such as PLA and PHA are now experiencing negative hype as more companies trial them. Time-temperature indicators, after more than 40 years, have moved up the slope of enlightenment but are unlikely to ever become more than a niche market. The retort pouch is approaching the plateau of productivity. It is to remember the changes through big innovations in food packaging such as MAP and aseptic packaging took 20–30 years before they reached the plateau of productivity.

23.5 Future Trends and Conclusion

Different factors, including global climatic changes and global trade demand for food resources, influence the development of bio-based packaging materials. The use of bio-based materials in food packaging improving mechanical and barrier properties, and decreasing packaging costs. Combinations of polymers and nanoclays (so-called bio-nanocomposites) are likely to be future products based on bio-based packaging materials to achieve the desired barrier, mechanical properties are on demand. Several researchers have been undertaken in this area and commercialized; in the next decade, the production of bio-nanocomposites for use in the food industry will be significant. Cost is also a limitation to the widespread use of bio-based packaging materials. Bio-based packaging puts upward pressure on the cost of raw materials. It must be produced on a much larger scale than any other aerobically produced microbial product. According to studies global bio-based plastic packaging demand is forecast to reach 0.884 Mt. (million metric tons) by 2020. From 2010 to 2015, a 24.9% compound annual growth rate (CAGR) is anticipated, slowing to 18.3% in 5 years to 2020. Fast-growing segments of GM and bioderivatives to account for a quarter of the total market demand for bioplastic packaging by 2020. PHAs are forecast to reach a 41% CAGR and 83% bio PE over the period. At the same time, traditional bio-based starch, cellulose, and polyester-based plastic packaging technologies are each predicted to show a decline in market share.

Europe was the largest market for bio-based plastic packaging in 2010, with demand representing more than 50% of the world's tonnage. While North America and Asia are currently lagging behind, growth rates are expected to outstrip and Europe in these two regions in during this period of 2020. Rigid packaging currently accounts for 52% of the bio-based plastics segment, but the report predicts that flexible packaging, which currently accounts for the remaining 48%, will account for an increasing share of the bio-based plastics segment by 2020.

The commercialization of bio PE and PHA and the wider availability and improvement of BOPLA film properties are likely to drive demand. An earlier report (Shen et al. 2009) indicated that global capacity of bio-based plastics will increase from 0.36 Mt. in 2007 to 3.45 Mt. in 2020. The most significant materials are projected to be starch plastics (1.3 Mt), PLA (0.8 Mt), bio-PE (0.6 Mt), and PHA (0.4 Mt). However, not all of these bio-based plastics will be used for packaging; a significant proportion will be used for man-made fibers. Bio-based plastics will grow to 884,000 tons in 2020 as per the reported studies. The quantity of maize needed to produce this quantity of bio-based plastics would be equal to 0.1% of the total annual global maize (maize) harvest of 800 million tones. Other raw materials such as sucrose, wheat, or rice starch will be used. Cellulosic materials will be available as monomers (lactic acid) and for bio-based plastics (PLA, PHA). Nanofilms, semi-rigid, and rigid materials will be used as per demand of products.

References

Fenn J, Raskino M (2008) Mastering the hype cycle: how to choose the right innovation at the right time. Harvard Business Press, Boston, MA

Morris SA (2011) Food and package engineering. Wiley-Blackwell, Chichester

Robertson GL (2009) Food packaging. In: Campbell-Platt G (ed) Textbook of food science and technology. CRC Press, New York

Shen L, Haufe J, Patel MK (2009) Product overview and market projection of emerging bio-based plastics. Commissioned by European polysaccharide network of excellence and European bioplastics. Utrecht, the Netherlands

Edible Packaging of Liquid Foods

24

Bhasha Sharma, Meenakshi Garg, Rajni Chopra, and Susmita Dey Sadhu

Abstract

The conventional food packaging materials manifest gigantic stumbling blocks in terms of exacerbating environmental pollution due to the extensive utilization of non-renewable sources of energy. The increasing quest and indagation diversion in sustainable packaging framework are attributed to the emerging patron predilection for stable, healthy foods and the apprehension concerning the detrimental effects of conventional packaging. Therefore, the proliferation of edible materials attributes to the escalating exigency to burgeon new-fangled, environmentally benign, and extortionate quality products could be a magic bullet in sustainability. Edible packaging and coatings do not dissimulate to replace conventional materials in packaging but to bestow an auxiliary benefit for the preservation of food and to alleviate the cost of traditional packaging materials. The edible food packaging is the cogent candidate for packaging competent for human consumption. Food coating and edible films are one of the transpired blueprints for the optimization of the quality of food. The pioneering of edible packaging perceives

B. Sharma
Applied Sciences and Humanities, University of Delhi, New Delhi, India

M. Garg (✉)
Department of Food Technology, Bhaskaracharya College of Applied Sciences, University of Delhi, New Delhi, India
e-mail: meenakshi.garg@bcas.du.ac.in

R. Chopra
Department of Food Science and Technology, National Institute of Food Technology Entrepreneurship and Management, New Delhi, India

S. D. Sadhu
Department of Polymer Science, Bhaskaracharya College of Applied Sciences, University of Delhi, New Delhi, India

A. Poonia, T. Dhewa (eds.), *Edible Food Packaging*,
https://doi.org/10.1007/978-981-16-2383-7_24

that plastics integrity is tough to replicate. To fabricate edible plastic proxies, scientists have turned them into robust and natural polymers extracted through plants by employing pre-emptive strategies. This intrinsic packaging is a virtuous matrix as the carrier of antimicrobial agents which enkindles numerous advantages against traditional coatings for instance better solubility, spreading, and diffusivity. The absolute edible packaging would be constructed by juxtaposing carbohydrates and proteins, the biological polymers produced from plant tissues which could be a felicitous strive for mankind. The edible coating efficiency for the prolongation and preservation of vegetables and fruits to augment the products shelf-life is dependent mainly on regulating the wettability and adhesion of the coating solutions on the surface. These packaging can be very commodious employed for microencapsulation of leveraging agents and food flavoring to coherently regulate their release and inclusion into the interior of liquid foods. The exploration will emphasize diversification in the fabrication of edible materials by elucidating active and smart food packaging with functional food coatings. This stalwart indagation will decipher standpoint on the classification of proteins, lipids, and carbohydrates with their desired properties and applications in the food and agricultural domain. Additionally, there is a desideratum for more fundamental comprehension to validate design to preferred specifications. The integration of nanotechnology in the concoction of edible polymeric materials to aggrandize the productivity and efficiency of these materials would be a benchmark in the myriad of food applications. Nanoscale revolution is proficient in detecting pathogens, barrier formation, and active packaging that is poised to ameliorate edible packaging to novel frontiers. Furthermore, aseptic packaging which infers exclusion or absence of unwanted organisms has also been delineated. Inadequate marketing, lack of cognizance, toxicity, and health reverberations have been addressed to discern the potential of edible films for liquid foods. This expedition will decipher consumer acceptance and legislation by divulging the impediment in employing edible packaging for liquid products. In a nutshell, to conserve the ecosystem and to mitigate the plight of increasing plastic wastes jeopardizing the environment, there is a dire need to employ innovative edible packaging for liquid products such as beverages, dairy products, etc., by incorporating active components.

Keywords

Food packaging · Edible materials · Biopolymers · Carbohydrates · Smart packaging

24.1 Standpoint

With the advent of exigency in food, it is imperative to employ specific materials to protect and impart the safety of food. The prime emphasis of packaging materials to protect food from environmental conditions and external influences. The considerable indagation has been explored to meliorate the applications of edible films and coatings in a myriad of the food industry as an alternative packaging system to provide safety and quality of food (Campos et al. 2011; Işık et al. 2013). The increasing progression and quest in edible packaging attributes to the escalating stable or healthy foods, and due to the increasing awareness of the obnoxious impacts of traditional packaging (Hassan et al. 2018). Edible packaging provides auxiliary assistance in reducing the cost and imparting food preservation. The packaging system acts as the carrier of antimicrobial agents and bestows numerous advantages against synthetic packaging to provide solubility, diffusivity, and better spreading (Ramos et al. 2012). Edible coatings are termed as skinny layers on foods for primary packaging with edible components. They are having properties like gas permeability and moisture in the food and can also be consumed with the food. Besides, the coating is a slightly thin layer of material which should be created on food products as a coating (Kang et al. 2013; Espitia et al. 2014). The function of edible coatings is to act as a blocker toward oxygen, moisture, and motion of solute from food without transforming its original characteristics (Krochta 2002). They are fabricated through renewable sources and are safe to eat. These coatings can be applied in liquid shapes through dipping, spraying, and electro-spraying that facilitate uniform and thin coating (Khan et al. 2012; Khan et al. 2013; Khan et al. 2017). The augmenting stipulation of patrons for the preservation of food in a way which has raised pathways for protection methodologies. For instance, the employment of edible polymers derived from renewable sources. To enhance the prolongation of food, edible packaging on food can be implemented. Besides, the prolongation and preservation of food products can be increased by diminishing respiration by application of edible coatings (Galus et al. 2020; Baldwin et al. 1995). The edible coatings are environmentally benign and biodegradable which helps reduce plastic waste pollution. Furthermore, employment of edible packaging bestows extension of shelf-life which also has an advantage in the economical aspect of food companies. Coatings are usually acted as carriers for preservatives and components which include antimicrobial and anti-browning agents like citric acid and ascorbic acid.

Edible packaging is a revolutionary proposition. For example, strolling around grocery stores where the food kinds of stuff are wrapped in edible skins. Hence, the person would be able to consume a protein bar or ice cream straight away from the shelf (wrapper and package included). These edible skins make consumers afraid of germs. Often, it makes us anxious to eat the food package as well as the food inside. Beguile innovations of edible packaging are escalating, numerous uncertainties restrict consumers to give try on these types of food products. Yet these innovations may impart sustainable products by reducing landfill waste created from conventional packaging. To comprehend this state of the art, let us focus on the lemons and potato skin which are delicacies and have water repellent properties that act as a protective layer on fruits, and impart flavor to savory and baked dishes. Furthermore, other synthetic foods like edible wrapper sausages, caramel candies, Japanese ice creams, etc., also employ this concept. Edible packaging can be fabricated from proteins, fat, carbohydrates based on the applications. Edible packaging can be categorized into several categories such as food paired with edible containers, food wrapped in food, container eaten with the beverage, disappeared packaging, and edible packaging served at restaurants.

24.1.1 Significance of Edible Packaging in Liquid Product

Attributable to the short shelf-life of most liquid foods such as sauces, juices, etc., therefore, there is a demand for intelligent food packaging across the globe. Customers are looking out for packaging which is easy to use and preserve the liquid food at the same time for a long duration. The edible coatings are employed to meliorate the organoleptic characteristics of packaged foods. They act as carriers of numerous components like flavor, coloration, antioxidants, antimicrobials, and sweeteners (Han 2005; Vásconez et al. 2009). These coatings are required to preserve moisture content of internal constituents and movement of solute in several foods like pizza, candies, and pies (Bourtoom 2008). Apart from effectiveness toward barrier to solute migration, gas, and moisture (Fig.24.1), these coatings can lessen the microbial growth in liquid products of food by reducing the diffusion rate of antimicrobials from coating substances to food products (Aloui et al. 2021).

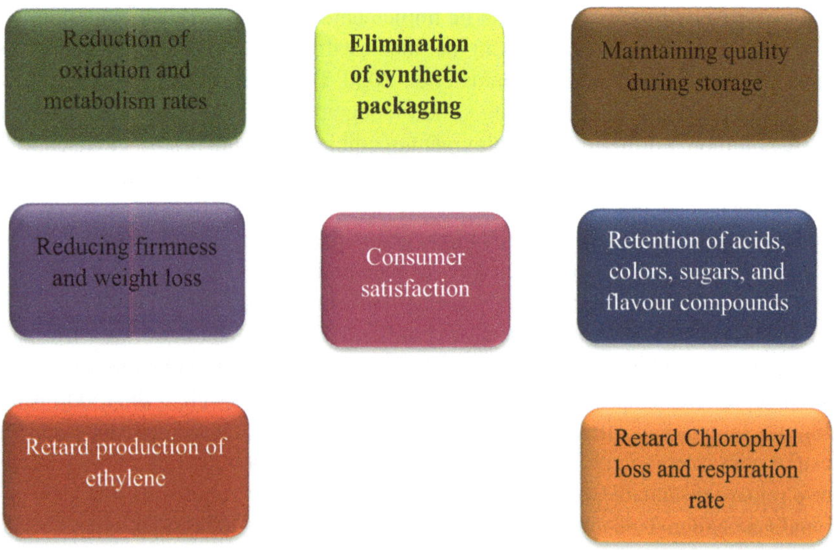

Fig. 24.1 Edible coating benefits

24.2 Functional Properties

24.2.1 Oxygen Barrier

The deterioration of food is majorly due to the oxidation, enzymatic browning of fresh food, and discoloration. The edible packaging can extend the prolongation and preserve the O_2 sensitive liquid foods quality with low oxygen permeability by eradicating the employment of non-recyclable plastics. Furthermore, the progression of edible films under certain storage conditions with defined permeability of gas can form a modified atmosphere which also diminishes the rate of respiration of food products or production of ethylene during distribution and storage (Rahman 2007).

24.2.2 Environment Barrier

The eminent process to employ edible packaging is to regulate the mass transfer between ambient atmosphere and food products. The most pertinent characteristic to tailor and select edible material in liquid packaging is permeability. The moisture content and temperature of biopolymeric materials entail permeability of the edible films (Janjarasskul and Krochta 2010).

24.2.3 Moisture Barrier

The edible packaging is utilized to obstruct the moisture exchange between the atmosphere and the food product. The modifications in packaged food (water activity) results in unwanted textural changes, problematic microbial growth, enzymatic reactions, and deteriorative chemical reactions. Hydrophobic and lipid compounds are frequently employed to form coatings of moisture barrier or to meliorate the moisture barrier characteristics of hydrocolloid derived films owing to the low polarity, low water affinity, and dense molecular structured matrices (Kalpana et al. 2019).

24.2.4 Oil Barrier

Grease and oil resistance can be proffered to any lipid-containing product with the aid of edible packaging. Albeit accustomed evaluation strategy, ASTM F119 for the oil entrance rate of adaptable obstruction materials is applicable to palatable films, quantitative information concerning porousness is restricted. Inborn carbohydrate- and protein-based hydrophilicity of polymeric films is required to deliver grease/oil-resistant attributes to these films (Aguirre-Joya et al. 2018).

24.3 Classification of Natural Polymers-Based Edible Films

24.3.1 Polysaccharides

The long-chain polymeric structures are derived from di- or monosaccharide repeating units formed by joining glycosidic bonds. The formation of hydrogen bonds facilitates film formation due to the existence of hydroxyl and hydrophilic moieties in the structure. It exhibits an oil barrier, oxygen, and moisture properties. Besides, it imparts structural integrity, strength, and slight resistance toward water migration. Notwithstanding, the fabrication of polysaccharide-derived coatings exhibits poor water vapor barrier attributes which therefore is an explanation of giving up advertisers postponing moisture misfortune from suppers (Hassan et al. 2018). Polysaccharide-based coatings are vapid, have minor caloric substance and sleek free appearance, and can be pertained to favorable to long the period of usability of organic products, vegetables, shellfish, or meat items by altogether lessening drying out, obscuring of the surface and oxidative rancidity (Mohamed et al. 2020). Chitosan, due to its physicochemical, mechanical, and thermal properties, was effective in the fabrication of edible films with enhanced solubility, moisture content, optical properties, and water vapor permeability (Homez-Jara et al. 2018). A robust review on starch and chitosan-derived edible films was recently published (Pelissari et al. 2019). Polysaccharides-based edible coatings can be classified in different ways (Fig. 24.2) An antimicrobial edible film for active packaging of foods was prepared to utilize carboxymethyl cellulose and corn starch embedded with the bacterium Lactococcus lactis against *Staphylococcus aureus* (Haghighi et al. 2020; Lan et al. 2020).

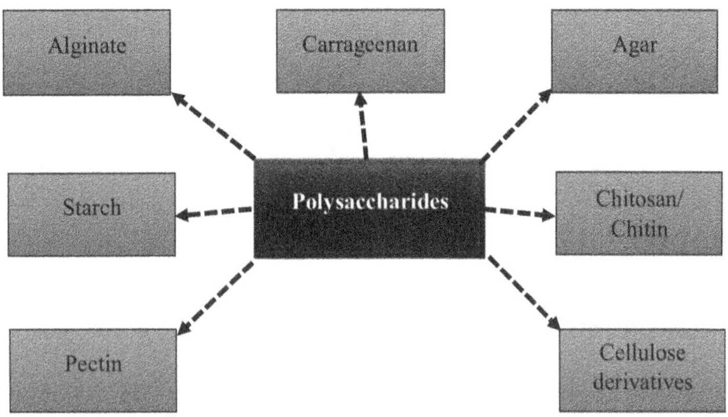

Fig. 24.2 Classification of polysaccharides-based edible coatings

24.3.2 Protein

Proteins usually prevail as either globular or fibrous proteins. Water insolubility and primary structure formation of animal tissues are the essential underlying properties of fibrous proteins, whereas water solubility and solubility in aqueous arrangements of salts, acids, or bases and better performance capacity for different purposes in living frameworks are the underlying properties exhibited by globular proteins (Dinika et al. 2020; Tkaczewska 2020; Calva-Estrada et al. 2019). Interconnect H-bonded fibers make up the fibrous proteins. Globular proteins fold into complex round structures, which are attributed to the presence of hydrogen, ionic, and covalent bonds. The physicochemical properties of proteins are completely subject to the course of action of amino acid pendant groups and their relative amount close to the polymeric chain. Protein dispersion or solution is utilized to fabricate coatings and films' solvent employed is ordinarily restricted to ethanol, water, or ethanol-water mixes (Chiralt et al. 2018).

The extra drawn out frameworks can be fabricated into films by the employment of heat, solvent, acid, or base that results in denaturation of the protein. On prolongation, hydrogen, ionic, and covalent holding connect the protein chain. The equivalent positioning of polar gatherings close by the polymeric chain improves the possibility of the necessary communications. Chain to chain communication decides the strength of the edible film; higher communications yield more grounded films with less permeability to fumes, fluids, and gases. In this way, protein-derived coatings or films are thought to be exceptionally successful blockers of oxygen even at low relative humidity (Mohamed et al. 2020; Chiralt et al. 2018). Diverse array of protein is employed in the fabrication of healthy to eat films/coatings that involves whey protein, corn zein, wheat gluten, soy protein, casein, and gelatin (Fig. 24.3). In a research, Wang et al. (2019) evaluated the function of pullulan polysaccharide, glycerol, and trehalose casted on gelatin-soluble edible membranes. The films demonstrated enhanced mechanical properties, texture comprehensive index, and solubility at lower concentrations, howbeit the properties decreased at higher concentrations. Whey protein isolates 96% protein ratio (at 8% 20 (w/v)), glycerol, lemon, and bergamot essential oils employed in the fabrication of edible

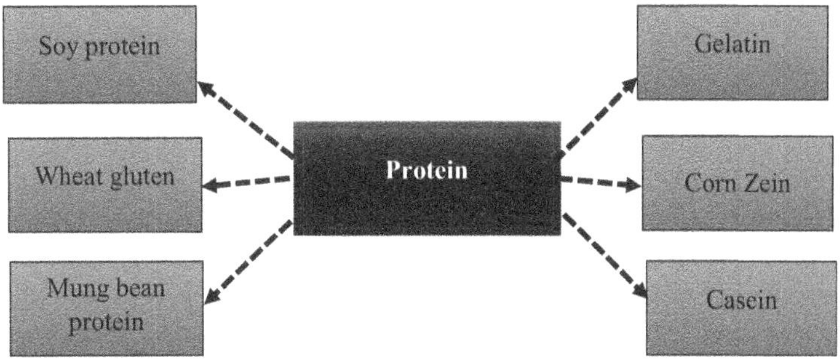

Fig. 24.3 Classification of protein-based edible coatings

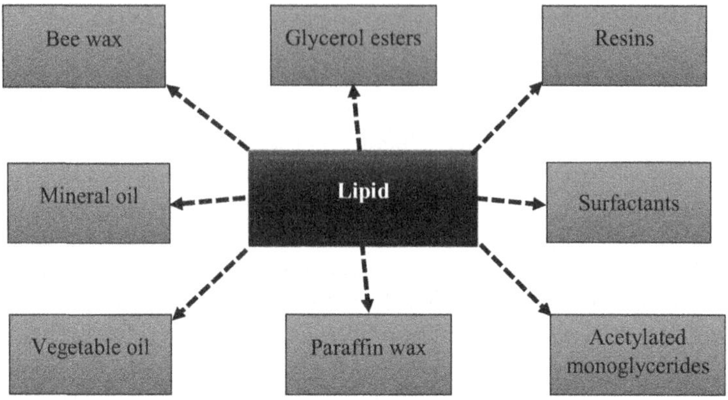

Fig. 24.4 Classification of lipid-based edible coatings

film demonstrated effective antimicrobial activity against E. coli and S. aureus with excellent retaliation in opposition to oxygen permeability and water vapor (Çakmak et al. 2020) Similarly, plum seed protein isolate and gum acacia conjugates demonstrated enhanced water vapor barrier property, mechanical properties, surface hydrophobicity, and thermal stability (Li et al. 2020).

24.3.3 Lipids

Lipids are brilliant hindrances against moisture relocation. Lipids, blended in with polysaccharides and proteins, deliver films with enhanced mechanical and hindrance characteristics. But, these composite structures may also demonstrate augmented moisture penetrability when contrasted with that of unadulterated lipids. Lipid-based mixes are utilized in protecting film/coating involving surfactants, herbal wax, and acetylated monoglycerides. The least complex lipid mixes are paraffin and beeswax. Lipid-derived films and coating are viewed as profoundly powerful to obstruct the conveyance of moisture because of their low extremity. Ordinarily, the films or, on the other hand, coatings produced using lipid are additional fragile and thicker because of their hydrophobicity. With expansion in the centralization of the hydrophobicity stage, the water fumes penetrability declined (Debeaufort and Voilley 2009). Notwithstanding, it was accounted for that lipid-accumulating films or coatings may harm the display and sparkle of covered food items. Lipid-based edible coating can be classified into bee wax, mineral oil, resin, surfactants, etc. (Fig. 24.4). Recently, flaxseed gum, oligomeric procyanidins, and lauric acid-derived films were evaluated for packaging of oil, salt, and vegetable. The films demonstrated remarkable seasoning packaging approximately 75-day storage period capacity (Liu et al. 2018).

24.4 Factors Involved in Aseptic Packaging

Four unique ways exist to achieve aseptic packaging (Von Bockelmann and Von Bockelmann 1986):

1. Utilization of pre-sterilized packages and filling hardware working in the arena devoid of any microbial contamination;
2. Utilization of pre-assembled, non-sterile packaging materials to be disinfected in the filler;
3. Pre-assembled, pre-sterilized packages to be filled in microbiologically spotless filling hardware;
4. Creation, disinfection, and filling of the packaging or food package material in the filler.

To keep up the elevated stage of microbial quality accomplished by ultrahigh-temperature handling of different food items, aseptic filling systems should be applied since "it doesn't bode well to deliver food items with a high hygienic norm if this is ruined by deficient bundling methods". Aseptic bundling infers three distinct advances; (1) cleansing of the bundling material in contact with the food surface; (2) making and keeping up a sterile encompassing in the territory where the sanitized item and the cleaned bundling material/bundle are united; (3) creation of compartments that are sufficiently tight to forestall passage of waste creatures. As far as adaptable and semirigid materials in packaging are concerned, the accompanying frameworks have taken entry into the merchandise of aseptic filling: (1) pockets or sacks; (2) pre-assembled cups; (3) structure fill-seal cups from move stock material; (4) plastic bottles; (5) pre-assembled, paper-derived overlaid containers; (6) containers created from move stock material of paper-based overlays (Murty 2020; Anderson et al. 2020).

Shockingly, in practical manner, all findings accessible on aseptic food packages have been distributed by the producers of such gear. Not many proclamations are accessible on the microbial execution of these filling techniques. It is, notwithstanding, protected to expect that frameworks that have effectively entered a region likewise meet the execution requests legitimate in that particular market (Sanjana et al. 2019).

24.5 Smart Food Packaging

24.5.1 Intelligent and Active Packaging

Food bundling innovation is constantly expanding over the most recent couple of years according to developing difficulties from an advanced society. The development has gone from safeguarding to the examination of the expansion of safe food added substances and adds to lessening natural contamination. Dynamic and astute bundling frameworks are an advanced development that goes past the customary capacities where exists an association between the item and its current circumstance to expand the timeframe of realistic usability of food, improving tangible properties keeping up the nature of the pressed food (Chen et al. 2020; Hammam 2019; Majid et al. 2018). Dynamic and insightful bundling ought not to be befuddled; dynamic bundling causes a change of the states of the pressed food to expand the timeframe of realistic usability, keeping up the nature of result while a shrewd bundling framework screens the state of bundled nourishments to give data about the quality items during transport and capacity. The two frameworks can work synergistically to acknowledge what is called savvy bundling, giving an absolute bundling arrangement utilizing every one of their points of interest. Other than this, dynamic, insightful, furthermore, shrewd bundling ideas are regularly utilized conversely in writing. A few sorts of brilliant bundling permit the controlled arrival of bioactive substances (cell reinforcements or antimicrobials) or can be appended with embodied mixes (Kalpana et al. 2019; Bhargava et al. 2020).

24.6 Role of Antimicrobial Agents in Protective Edible Packaging

Safe-clean food is a worldwide need and an indispensable goal of the present food enactment due to microbial hazards of food items that are among the primary wellsprings of food-borne illness. Generally, antimicrobial specialists are straightforwardly blended into the underlying food definitions; however, inordinate sums might alter the flavor of the food brought about by the direct expansion of antimicrobial segments, which may alter about an inactivation because of food parts or, on the other hand, weakening beneath dynamic fixation or vanishing of these mixes (Motelica et al. 2020). The primary explanation behind fusing this compound into the package is to forestall surface development in nourishments where a huge part of deterioration, as well as defilement, happens that may decrease the expansion of incredible amounts of antimicrobials. A potential answer for regulating the development of phytopathogens in organic products during the postharvest period of usability, and to broaden the security and rack life of prepared to eat nourishment, is the utilization of antimicrobials mixes. The antimicrobial packaging can be compelling during the capacity time frame, taking care of transport, what's more, when the package is opened, the antimicrobial film will become dynamic (Kadzińska et al. 2019). The antimicrobials installed in films can likewise be moved to the surface of food for additional activity, and generally, low sums are needed to accomplish an objective rack life. The general moist climate and the moisture substance of the item can impact the antimicrobial characteristics of palatable packaging. This can initiate the buildup of moisture on the surface of food, expanding the opportunities for the development of microbes (Guo et al. 2017). Eatable packagings, when fused with antimicrobial mixes, can be utilized as antimicrobial dynamic bundling to control food-borne microbes and deterioration microorganisms, accordingly upgrading sanitation. The specific activity of such films comprises the delivery during food stockpiling of a suitable dynamic substance that influences microorganisms (Suhag et al. 2020).

The edibility of coatings and films is possibly accomplished when all parts including plasticizers, biopolymers, and different added substances are food-grade fixings, while the entirety of the elaborate cycles and gear ought to be likewise satisfactory for food handling. The primary coating strategies utilized in these materials are depicted beneath (Valdés et al. 2017).

24.6.1 Use of Spraying

A spray is basically an assortment of mobile beads as the consequence of atomization cycles to separate mass fluids into beads, including the accompanying contemplations:

- Augmentation in the fluid surface zone is a significant problem in cycles where fast vaporization is needed. Indeed, in applications that require/involve antimicrobial activity, it is essential to acquire uniform coatings where the added substance is accessible to deliver rapidly to the general climate.
- The development of a uniform surface, as the beads scattering produces coatings with uniform spatial examples and controlled thicknesses, is basic to assess the energy arrival of the antimicrobial added substance.
- Decrease in cost, as spraying procedures are generally quick and productive cycles as far as dissolvable and material utilization (Andrade et al. 2012).

Distinctive spraying procedures have been propounded:

- Air-spray atomization: Here, the liquid rises out of a spout at very low speed which is encircled by a fast stream of compacted air (up to 8 bar). Atomization is caused by the quickening of rubbing among the air particles and fluid that upsets the liquid stream.
- Pressure (airless) atomization: High weights (34–340 bar) power the liquid from a little spout (spray tip) to arise as a sheet. The rubbing among the liquid and the air atoms upsets the stream, breaking it at first into sections and at last into beads. The quick-moving, high-pressure fluid stream gives enough energy to beat the liquid's consistency and surface strain by framing little beads.
- Air-helped airless atomization: This strategy consolidates the highlights of air splashing and airless strategies. It depends on the guideline of the airless atomization with the expansion of a concentrated wind stream to acquire beads in a more controlled manner.

24.6.2 Use of Dipping

The technique has been utilized to frame coatings onto vegetables, organic products, and meat, among other food items. Characteristics, for example, viscosity, thickness, and surface pressure of the covering arrangement, are essential to decide the thickness of the film, as dipping procedures can shape thick covering layers. A membranous film is shaped over the item surface by straightforwardly plunging the item into the fluid covering detailing and additionally air-drying (Lu et al. 2010). This cycle might be isolated into three phases:

- Immersion and staying: The substrate is inundated into the antecedent arrangement at a steady speed followed by staying to guarantee that collaboration of the substrate with the covering arrangement which is enough for complete wetting.
- Deposition: A dainty layer of the forerunner arrangement is framed on the food surface.
- Evaporation: The dissolvable overabundance dissipates from the liquid, shaping the slim film (Dhanapal et al. 2012).

24.6.3 Use of Spreading

Spreading, otherwise called brushing, comprises the regulated suspension spreading onto the material surface to be additionally dried. This strategy is viewed as a legitimate option for fabrication of films with measurements bigger in comparison to those readied by projecting systems (McIver et al. 2020). The covering suspension thickness is constrained by an edge connected to the spreading gadget's lower part. Furthermore, the drying of film is hung by a course of hot air. The strategy can be implemented to the creation of protein and polysaccharide-based films. Two boundaries can be utilized to portray the fluid beads spreading: the rate of spreading and wetting degree. Therefore, contact point estimations are generally employed to assess the level of wettability or spreading on the surface by a specific fluid. Evidently, the spreading is influenced by a few components, for example, the substrate properties (surface unpleasantness and calculation), framework conditions, for example, relative moistness, temperature, and fluid properties (surface pressure, consistency, and thickness). Thickness was observed to demonstrate a significant impact as it characterizes the opposition of fluid to spreading on strong surfaces. Subsequently, the spreading of exceptionally gooey fluids is more troublesome than fluids with low viscosity (Ribeiro et al. 2020).

24.7 Lack of Knowledge and Consumer Affirmation

The exigency of consumers is transforming concerning purchasing and awareness of repercussions of food packaging (Fig. 24.5). Implications about the environment, food safety, sustainability, ethics, product cost, and food quality are eminent factors

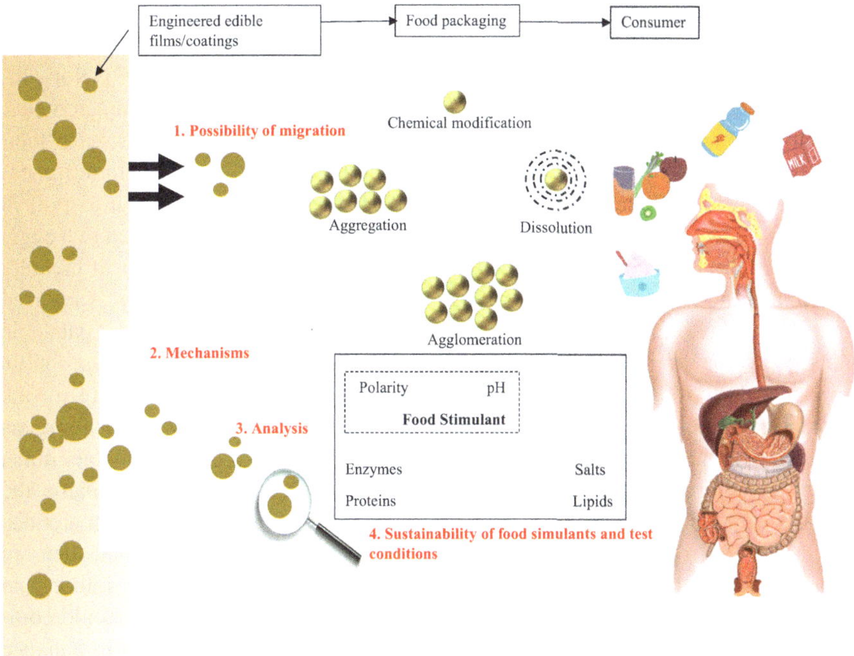

Fig. 24.5 Consumer perceptions and questions about the use of edible packaging in liquid food products

for present-day consumers when buying food products. Besides, changes in exigency of consumers are being enlightened by continual drip feeding of fatalistic information about traditional packaging (Guillard et al. 2018).

Customers acknowledge palatable films just on the off chance that they feel it is protected. Even though analysts guarantee that palatable films/coatings are edible alongside food, just restricted examinations were upheld with lucid proof on edibility as well as biodegradability of palatable films. Ongoing examinations bestow that the film framing technique, film shaping materials, kind of plasticizers, added substances, structural attributes, piece, chain length, subatomic course of action, furthermore, crystallinity file can essentially influence the edibility and digestion rates (Kowalska et al. 2020). Fundamental oils, which are commonly viewed as protected, are commonly utilized in edible films/coatings to enhance the antimicrobial characteristics. While low convergence of fundamental oils is helpful, higher centralization of basic oils may produce genuine toxicological impacts and hypersensitive responses. A study on assessing the edibility and biodegradability and guaranteeing well-being is infrequently observed in the literature, and broad examinations are needed to assess these attributes. The consumer adequacy of eatable films not just relies upon the utilitarian properties, yet additionally different elements, for example, film appearance, organoleptic properties, showcasing, cost, and so forth palatable films, ought not to influence the tangible properties and dietary benefits of the contained food. They ought to seem straightforward, dismal, unscented, dull, polished, and so on, and help food shading, flavor, and centralizations of flavors, corrosive, sugar, or salt (Hasan and Khan 2009). Notwithstanding, much consideration is required to enhance the organoleptic characteristics. The absence of mindfulness and dread about palatable films can decrease their acknowledgment. Blemish keting procedures, for example, directing mindfulness programs, value dabchicks, alluring offers, and notices, may be useful to pull in customers. The creature inferred palatable films may not be acknowledged by veggie lovers and the individuals from religions that do not permit utilization of creature determined items. Individuals, who wish to keep away from the creature's inferred food items, may likewise have worries to utilize eatable films. Moreover, if palatable films comprise any sorts of allergens that may result in unfavorably susceptible responses. Individuals might not acknowledge the palatable films dreading for the existence of such allergens. Appropriate naming of any such realized allergens would aid in ameliorating the consumer acknowledgment of the eatable films. Hence, the administrative bodies ought to underscore food makers to name the necessary data about allergens and the existence of creature determined materials. Purchasers deem the film properties from the individual perspective, and they are improbable to judge the natural or mechanical advantages. Henceforth, rather than featuring the ecological advantages, promoting and item development should comprehend the palatable films from the customers' perspective, and ought to enhance the properties of film to make the palatable films to be received. Various nations follow various guidelines on food bundling materials. These varieties can essentially influence the measure of information needed for deciding if a material is good enough for food bundling application. European Directive and US guidelines order films and coatings that are edible as food items, food fixings, food advertisement additives, food contact substances, or food bundling materials. Since eatable films are contemplated as the vital constituent of food, they should cling to the guidelines identified with food items. Consequently, for the model, all fixings utilized for fabricating an edible film should accomplish for the generally recognized as safe (GRAS) status, according to Food and Drug Organization (FDA) guidelines. The edible packaging materials and additives must be employed following good manufacturing practice (GMP). Agreeing to EU guideline (EC) No. 1935/2004, all food contact materials should meet the accompanying four essential prerequisites: (1) They will not imperil human well-being, (2) will not alter the composition of the food in an unsuitable manner, (3) will not cause any alteration in taste, scent, or surface of the food, and (3) will be made as indicated by great assembling practice. As per this guideline, materials will be approved for food bundling just if no danger to human well-being is confirmed. Be that as it may, security assessment on the incorporation of nanoparticles in food bundling and their toxicological impacts are not referenced obviously (Kouhi et al. 2020; Sason and Nussinovitch 2021).

What's more, every nation may have a distinctive rundown of affirmed materials appropriate for palatable film formation. The material utilized in one nation may not be demonstrated in different nations. Subsequently, the food producers delivering to other nations ought to think about the guidelines of different nations, what's more, the film should be shaped with the materials recorded in the nation's affirmed rundown of food materials. At the point when food makers produce the palatable films, they ought to incorporate all the fixings utilized for the film arrangement on the names of their food items. Notwithstanding, on the off chance that they utilize palatable film created by other providers, the consumable film providers ought to acquire freedom (no complaint authentication) from the approving organizations before utilizing them for bundling with the legitimate marking of materials, nourishing data, and conceivable allergenicity (Umaraw et al. 2020).

Lastly, the price of the food packaging material is a significant driving element for the client's acknowledgment of eatable films. Presently, the expense of edible films is more costly, being 10–50 times higher than the petroleum-derived plastic derivatives. Notwithstanding, as the creation of palatable films is in the turn of events stage, and fewer amounts are delivered, the significant expense of palatable films cannot be taken as a negative point as of now. The complete expense of the bundling should be under 10% of the item cost. Legitimate money-saving advantage examinations are necessary for the justification of the variation of eatable films. The expense of the edible films should be lower than or equivalent to the oil-based plastics to pull in the clients (Jeya et al. 2020).

24.8 The Apprehension of Safety Perspective and Legislations

With the developing utilization of prepared to eat items, pre-cut foods grown from the ground, and insignificantly handled nourishments, there is worry about its security and poisonousness issues. Moreover, well-being relates to general well-being and protection from bioterrorism. Any food-borne illnesses and food quality alterations should be evaded from the natural pecking order. Any guideline on materials concomitant to food ought to guarantee that no compound response generated by grocery contact materials can alter their organoleptic or synthesis qualities. Food added substances consolidated into edible polymers definition should be presented under explicit virtue standards. Plan and creation of natural cordial matcrials is a moderately avant-garde pattern in the arena of consumable polymers. The bio-based plan assists with changing society giving materials in a repeating or helpful way, prompting a roundabout economy. Supplanting the manufactured oil materials with palatable polymers may decrease the aggregate sum of different materials utilized, with a higher measure of recyclable and reusable materials (Trinetta 2016).

In packaging enterprises, planning new materials because of composites of polysaccharides, proteins, and lipids may progress because of their potentiality to conquer the downside of every individual part. Utilization of high-boundary materials for bundling may diminish material taking care of, dissemination, and transportation costs, just as squander. Accommodation in assembling, circulation, transportation, deals, promoting, utilization, and garbage removal is another future pattern for palatable polymers in food bundling. Arising utilizations of bio-based polymers as a gel that can be changed to old food, drinkable jams, 3D-printed nourishments, or even fat mimetics are normal in the years to come. Burgeoning curiosity in the utilization of basic oils as generally perceived as safe (GRAS) segments for food conservation is also observed. Other than giving antimicrobial/ cell reinforcement movement, they can lessen water vapor permeability of palatable coatings and films. Examination, creation, and commercialization of consumable materials for various implications are still developing, in light of interest in investigating new wellsprings of crude common materials, nearly illimitable utilization, specific exercises by requests of purchasers, and changing the current pattern in customary plastics employments (Debeaufort et al. 2000).

24.9 Shortcomings of Edible Packaging

Edible packaging has few limitations owing to the inferior physical properties. For instance, lipid-derived films have excellent moisture barrier properties but poor mechanical strength. The formation of laminated structures can overcome these disadvantages by engineering edible films with several functional layers. As food parts, edible film and coatings, as a rule, must be as bland as conceivable all together not to be recognized during the utilization of the consumable bundled food item. At the point when edible packages have a huge or specific taste and flavor, their sensorial attributes must be viable with those of the food. Additionally, one rule impediment of utilizing edible coating on the new product is the expected advancement of off-flavors if the restraint of O_2 and $C0_2$ trade results in anaerobic respiration. What's more, unfortunate tangible characteristics may create on a few covered items, where nonuniform, white, and tacky surfaces become unattractive to customers.

The major limitations and challenges facing the edible food packaging industry can be summarized as:

1. Unplasticized edible films are fragile and demonstrate mediocre film attributes than oil inferred plastics. Notwithstanding, properties including heat and gas obstruction are tantamount to those of plastics. While the fuse of a plasticizer enhances the film adaptability, the production of multilayers, composites, and nanocomposite films ameliorate the film properties. Additional exploration in the field is as yet needed to deliver exceptional film attributes.
2. Present lab-scale creation of edible films is unsatisfactory for scaling up to mechanical scale because of issues, for example, the powerlessness of making constant films, high energy utilization, long drying time and off base thickness control, and significant expense. Future exploration should deliver these issues to scaling up creation.
3. Most exploration zeroed in on just not many of the film attributes, and numerous different characteristics were not researched. The future palatable film would be multifunctional and viable to the cutting-edge bundling technologies. Notwithstanding, a broad examination is needed on the fundamental examination of the overlooked factors and film properties.

4. Lack of proof on edibility and biodegradability, organoleptic viewpoints, lacking lawful perspectives, the dread of toxicological and well-being impacts, deficient advertising, public mindfulness, and social issues, and so on can influence the shopper acknowledgment of palatable films. Future research on edible films ought to likewise think about these perspectives to improve commercialization achievement.

24.10 Conclusions

Studies have perceived edible packages as a well-being wellspring of food insurance from different components as they are naturally happening, modest, and sustainable. These films are additionally a decent transporter of basic supplements which we consume through our food. As of now, the creation of palatable films is basically at the research center scale and viewed as expensive contrasted and engineered plastic films. Examination on cost decrease and creation on bigger scales are important to advance the attainability of popularized consumable bundling. The practicality of popularized frameworks relies upon the multifaceted nature of the creation cycle, size of venture for film creation or covering gear, possible clashes with traditional food bundling frameworks, and producer protection from the utilization of new materials. Moreover, food makers request a long-time frame of realistic usability for items in interstate or international trade. Palatable bundling materials are themselves naturally powerless to biodegradation, accordingly, their defensive capacities are steady for more limited terms than traditional bundling. Consequently, the steadiness and well-being of consumable bundling under the planned stockpiling/use conditions require examination. Despite the numerous advantages proffered by edible packaging, the major stumbling blocks such as enhanced water permeability, cost-effectiveness, and technical methods should be overcome. The complete application and production are still far away from their complete exploitation.

Acknowledgments The authors would like to thank Bhaskaracharya College of Applied Science, University of Delhi, India.

References

Aguirre-Joya JA, De Leon-Zapata MA, Alvarez-Perez OB, Torres-León C, Nieto-Oropeza DE, Ventura-Sobrevilla JM, Aguilar MA, Ruelas-Chacón X, Rojas R, Ramos-Aguiñaga ME, Aguilar CN (2018) Basic and applied concepts of edible packaging for foods. In: Food packaging and preservation. Academic Press, pp 1–61

Aloui H, Deshmukh AR, Khomlaem C, Kim BS (2021) Novel composite films based on sodium alginate and gallnut extract with enhanced antioxidant, antimicrobial, barrier and mechanical properties. Food Hydrocoll 113:106508

Anderson NM, Benyathiar P, Mishra DK (2020) Aseptic processing and packaging. In: Food safety engineering. Springer, Cham, pp 661–692

Andrade RD, Skurtys O, Osorio FA (2012) Atomizing spray systems for application of edible coatings. Compr Rev Food Sci F 11(3):323–337

Baldwin EA, Nisperos-Carriedo MO, Baker RA (1995) Edible coatings for lightly processed fruits and vegetables. Hort Sci 30(1):35–38
Bhargava N, Sharanagat VS, Mor RS, Kumar K (2020) Active and intelligent biodegradable packaging films using food and food waste-derived bioactive compounds: a review. Trends Food Sci Technol
Bourtoom T (2008) Edible films and coatings: characteristics and properties. Int Food Res J 15(3):237–248
Çakmak H, Özselek Y, Turan OY, Fıratlıgil E, Karbancioğlu-Güler F (2020) Whey protein isolate edible films incorporated with essential oils: antimicrobial activity and barrier properties. Polym Degrad Stab 179:109285
Calva-Estrada SJ, Jiménez-Fernández M, Lugo-Cervantes E (2019) Protein-based films: advances in the development of biomaterials applicable to food packaging. Food Eng Rev 11(2):78–92
Campos CA, Gerschenson LN, Flores SK (2011) Development of edible films and coatings with antimicrobial activity. Food Bioprocess Technol 4(6):849–875
Chen S, Brahma S, Mackay J, Cao C, Aliakbarian B (2020) The role of smart packaging system in food supply chain. J Food Sci 85(3):517–525
Chiralt A, González-Martínez C, Vargas M, Atarés L (2018) Edible films and coatings from proteins. In: Proteins in food processing. Woodhead Publishing, Cambridge, pp 477–500
Debeaufort F, Voilley A (2009) Lipid-based edible films and coatings. In: Edible films and coatings for food applications. Springer, New York, NY, pp 135–168
Debeaufort F, Quezada-Gallo JA, Delporte B, Voilley A (2000) Lipid hydrophobicity and physical state effects on the properties of bilayer edible films. J Membr Sci 180(1):47–55
Dhanapal A, Sasikala P, Rajamani L, Kavitha V, Yazhini G, Banu MS (2012) Edible films from polysaccharides. Food Sci Qual Manag 3:9
Dinika I, Verma DK, Balia R, Utama GL, Patel AR (2020) Potential of cheese whey bioactive proteins and peptides in the development of antimicrobial edible film composite: a review of recent trends. Trends Food Sci Technol 103:57–67
Espitia PJ, Du WX, de Jesús A-BR, Soares ND, McHugh TH (2014) Edible films from pectin: physical-mechanical and antimicrobial properties—a review. Food Hydrocoll 35:287–296
Galus S, Arik Kibar EA, Gniewosz M, Kraśniewska K (2020) Novel materials in the preparation of edible films and coatings—a review. Coatings 10(7):674–687
Guillard V, Gaucel S, Fornaciari C, Angellier-Coussy H, Buche P, Gontard N (2018) The next generation of sustainable food packaging to preserve our environment in a circular economy context. Front Nutr 5(121):1–13
Guo M, Yadav MP, Jin TZ (2017) Antimicrobial edible coatings and films from micro-emulsions and their food applications. Int J Food Microbiol 263:9–16
Haghighi H, Licciardello F, Fava P, Siesler HW, Pulvirenti A (2020) Recent advances on chitosan-based films for sustainable food packaging applications. Food Packag Shelf 26:100551
Hammam AR (2019) Technological, applications, and characteristics of edible films and coatings: a review. SN Appl Sci 1(6):632
Han JH (ed) (2005) Innovations in food packaging. Elsevier, Amsterdam
Hasan SA, Khan MZ (2009) The impact of packaging characteristics on consumer brand preference. SAJMS 3(1):1–10
Hassan B, Chatha SA, Hussain AI, Zia KM, Akhtar N (2018) Recent advances on polysaccharides, lipids and protein based edible films and coatings: a review Int. J Biol Macromol 109:1095–1107
Homez-Jara A, Daza LD, Aguirre DM, Muñoz JA, Solanilla JF, Váquiro HA (2018) Characterization of chitosan edible films obtained with various polymer concentrations and drying temperatures. Int J Biol Macromol 113:1233–1240
Işık H, Dağhan Ş, Gokmen S (2013) A study on edible coatings used in food industry.Electronic J. Food Technol 8(1):26–35
Janjarasskul T, Krochta JM (2010) Edible packaging materials. Annu. Rev. Food Sci 1:415–448

Jeya J, Chandrasekaran M, Venkatesan SP, Sriram V, Britto JG, Mageshwaran G, Durairaj RB (2020) Scaling up difficulties and commercial aspects of edible films for food packaging: a review. Trends Food Sci Technol 100:210–222

Kadzińska J, Janowicz M, Kalisz S, Bryś J, Lenart A (2019) An overview of fruit and vegetable edible packaging materials. Packag Technol Sci 32(10):483–495

Kalpana S, Priyadarshini SR, Leena MM, Moses JA, Anandharamakrishnan C (2019) Intelligent packaging: trends and applications in food systems. Trends Food Sci Technol 93:145–157

Kang HJ, Kim SJ, You YS, Lacroix M, Han J (2013) Inhibitory effect of soy protein coating formulations on walnut (Juglans regia L.) kernels against lipid oxidation. LWT Food Sci Technol 51(1):393–396

Khan MK, Schutyser MA, Schroën K, Boom R (2012) The potential of electrospraying for hydrophobic film coating on foods. J Food Eng 108(3):410–416

Khan MK, Mujawar LH, Schutyser MA, Schroën K, Boom R (2013) Deposition of thin lipid films prepared by electrospraying. Food Bioprocess Technol 6(11):3047–3055

Khan MK, Nazir A, Maan AA (2017) Electrospraying: a novel technique for efficient coating of foods. Food Eng Rev 9(2):112–119

Kouhi M, Prabhakaran MP, Ramakrishna S (2020) Edible polymers: an insight into its application in food, biomedicine and cosmetics. Trends Food Sci Technol

Kowalska M, Janas S, Kudełka W (2020) Consumer preferences and physical evaluation of shortbread stored in different packaging. Int J Consum Stud 00:1–12

Krochta JM (2002) Proteins as raw materials for films and coatings: definitions, current status, and opportunities. Protein Based Films Coat 1:1–40

Lan W, Zhang R, Ji T, Sameen DE, Ahmed S, Qin W, Dai J, He L, Liu Y (2020) Improving nisin production by encapsulated Lactococcus lactis with starch/carboxymethyl cellulose edible films. Carbohydr Polym 251:117062–117072

Li C, Pei J, Xiong X, Xue F (2020) Encapsulation of grapefruit essential oil in emulsion-based edible film prepared by plum (Pruni Domesticae semen) seed protein isolate and gum acacia conjugates. Coatings 10(8):784–802

Liu R, Cong X, Song Y, Wu T, Zhang M (2018) Edible gum–phenolic–lipid incorporated gluten films for food packaging. J Food Sci 83(6):1622–1630

Lu F, Ding Y, Ye X, Liu D (2010) Cinnamon and nisin in alginate–calcium coating maintain quality of fresh northern snakehead fish fillets. LWT Food Sci Technol 43(9):1331–1335

Majid I, Nayik GA, Dar SM, Nanda V (2018) Novel food packaging technologies: innovations and future prospective. J Saudi Soc Agric Sci 17(4):454–462

McIver A, Eydelnant I, Geller B, Jen CK (2020) U.S. Patent No. 10,610,035. Washington, DC: U.S. Patent and Trademark Office

Mohamed SA, El-Sakhawy M, El-Sakhawy MA (2020) Polysaccharides, protein and lipid-based natural edible films in food packaging: a review. Carbohydr Polym 116178

Motelica L, Ficai D, Ficai A, Oprea OC, Kaya DA, Andronescu E (2020) Biodegradable antimicrobial food packaging: trends and perspectives. Foods 9(10):1438–1454

Murty R (2020) Aseptic packaging and labeling. Aseptic pharmaceutical manufacturing II: applications for the 1990s, 337

Pelissari FM, Ferreira DC, Louzada LB, dos Santos F, Corrêa AC, Moreira FK, Mattoso LH (2019) Starch-based edible films and coatings: An eco-friendly alternative for food packaging. In: Starches for food application. Academic Press, New York, pp 359–420

Rahman MS (ed) (2007) Handbook of food preservation. CRC Press, Boca Raton, FL

Ramos ÓL, Santos AC, Leão MV, Pereira JO, Silva SI, Fernandes JC, Franco MI, Pintado ME, Malcata FX (2012) Antimicrobial activity of edible coatings prepared from whey protein isolate and formulated with various antimicrobial agents. Int Dairy J 25(2):132–141

Ribeiro AM, Estevinho BN, Rocha F (2020) Preparation and incorporation of functional ingredients in edible films and coatings. Food Bioprocess Technol 14:209–231

Sanjana MC, Hemegowda R, Sushma RE (2019) Aseptic packaging–a novel technology to the food industry

Sason G, Nussinovitch A (2021) Hydrocolloids for edible films, coatings, and food packaging. In: Handbook of hydrocolloids. Woodhead Publishing, pp 195–235
Suhag R, Kumar N, Petkoska AT, Upadhyay A (2020) Film formation and deposition methods of edible coating on food products: a review. Food Res Int 136:109582–109597
Tkaczewska J (2020) Peptides and protein hydrolysates as food preservatives and bioactive components of edible films and coatings-a review. Trends Food Sci Technol 106:298–311
Trinetta V (2016) Edible packaging
Umaraw P, Munekata PE, Verma AK, Barba FJ, Singh VP, Kumar P, Lorenzo JM (2020) Edible films/coating with tailored properties for active packaging of meat, fish and derived products. Trends Food Sci Technol 98:10–24
Valdés A, Ramos M, Beltrán A, Jiménez A, Garrigós MC (2017) State of the art of antimicrobial edible coatings for food packaging applications. Coatings 7(4):56–79
Vásconez MB, Flores SK, Campos CA, Alvarado J, Gerschenson LN (2009) Antimicrobial activity and physical properties of chitosan–tapioca starch based edible films and coatings. Food Res Int 42(7):762–769
Von Bockelmann BA, Von Bockelmann IL (1986) Aseptic packaging of liquid food products: a literature review. J Agric Food Chem 34(3):384–392
Wang B, Yang C, Wang J, Xia S, Wu Y (2019) Effects of combined pullulan polysaccharide, glycerol, and trehalose on the mechanical properties and the solubility of casted gelatin-soluble edible membranes. J Food Process Preserv 43(1):e13858

Nanoedibles: Recent Trends and Innovations

25

Anu Kumar, Mrinal Samtiya, and Tejpal Dhewa

Abstract

Presently nanotechnology is most widely applicable in industrial environments and rapidly emerging as a promising future technology. Nanotechnology is utilized in food industries in order to elevate quality as well as safety. Several studies have been conducted in many areas such as food processing, food additives, and food packaging. Nanotechnology has attracted much interest toward food and beverage industries also. Edible packaging is nowadays becoming an area of interest as it has a potential to compensate the problems with plastic packaging. Edible packaging is eco-friendly and can improve shelf life of fruits and vegetables. The objective of this review chapter is to explore the recent work on nanoedible films and about the use of nanotechnology for applications in the food packaging process.

Keywords

Nanotechnology · Nanoedibles and food packaging

25.1 Introduction

Storage of food items through food packaging is a cost-effective way and this is one of the best methods to protect the food items from damage during transportation and other environmental factors. It is also a promising approach for the maintenance of

A. Kumar
University Institute of Biotechnology, Chandigarh University, Mohali, Punjab, India

M. Samtiya · T. Dhewa (✉)
Department of Nutrition Biology, Central University of Haryana, Mahenderga, Haryana, India
e-mail: tejpaldhewa@cuh.ac.in

A. Poonia, T. Dhewa (eds.), *Edible Food Packaging*,
https://doi.org/10.1007/978-981-16-2383-7_25

quality of food during the packaging of food and at the time of consumption too (Marsh and Bugusu 2007; Petersen et al. 1999). Plastic packaging has been found as one of the most dominant packaging materials among the above mentioned is because of their high cost-to-performance ratio and light weight (FICCI 2016). During 2003–2009, approximately 54% of plastic packaging were consumed by food and beverages sectors (World Packaging Organization 2008). Very minute quantity of plastics is recovered through recycling; if not recycled, it is burnt and that releases harmful gases leading to air pollution. Dependency on petroleum reserves is another disadvantage with plastics (Hopewell et al. 2009).

Thin layer in edible packaging that is known as film or coating is an inherent component of food products. This thin layer can be wrapped on the surface of food (film) or may be formed directly on the food surface (coating). This is called edible coating or edible packaging. Edible composites and polymers are able to constitute such continuous inseparable structures, and qualify as edible packaging materials (Guilbert et al. 1995). One advantage with this type of packaging is that nothing will be remaining for disposal as such packages can be consumed along with food products. In comparison to synthetic packaging material, degradation of edible packages is easier and more reliable. Because of this, such materials for utilization in edible packaging are becoming a better choice of the researchers nowadays, which has ultimately a remarkable ability to be a remarkable alternative to synthetic as well as biodegradable plastics (Shit and Shah 2014). The food and beverage industries are responding with more interest to explore the potential advantages of nanotechnology. Food products can be prevented from deterioration and their shelf life can be extended with the help of nanotechnology. Nanotechnology-based packaging materials are lighter in weight as compared to synthetic material. This type of material is available with strong barrier properties and it can protect the quality of food during their consumption and transportation; for example, poultry products or meat can be prevented from food spoilage.

Many kinds of packaging materials can be prepared by using nanomaterials of size ranging from 10 to 100 nm. These can be incorporated into different kinds of packaging materials, and effective and more convenient packing materials with improved mechanical strength or barrier properties can be produced (Mellinas et al. 2016). Selection of right nanomaterials for incorporating into packaging material is an important matter of concern and it can offer various benefits, such as thermal stability, mechanical strength, and better optical properties, in comparison to conventional materials used for packaging purpose (Cerqueira et al. 2018). Some related studies in the area of food packaging have been conducted earlier by many researchers (Vasile 2018). Overview of applications of nanotechnology in food packaging is schematically shown below (Fig. 25.1).

This chapter is mainly focused on exploring the use of nanosystems in edible packaging and to conclude the different significantly related studies on nanoedible coatings in the recent period in order to solve the problems associated with such coatings, which could lead to better opportunities for future research.

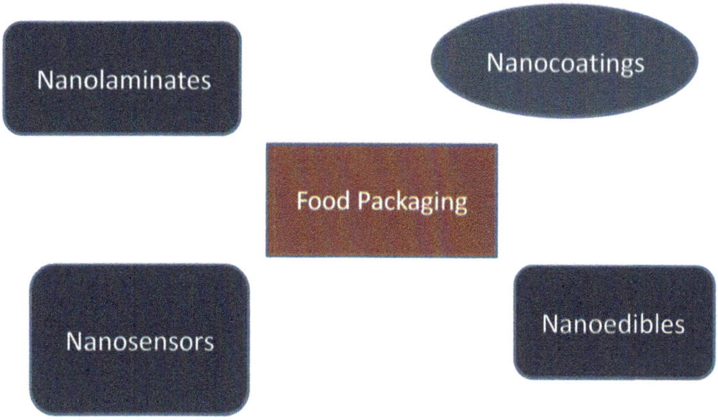

Fig. 25.1 Overview of applications of nanotechnology in food packaging

25.2 Basic Components of Edible Coatings

Various examples of nanosystems, which are evident in the advances made in the incorporation of ingredients as a component of food products, lead to the possibility of exploring more insights into edible coatings including the functional modifications from time to time. By combining two or more materials in which at least one should be based on nanometric scale which can ultimately improve the properties of a nanosystem (Mallakpour and Sadaty 2016). For example, nanocomposites, has enables to be used as "temporal distribution systems". Such systems release bioactive substances from coating or film to the food, to improve conservation (Liu et al. 2017).

25.2.1 Nanoemulsion

Nanoemulsions are made up of emulsifying molecules that basically consist of oil in aqueous phase, in which a thin interfacial layer surrounds an oil drop. Nanoemulsions are having particle size in the range of 50–500 nm (Ranjan et al. 2014). Out of the two kinds of nanoemulsions that are available, for example, oil/water (o/w) or water/oil (w/o), o/w systems are of much interest and it can permit the incorporation of various lipophilic components. This can enhance the possibility of addition of antibacterial, antifungal, or antioxidant potential into a polymeric matrix of hydrophilic nature. Bioactivity extension can only be possible by minimizing the size of oil drops to nanoscale range. This is one of the greatest advantages of utilizing such nano-based systems. In the present scenario, researchers are attracted toward the use of nanoedibles in food packaging. Area of research to explore the use of nanoemulsions in the formulation of the edible coatings is becoming popular just

Table 25.1 Potential application of nanoemulsions as edible coatings

Bioactive substance	Biopolymer matrix	Food product	Potential	Findings
α-Tocopherol	Nopal mucilage	Fresh-cut apples	Shelf-life enhancer and enzymatic activity	Zambrano-Zaragoza et al. (2016)
Essential oil (oregano)	Mandarin fiber	Cheese of low fat	Antimicrobial	Artiga-Artigas et al. (2017)
Lemongrass oil	Chitosan	Grape berry	Antimicrobial and antioxidant	Oh et al. (2017)
Clove bud and oregano essential oils	Methylcellulose	Sliced bread	Antimicrobial and shelf-life extender	Otoni et al. (2014a, b)
Cinnamaldehyde	Pectin (low and high methyl ester)	Edible films (in vitro)	Antimicrobial	Otoni et al. (2014a, b)
Carvacrol	Chitosan	Cucumber	Antimicrobial	Tastan et al. (2017)
Carvacrol	Chitosan	Cabbage	Antimicrobial	Sow et al. (2017)

because of their characteristics. Also, the inclusion in a polymeric matrix in the range of 100–1000 nm allows sustained control over the release of active ingredients (McClements 2015). Table 25.1 shows a list of functional bioactive substances and applications of nanoemulsions in the formulation of edible coatings in different foods.

25.2.2 Polymeric Nanoparticles

Nanocapsules and nanospheres are two well-studied examples of polymeric nanoparticles that are formed by a dense polymeric matrix (Yousuf et al. 2018). After the polymerization reaction, release of some toxic compounds was reported in polymeric nanoparticles, for example, monomers, oligomers, and initiators. Due to this reason, the concerned technique is usually avoided. More interestingly, polymeric nanoparticles were also reported in the pharmaceutical field for their utilization in targeted drug delivery systems. Such systems basically can be utilized in order to protect the food products such as dairy products, bakery products, and nutraceuticals from degradation. Delivery and controlled release of nutraceuticals across edible coatings improved. Also such systems can enhance the shelf life of food products if explored in the preparation of edible packaging. Through this, antioxidants or antimicrobial components can also be encapsulated into polymeric nanoparticles so that active edible coatings can be achieved. Edible coatings are packing systems that can include polymeric nanoparticles and ensure the well dispersion over the surface of treated food (Zaragoza et al. 2018). One earlier, similar study on hybrid nanodispersion including curcumin-loaded, lipid-polymer, hybrid, nanoparticles depicted that such edible coatings can enhance the freshness of fruits and vegetables. Researchers revealed that synthesized nanoparticles showed an excellent antitumor activity and nontoxicity. Such nanosystems were found better candidates for edible coating and employed for fresh apples and tomatoes (Joshy et al. 2020).

25.2.3 Nanotubes

Nanotubes and nanofibers have been regarded as potential carriers of antimicrobials or antioxidants because of their utility in edible coatings. Main function of these nanoedibles is to improve controlled release and in the protection of fresh food. These nanosystems also assisted in safe transportation of substances during food storage and also in the improvement of function during consumption time (Zaragoza et al. 2018).

Carbon nanotubes are most widely used but basically in development of containers. One advantage of these nanotubes is that these are of nanosizes, that is, 8 nm in diameter; and different active materials can be encapsulated into food by utilizing these nanosystems. Also, these can be incorporated into polymeric matrices so that their tensile strength and elasticity can be improved. Controlled release through coatings is possible by encapsulating the material on the support polymer used along with nanotubes (Ramos et al. 2017).

25.2.4 Nanofibers

This is another choice among nanosystems as the best components of edible coatings. These have been well explored in food preservation because of their potential in the encapsulation of antioxidants and antimicrobial substances, so that this approach can enhance the quality, safety, and efficacy of food. Nanofibers are known as fibrous scaffolds available in the range of 100–500 nm and have been utilized in the immobilization of enzymes. These can also improve the properties of the film or coatings so that better encapsulation of active ingredients can be possible. In the upcoming future, this seems to be an alternative for the development of edible film or coatings (Fahami and Fathi 2018).

25.2.5 Nanolaminate

Food scientists are nowadays attracted toward a novel research area, that is, utilizing nano-based films. Novel laminate films can be created using nanotechnology approaches and such films are appropriate for their utilization in food industries. There are many advantages of utilizing nanolaminates, which can open various ways for food scientists to use them in various sectors within food industries. Edible films and coatings are employed on a broad variety of foods including French fries, apples, tomatoes, meats, bakery products, etc. (Assis et al. 2017). These coatings or films could be a better gas, moisture, or lipid barrier. Moreover, these could modify the texture and other relevant properties of foods or act as coloring and flavoring agent. These coatings also serve as a source of nutrition, antioxidants, and antimicrobials, and these properties depend upon the quality of material used for film preparation. Film-forming materials must be a good moisture barrier and must be resistant to gas transfer. Generally, both lipid-based films and biopolymer-based films are of good choice and are good moisture barriers, but out of these two former is a good oxygen carrier. Comparatively lipids-based films offer very minute resistance to gas transfer and have improper mechanical characteristics (Zaragoza et al. 2018). Extremely thin layer of nanolaminate makes them more suitable to be used in direct coating on the surface of food products (Brum et al. 2017). Packaging involves natural antimicrobial agents is a new promising approach for the fulfillment of demands of consumers. This packaging is known as active packaging (AP) and is a better candidate for upgrading the shelf life of food products (Bahrami et al. 2020).

25.3 Conclusions

Food with potential characteristics is in more demand. Nanotechnology has the potential to make food healthier, tastier, and, in fact, more nutritious. Texture and flavor of food can be enhanced through nanotechnology. Nanotechnology can be used to reduce fat content and in the encapsulation of nutrients, for example, vitamins, so that their shelf life can be improved in order to prevent their degradation. In addition to this, freshness of the product can also be enhanced by the use of nano-based packaging. Many types of nanomaterials can be used to achieve this purpose. Incorporating nanosensors into packaging process can keep consumers updated with information on the present status of the food inside. Nanoparticles-based food packaging can alert consumers regarding the safety of a product. When a food product is no longer safe for eating purpose, the nanosensors can alert consumers. Sensors can warn us before rotting food and simultaneously we can also get information related to nutritional value of food products.

References

Artiga-Artigas M, Acevedo-Fani A, Martín-Belloso O (2017) Improving the shelf life of low-fat cut cheese using nanoemulsion-based edible coatings containing oregano essential oil and mandarin fiber. Food Control 76:1–12

Assis RQ, Lopes SM, Costa TMH, Flôres SH, de Oliveira RA (2017) Active biodegradable cassava starch films incorporated lycopene nanocapsules. Ind Crop Prod 109:818–827

Bahrami A, Delshadi R, Assadpour E, Mahdi Jafari S, Williams L (2020) Antimicrobial-loaded nanocarriers for food packaging applications. Adv Colloid Interf Sci 278:102140

Brum AAS, dos Santos PP, da Silva MM, Paese K, Guterres SS, Costa TMH, Pohlmann AR, Jablonski A, Flôres SH, de Oliveira RA (2017) Lutein-loaded lipid-core nanocapsules:

physicochemical characterization and stability evaluation. Colloids Surf A Physicochem Eng Asp 522:477–484
Cerqueira MA, Vicente AA, Pastrana LM (2018) Nanotechnology in food packaging: opportunities and challenges. Nanomater Food Packag. https://doi.org/10.1016/b978-0-323-51271-8.00001-2
Fahami A, Fathi M (2018) Fabrication and characterization of novel nanofibers from cress seed mucilage for food applications. J Appl Polym Sci 135:45811–45816
Federation of Indian Chambers of Commerce and Industry (FICCI) (2016) A report on plastic industry. In: 2nd national conference on plastic packaging—the sustainable choice, pp. 1–32
Guilbert S, Gontard N, Cuq B (1995) Technology and applications of edible protective films. Packag Technol Sci 8:339–346
Hopewell J, Dvorak R, Kosior E (2009) Plastics recycling: challenges and opportunities. Philos Trans R Soc 364:2115–2126
Joshy KS, Augustine R, Li T, Snigdha S, Hasan A, Komalan C, Kalarikkal N, Thomas S (2020) Carboxymethylcellulose hybrid nanodispersions for edible coatings with potential anti-cancer properties. Int J Biol Macromol 157:350–358
Liu R, Liu D, Liu Y, Song Y, Wu T, Zhang M (2017) Using soy protein SiOx nanocomposite film coating to extend the shelf life of apple fruit. Int J Food Sci Technol 52:2018–2030
Mallakpour S, Sadaty MA (2016) Thiamine hydrochloride (vitamin B1) as modifier agent for TiO2 nanoparticles and the optical, mechanical, and thermal properties of poly(vinyl chloride) composite films. RSC Adv 6:92596–92604
Marsh K, Bugusu B (2007) Food packaging—roles, materials, and environmental issues. J Food Sci 72:39–55
McClements DJ (2015) Food emulsions: principles, practices, and techniques. CRC Press, Boca Raton, FL
Mellinas C, Valdés A, Ramos M, Burgos N, Garrigos MD, Jiménez A (2016) Active edible films. Current state and future trends. J Appl Polym Sci 133:1–15
Oh YA, Oh YJ, Song AY, Won JS, Song KB, Min SC (2017) Comparison of effectiveness of edible coatings using emulsions containing lemongrass oil of different size droplets on grape berry safety and preservation. LWT Food Sci Technol 75:742–750
Otoni CG, de Moura MR, Aouada FA, Camilloto GP, Cruz RS, Lorevice MV, Soares NDF, Mattoso LH (2014a) Antimicrobial and physical-mechanical properties of pectin/papaya puree/cinnamaldehyde nanoemulsion edible composite films. Food Hydrocoll 41:188–194
Otoni CG, Pontes SFO, Medeiros EAA, Soares NDF (2014b) Edible films from methylcellulose and nanoemulsions of clove bud (Syzygium aromaticum) and oregano (Origanum vulgare) essential oils as shelf life extenders for sliced bread. J Agric Food Chem 62:5214–5219
Petersen K, Nielsen PV, Bertelsen G, Lawther M, Olsen MB, Nilsson NH, Mortensen G (1999) Potential of biobased materials for food packaging. Trends Food Sci Technol 10:52–68
Ramos OL, Percira RN, Martins A, Rodrigues R, Fuciños C, Teixeira JA, Pastrana L, Malcata FX, Vicente AA (2017) Design of whey protein nanostructures for incorporation and release of nutraceutical compounds in food. Crit Rev Food Sci Nutr 57:1377–1393
Ranjan S, Dasgupta N, Chakraborty AR, Melvin Samuel S, Ramalingam C, Shanker R, Kumar A (2014) Nanoscience and nanotechnologies in food industries: opportunities and research trends. J Nanopart Res 16:2464–2468
Shit SC, Shah PM (2014) Edible polymers: challenges and opportunities. J Polym Article ID 427259:1–13
Sow LC, Tirtawinata F, Yang H, Shao Q, Wang S (2017) Carvacrol nanoemulsion combined with acid electrolysed water to inactivate bacteria, yeast in vitro and native microflora on shredded cabbages. Food Control 76:88–95
Tastan Ö, Pataro G, Donsì F, Ferrari G, Baysal T (2017) Decontamination of fresh-cut cucumber slices by a combination of a modified chitosan coating containing carvacrol nanoemulsions and pulsed light. Int J Food Microbiol 260:75–80
Vasile C (2018) Polymeric nanocomposites and nanocoatings for food packaging: a review. Materials 11:1–49. https://doi.org/10.3390/ma11101834

World Packaging Organization (2008) Market statistics and future trends in global packaging. https://www.worldpackaging.org, pp 1–44. Accessed 7 Feb 2019

Yousuf B, Qadri OS, Srivastava AK (2018) Recent developments in shelf-life extension of fresh-cut fruits and vegetables by application of different edible coatings: a review. LWT Food Sci Technol 89:198–209

Zambrano-Zaragoza ML, Gutiérrez-Cortez E, Del Real A, González-Reza RM, Galindo-Pérez MJ, Quintanar-Guerrero D (2016) Fresh-cut red delicious apples coating using tocopherol/mucilage nanoemulsion: effect of coating on polyphenol oxidase and pectin methylesterase activities. Food Res Int 62:974–983

Zaragoza Z, María L, González-Reza R, Mendoza-Muñoz N, Miranda-Linares V, Tania F, Couoh B, Mendoza-Elvira S, Quintanar-Guerrero D (2018) Nanosystems in edible coatings: a novel strategy for food preservation. Int J Mol Sci 19(3):705

26 Edible Packaging: A Vehicle for Functional Bioactive Compounds

Vandana Chaudhary, Neha Thakur, and Suman Bishnoi

Abstract

Packaging plays a very important part in the food supply chain as well as it is a consequential section of the food manufacturing process. Edible packaging is a swiftly emerging art of science in which edible biopolymers like lipids, polysaccharides, proteins, resins etc., and other consumable constituents extracted from various non-conventional sources are used alone or imbibed together. These edible packagings are indispensable and are meant to be consumed with the food. These are biodegradable, environment friendly or sustainable in nature. The physical and chemical properties of the components influence the properties of the edible packaging. Commonly edible packaging includes films, coatings, edible containers or biodegradable films. Edible packaging can also be employed as a carrier of various antioxidants and antimicrobial compounds, flavour or aroma blends or nutritive additives because of their distinctive properties, such as their capability to provide protection to food products (as a result of their mechanical and barrier properties), elevation of sensorial characteristics, sustained and controlled release of functional ingredients, and controlled mass transfer. Edible packaging can also serve as a carrier of certain drugs and medications subjected to regulations.

V. Chaudhary (✉)
College of Dairy Science and Technology, Lala Lajpat Rai University of Veterinary and Animal Sciences, Hisar, Haryana, India

N. Thakur · S. Bishnoi
Department of Livestock Product Technology, Lala Lajpat Rai University of Veterinary and Animal Sciences, Hisar, Haryana, India

A. Poonia, T. Dhewa (eds.), *Edible Food Packaging*,
https://doi.org/10.1007/978-981-16-2383-7_26

Keywords

Bioactive compounds · Functional · Edible packaging · Mechanical properties · Active packaging

26.1 Introduction

Packaging plays a very imminent part in the food supply chain as well as it is a consequential section of the process of food manufacturing. Edible packaging is a swiftly emerging art of science in which edible biopolymers like lipids, polysaccharides, proteins, resins etc., and other consumable constituents extracted from various non-conventional sources are used alone or imbibed together. This may be applied directly to food or applied to it in the form of a film or coating without altering the originality of food or its products. These edible packagings are indispensable and are meant to be consumed with the food. These are biodegradable, environment friendly or sustainable in nature. For the films to be edible in nature, the film-forming process should involve processes like alteration of pH, warming, addition of salt, modifications carried out by enzymes or other food grade processes. It is very essential to monitor the fabrication process, as any change associated with it may lead to a change in the mechanism of reaction or its kinetics.

Usefulness of edible packaging usually relies on its capacity to retain the quality of the product being packed, its ability to enhance the storage life and its contribution to the economic efficiency of packaging materials. Edible packaging helps to achieve superior quality of products and also helps in protection against biological, chemical and physical deterioration. By applying edible coatings or films on food products, physical properties of the food like strength can be enhanced, which also helps in improving the aggregation of particles and appearance. In addition, it acts as a barrier against moisture, oil, oxygen, microorganisms and light, thereby enhancing shelf life and ensuring food safety.

The physicochemical features of the components influence the properties of edible packaging. Edible packaging with antimicrobial components had led to the development of the hypothesis of active packaging. This helps in reduction, inhibition or complete arrest of growth of microorganisms on the surface of food. A number of essential oils from spices like cloves, cinnamon, coriander, etc.; vegetables such as garlic, onion and citrus fruits having antimicrobial properties can be incorporated into edible packaging to serve the purpose. Other naturally occurring substances having antimicrobial activity include polypeptides and chitosan. Antioxidants may also be used in edible packaging, so as to provide protection against oxidative changes in food like rancidity, discolouration and degradation. Acidic compounds like ascorbic acid, citric acid or phenolic compounds such as butylated hydroxytoluene (BHT) and butylated hydroxyanisole (BHA) serve as antioxidants.

However, the usage of edible packaging may be constrained due to issues pertaining to consumer acceptance, safety issues, religious and cultural restrictions.

26.2 Functional Properties of Edible Packaging

26.2.1 Biodegradable and Edible

The utmost important functional properties of edible packaging are that it is edible and biodegradable. To ensure that the packaging is edible, all the constituents of packaging should be food grade, biodegradable in nature and environment friendly. The use of solvents must be limited to ethanol and water (Zaritzky 2011).

26.2.2 Mechanical Properties

Edible packaging should provide strength to withstand transport or handling. It should protect food products against pressure, physical impacts, vibrations, etc. A number of mechanical tests like tensile strength, elongation, tear resistance, puncture resistance, impact resistance, burst resistance, adhesion resistance, stiffness, compression strength and others may be used to check the mechanical strength of the edible packagings. Mechanical properties depend upon the kind of ingredients used to prepare the film, structural cohesion, further controlled by the kind and amount of film producing constituents used, film manufacturing process and the experimental conditions. For instance, alginate films show lower water barrier properties but have good tensile strength (Gheorghita Puscaselu et al. 2020). Protein-based packaging has reduced tensile strength than polysaccharide-based films. Protein- and polysaccharide-based packaging material is brittle, fragile and vulnerable to crack because of strong bonding between the two polymers. In order to solve this obstacle, plasticizers are frequently supplemented to escalate the elasticity and flexibility of the edible packaging material, as these supplements minimize the intermolecular hydrogen bonding that expands the spacing in between the biopolymer chain. In an experiment by Saberi et al. 2017, it was confirmed that tensile strength and Young's modulus decreased and elongation at break increased with rise in molecular weight of plasticizers and relative humidity (RH) for all the studied formulations. Alvarado et al. 2015 examined the outcome of relative humidity on the tensile strength of composite films made by fish gelatin, chitosan and microbial transglutaminase. They found that with escalation in relative humidity, the tensile strength showed a downtrend and an increase in deformation. Hence, it can be deduced that the laboratory and environmental conditions have a direct impact on strength p characteristics of the packaging material.

26.2.3 Barrier Properties

Regulation of mass transfer in between the packed product and the outer environment is a paramount function of the packaging material. There are various factors that affect the quality of packed food like loss of flavour and colour, interaction of packaging material with food, loss or gain of moisture and so on.

Permeability of films is an important criterion during the selection of raw materials for edible packaging. In general, edible packaging materials are quiet sensitive to changes in moisture. As the relative humidity is high, their physical strength de-escalates because the moisture absorbed by the packaging material functions like a plasticizer (Han 2014).

Polysaccharide-based packaging is designated by poor moisture and gas barrier properties. Protein-based packaging is considered to be recognized for the proteins' ability to form a film, as polysaccharides have superb mechanical as well as barrier properties, exception being they possess reduced moisture resistance characteristics. Edible packaging having protein as the main constituent has excellent mechanical as well as barrier properties, contrary to oil, aroma and oxygen. But they show limited resistance to water vapours (Raajeswari and Pragatheeswari 2019). Proteins are not completely hydrophobic and they primarily contain amino acid residues having high water affinity, thus restraining their moisture migration properties. Edible packaging based on lipids has superb moisture and oxygen barrier properties. A study by Santosa et al. (2017) proposed that with increase in the quantity of wax it leads to a significant improvement in its barrier characteristics. It was noted that with 30% wax, there was 63% reduction in water vapour transmission rate (WVTR). This decline in WVTR may be attributed to surge in contact angle with the addition of wax, illustrating the transformation of the hydrophilic to the hydrophobic feature of the biofilm (Santosa et al. 2017; Goslinska and Heinrich 2019).

Oxygen permeability is the second most widely reviewed feature of edible biopolymer films after moisture transmission. Most of the food degradation is a result of fat degradation or lipid oxidation alike browning observed in cut fruits and myoglobin discolouration in meat. By choosing edible packaging with low oxygen permeability, we will be able to retain quality together with enhancing the shelf life of the food product. Recent advances in edible packaging with established gas permeability under specified keeping situation can be helpful in creating a customized atmosphere, thereby curbing the rate of respiration of fruits and vegetables or the ethylene gas production of physiologically active climacteric produce in the course of storage and marketing (Janjarasskul and Krochta 2010).

Protein-based edible packaging appears to have reduced oxygen accessibility, as compared to non-ionic polysaccharide-based packaging. It can be attributed to their polar nature and aliphatic (non-ring) structure, contributing to greater cohesive energy density and small free volume. It was noticed by McHugh and Krochta 1994 that milk proteins at low to moderate relative humidity (RH) showcased as excellent gas barriers.

Oxygen transmission of whey protein isolate (WPI) packaging exhibited an exponential effect on RH, thus revealing that WPI-based packaging has a better potential in preventing oxygen migration into a food system which is kept at intermediate or low RH. One more study specified that WPI-based packagings have exceptional gas-resistant property; however, they are impacted by the relative humidity that affects the opposition of oxygen and carbon dioxide gas. As the relative humidity drops, defiance to the transfer of gas enhances. It was perceived that at lower RH, the oxygen concentration diminishes and carbon dioxide gets enhanced in fruits with edible coatings. When the relative humidity values ranged from 70 to 80%, anaerobic metabolism was activated because of the low levels of oxygen (Zevallos and Krochta 2003).

In order to avoid the loss of volatiles or the imbibition of external flavour or aroma during the period of storage, aroma barrier properties of the edible packaging are very prominent. When the affinity of packaging materials is low for the aroma volatile compounds and lower diffusivity is observed, the aroma barrier efficacy of the packaging is optimized. Hence, the values of diffusivity, permeability and solubility of the edible packaging material should be inadvertently low.

26.2.4 Shelf Life Enhancement

Increasing the consumer awareness about convenient, safe and stable foods in addition to the adverse effect of non-biodegradable packaging materials has reinforced the interest of researchers in edible packaging. Edible packaging forms a protective layer around the food, thus enhancing shelf life of the food while sustaining its sensory as well as nutritional quality. It was concluded by Mali and Grossman in 2003 that shelf life of strawberries could be extended to 21 days by affixing yam starch films. Reduction in spoilage microorganisms was also reported (Mali and Grossmann 2003). Similar trends were addressed by Mahfoudhi and Hamdi 2015 who used an edible coating made from gum arabic and almond gum on cherries. They stated that the coated fruits exhibited delayed ripening that extended shelf life with no off-flavour development.

26.2.5 Food Appearance Enhancer

Tangible properties such as smooth surface, tenacious/non-greasy surface and sensory attributes including visual characters likc colour and glossiness can be refined by the application of edible packaging. In order to give a glossy finish on fruits, a lipid-based edible packaging incorporated with shellac and wax can be a good alternative. Maize protein zein and ethanol-based shellac packagings are commonly used polishing agents on candy and confectionary items (Krochta 2002).

Edible packaging formed from methylcellulose is observed to have very low fat permeability, even less than plastics, and is capable of minimizing fat transfer, thereby restraining fat bloom in chocolates (Debeaufort et al. 1998). Bilayer films made of hydrocolloid possessing film-forming ability and wax that has enhanced adhesiveness were applied on chewing gums to lengthen the storage and to depreciate the loss of moisture. This formulation was created and patented by Wrighley JR Company (Meyers 1994).

26.2.6 Carrier of Active Substances

The potentiality of edible packaging material to act as carrier of active substances and the controlled release of these compounds is a very promising field of science which is attracting a lot of attention from food and pharmaceutical manufacturers.

Edible packaging should be conceived in such a way that it should control the pace of release of encompassing active substances to accomplish maximum adequacy of the functions of active compounds. It is also essential to take into consideration the interaction between the active compounds and the packaging constituents as well as with the environment. Many researchers have devised edible packaging material having antimicrobial and antioxidant properties (Lopez et al. 2014; Bie et al. 2013; Souza et al. 2013; Jiménez et al. 2013; Acevedo-Fani et al. 2015; Gómez-Estaca et al. 2014).

26.3 Active Packaging—Embodiment of the Active Substances within Edible Packaging

26.3.1 Antioxidants

One of the major issues influencing the quality of food is fat oxidation. Synthetic antioxidants, like butylated hydroxytoluene (BHT), tertiary butylhydroquinone (TBHQ), propyl gallate (PG) or butylated hydroxyanisole, are used generally to prevent food from oxidation. During recent times, considering the increasing awareness among consumers and their demand for less use of chemicals and minimally processed foods, major consideration is being paid to find natural antioxidants. As a result, the food products preserved with natural importance are getting popular with masses.

Some of the important natural antioxidants are tocopherols, essential oils, volatile phenols, flavanols, and organic acids and so on. Scientists have explored the incorporation of natural active substances into a packaging material (Tian et al. 2013). The imbibition of antioxidants in edible packaging helps to improve the storage life of a food product by protecting it against oxidation, oxidative rancidity and colour distortion.

During the making process of edible packaging, a judicious selection of active compounds is very crucial. In contemplation to accomplish homogeneous dispensation of antioxidants, the natural antioxidants and packaging constituents should be compatible with each other. However, some disadvantages may be associated while incorporating these compounds in edible packaging like susceptibility to increased temperature and light, high volatility, restricted solubility and displeasing taste, and these attributes result in limitation on their application part.

For example, there are certain phenols and flavones which have restricted mixibility in fat-based formulations (Viskupicova et al. 2010). There are also certain polyphenols (thymol, quercetin, eugenol) and carotenoids, which offer astringent and obnoxious taste. One natural oxidant named ferulic acid is volatile at raised temperature and hence cannot show its influence at elevated temperatures (Nystrom et al. 2007). Likewise vitamins E, C and K show high sensitivity to ultraviolet radiations, thereby limiting their usage in food products under these conditions (Durand et al. 2010).

26.3.2 Antimicrobials

The biggest challenge in food processing industry is the restricted shelf stability of the food products due to microbial spoilage. An edible packaging with assimilated antimicrobial substances aimed to subjugate surface contamination of food products, to increase safety from microorganisms and to enhance keeping quality is a new alternative for the food industry. The natural antimicrobial compounds used in edible packaging can be assorted as animal- or plant based like bacteriocins, enzymes and plant extracts.

Essential oils derived from plants and extracts had been used from a long time as food supplements not only to augment taste but also to enhance storage life by restricting oxidative rancidity and managing microbial spoilage. These oils are actually able to minimize or control the growth of spoilage microorganisms because of higher levels of secondary metabolites, chiefly phenols, terpenes, aldehydes, ketones and iso-flavanols along with others (Tiwari et al. 2009).

Several studies have exhibited that polysaccharide-based edible packagings can be utilized as conveyor of natural antimicrobial agents, predominantly plant-based extracts to complement the quality and microbial stability of fresh fruits and vegetables (Vieira et al. 2016; Gol et al. 2013; Jin et al. 2016; Treviño-Garza et al. 2015).

Falagán et al. (2018) also narrated that aggregation of alginate-based edible coating and natural antimicrobial compounds helped in preserving the quality characteristics of fresh-cut peaches, covering either sensory, microbiological or functional aspects. It was displayed by Postma et al. 2009 that when chitosan and *Lysobacter enzymogenes* were used collectively in cucumber in edible packaging, it exhibited improved control in the development of *Pythium aphanidermatum* in cucumber.

Meat and meat products are very rich source of nutrients, providing high biological value (BV) proteins, and are also a very good supplier of essential amino acids. But these products are more vulnerable to the attack of spoilage microorganisms being nutrient dense, thereby posing a great health risk for consumers. Edible packaging with incorporated antimicrobial agents has been scrutinized by the meat processing industry for being able to enhance microbial stability and increase the storage life of products by inhibiting moisture and oxygen transfer, by reducing microbial spoilage and by decreasing colour losses.

Thyme and propionic acid at 0.5%, w/w have been found to be effective antimicrobial agents after being imbibed into edible packaging based on sorbitol and sodium alginate. It was observed that this combination improved the shelf stability of fresh chicken from *E. coli* and *L. innocua* (Matiacevich et al. 2015). It was revealed that essential oil derived from oregano when used at 1.5% v/v reduced the total viable count by 2 log cycles in smoked salmon after 28 days of storage in a refrigerator.

In addition, galactomannan-based edible packaging impregnated with nisin was found to minimize post-contamination of cheese products by *L. monocytogenes* during storage (Martins et al. 2010). When natamycin was introduced in tapioca starch edible packaging, it helped in limiting the proliferation of *S. cerevisiae* during storage of Port Salut cheese (Ollé Resa et al. 2014). These perceptions affirmed the findings determined by Ture et al. 2011 that when natamycin is added to gluten (wheat)-based packaging, it almost stopped the surge of *A. niger* on the surface of fresh kashar cheese.

26.3.3 Antibrowning Agents

Colour is considered to be an important parameter in relation to quality. Minimally processed fruits and vegetables undergo browning when kept for certain period of time, making them consumer nonacceptable. The main reason behind this phenomenon is enzymatic or non-enzymatic browning as a result of oxidation of phenolic compounds. During minimal processing of fruits and vegetables, internal cellular structure of the food is disrupted, thus increasing the rate of browning reaction. To initiate browning reaction, polyphenol oxidase (PPO) requires oxygen, therefore choosing edible packaging offering excellent oxygen barrier properties will do the needful. In a research study by Oms-Oliu et al. (2008), it was reported that the edible packaging with N-acetylcysteine and glutathione in polysaccharide-based packaging helped in averting browning in fresh-cut pears for a period of 2 weeks without altering the textural properties of the fruit.

Non-enzymatic-assisted browning can be repressed in peeled and blanched potatoes by coating them with a solution comprising of gums, starch, gelatin and calcium chloride (Mazza and Qi 1991). Indeed, coatings based on corn protein zein when applied on tomatoes or sucrose polyester "semperfresh" when coated on apples postponed colour, weight and textural changes (Park et al. 1994).

26.3.4 Nutraceuticals

Edible films can act as an excellent transport vehicle of nutraceuticals. For the incorporation of nutraceuticals in packaging, basic properties like mechanical and barrier properties should be studied carefully.

A study in 2003 by Mei and Zhao appraised the viability of casein-based edible packaging to convey increased amount of vitamin E and calcium. It was inferred from the results that both calcium caseinate and whey protein isolate based packagings were successful in transporting the calcium and vitamin E but it led to a reduction in mechanical strength of the film. Films containing blackberry pulp developed by Gutierrez 2017 displayed even, unwrinkled and condensed morphological characters, which may be attributed to the lower in vitro digestibility rate and more concentration of resistant starch. Moreover, the film produced displayed higher anti-inflammatory activity, higher cell viability and approval by a group of panelists, signalling potential health benefits for persons with special dietary requirements like having high sugar levels, being overweight and having celiac disease.

Calcium and vitamin E were added to chitosan-based edible packaging (Han et al. 2004) with an aim to enhance keeping the quality as well as the nutritional value of fresh and frozen strawberries and red raspberries. Chitosan-based packaging served as a good carrier of increased amount of calcium and vitamin E, thus remarkably improving the nutrient content of the two in fresh as well as frozen fruit. Iron-fortified rice premix (IFRP) was prepared by Mridula and Jha (2014) using soaking and spraying process, accompanied by layering it with hydroxypropyl methylcellulose (HPMC), methyl cellulose (MC), HPMC and MC mixture, zein, palmitic acid (PA) and stearic acid (SA). Thiamine- and niacin-fortified rice were developed by Chatiyanont and Wuttijumnong 2007 by using pectin-based packaging. Vitamin E content of baby carrots could be enhanced by xanthan gum based packaging containing alpha-tocopherol acetate (Mei et al. 2002).

26.3.5 Probiotics

Nowadays, with the consumer being more aware about their health and well-being, the consumption pattern of foods having probiotics is increasing. In order to confer health benefits and get the desired outcome from probiotics, a dosage of 10^8–10^9 viable cells is usually recommended. It is quiet challenging to meet the desirable dose of probiotics due to their sensitivity to the change in environmental conditions. The imbibition of probiotics into edible films was first exercised by Tapia et al. 2007. In reference to probiotic containing edible films, López de Lacey et al. 2012 proclaimed that the bacterial strains, namely *Bifidobacteria bifidum* and *Lactobacillus acidophilus*, could be immobilized in gelatin-based matrix. This helped in protecting the loss of viability of probiotic strains for not less than six days when stored at 2 °C. An interesting development to deliver probiotics in bread with the help of edible films and anhydrobiotics was devised by Altamirano-Fortoul et al. 2012. Consecutive layers of starch-based edible coating and encapsulated strains of probiotics were spread on the partly baked breads. After the application, the breads were baked for very short time. To their surprise, it was noticed that a large number of viable bacteria could be retained after baking.

Starches derived from tapioca, corn, potato and pullulan were utilized for the development of probiotics-incorporated edible films (*Lactobacillus reuteri* ATCC 55730, *L. rhamnosus* GG ATCC 53103 and *L. acidophilus* DSM 20079) (Kanmani and Lim 2013). Pure pullulan and various combinations of starches and pullulan were used for film formation and stored at two temperatures that is at 25 and 4 °C. After 10 days of storage at 25 °C, it was found that the viability of probiotics in pure pullulan film was around 80%. However, the edible films made from starches showed decreased viability.

26.4 Methods of Film Formation

Commonly edible packaging includes films, coatings, edible containers or biodegradable films. Edible film is a free-standing sheet that can be placed on or between the layers of food constituents. In general, the thickness of edible films varies from 25 μm to 140 μm. Depending on the raw materials employed and the manufacturing method used, the transparency, translucency and opaqueness of the film vary. Edible coating is often described as a thin layer of edible component, but it is administered on the exterior surface of the product as a viscous liquid by making use of techniques like brushing, spraying, dipping and so on.

Techniques employed to form films and the conditions under which coating takes place have a paramount effect on the physical characteristics of the film. Consistent, homogeneous and a film with no defects like presence of air bubbles or any other kind of damage is extremely critical to optimize the functionalities of the film.

Edible films can be produced by two main processes, that is, wet and dry. In the wet process, the biopolymers are disseminated or solubilized in a film-forming solution (solution casting) followed by the solvent evaporation process. Film-forming solutions commonly contain water-soluble polymers, plasticizers, fibres, colours, flavours and sweeteners. Drying is a very crucial step during the process of manufacturing of edible films. Water is the primary solvent that is removed, although small quantities of volatile flavours and other additives may also be removed. Film drying is comprised of three stages: transition, constant rate and falling rate. Wet process facilitates the administration of viscous coating straightforwardly on products either by brushing, dipping or by spraying, etc.

Dry process depends upon the thermoplastic behaviour expressed by polymers like polysaccharides and proteins, particularly when they are heated at temperatures more than their glass transition temperatures at low moisture levels. The main drawback of extruded films is that it is not practical to use them for coating uneven surfaces.

26.4.1 Solvent Casting Method

The most widely used technique for film forming is the casting or solvent casting method. This method encompasses three steps: (i) dispersion of biopolymers in an appropriate solvent, (ii) mould casting, followed by (iii) drying. The process of edible film formation starts with the selection of the biopolymers. This is followed by the dispersion of selected biopolymer/s in an appropriate solvent. This process is also known as solubilization. Formation of film in the method relies on biopolymer solubility, instead of melting (Koide et al. 2013). Solution thus derived is glugged onto plates made from glass, Teflon, polyethylene and other materials. This is followed by evaporation of the solvent by air-drying or oven drying, vacuum drying and tray drying. The film is then peeled off from the mould. Rapid drying of the film should be avoided, as the concentration of solvent will decrease at a faster rate restricting the biopolymer chain mobility and leading to the formation of intermolecular relationships in the film (Velaga et al. 2018).

Primary benefits of film formation by the solvent casting method are:

1. Simplicity of the method without usage of any specialized instruments.
2. The cost of production is low.
3. As it is a wet process, there is better interaction between the constituent particles.
4. Fewer defects.
5. Films are homogeneous in nature.
6. Can be moulded at lower temperature.

The main disadvantages of the casting method are:

(a) Can be moulded into sheets and tube forms only.
(b) May lead to denaturation of biopolymers by solvents.
(c) Time-consuming process.
(d) Restriction in the amount of film produced.
(e) Trapping of solvent within the biopolymer matrix.
(f) A little change in processing conditions like temperature and relative humidity at commercial scale could change the properties of the film.

26.4.2 Extrusion Process

Extrusion process of film formation relies on the thermoplastic properties of the biomolecules. This involves change in the structure of the polymer and helps in refining their physical and chemical properties. Macromolecules or biopolymers in combination with plasticizers, at less humidity and elevated temperatures under pressure or shear forces, obtain a viscoelastic nature that enables them to be formed into a variety of shapes and sizes such as films, cups, lids, trays and so on. Among the macromolecules, proteins obtained from different sources are regarded as the preferred constituent for film formation as they get denatured upon heating, leading to conformational changes, revealing their concealed functional groups, thereby allowing the formation of new linkages leading to improvements in functional properties of edible films.

Extrusion-developed films are found to exhibit superior barrier, mechanical and microstructural properties in comparison to those obtained by the casting method (Hernandez-Izquierdo and Krochta 2008). The extrusion process is completed in three steps: (i) feeding, (ii) kneading and (iii) heating. The constituents of the film are fed into the hopper and compressed with air. During this phase, there should be minimum solvent or water, hence it is known as the dry process. Plasticizers are required to enhance the flexibility of the film. When this mixture is forwarded to the kneading zone, their density, temperature and strain increase. Multilayered films can be developed by using the coextrusion method. This helps in obtaining edible films with desired properties (Winotapun et al. 2019).

26.5 Methods of Application of Coatings

Thin layer of edible material deposited on the surface of food in liquid form is known as edible coating. The method by which edible coating is applied and the tendency of it to adhere to the surface of food product are the two main distinctive features of edible coating. Edible coating can be done by using different application methods such as dipping, dripping, foaming, panning, spraying, electrostatic coating and fluidized bed coating. Choice of a suitable method relies on the characteristics of the food, the coating material, the expected effect of the coating and the cost.

26.5.1 Dipping

It is the most common method employed for depositing coating on the food product. In this process, the food product to be coated is immersed in the solution, dispersion or emulsion of coating material, so that the required amount of coating material is deposited on the surface as well as to ensure that the interaction between the food and the coating material is complete. After this, the sample is withdrawn from the dispersion. Excess of coating material is drained off from the product. Sometimes, the products may be subjected to a second immersion, also to achieve proper gel formation and to drain out any excessive coating material. Thereafter, the solvent and the excess material are evaporated by heating which may be carried out at room temperature. It has been seen that to apply edible coating on the cut surface of a fruit is quiet difficult because of the hydrophilic nature of the surface which does not allow the coating material to gel properly (Skurtys et al. 2010). Multilayer coating or the layer-by-layer (LbL) technique was used to overcome this problem. In this process, the food product is dipped one by one into various coating solutions that contain oppositely charged polyelectrolytes to achieve physical and chemical bonding with each other (Sipahi et al. 2013).

As the dipping method is very simple and cost effective, this makes it the most preferable method in most of the research studies. Dipping ensures complete and uniform coverage of the food product. However, it is associated with certain difficulties also like dilution of the coating solution, growth of microorganisms in the vat containing coating solution, adherence of dust and dirt on the wet surface. Sometimes, the coating may detach itself from the surface of the food product leading to deterioration of its quality.

26.5.2 Spraying

Spraying is another commonly used method for the coating of food products. It increases the surface area of the coating liquid by converting it into fine droplets with the help of nozzles. The coating solution in this technique should be of low viscosity, so as to ensure better and uniform coverage, and better control of the thickness of coating. With this technique, multilayer coating can also be exercised. This also enables to work on a bigger area. Many factors like temperature, surface tension, viscosity, density of the coating solution; pressure and flow rate; design of the nozzle and spray angle play a crucial role in the spraying technique. A study by Zhong et al. 2019 presented that the application of edible coating by the spraying method helped in improving the appearance and quality of food products reported using the spraying technique.

26.5.3 Fluidized Bed Coating

It is a process by which a very thin layer is applied on dry particles of very low density or small size. It was initially developed as a pharmaceutical coating technique but it is now being gradually applied in the food industry. In this process, the coating solution is deposited by spraying on the fluidized powder surface with the help of nozzles, resulting in the formation of shell on the food outer surface. It can be accomplished by top or bottom or rotary spraying.

Fluidization happens when a liquid proceeds upwards across a particle bed that achieves ample velocity to sustain particles without taking them away in the fluid stream. Subsequently, the particle bed acquires the attributes of a gurgling liquid, thereby termed as fluidization. Dimensions of fluidized bed particulate matter should not be less than 100 μm because particles with small size do not fluidize stably or may lead to the formation of agglomerates (Chen et al. 2009).

26.5.4 Panning

Panning has been practiced since the ninth century when the Greeks used to coat drugs. The food to be coated is placed in large enclosed pans that are perforated along the side panels. A pump is used to distribute the coating solution to drizzlers or sprinklers installed in different sections of the rotating pan. Thus, action helps in coating the food product uniformly. After coating, air at ambient or elevated temperatures is allowed to enter the pan for drying. In the course of this process, heat is produced due to friction which should be removed with the help of cool air to maintain the quality characteristics. Considerably, extruded food products are best suited for panning, as the size and shape of these products can be controlled to coat them uniformly.

26.5.5 Foaming

Foam application is used for some emulsion coating. A foaming agent is added to the coating solution and then air under pressure is blown into the applicator tank. Comprehensive spiraling action is necessary to break the foam for consistent application of coating. The agitated foam is applied to the food commodities moving on rollers, thereby spreading the foam coating solution on products. As this type of emulsion contains very little water, therefore the coating dries very quickly, but inadequate coverage is usually a drawback of this application.

26.5.6 Electrostatic Coating

This process employs charged particles (by electric field) to efficiently coat a surface. Because of the same charge on these particles, there is repulsion between them, thus forming an even cloud around the surface to be coated. Atomized liquid or powdered particles are projected towards a conductive surface by spraying and thereafter accelerated towards the surface to be coated by a powerful electrostatic charge. These particles are deposited due to the difference in charges of the coating material and the food surface to be coated (Khan et al. 2012).

26.6 Raw Materials for Edible Packaging

With the advancement of processing technologies, knowledge of material science edible packaging continues to gain importance in the coming times. This sector has a great potential and scope in view of the environment-friendly nature of these edible films and coatings (Galus et al. 2020).

Hydrocolloid is made of two words, hydro meaning "water" and kolla meaning "glue" and contains a varied group of heterogeneous, long-chain biopolymers. Hydrocolloids are hydrophilic polymers constituting long chains of either proteins or polysaccharides. Colloidal dispersions and/or liquid gel is formed from these biopolymers (Milani and Maleki 2012; Razavi 2019). Basically, the natural raw materials of biopolymers can be obtained from the following four groups, namely, plant, animals, microbial and seaweeds (Bisht et al. 2020).

The requirements of materials for edible coatings and edible films are precisely based on the nature and properties of the food which is to be packaged and their application (Mohamed et al. 2020). The main film-forming materials either used alone or in combination include biopolymers such as proteins, polysaccharides (pectin, starch, carrageenan, alginate and xanthan), gums, resins and lipids (Fig. 26.1). Vegetable and fruit purees can also be utilized for the manufacture of edible films (Azeredo et al. 2009). Composite film ingredients could be achieved, either as bilayer or as emulsions. In a bilayer film structure, firstly, a thin layer of polysaccharide is to be formed, followed by a second layer of dispersed lipids (Galus and Kadzińska 2015).

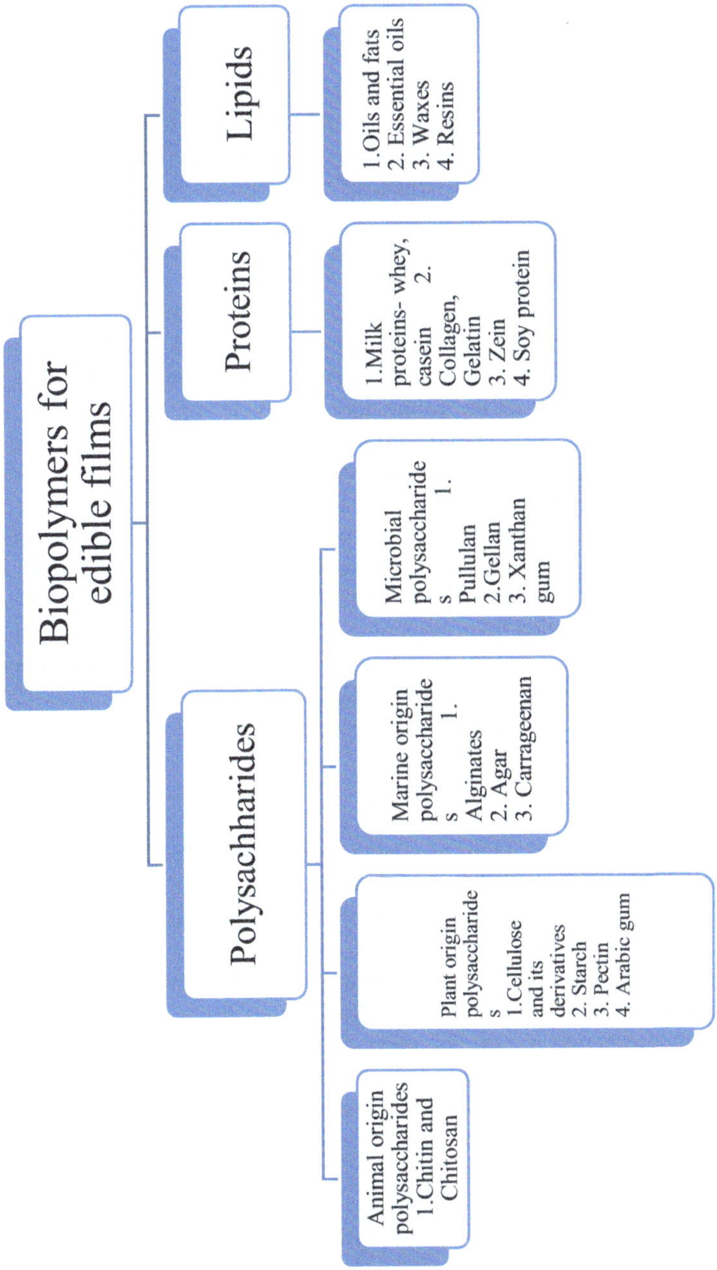

Fig. 26.1 Biopolymers of edible packaging

26.6.1 Proteins

Proteins used for film-forming substances are derived from renewable sources and can easily be degraded, as compared to their plastic analogues. Proteins are macromolecules having precise amino acid sequences joined by amide linkage and molecular arrangement, which can be degraded by proteases (Clarinval and Halleux 2005). They are frequently used as film-forming substances. Proteins possess an additional benefit of their amphiphilic nature, besides electrostatic charge and denaturation properties (Han 2014). Various changes can be brought about in the secondary, tertiary and quaternary structures of proteins to suit the needs of film-forming substances. These variations can be done by the utilization of heat, irradiation, chemical, mechanical treatment, pressure and enzymatic applications.

The most commonly used proteins in edible film and coating formulations include proteins derived from milk (casein and whey protein), plant sources (soy protein and corn zein), and wheat gluten, pea protein, rice bran protein collagen, egg albumin from eggs, myofibrillar protein of fish and keratin (Ramos et al. 2012). Total milk proteins can be used in the formulation of edible films or the components thereof (Shendurse et al. 2018). Edible films made from whey proteins are nutritious, bland in taste and have fine mechanical strength and potent gas, aroma and lipid barrier properties (Azevedo et al. 2018; Lara et al. 2019). Considering the fibrous proteins, collagen is the protein of interest for the manufacture of edible films and generally, the protein films are formed from protein solutions and dispersions, either by surface film formation or by deposition methods (Chiralt et al. 2018). Gelatin is obtained from the controlled hydrolysis of collagen and has an exclusive feature of melting near the body temperature, which is noteworthy in the pharmaceutical industry and for edible packaging.

26.6.2 Polysaccharides

The polysaccharides are made up of simple monosaccharide units but have larger molecular weights due to complicated and unpredictable conformational structural changes. They occur in natural world in diverse forms and are newly included biopolymers for edible films and coatings (Mellinas et al. 2016). Starch-based packaging is promising, due to desirable attributes like flexibility, transparency, environment-friendly nature and low cost (Suput et al. 2015). Due to firm configuration of polysaccharide molecules, they are good in blocking the diffusion of oxygen and carbon dioxide gases. But, these biomolecules are prone to water transmission through films, therefore they are coupled with lipids for counteracting this limitation. Polysaccharides can be obtained either from plant sources (chitin and chitosan). The various examples of polysaccharides cover the different starch types, modified cellulose (carboxymethyl cellulose; methylcellulose and hydroxypropyl cellulose; hydroxypropyl methylcellulose) non-starch carbohydrates, gellan gums and fibres (Han 2014). The carbohydrates are mainly neutral in nature but can be charged negatively or positively which is uncommon.

The neutral carbohydrates have hydroxyl groups in their structure. These groups or the hydrophilic moieties have strong affinity for hydrogen bond formation which is responsible for the film-forming properties. Alginate, pectin and carboxymethyl cellulose are negatively charged gums that illustrate different rheological characteristics in diverse mediums like acidic, alkaline or neutral.

26.6.2.1 Broad Categories of Polysaccharides

Natural polysaccharides comprise mainly of the following groups:

Chitin and chitosan-chitin is considered as the second most plentiful occurring polysaccharide following cellulose and it is originally found in the exoskeleton of crustaceans and the cell wall of fungi. Deacetylation reaction of chitin furnishes chitosan (Mohamed et al. 2020).

Pectin, an ionic biopolymer group, used for thickening, stabilizing and gel formation is present in the fruits and vegetables (Dhanapal et al. 2012).

Alginates are considered to be the polysaccharides derived from brown algae present in marine water and can prove to be a good and strong film-forming substrate when calcium addition is done to increase the water barrier properties (Rangel-Marrón et al. 2013).

Pullulan, gellan and xanthan gum are the sources of polysaccharides acquired from microbial sources. These biopolymers have the ability of gel formation, so they are used to make safe and transparent edible films.

26.6.3 Lipids

The majority of lipids and resins are edible, biodegradable and possess good film-forming properties. At room temperature, they are soft solids and at transition temperatures, their phase can be altered reversibly. Their outline can be changed by giving heat treatment and then using casting moulding system. The hydrophobic nature of lipids is responsible for their high water resistance and low surface energy. A range of biopolymers like oils, free fatty acids, bee waxes, carnauba wax, paraffin wax, shellac resin, terpene resin and acetogylcerides are included in lipids (Suput et al. 2015). Oils and fats are basically mixtures, the former is obtained from plant sources and animals are source for the latter. Waxes having high molecular weight with protection being major function have animal as well as plant origin. They result from long-chain acid esterification and/or alcohol formation (Sikorski and Kolakowska 2010). Lipids can be used in combination with polysaccharides and proteins, so as to attain better water-resistant properties. As multilayer plastics are available, likewise they can also be constructed by use of biopolymers and their combination. Lipids along with proteins can be used as either emulsion or multilayer coatings for improvement and to bring about desirable changes to the structure of the film formed (Han 2014). In view of improving resistance to water penetration, lipids are generally combined with other biopolymers like proteins and polysaccharides (Mehyar et al. 2012).

26.6.4 Additives

Additives are the compounds which can be added to the edible films in adequate amount to cause desirable effect without compromising the various physical, chemical and mechanical properties of the coatings and films. Active agents, for instance, antioxidants, nutraceuticals, emulsifiers, antimicrobials, flavours, pigments, light absorbers and colourants can be incorporated in the edible films. These additives play an active role in maintaining, improving and protecting the quality and safety of food. Antioxidants are added to films to regulate the changes caused due to oxidative rancidity of lipids and extend the shelf life (Vargas et al. 2017). Plant extracts are the source of many natural antioxidants and can be obtained from various parts like fruits, leaves, seeds and oil, sometimes having activity as good as synthetic antioxidants (Okonogi et al. 2007). Microbial degradation can be controlled by the adjoining of antimicrobial compounds in the coatings and biofilms. The product shelf life and safety can be improved by the addition of antimicrobials such as curing agents (e.g. sodium chloride and sodium nitrite and nitrates), benzoates, sorbates, parabens, acidifying agents (e.g. acetic and lactic acids), propionates, bacteriocins and natural preservatives (e.g. essential oils, lysozyme and liquid smoke) in the edible films (Cagri et al. 2004).

Emulsifiers, surface-active agents, work to stabilize an emulsion. Due to their amphiphilic nature, they have a vital function of reducing the surface tension at the interface of either air–water or lipid water and improving the adhesion and wettability of the formed film (Ramos et al. 2012). So, the addition of emulsifiers is required in lipid films, although films made of proteins may not require emulsifiers as protein structures are also amphiphilic. The sensory attributes can be improved by incorporating flavours and colourants as additives in the biopolymers of films. A range of medications can be delivered by inclusion of pharmaceutical and nutraceutical compounds in biodegradable films and coatings.

26.7 Current Scenario

Packaging of foods and food products is done to protect them from external physical, chemical and biological conditions. Besides, packaging also provides valuable information about the food products. Usually, fossil raw materials are used increasingly to produce synthetic plastics which are non-eco-friendly, owing to their nature being non-biodegradable (Rodríguez et al. 2020). Consequently, the use of natural biopolymers for the production of biofilms and coatings was necessitated to replace the plastic analogues (Ahmed et al. 2018).

The edible packaging has been acknowledged by food, pharmaceutical and biotech sectors as a better substitute to synthetic packaging in reducing waste and to craft novel application with desired features for improving a product, for example, safety, stability, quality, variety and convenience for customers (Aguirre-Joya et al. 2018). The barrier properties (barriers to oxygen, light, moisture and flavours) of these edible films are not alike those of petrochemical-based films but the properties can be improved by making use of suitable other biopolymers for composite films and blends (Qamar et al. 2020).

Proteins, lipids and polysaccharides are the main natural edible polymers which when made into films can provide safety to fresh foods (Mohamed et al. 2020). Different additives with antibacterial and antifungal properties, biobased materials, for example, plant essential oils from plants and natural additives are being incorporated in biofilms to enhance the functionality of these films (Qamar et al. 2020).

26.8 Challenges and Opportunities

Though, edible packaging presently covers a small place in the markets it has noteworthy potential as packaging material in the times to come. The possibilities of various applications of these coatings and films are exciting and restricted only by one's thoughts.

26.8.1 Opportunities

- The opportunities of edible films and coatings lie with benefits of addition of numerous functional agents like antioxidants, nutraceuticals, antimicrobial agents, vitamins, probiotics, flavouring and colouring agents.
- The replacement of the synthetic plastics with edible films offers a great solution to waste generated because of the non-biodegradable nature of these petrochemical films.
- The edible films and coatings have a wide range of applications in diverse range of food types including meat and meat products.
- The use of polysaccharides to form edible films offers newer prospects to develop biodegradable films.
- The different antioxidant compounds can be nano-encapsulated on edible films as lipophilic particles which offer to be an appealing strategy.
- As the name suggests, these biofilms are edible in nature, so can be consumed along with the product.

26.8.2 Challenges

Lack of knowledge: The researchers have studied about the various film-forming substances, plasticizer concentration and about the properties of formed films, but inadequate data are available on the effects of different factors such as temperature and pressure, free volume, acid and base concentrations, orientation etc. which might affect the films differently.

Consumer acceptance: Although, edible films can be eaten along with the food, consumers agree to it only when they feel it as safe and their acceptance is governed by a large number of factors like the appearance of film, sensory attributes as well as cost.

Cheap and complete biodegradability methods of disposal of the biopolymers still need to be focused upon.

Till date, edible films cost higher than the corresponding traditional plastics, so cost has also been a major constraint in adoption of these films. Although, addition of these compounds justifies the continued research cost-effective methods of production of these films and coatings still need to be focused upon.

Application possibilities are exciting and limited only by one's imagination. Deep-seated research of edible films is desirable to guarantee a thorough knowledge of the molecular basis of functional properties and is helpful in the revelation of possibilities.

26.9 Conclusion

Biopolymers required for making edible films and coatings can be obtained from polysaccharides, such as pectin, starch and alginate, as well as proteins derived from both plant and animal sources and lipids. These biopolymers help to replace and reduce the waste generated due to conventional plastics made from fossils. Not a single biopolymer is capable of fulfilling all the properties required for edible films. So, different biopolymers can be combined with suitable plasticizers to attain desirable effects. These materials are novel, natural, safe, eco-friendly and have an added advantage of the fact that they could be made from renewable sources.

References

Acevedo-Fani A, Salvia-Trujillo L, Rojas-Graü MA, Martín-Belloso O (2015) Edible films from essential-oil-loaded nanoemulsions: physicochemical characterization and antimicrobial properties. Food Hydrocoll 47:168

Aguirre-Joya JA, De Leon-Zapata MA, Alvarez-Perez OB, Torres-León C, Nieto-Oropeza DE, Ventura-Sobrevilla JM, Aguilar MA, Ruelas-Chacón X, Rojas R, Ramos-Aguiñaga ME, Aguilar CN (2018) Basic and applied concepts of edible packaging for foods. In: Food packaging and preservation. Academic Press, Cambridge, MA, pp 1–61

Ahmed T, Shahid M, Azeem F, Rasul I, Shah AA, Noman M, Hameed A, Manzoor N, Manzoor I, Muhammad S (2018) Biodegradation of plastics: current scenario and future prospects for environmental safety. Environ Sci Pollution Res 25(8):7287–7298

Altamirano-Fortoul R, Moreno-Terrazas R, Quezada-Gallo A, Rosell CM (2012) Viability of some probiotic coatings in bread and its effect on the crust mechanical properties. Food Hydrocoll 29(1):166–174

Alvarado S, Sandoval G, Palos I, Tellez S, Aguirre-Loredo Y, Velazquez G (2015) The effect of relative humidity on tensile strength and water vapor permeability in chitosan, fish gelatin and transglutaminase edible films. Food Sci Technol 35(4):690–695

Azeredo HM, Mattoso LH, Wood D, Williams TG, Avena-Bustillos RJ, McHugh TH (2009) Nanocomposite edible films from mango puree reinforced with cellulose nanofibers. J Food Sci 74(5):N31–N35

Azevedo VM, Dias MV, de Siqueira Elias HH, Fukushima KL, Silva EK, Carneiro JD, Soares ND, Borges SV (2018) Effect of whey protein isolate films incorporated with montmorillonite and citric acid on the preservation of fresh-cut apples. Food Res Int 107:306–313
Bie P, Liu P, Yu L, Li X, Chen L, Xie F (2013) The properties of antimicrobial films derived from poly (lactic acid)/starch/chitosan blended matrix. Carbohydr Polym 98:959–966
Bisht B, Lohani UC, Kumar V, Gururani P, Sinhmar R (2020) Edible hydrocolloids as sustainable substitute for non-biodegradable materials. Crit Rev Food Sci Nutr 8:1–33
Cagri A, Ustunol Z, Ryser ET (2004) Antimicrobial edible films and coatings. J Food Prot 67(4):833–848
Chatiyanont L, Wuttijumnong P (2007) Application of pectin coating in the production of vitamin fortified rice. Kasetsart J Nat Sci 4:356–362
Chen Y, Yang J, Mujumdar A, Dave R (2009) Fluidized bed film coating of cohesive Geldart group C powders. Powder Technol 189(3):466–480
Chiralt A, González-Martínez C, Vargas M, Atarés L (2018) Edible films and coatings from proteins in: proteins in food processing. Woodhead Publishing, Cambridge, pp 477–500
Clarinval AM, Halleux J (2005) Classification of biodegradable polymers. In: Biodegradable polymers for industrial applications. Woodhead Publishing, Cambridge, pp 3–31
Debeaufort F, Quezada-Gallo J-A, Voilley A (1998) Edible films and coatings:Tomorrow's Packagings: a review. Crit Rev Food Sci Nutr 38(4):299–313
Dhanapal A, Sasikala P, Rajamani L, Kavitha V, Yazhini G, Banu MS (2012) Edible films from polysaccharides. Food Sci Qual Manag 3:9
Durand L, Habran N, Henschel V, Amighi K (2010) Encapsulation of ethylhexyl methoxycinnamate, a light-sensitive UV filter, in lipid nanoparticles. J Microencapsul 27(8):714–725
Falagán N, Artés F, Aguayo E (2018) Antimicrobial properties of edible coatings enriched with natural additives on fresh-cut peaches. Acta Hortic 1194:519–526
Galus S, Kadzińska J (2015) Food applications of emulsion-based edible films and coatings. Trends Food Sci Technol 45(2):273–283
Galus S, Arik Kibar EA, Gniewosz M, Kraśniewska K (2020) Novel materials in the preparation of edible films and coatings—A review. Coat 10(7):674
Gheorghita Puscaselu R, Amariei S, Norocel L, Gutt G (2020) New edible packaging material with function in shelf life extension: applications for the meat and cheese industries. Foods 9(5):562
Gol NB, Patel PR, Rao TVR (2013) Improvement of quality and shelf-life of strawberries with edible coatings enriched with chitosan. Postharvest Biol Technol 85:185–195
Gómez-Estaca J, Montero P, Giménez B, Gómez-Guillén MC (2007) Effect of functional edible films and high pressure processing on microbial and oxidative spoilage in cold-smoked sardine (Sardina pilchardus). Food Chem 105(2):511–520
Gómez-Estaca J, López-de-Dicastillo C, Hernández-Muñoz P, Catalá R, Gavara R (2014) Advances in antioxidant active food packaging. Trends Food Sci Technol 35:42–51
Goslinska M, Heinrich S (2019) Characterization of waxes as possible coating material for organic aerogels. Powder Technol 357:223–231
Gutierrez TJ (2017) Surface and nutraceutical properties of edible films made from starchy sources with and without added blackberry pulp. Carbohydr Polym 165:169–179
Han JH (2014) Edible films and coatings in: innovations in food packaging, 2nd edn, pp 277–294
Han JH (2019) Edible films and coatings: a review. Innov Food Packag 1:213–255
Han C, Zhao Y, Leonard SW, Traber MG (2004) Edible coatings to improve storability and enhance nutritional value of fresh and frozen strawberries (Fragaria x ananassa) and raspberries (Rubus ideaus). Postharvest Biol Technol 33:67–78
Hernandez-Izquierdo VM, Krochta JM (2008) Thermoplastic processing of proteins for film formation—a review. J Food Sci 73(2):R30–R39
Janjarasskul MT, Krochta JM (2010) Edible packaging. Annu Rev Food Sci Technol 1:415–448
Jiménez A, Fabra MJ, Talens P, Chiralt A (2013) Physical properties and antioxidant capacity of starch-sodium caseinate films containing lipids. J Food Eng 116:695–702

Jin TZ, Huang M, Niemira BA, Cheng L (2016) Shelf life extension of fresh ginseng roots using sanitiser washing, edible antimicrobial coating and modified atmosphere packaging. Int J Food Sci Technol 51:2132–2139

Kanmani P, Lim ST (2013) Development and characterization of novel probiotic-residing pullulan/starch edible films. Food Chem 141:1041–1049

Khan M, Schutyser M, Schroën K, Boom R (2012) Electrostatic powder coating of foods – state of the art and opportunities. J Food Eng 111

Koide Y, Ikake H, Muroga Y, Shimizu S (2013) Effect of the cast-solvent on the morphology of cast films formed with a mixture of stereoisomeric poly (lactic acids). Polym J 45:645–650

Krochta JM (2002) Protein as raw materials for films and coatings: definitions, current status, and opportunities. In: Gennadios A (ed) Protein-based films and coatings, vol 1. CRC Press, New York, pp 1–40

Lara BR, Araújo AC, Dias MV, Junior MG, Santos TA, Ferreira LF, Borges SV (2019) Morphological, mechanical and physical properties of new whey protein isolate/polyvinyl alcohol blends for food flexible packaging. Food Packag Shelf 1(19):16–23

López de Lacey ME, López-Caballero, Gómez-Estaca J, Gómez-Guillén MC, Montero P (2012) Functionality of Lactobacillus acidophilus and Bifidobacterium bifidum incorporated to edible coatings and films. Innov Food Sci Emerg Technol 16:277–282

Lopez O, Garcia MA, Villar MA, Gentili A, Rodriguez MS, Albertengo L (2014) Thermocompression of biodegradable thermoplastic corn starch films containing chitin and chitosan. LWT Food Sci Technol 57:106–115

Mahfoudhi N, Hamdi S (2015) Use of almond gum and gum arabic as novel edible coating to delay postharvest ripening and to maintain sweet cherry (Prunus avium) quality during storage. J Food Process Preserv 39:1499–1508

Mali S, Grossmann MVE (2003) Effects of yam starch films on storability and quality of fresh strawberries (Fragaria ananassa). J Agric Food Chem 51:7005–7011

Martins JT, Cerqueira MA, Souza BW, Carmo Avides M, Vicente AA (2010) Shelf life extension of ricotta cheese using coatings of galactomannans from nonconventional sources incorporating nisin against Listeria monocytogenes. J Agric Food Chem 58:1884–1891

Matiacevich S, Acevedo N, López D (2015) Characterization of edible active coating based on alginate-thyme oil-propionic acid for the preservation of fresh chicken breast fillets. J Food Process Preserv 39:2792–2801

Mazza G, Qi H (1991) Control after-cooking darkening in potatoes with edible film-forming products and calcium chloride. J Agric Food Chem 39(10):2163–2166

McHugh TH, Krochta JM (1994) Permeability properties of edible films In: JM Krochta, EA Baldwin, M Nisperos-Carriedo (ed) Edible coating and films to improve food quality, Lancaster PA: Technomic Publishing Co., Inc 139–187

Mehyar GF, Al-Ismail K, Han JH, Chee GW (2012) Characterization of edible coatings consisting of pea starch, whey protein isolate, and carnauba wax and their effects on oil rancidity and sensory properties of walnuts and pine nuts. J Food Sci 77(2):E52–E59

Mei Y, Zhao Y (2003) Barrier and mechanical properties of milk protein-based edible films

Mei Y, Zhao Y, Yang J et al (2002) Using edible coating to enhance nutritional and sensory qualities of baby carrots. J Food Sci 67(5):1964–1968

Mellinas C, Valdés A, Ramos M, Burgos N, Garrigos MD, Jiménez A (2016) Active edible films: current state and future trends. J Appl Polym Sci 10:133(2)

Meyers MA (1994) Use of edible film to prolong chewing gum shelf life. U.S. Patent No. 5 286 502

Milani J, Maleki G (2012) Hydrocolloids in food industry. Food Ind Process Methods Equip 22: 22–37

Mohamed SA, El-Sakhawy M, El-Sakhawy MA (2007) Polysaccharides, protein and lipid-based natural edible films in food packaging: a review. Carbohydr Polym 20:116178

Mohamed SAA, El-Sakhawy M, El-Sakhawy MAM (2020) Polysaccharides, protein and lipid -based natural edible films in food packaging: a review. Carbohydr Polym: 116178

Mridula D, Jha P (2014) Preparation of iron-fortified rice using edible coating materials. Int J Food Sci Technol 49(1):246–252

Nystrom L, Achrenius T, Lampi AM, Moreau RA, Piironen V (2007) A comparison of the antioxidant properties of steryl ferulates with tocopherol at high temperatures. Food Chem 101(3):947–954

Okonogi S, Duangrat C, Anuchpreeda S, Tachakittirungrod S, Chowwanapoonpohn S (2007) Comparison of antioxidant capacities and cytotoxicities of certain fruit peels. Food Chem 103(3):839–846

Oms-Oliu G, Soliva-Fortuny R, Martin-Belloso O (2008) Edible coatings with antibrowning agents to maintain sensory quality and antioxidant properties of fresh-cut pears. Postharvest Biol Technol 50:87–94

Park HJ, Chinnan MS, Shewfelt RL (1994) Edible coating effects on storage life and quality of tomatoes. J Food Sci 59(3):568

Postma J, Stevens LH, Wiegers GL, Davelaar E, Nijhuis EH (2009) Biological control of Pythium aphanidermatum in cucumber with a combined application of Lysobacter enzymogenes strain 3.1T8 and chitosan. Biol Control 48(3):301–309

Qamar SA, Asgher M, Bilal M, Iqbal HM (2020) Bio-based active food packaging materials: sustainable alternative to conventional petrochemical-based packaging materials. Food Res Int 14:109625

Raajeswari PA, Pragatheeswari R (2019) Edible packaging and market overview. Food Marketing & Technology, Indian addition

Ramos ÓL, Fernandes JC, Silva SI, Pintado ME, Malcata FX (2012) Edible films and coatings from whey proteins: a review on formulation, and on mechanical and bioactive properties. Crit Rev Food Sci Nutr 52(6):533–552

Rangel-Marrón M, Montalvo-Paquini C, Palou E, López-Malo A (2013) Optimization of the moisture content, thickness, water solubility and water vapor permeability of sodium alginate edible films. Rec Adv Ch Eng Biochem Computat Chem 10:29–31

Razavi SM (2019) Editor emerging natural hydrocolloids: rheology and functions. John Wiley & Sons, Chichester

Resa CPO, Gerschenson LN, Jagus RJ (2014) Natamycin and nisin supported on starch edible films for controlling mixed culture growth on model systems and port Salut cheese. Food Control 44: 146–151

Rodríguez GM, Sibaja JC, Espitia PJ, Otoni CG (2020) Antioxidant active packaging based on papaya edible films incorporated with Moringa oleifera and ascorbic acid for food preservation. Food Hydrocoll 103:105630

Saberi B, Chockchaisawasdee S, Golding JB, Christopher JS, Costas ES (2017) Physical and mechanical properties of a new edible film made of pea starch and guar gum as affected by glycols, sugars and polyols. Int J Biol Macromol 104(Pt A):345–359

Santosa FKG, Silvab KNO, Xaviera TDN, Leitea RHL, Arouchaa EMM (2017) Effect of the addition of carnauba wax on physicochemical properties of chitosan films. Mater Res 20(2):479–484

Shendurse AM, Gopikrishna G, Patel AC, Pandya AJ (2018) Milk protein based edible films and coatings–preparation, properties and food applications. J Nutr Health Food Eng 8(2):219–226

Sikorski ZZ, Kolakowska A (eds) (2010) Chemical, biological, and functional aspects of food lipids. CRC Press, Boca Raton, FL

Sipahi RE, Castell-Perez ME, Moreira RG, Gomes C, Castillo A (2013) Improved multilayered antimicrobial alginate-based edible coating extends the shelf life of fresh-cut watermelon (Citrullus lanatus). LWT Food Sci Technol 51:9–15

Skurtys O, Acevedo C, Pedreschi F, Enrione J, Osorio F, Aguilera JM (2010) Food hydrovolloid edible films and coatings IN: food hydrocolloids: characteristics, properties and structures. Nova science publishers, Inc, pp 41–80

Souza AC, Goto GEO, Mainardi JA, Coelho ACV, Tadini CC (2013) Cassava starch composite films incorporated with cinnamon essential oil: antimicrobial activity, microstructure, mechanical and barrier properties. LWT Food Sci Technol 54:346–352

Suput DZ, Lazić VL, Popović SZ, Hromiš NM (2015) Edible films and coatings: sources, properties and application. Food Feed Res 42(1):11–22

Tapia MS, Rojas-Graü MA, Rodríguez FJ, Ramírez J, Carmona A, Martin-Belloso O (2007) Alginate and gellan-based edible films for probiotic coatings on fresh-cut fruits. J Food Sci 72:190–196

Tian F, Decker EA, Goddard JM (2013) Controlling lipid oxidation of food by active packaging technologies. Food Func 4(5):669–680

Tiwari BK, Valdramidis VP, O'Donnell CP, Muthukumarappan K, Bourke P, Cullen P (2009) Application of natural antimicrobials for food preservation: a review. J Agric Food Chem 57: 5987–6000

Treviño-Garza MZ, García S, Flores-González MS, Arévalo-Niño K (2015) Edible active coatings based on pectin, pullulan, and chitosan increase quality and shelf life of strawberries (Fragaria ananassa). J Food Sci 80:1823–1830

Ture H, Eroglu E, Ozen B, Soyer F (2011) Effect of biopolymers containing natamycin against Aspergillus Niger and Penicillium roqueforti on fresh kashar cheese. Int J Food Sci Technol 46: 154–160

Vargas CG, Costa TM, de Oliveira RA, Flôres SH (2017) Comparative study on the properties of films based on red rice (Oryza glaberrima) flour and starch. Food Hydrocoll 65:96–106

Velaga SP, Nikjoo D, Vuddanda PR (2018) Experimental studies and modeling of the drying kinetics of multicomponent polymer films. AAPS Pharm Sci Tech 19:425–435

Vieira JM, Flores-López ML, de Rodríguez DJ, Sousa MC, Vicente AA, Martins JT (2016) Effect of chitosan–aloe vera coating on postharvest quality of blueberry (Vaccinium corymbosum) fruit. Postharvest Biol Technol 116:88–97

Viskupicova J, Danihelova M, Ondrejovic M, Liptaj T, Sturdik E (2010) Lipophilic rutin derivatives for antioxidant protection of oil-based foods. Food Chem 123(1):45–50

Winotapun C, Phattarateera S, Aontee A et al (2019) Development of multilayer films with improved aroma barrier properties for durian packaging application. Packag Technol Sci 32: 405–418

Zaritzky N (2011) Edible coating to improve food quality and safety. In: Aguilera JM, Barbosa-Canovas GV, Simpson R, Welti-Chanes J, Bermudez Aguirre D (eds) Food engineering interfaces, Food engineering series. Springer, New York

Zevallos CL, Krochta JM (2003) Whey protein coatings for fresh fruits and relative humidity effects. J Food Sci 68(1):176–181

Zhong Y, Zhuang C, Gu W, Zjao Y (2019) Effect of molecular weight on properties of chitosan films prepared using electrostatic spraying technique. Carbohydr Polym 212:197–205

27 Antimicrobial Edible Packaging: Applications, Innovations, and Sustainability

A. Poovazhahi and Monika Thakur

Abstract

The edible packaging is the recent interdisciplinary approach, which primarily focus on the sustainability and food safety. The antimicrobial edible packaging is a potential alternative method to protect the food products during their production process like preparation, storage, and distribution to improve the shelf life of the food product by reducing microbial activity. It is a technique where the antimicrobial microbes (like gram-positive bacteria, gram-negative bacteria, molds, and yeasts) are impregnated into the packaging film, which in turn extend the shelf life of the product by the controlled release of antimicrobial agents to hinder the growth of harmful bacteria. The polymer film suppresses the contaminating microbial activity growth with antimicrobial agents. The primary focus of the packaging technique is to protect the food from any physical damage and to preserve it for many days. The antimicrobial edible packaging is specially contrived to promote the shelf life of the product by using edible biopolymer in the packaging material and it also aims at maintaining the quality and safety assurance of the food product. The edible biopolymer used in the antimicrobial packaging is derived from the edible biomass like polysaccharides, protein, lipids, etc. This packaging extends the shelf life by controlling the gas exchange, moisture permeation, lipolysis, and oxidation process in the food product. The recent global foodborne microbial outbreaks have led to the search for more innovative ways to inhibit microbial growth in foods that will maintain their quality and safety. This chapter gives a detailed description about the various

A. Poovazhahi
Cranfield University, Cranfield, UK

M. Thakur (✉)
Amity Institute of Food Technology, Amity University, Noida, Uttar Pradesh, India
e-mail: mthakur1@amity.edu

A. Poonia, T. Dhewa (eds.), *Edible Food Packaging*,
https://doi.org/10.1007/978-981-16-2383-7_27

antimicrobial agents, which aim to enhance the shelf life of the food products, their commercial application in the industry, and sustainability also.

Keywords

Active packaging · Edible biopolymers · Antimicrobial agents · Sustainability

27.1 Introduction

In recent scenario, food preservation, quality parameters, and food safety are the major key challenges in the food industry. Consumers are facing lots of issues in the delivery of safe level of products with stringent regulations to prevent food-borne infections and diseases in the developing world (Thakur and Modi 2020). Antimicrobial packaging paves the way to most safe packaging with dual purpose of preserving and protecting the food from pathogens.

Antimicrobial packaging system is a novel development which incorporates antimicrobial agent into a polymer film to suppress the activities of targeted microorganisms that are contaminating foods. The food-borne pathogens endangering the health of the consumers have to be dealt with by the multifunctional bio-based antimicrobial packaging agents that provide enhanced food safety and stability (Sung et al. 2013). It is also defined as the system of packaging that incorporates antimicrobial agents into the matrix, which in turn prevents the microbial growth and, thereby, extends the food product life. This mode of packaging with antimicrobial agents in its structure is generally biodegradable and it also promotes toward the sustainability in the food supply chain. These antimicrobials agents are usually incorporated into the biopolymer in the packaging material or it is added as an additive in packaging system, and each antimicrobial agent have different effects toward the target microorganisms. The rate of antimicrobial activity of the packaging material is directly proportional to the antimicrobial property of the component. The antimicrobial packaging takes up the benefits of various hurdles like water activity, pH, RH, etc. The combination of different processing techniques to create a hurdle environment for the microbes is a key line technique for antimicrobial packaging. Incorporating this technique into packaging system will help to maintain the organoleptic characteristics of the food (Thakur et al. 2020). The antimicrobial activities in the food products will be achieved by addition of various antimicrobial things in the packaging system to extend the lag period of microbial growth curve and decrease their viable counts (Han 2003).

But, nowadays, food security is also one of the biggest issues and both food safety and food security are to be worked upon simultaneously; therefore, the antimicrobial packaging system has been specifically designed to control the growth of microorganisms that severely affect the above-mentioned three main goals.

27.2 Antimicrobial Packaging System

Antimicrobial packaging is one of the applications of active packaging and it prevents the surface growth of pathogenic microbes in food by use of various antimicrobial agents where large portion of spoilage and contamination occurs. Basically, this system allows a controlled release of antimicrobial agents with sound barrier and active layers outside the food surface during storage and distribution.

The basic conventional food packaging system aims at three basic concepts:

1. Extension of shelf life.
2. Quality maintenance.
3. Safety assurance.

The antimicrobial packaging is specifically designed to control the growth of microorganism that generally affects the above-mentioned goals adversely. The products which are not affected by microbial contamination, may not need the antimicrobial packaging, but, since most of the food products belongs to perishable category, are much susceptible to microbial contaminations.

Therefore, the primary goal of an antimicrobial packaging system is as follows:

1. Safety assurance.
2. Quality maintenance.
3. Extension of shelf life.

Therefore, we observed that the role of antimicrobial packaging is reverse of the primary goals of conventional packaging systems (Fig. 27.1).

Antimicrobial packaging systems consist of two main components:

1. **Antimicrobial agents:**
   (a) Natural: plant (spices/oils/herbs), animals (chitosan), or microbial (bacteriocins/enzymes) sources.
   (b) Chemical/artificial: benzoic acid and sorbic acid.
2. **Films:**
   (a) LDPE, LLDPE, HDPE, PET, polyolefin, EVA, MC, WPI, SPI, egg albumen, methyl cellulose and corn zein films.

This method of packaging involves use of three modus operandi to create a challenging environment for the microbial growth. It actually delays the lag phase of microbes in the growth curve, thereby inhibiting its favorable environment with the addition of antimicrobial agents in its structure. The three working modes include (DSO) disentangle, subsume, and obstruction techniques.

Fig. 27.1 Order of the primary goals of antimicrobial and conventional packaging systems

1. **Disentangle mode:** It is the mode where the hurdle environment for the microbes is created by the controlled release of antimicrobial agents into the headspace of the packed products and thereby inhibiting the growth of microorganisms.
2. **Subsume**: It is the process where the positive environment factors responsible for the microbial growth are absorbed by the antimicrobial agent in the packaging system, thereby creating a hurdle for the growth of microbes. The favorable conditions for the microbial growth include water activity, carbon dioxide, oxygen, and pH.
3. **Obstruction:** It is the process where the antimicrobial agents will suppress the growth of microbes in the surface of the packaging system and it is very effective for liquid foods.

Example: Glucose oxidase enzyme in the polymer package acts as a suppress system for the microbial growth (Fields et al. 1986). There is no universal antimicrobial agent that can be used against array of microbes and the act of choosing the best antimicrobial agents is purely dependent upon the target microorganisms and the packaging material. The active packaging is the precursor for antimicrobial packaging (Brody et al. 2001) and it diversified the applications in the food industry (Muthukumarasamy and Holley 2003; Vermeiren et al. 2002).

The elements of antimicrobial packaging involve packaging material, antimicrobial agents, in-house packaging atmosphere, and food product. The carrier for the antimicrobial agents to enter the packaging system is the packaging material, and the choice of choosing the packaging material depends on the food product, cost involved, and it is purely a manufacturer's choice. The two most common routes of antimicrobial agent carrier in food industry are biodegradable and nonbiodegradable.

1. Biodegradable packaging can be degraded into the landfill when disposed by the action of fungi, algae, and bacteria, and thereby helping us to achieve sustainability. Edible coatings, films from protein, lipids, starch, polylactic acid, polyhydroxyalkanoate, and polyhydroxy butyrate can be used as a biodegradable packaging using antimicrobial agents as a carrier (Dawson et al. 2003).
2. Nonbiodegradable materials like high-density polyethylene (HDPE) and low-density polyethylene (LDPE) not only possess low capability to degrade and cause landfill depletion, pollution, etc. but also possess high mechanical strength, sealability, and printability, which make it more advantageous than other biodegradable packaging materials.

27.3 Aspects of Engineering Properties of Antimicrobial Packaging Material

The method of incorporating antimicrobial agent in the packaging material requires a keen understanding of the engineering and design parameters in choosing the packaging material. The thermomechanical characteristics of the packaging material in the design consideration include tensile strength, elastic modulus, durability, elongation length, etc. The material used in the packaging should also act as a barrier such as permeability to maintain food quality throughout the process. The reciprocity between the packaging polymer and the antimicrobial agent is the baseline criteria in determining the successful packaging, and it is purely dependent on the molecular weight of the antimicrobial agents imbibed. Cellulose- or gelatin-based polymer in combination with synthetic polymer is the most recommended packaging material.

The solubility of antimicrobial agent also affects the thermomechanical characteristics of the packaging material. Techniques like lamination, blending, spraying, and coating also possess positive impact in incorporating these antimicrobial components, which greatly affects the tensile characteristics of the material. Recent studies stated that antimicrobial agents—nisin and natamycin—have influenced the engineering properties and reduced the resistance toward the elongation length values and resulted in fourfold decrease in the control films (Coma et al. 2001). In a commercial study, where the antimicrobial agents like sodium citrate and sodium metabisulfite imbibed with gelatin-based films, they have resulted in positive impact of the mechanical properties like tensile strength etc. The choice of antimicrobial agent and the system is the key factor which ensures the efficiency of the antimicrobial packaging.

The key criterion that is most crucial in food packaging includes gas permeability properties of the packaging material because improper barrier characteristics of the material will tend to increase the spoilage mechanisms in food. Gas barrier characteristics of the packaging material are directly dependent on the choice of antimicrobial agents. In some cases, the antimicrobial substances are incorporated with polymer which results in decrease in the permeability of water vapor, whereas in some cases, incorporating silver into the polymer film increased the permeability of water vapor and hydrophobicity and this is because of its ability to create pinhole in the film matrix. The imbibing phenomenon of antimicrobial components in the packaging material also influences the barrier property of the films. The recent techniques like immersion and supercritical solvent impregnation increased the hydrophobicity of film and resulted in better barrier properties.

The antimicrobial agents in the packaging system can be released in the controlled phase through the single/double-layer food packaging through migration, diffusion, and immobilization, whereas in multilayer packaging system, the antimicrobial agents are released as a volatile into the headspace by evaporation or introduced into the matrix layer and then released back into the headspace. The other techniques of antimicrobial agents in the packaging system involves coating the product and introducing the antimicrobial pouches and sachets in the packaging. The antimicrobials can also be added as a food ingredient in the formulation.

27.4 Mode of Antimicrobial Action and Its Volatility

The antimicrobial components are usually introduced into the surface of the packaged foods because the exterior surface is prone to microbial growth than the inner surface. The antimicrobial activity in packaging material can be introduced into the headspace and package atmosphere, and it can be incorporated into the surface to inhibit the microbial activity. The impregnation of antimicrobial agent and the method of embodiment in packaging material help to achieve effective antimicrobial packaging.

The edible coating has numerous advantages because of its biodegradable properties in nature, and involves dry coating and wet coating. Dry coating is the process of mixing antimicrobial agents which itself acts as a microbial barrier, whereas wet coating involves the utilitarian agent like probiotics and an external binding cover. The predominantly used probiotics in wet process coating is lactic acid bacteria and it finds application in fresh produced food products, meat industry, etc. (Baron and Sumner 1993; Daeschul 1989).

The chemical gridlock in packaging material covalently adheres the antimicrobial agents, which thereby not only hinders the migration route of harmful chemical agents in the food surface but also helps in inhibition of microbial growth in packed foods. There are basically three different methods to incorporate antimicrobial properties into packaging system as explained below:

27.4.1 Passage of Nonvolatile Agents

The nonvolatile antimicrobial agents are usually introduced in the packaging system via spraying onto the food surface or by sandwiching in between the food surface and packaging material. The mass transfer kinetics, diffusion coefficient, partition coefficient/solubility, and concentration distribution activity are the prime phenomenon for building the effective antimicrobial packaging. Diffusional migration is highly dominated by the mass transfer of nonvolatile antimicrobials on to the food surface. If the antimicrobial agents are sprayed on to the food surface, the first initial surface concentration will be in peak and tends to reduce due to its diffusivity and partition coefficient. If the antimicrobial components are incorporated into the packaging material, its action begins at run-off from the packaging material and then released into the food produce. Therefore, the diffusivity and the solubility agents are the key characteristics in mass transfer of the antimicrobial agents. This mode of nonvolatile antimicrobials agents prefers continuous matrix food, for example, one-piece solid (fermented meats), sausages battered with antimicrobial agents, cheeses spayed with potassium sorbate, antimicrobial wax coating in fresh produce, etc. (Cooksey et al. 2000; Devlieghere et al. 2000; Ming et al. 1997).

27.4.2 Passage of Volatile Antimicrobial Agent

The other method of incorporating the antimicrobial properties into the packaging system involves addition of volatile agents which improves the packaging efficiency by direct contact with the antimicrobial agent. The equilibration between the headspace concentration and the food surface concentration relies on the impact in determining the volatile agent concentration in the packaging system. It is very important to determine the headspace concentration of the volatile antimicrobial agent. The volatile agent is introduced in the packaging material, where it is vaporized into the headspace, and it reaches the food surface, and finally, it is absorbed by the food. The rate of volatile agent release is directly proportional to the volatility of the antimicrobial agent and this can be stabilized and controlled with the help of microencapsulation, cyclodextrin. The food substance also affects the absorption characteristics of the volatile components, and its chemical reaction kinetics with the volatile components. This finds applications in porous foods and powdered foods, such as grated cheese, beef, fruits, vegetables, etc. (Cooksey et al. 2000).

27.4.3 Passage of Nonmobilization of Volatile Agents

In this method, the antimicrobial agent is not diffused in the surface of the food rather it is attached to the polymer linkage. The bioactive agents like enzymes, proteins, etc. interlinked to the polymer via the cross linkers and its activity are limited to the surface contact (Dawson et al. 2003). This system of nonmobilization of volatile composition can be effectively used in the process to change any existing substrates into bioavailable compounds using immobilized enzymes. The antimicrobial agents involved in this system are immobilized and it have unique advantage in marketing and regulation because the food does not contain any chemical antimicrobial agent.

27.5 Affinity Between the Process and Food Substrate

The antimicrobial effectiveness is highly dependent on the process methodology. In extrusion technology, the temperature and force applied as input play a significant role in determining the effectiveness of antimicrobial activity. The antimicrobial agent is not heat resistant, whereas it is subjected to thermal degradation in the process of extrusion. In solvent casting films, the solubility and diffusivity of antimicrobial agents are the critical parameters in designing antimicrobial packaging. The homogenous and heterogeneous characteristics of food substrate will hinder the solubility of antimicrobial agents.

Water-soluble antimicrobial agents are imbibed into plastic resins, and then, it is extruded as films with antimicrobial agents incorporated into it, without losing the physical integrity of the film. The choice of hydrophilic agent and hydrophobic plastic resin will affect the transparency of the film.

The pH in the packaging material also affects the chemical structure of the antimicrobial component and the food composition. The time-temperature relationship affects the growth rate of microbial growth, chemical reaction of antimicrobial agents, and the storage. In MAP with antimicrobial gas, the gas permeability is greatly affected by storage and distribution. When the gas in the modified atmosphere packaging is altered in the antimicrobial system environment, it will lead to the spoilage of food produce.

27.6 Affinity Between Packaging Material and Food Substrate

The technique of incorporating antimicrobial components also affects the packaging material, and its composition deteriorates the physical and mechanical integrity of the material. The polymer package in packaging material and the imbibed antimicrobial agent relate to the strength of the packaging material. Polymer-polymer interaction helps in designing the amorphous structure of the packaging material. The level of antimicrobial agent in the packaging material reflects the level of damage to the integrity of the components of the material.

The implication involved in designing of antimicrobial packaging system is the careful consideration of migration characteristics of antimicrobial components and the growth kinetics of the microorganisms. If the diffusion kinetics of antimicrobial agent is higher than the growth kinetics of the target microbes, it will lead to the decomposition of concentration of the antimicrobial component and the packaging will no longer be effective to the microbial load. When the diffusion rate of antimicrobial agent is lower than the growth kinetics of microbes, it will lead to the microbial growth before the inhibition process acts upon the packaging system. The release rate of antimicrobial components into the packaging system is directly proportion to the growth rate of the target microbial load. In case of coating of the antimicrobial agent in food, the mass transfer of the component should be minimum to retain the coating layer in the food layer for a longer period to inhibit the growth rate of microorganism. The migration and mass transfer should be properly designed to increase the antimicrobial effectiveness of the packaging system (Vojdana and Torres 1990).

The rate of diffusion of antimicrobial agent in the packaging system is also controlled by the permeability of the packaging material and in case of liquid produce; the release rate is governed by the base layer. Certain volatile antimicrobial agent includes horseradish oil which contains isothiocyanate as the major antimicrobial agent in its composition and this will create a space between oil and headspace concentration.

27.7 Application of Antimicrobial Films

Antimicrobial edible packaging is one of the major applications of active packaging in the food products. It is one of the emerging technologies with very significant impact on extending the shelf life of perishable food products by reducing, inhibiting, or retarding the growth of spoilage microflora. Nowadays, food security is a big issue and antimicrobial packaging could play a pivotal role in food security assurance to every common consumer. Various advantages of using antimicrobial packaging systems are mentioned below:

1. Multifunctional role by reducing harmful microbial load in food.
2. Help to increase food safety.
3. Reduce the food wastage and improve shelf life.
4. Bio-based antimicrobial agents in packaging provide extra safety for consumer's health.

The application of antimicrobial films is multifold. Some of the applications are explained below:

1. The method of articulating antimicrobial agents in antimicrobial films are generally classified as natural agents and chemical agents. The antimicrobial agents used in food industry include benzoic acids, sorbates, lysozymc, nisin, EDTA, citrate, and silver zeolite (Appendini and Hotchkiss 1997; Chen et al. 2003; Dobias et al. 2000). The antimicrobial agents incorporated into the packaging material are dispensed as vapor into the headspace.
2. Potassium sorbate is commonly used in food industry because it possesses a strong antimicrobial property against the fungi (Takeuchi and Yuan 2002).
3. The zinc nanoparticles have recent application in antimicrobial packaging where it affects the cell wall permeability, respiration, etc. Metallic substances like silver and titanium oxide show antimicrobial activity after UV excitement.
4. Antimicrobial agents are sometimes used in the gaseous form and to implement these gases like CO_2, SO_2, ClO_2, C_2H_5OH, and O_3 find applications in antimicrobial packaging.
5. Certain bacteria like LAB produce bacteriocins and nonpeptide growth-inhibiting chemicals (reuterin), which will retard the bacterial growth in the substance (Kim

et al. 2000; Cutter et al. 2001). Certain bacteriocins like pediocin, lacticin, propionicin, nisin, and EDTA provide antimicrobial activity against bacteria, especially gram-positive bacteria like *Listeria monocytogens* (Coma et al. 2002; Cutter et al. 2001; Kim et al. 2002).

6. The probiotics is a source of beneficial bacteria and it is rich in antimicrobial probiotics. Researches are carried out to find the possibility of using probiotic antimicrobials in antimicrobial packaging.
7. The inorganic particles, especially nanosized molecules, inhibit the wide spectrum of microorganism by generating the radical scavengers and ROS (reactive oxygen species) which will oxidize the cell membrane component phospholipids.
8. The most inorganic antimicrobial agent used includes titanium dioxide and zinc oxide. Studies revealed that titanium dioxide coated with polypropylene film reduced the *Escherichia coli* count in lettuce (Cha et al. 2002). Further treatment of polypropylene film exposed to UV radiation for 1 h drastically improved the antimicrobial activity of the film against *Escherichia coli* and *Staphylococcus aureus* (Chawengkijwanich and Hayata 2008).
9. Polyvinyl alcohol (PVA) film coated with silver shows inhibition toward the microbes, *Listeria innocuous* and *E. coli*, and two fungi, *Aspergillus niger* and *Penicillium expansum* (Chen et al. 2003; Thakur et al. 2020). Also the silver coating with nanoparticles retards the meat spoilage bacteria.

Various antimicrobial agents are incorporated into the food packaging system and listed in Table 27.1.

27.8 Marketing Perspective

Antimicrobial agents are generally categorized as additives of packaging material, where they immobilize the antimicrobial components in the food surface to inhibit the microbial load. The natural antimicrobial agent like spices and herbs is the best alternative than the chemical antimicrobial agents (Thakur and Kale 2016).

In large-scale implementations of antimicrobial packaging systems, the supply chain, cost involved, and distribution are the major parameters involved. The antimicrobial agents can be implemented in existing packaging line with minimal cost. The level of consumer acceptance to the upgradation of packaging system is a challenge faced while implementing the antimicrobial packaging (Suppakul et al. 2003). The antimicrobial agent should not affect the organoleptic properties of the final produce. The three prime goals of antimicrobial packaging system include food safety assurance, quality control, and shelf-life improvement (Kelly et al. 2003; Steven and Hotchkiss 2003). There is a lot of market potential in the present scenario, where the whole world is suffering from various microbial infections. Therefore, while working on the development of any antimicrobial packaging systems, we have to focus on these three prime goals only and have immense marketing perspective.

Table 27.1 List of most commonly used antimicrobial agents in food packaging

	Antimicrobials agents	Food applications	Packaging carrier	Target microorganism
1	**Organic acids** 1. Benzoic acid 2. Sorbates 3. Sorbic anhydride	Fresh meat and fillets	PE and LDPE	Total bacteria migration test
		Cheese and poultry	LDPE, chitosan, and starch/ glycerol	Yeast and mold migration test
		Fruits, vegetables, and jam	PE	*Saccharomyces cerevisiae* and molds
2	**Enzymes** 1. Lysozyme, EDTA, and Nisin	Fish, ham, meat balls, and cheese	SPI and Zein	*Escherichia coli* and *Lactobacillus plantarum*
	2. Lysozyme immobilized	Fish	PVOH, nylon, and cellulose acetate	Lysozyme activity test
3	**Bacteriocins** 1. Nisin 2. Lauric acid	Beef and sausage	PE	*Brochothrix thermosphacta* and *Staphylococcus aureus*
		Fish and shellfish	PE	Migration test
4	**Fungicides** 1. Benomyl 2. Imazalil	Strawberry	Ionomer	Molds
		Bell pepper cheese	LDPE and PE	Molds
5	**Polymers** 1. Chitosan 2. Chitosan/ herb extract	Strawberry	Chitosan/ paper	*E. coli*
		Citrus and berries	LDPE	*E. coli, lactobacillus plantarum, Fusarium oxysporum* and *S. cerevisiae*
6	**Natural extract** 1. Grapefruit seed extract 2. Clove extract 3. Eugenol and cinnamaldehyde 4. Horseradish extract	Ground beef	LDPE and nylon	*Aerobes sp. and* coliforms
		Fresh meat, noodles, and pasta	LDPE	*E. coli, L. plantarum, Fusarium oxysporum,* and *S. cerevisiae*
		Bologna and ham	Chitosan	*Enterobacter* and *Serratia sp.*
		Ground beef	Paper	*E. coli*
7	**Oxygen absorber** 1. Ageless	Bread	Sachets	Molds
8	**Gas** 1. Ethanol	Fresh produce	Silica gel sachets	Molds

(continued)

Table 27.1 (continued)

	Antimicrobials agents	Food applications	Packaging carrier	Target microorganism
9	**Others** 1. Silver zeolite 2. Antibiotics	Nuts, jam, and jelly	LDPE	*E. coli, Staphylococcus aureus, S. cerevisiae*, and *Salmonella typhimurium*
		Fermented vegetables and yogurt	PE	*E. coli, Staphylococcus aureus, S. cerevisiae, Sal. Typhimurium*, and *Klebsiella pneumoniae*

PE polyethylene, *HDPE* high-density polyethylene, *LDPE* low-density polyethylene, *HPMC* hydroxypropyl methylcellulose, *MC* methylcellulose

27.9 Innovations and Sustainability in Antimicrobial Packaging

The sustainability for active antimicrobial packaging with the potential to increase the food safety with the prevention of resistant strains of pathogenic microbes is the need of the hour. A lot of research is going on in the development of new antimicrobial packaging films as the other technologies can passively protect the food products but, by use of antimicrobial technology, we can inhibit the growth of pathogenic microbes in the food products and also extend the shelf life of food products while combating various environmental factors. New innovations are used in the development of antimicrobial-based nanocomposite food packaging films having enhanced thermal, physicochemical, optical, and mechanical properties as well. In the current scenario, the nano-based films with antimicrobial strains are in use and they also improve the efficacy of the delivery of antimicrobial agents very well Tunç and Duman 2011).

The use of the material in the development of these antimicrobial films is safe for consumption and use, but still we have to follow the regulatory guidelines. It has been observed that few components are allowed as per EU and have GRAS certifications, but the same components are banned in other countries due to toxicological reasons. On a broader scale, these antimicrobial components do not affect the organoleptic properties of the food products, but, many times, there has been change in color, texture and appearance of the food products due to release of many essential oils (Burt 2004; Tongnuanchan et al. 2012). The use of antimicrobial film strategies to extend the shelf life of food products allows efficiency in the food supply chain and also enhances the sustainability of the food production. Since the antimicrobial films are sustainable and biodegradable, they are able to meet the growing demand for "Green Packaging" from the flourishing food sector.

27.10 Conclusion

Antimicrobial packaging has achieved great potential in recent times because of its ability to inhibit the microbial growth. Researches and studies have revealed the benefits of incorporating harmless antimicrobial agents to improve the shelf life, quality, and sensory properties of the food products. In designing antimicrobial films, the mechanical and barrier properties should be analyzed to achieve maximum antimicrobial efficiency. It is really an emerging technology that helps to satisfy the consumer demands for safe, convenient, wholesome products in the current times. This technique serves a dual purpose of increasing the shelf life and improving the food safety. Thus, knowing the inherent unimaginable potential of antimicrobial agents in food products, a cutting-edge research is still required on the development of antimicrobial packaging systems, and we can further explore other underexploited products for their developments.

References

Appendini P, Hotchkiss JH (1997) Immobilization of lysozyme on food contact polymers as potential antimicrobial films. Packag Technol Sci 10(5):271–279

Baron JK, Sumner SS (1993) Antimicrobial containing edible films as an inhibitory system to control microbial growth on meat products. J Food Protect 56:916

Brody AL, Strupinsky ER, Kline LR (2001) Active packaging for food applications. Technomic Publishing Co., Lancaster, PA, pp 131–196

Burt S (2004) Essential oils: their antibacterial properties and potential applications in foods—a review. Int J Food Microbiol 94:223–253

Cha DS, Choi JH, Chinnan MS, Park HJ (2002) Antimicrobial films based on Na-alginate and kappa-carrageenan. Lebensm Wiss Technol 30(8):715–719

Chawengkijwanich C, Hayata Y (2008) Development of TiO_2 powder coated food packaging film and its ability to inactivate *Escherichia coli in vitro* and in actual tests. Int J Food Microbiol 123:288–292

Chen S, Wu JG, Chuo BY (2003) Antioxidant and antimicrobial activities in the catechins impregnated PVA-starch film. In: Book of Abstracts, 2003 IFT Annual Meeting. Institute of Food Technologists, Chicago, IL

Coma V, Sebti I, Pardon P, Deschamps A, Pichavant RH (2001) Antimicrobial edible packaging based on cellulosic ethers, fatty acids, and nisin incorporation to inhibit *Listeria innocua* and *Staphylococcus aureus*. J Food Protect 64(4):470–475

Coma V, Martial-Gros A, Garreau S, Copinet A, Salin E, Deschamps A (2002) Edible antimicrobial films based on chitosan matrix. J Food Sci 67(3):1162–1169

Cooksey DK, Gremmer A, Grower J (2000) Characteristics of nisin-containing com zein pouches for reduction of microbial growth in refrigerated shredded Cheddar cheese. In: Book of Abstracts. Institute of Food Technologists, Chicago, IL

Cutter CN, Willett JL, Siragusa GR (2001) Improved antimicrobial activity of nisin-incorporated polymer films by formulation changes and addition of food grade chelator. Lett Appl Microbiol 33(4):325–328

Daeschul MA (1989) Antimicrobial substances from lactic acid bacteria for use as food preservatives. Food Technol 43(1):164–167

Dawson PL, Hirt DE, Rieck JR, Acton JC, Sotthibandhu A (2003) Nisin release from films is affected by both protein type and film-forming method. Food Res Ind 36:959–965

Devlieghere P, Vermeiren L, Bockstal A, Debevere J (2000) Study on antimicrobial activity of a food packaging material containing potassium sorbate. Acta Alimen 29(2):137–146

Dobias J, Chudackova K, Voldrich M, Marek M (2000) Properties of polyethylene films with incorporated benzoic anhydride and/or ethyl and propyl esters of 4-hydroxybenzoic acid and their suitability for food packaging. Food Add Contam 17(12):1047–1053

Fields C, Pivamick LP, Barnett SM, Rand A (1986) Utilization of glucose oxidase for extending the shelf life of fish. J Food Sci 51(1):66–70

Han JH (2003) Antimicrobial food packaging. Food Technol 54:56–65
Kelly M, Steele R, Scully A, Rooney M (2003) Enhancing food security through packaging. Intl Rev Food Sci Technol 11:115–117
Kim DS, Lee YM, Paik SB, Lee DS (2000) Antimicrobial low density polyethylene film coated with bacteriocins in binder medium. Food Sci Biotechnol 9(1):14–20
Kim YM, Lee NK, Paik HD, Lee S (2002) Migration of bacteriocin from bacteriocin-coated film and its antimicrobial activity. Food Sci Biotechnol 9(5):325–329
Ming X, Weber GH, Ayres JW, Sandine WE (1997) Bacteriocins applied to food packaging materials to inhibit *Listeria monocytogenes* on meats. J Food Sci 62(2):413–415
Muthukumarasamy, Han JH, Holley RA (2003) Bactericidal effects of *lactobacillus reuteri* and allyl isothiocyanate on *Escherichia coli* 0157:H7 on refrigerated ground beef. J Food Protect 66 (11):2038–2044
Steven MD, Hotchkiss JH (2003) Non-migratory bioactive polymers (NMBP) in food packaging. In: Novel food packaging techniques. Woodhead Publishing Ltd., Cambridge
Sung SY, Sin LT, Tee TT, Bee ST, Rahmat AR, Vikhraman M (2013) Antimicrobial agents for food packaging applications. Trends Food Sci Technol 33:110–123
Suppakul P, Miltz J, Sonneveld K, Bigger SW (2003) Active packaging technologies with an emphasis on antimicrobial packaging and its applications. J Food Sci 68(2):408–420
Takeuchi K, Yuan J (2002) Packaging tackles food safety: a look at antimicrobials. Industrial case study 'Industrial Case Studies: Meats, Poultry, Produce, Juice, Smoked Salmon'. Institute of Food Technologists, Chicago, IL
Thakur M, Kale R (2016) Studies on biology and functional components of the Indian spices. Lambert Academic Publishing, Saarbrucken
Thakur M, Modi VK (2020) Emerging technologies in food science- focus on the developing world. Springer Nature, Singapore
Thakur M, Modi VK, Khedkar R, Singh K (2020) Sustainable food waste management: concepts and innovations. Springer Nature, Singapore
Tongnuanchan P, Benjakul S, Prodpran T (2012) Properties and antioxidant activity of fish skin gelatin film incorporated with citrus essential oils. Food Chem 134:1571–1579
Tunç S, Duman O (2011) Preparation of active antimicrobial methyl cellulose carvacrol/montmorillonite nanocomposite films and investigation of carvacrol release. Food Sci Tech 44:465–472
Vermeiren L, Devlieghere F, Debevere J (2002) Effectiveness of some recent antimicrobial packaging concepts. Food Add Contam 19:5163–5171
Vojdana F, Torres JA (1990) Potassium sorbate permeability of methylcellulose and hydroxypropyl methylcellulose coatings: effect of fatty acid. J Food Sci 55(3):841–846

Antioxidant-Rich Edible Packaging

28

Himanshi Garg, Soumya Ranjan Purohit, Vasudha Sharma, and Jatindra Kumar Sahu

Abstract

Edible packaging is a recent trend in food packaging industry. In last few years, it has gained prime importance due to potential functionality apart from basic role of packaging. Antioxidant-rich edible packaging is one such example which has well acknowledged and recognized in food packaging industries. The present chapter deals with edible food packagings incorporated with various antioxidants. The chapter outlines studies related to incorporation of antioxidants in edible films and coatings and shelf-life extension. Further, applications, advantages and disadvantages of antioxidant-based edible packaging are discussed in this chapter. The approach of this chapter is to focus more on applications of nature-based antioxidant in edible packaging, which includes essential oils, spice extracts, green tea extracts and alpha tocopherol among others. Further, regulatory aspects are reviewed and future aspects are also outlined.

Keywords

Antioxidant · Edible packaging · Food deterioration · Sustainability

H. Garg · S. R. Purohit (✉)
Amity Institute of Food Technology, Amity University Uttar Pradesh, Noida, Uttar Pradesh, India

V. Sharma
Department of Food Technology, Jamia Hamdard, New Delhi, India

J. K. Sahu
Centre for Rural Development and Technology, IIT Delhi, Hauz Khas, New Delhi, India

A. Poonia, T. Dhewa (eds.), *Edible Food Packaging*,
https://doi.org/10.1007/978-981-16-2383-7_28

28.1 Introduction

Food packaging not only protects but also advertises a food product. But with the advancement in food packaging industry, new dynamic concepts are outlined which includes active and intelligent packaging. The latest packaging demands the package to not only contain the product but also significantly helps in increasing the shelf-life of the product simultaneously maintaining food quality and monitoring safety. In similar context, natural and biodegradable packaging materials has become a trend these days. Consumer wants fresh and natural products enriched with nutrients and increased shelf life. And this shelf life of food product is dependent on how long it can resist the change in quality and safety but food deteriorates fast as it has to go through a long chain to reach to table. There are many phenomena responsible for food deterioration based on its composition but lipid oxidation is a well-known phenomenon responsible for food deterioration which ultimately results in:

1. Nutritional value loss (due to the destruction of essential fatty acids, proteins and lipid-soluble vitamins).
2. Decreased energy content.
3. Off flavour and odour production.
4. Toxic substances formation.
5. Change in colour (which influences consumers' behaviour).

Oxidation of food products and after results are major challenges for food industry. This is because all these challenges greatly impact not only the economy and consumer acceptance but also the environment. They demand the industry to look for some alternatives and if it could be served by packaging, then it is like cherry on cake. In past studies, vacuum or modified atmosphere packaging has been reported to be most effective in controlling oxidation reactions. However, these methods have certain limitations. Hence, direct incorporation of antioxidants to food and food packages was favoured by food industries for complete elimination of oxygen and enhancing the product life.

But recently, consumer and the food industry are looking for biodegradable and compostable packaging systems which not only ensures food safety by delaying deterioration but also somewhere reduce burden on environment. All these have led to a great shift in food packaging industry from normal plastic food packaging to active antioxidant packaging and now to packaging which not only curbs pollution but also can be eaten with food itself and serves with functional food requirements as well, that is, edible packaging enriched with antioxidants. So, the emerging nutraceutical industry, changing consumer mindset and industrial perception have created a platform for antioxidant-rich edible packaging to boom. From past 5 years, there is considerable increase in marketing of edible packaging.

However, this chapter will try to give a brief insight of antioxidant-rich edible packaging where firstly an introduction to antioxidants, their working will be given, following which introduction to edible packaging with its importance or why it came into existence will be mentioned. Further, this chapter will provide a detailed information about antioxidant-enriched edible packaging, studies based on edible packaging enriched with antioxidants, advantages and disadvantages of this kind of packaging, ways to enhance the film functional properties, some future trends and in end regulatory aspects will be discussed.

28.2 What Are Antioxidants?

Antioxidants are the substances which delays the onset or slow down the rancidity or oxidation reaction rate. However, the antioxidant structure influences the course of the food oxidation. Examples include α- tocopherol, butylated hydroxyl anisole (BHA), etc.

28.2.1 Antioxidants Work in the Following Manner

1. It may inhibit the chain reaction by acting as hydrogen donor or free radical acceptor. Example: phenols present in essential oils (EOs).
2. Decomposing peroxides, producing stable substances that are unable to produce radicals, such as tioethers, methionine, tiodipropionic acid, or even some enzymes such as glutathion peroxidase (Nerín 2010).
3. Forming complexes with metals that catalyse oxidation reactions. Examples of such are tartaric acid, citric acid, oxalic acid, ascorbic acid, succinic acid, oxalate, phosphate, gluconic acid and ethylenediaminetetraacetic acid (EDTA) (Nerín 2010).
4. It may prevent the reactions caused by reactive oxygen (Tkaczewska 2020). Examples include tocopherol and carotenoids.
5. Inhibiting oxidative enzymes (especially lipoxygenases). Examples include flavanoids, phenolic acids and gallates (Jamróz et al. 2021; Tkaczewska 2020).

So far, artificial antioxidants, such as butylated hydroxy anisole (BHA), butylated hydroxy toluene (BHT), tertiary butyl hydroquinone (TBHQ) and propyl gallate (PG) have been used as direct incorporation into food. But today's consumer is interested in consuming more natural and functional food that is health promoting, while industry desires to meet consumer demands with sustainability. The solution to this is none other than antioxidant-rich edible packaging which is the target of researchers these days.

Before going on to what all researches have been done, what all positive or negative impacts this kind of packaging has; let's get introduced with edible packaging and before that what all other methods are there to preserve the food from lipid oxidation or oxidative deterioration.

28.3 Edible Packaging

Lipid oxidation of food is an irreversible change which generates the deteriorative changes in food especially in terms of nutritional and sensorial qualities. The lipid oxidation of food occurs in three stages considering specifically auto-oxidation, that is, initiation, propagation and termination. The reactions involved in it has been depicted in Fig. 28.1.

Besides this, enzymatic oxidation of fruits and vegetables because of polyphenoloxidase (PPO) results in serious colour changes. This all has led to look for methods to eliminate oxygen from food products.

The methods for preventing deteriorative changes in food were traditionally focused on oxygen elimination from food package via either vacuum packaging or modified atmosphere packaging.

- *Vacuum packaging*: In this, vacuum is created around the product and then it is packaged in a packaging material with great oxygen-barrier properties. However, oxygen entrapped within food layers may get diffused which is controlled by lowering temperature. This may result in tissue drying and separation of lipid fractions in natural emulsions (Nerín 2010).
- *Modified atmosphere packaging (MAP)*: In this the atmospheric conditions around the commodity are changed from that of air to extend the shelf life of product to some extent. Examples: meat, fruits and vegetables, etc.

Later on, several formats of active packaging too have been developed and commercialized which acts as oxygen absorbers and oxygen scavengers like reduced iron salts (moisture in food packages led the iron oxidized to a ferric state) and ascorbic acid respectively.

But all the above-said techniques are partially effective in oxidation prevention which led researchers to look for alternatives and that is none other than edible

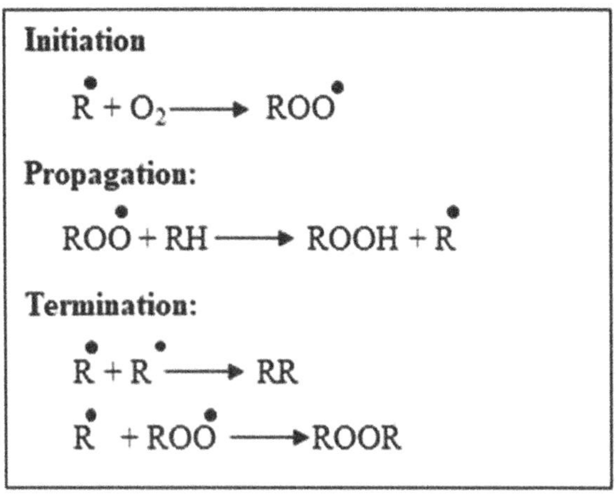

Fig. 28.1 Reactions involved in lipid oxidation

packaging which is preferred because instead of adding antioxidants to food they can be added to packaging material hence keeping food safe and chemical free.

Edible packaging is kin to active packaging technology. Further, edible coatings not only protect the containment from oxygen, moisture and light but also serves functional compounds required for human health and well-being in incorporated form like phytochemicals, antioxidants, flavours, essential oils, nutraceuticals, etc.

Well before going on to incorporation of antioxidants, it is necessary to be aware of what is edible packaging, what are the advantages of using it, the basic components involved in edible package making, etc. Therefore, any kind of edible material used for food packaging to enhance the shelf life of the product while maintaining appropriate organoleptic quality which can be eaten with the product itself is termed as edible package and it is of two types on the basis of how it has been applied:

1. *Coatings*: Spraying or brushing film forming solution and drying it to form a uniform layer.
2. *Films*: It is a dried film forming solution applied on food product (Ganiari et al. 2017).

An edible film or coating does not act as a package itself (Bhardwaj et al. 2019). However, edible coatings and films performs the following functions:

- Acting as barrier against oxygen, moisture, CO_2, oils and other aromatic substances (Tkaczewska 2020).
- Improving visual and tactile features of food (Tkaczewska 2020).
- Potentiality to envelope active substances which includes flavours, antimicrobial, anti-browning agents, antioxidants, nutraceuticals, colourants and spices.
- Organoleptic properties and nutritional value of the food get enhanced (Ganiari et al. 2017).
- Lowers impact on environment being biodegradable as well as consumable (Ganiari et al. 2017).
- Limits the usage of additional outer packaging.

Edible films are usually made from:

- *Animal proteins* like casein, creatine, gelatin and collagen.
- *Vegetable proteins* like soy proteins, gluten and zein.
- *Polysaccharides* like cellulose and its derivatives, chitosan, pectin, carrageenan and starch.
- *Fats and waxes* (Tkaczewska 2020).

Studies have shown that protein films exhibit good oxygen-barrier properties while polysaccharide films exhibit moderate mechanical properties (Salgado et al. 2015; Tkaczewska 2020). Similarly, evidences have been found that by mixing proteins and polysaccharides, composite films of multilayer or single layer can be manufactured which can exhibit the best properties of each component by avoiding individual film disadvantages (Salgado et al. 2015).

However, film efficiency is strongly dependent on gas reservation, water vapour, oxygen permeability (OP) and odour, which sequentially depends on the composition of the product, polymers structure, product characteristics and storage conditions (Skurtys 2014).

Literature has some evidences which make it necessary to evaluate the antioxidant capacity of the antioxidant to be added to films or coatings. Other properties which influence antioxidant capacity includes retention power of the film, flavour, colour and chemical modifications of the food product. Moreover, chemical additives adversely affect the food taste hence, addition of natural food additives with antimicrobial and antioxidant properties have gained more importance these days, forming a platform for biodegradable and edible packaging.

Further, it has also been investigated that film-forming materials and material incorporated must be compatible so as to ease the release (Benbettaïeb et al. 2019).

In recent decade, many film-forming solutions containing antioxidants and antimicrobials agents have been found to be compatible enough to form films and protect food (Benbettaïeb et al. 2019; Halim et al. 2016; Tkaczewska 2020). Some of these are discussed in the next section.

28.4 Antioxidant-Rich Edible Packaging

The consumer safety and regulatory concerns associated with plastic packaging materials and usage of synthetic antioxidants have restricted its use in food packaging direct or indirect in terms of food protection and preservation. This issue has made researchers to develop nature-based active edible films containing antioxidant agents. Natural agents can be natural extracts (De'Nobili et al. 2013; Ganiari et al. 2017; Murcia and Martínez-Tomé 2001); pure compounds, which includes citric acid, ascorbic acid, tocopherol or resveratrol (Ganiari et al. 2017), natural tea extracts (Das et al. 2013; Ganiari et al. 2017; Li et al. 2014), fruits and vegetables extracts, (Akhtar et al. 2012; Ganiari et al. 2017; Supapvanich et al. 2012), ginseng extract (Ganiari et al. 2017; Norajit et al. 2010), plant extracts (Ganiari et al. 2017; Gómez-Estaca et al. 2009) and propolis (Ganiari et al. 2017; Pastor et al. 2011). These compounds are chosen because they not only help in forming antioxidant models which ensures reduced oxidation and extended shelf life but also act as functional food, simultaneously preserving the sensory and nutritive quality of the food (León and Rojas 2007), without impacting the product integrity.

On the other hand, talking from consumer perspective then today's consumer is more aware. They have been demanding product to be more natural, nutrient-enriched and health-promoting which makes the search for nature-based antioxidants packages. So, the trend towards usage of natural antioxidants in edible film formation has advanced the research in this field, and materials used and studied so far includes essential oils, spices extract, etc.

Essential oils (EOs) are mixture of terpenes, aromatic, phenolic and various other aliphatic and antimicrobial substances acting as natural antioxidant, having potential to slow down lipid degradation and extending the food life (Perdones et al. 2014; Tongnuanchan et al. 2013).

Similarly, spices extract also exhibited antioxidant properties, which make it an interesting direct additive in food. Further, these extracts contain high concentrations of phenolic compounds. Phenolic compounds present in spice extracts exhibit strong H-donating activity which plays a great role in preventing oxidation, hence spice extract incorporation in packaging material is also widely studied and has presented positive results.

The most studied spices and tea extracts include oregano, green tea, rosemary, sage, thyme, marjoram, savoury and many other herbs belonging to the *Lamiaceae* family and *Theaceae* family (Ganiari et al. 2017).

Oregano and rosemary extracts are found to be potent antioxidants for fish and meat products. In the similar context, gelatin or gelatin–chitosan-based films were prepared using water extracts of both herbs and added to pack smoked sardines, the result revealed these extracts-enriched film to be effective in retarding oxidation (Ganiari et al. 2017; Gómez-Estaca et al. 2007). Similarly, ethanol-based rosemary extracts incorporated with carboxy methyl cellulose (CMC) coating was found to be effective in protecting smoked eel fillets from oxidation (Ganiari et al. 2017). In another study, thyme and basil essential oils were incorporated in CMC coating for sunflower seeds. Results indicated that CMC-based coating with added antioxidants showed improved sensorial stability while solely thyme-oil-based coating showed chemical stability (Riveros et al. 2016; Sahraee et al. 2019).

Another widely studied extract in edible package manufacturing is green tea extract (Ganiari et al. 2017). Green tea extract has exhibited great antioxidant activity when incorporated in films like chitosan and ethylene vinyl alcohol. Further, the antioxidant potential of green tea-based films is found to effective in case of pork sausages and food simulants. Moreover, a group of researchers analysed tea extracts and gelatin-based film through Fourier transform infrared (FTIR) and found green tea extracts to be effective in reducing free hydrogen (Li et al. 2014).

In contrast, cashew nuts packed in mango leaf extract-based chitosan films exhibited the extended nuts shelf life (Rambabu et al. 2019; Sahraee et al. 2019). In another research, extra virgin oil was packed in gelatin and corn starch-based film blended with phenolic and ascorbic acid enriched guabiroba pulp (Malherbi et al. 2019; Sahraee et al. 2019). It was found to be effective in preventing acidity and peroxide value (PV) of the oil from exceeding the legislated maximum limit for 15 days (Sahraee et al. 2019). The results from the above-mentioned studies ensured that green tea extract in edible packaging is a potent material for manufacturing antioxidant-rich edible packaging for food.

It will be worth mentioning here that the incorporation of some natural and synthetic antioxidant compounds to film is not well accommodated in literature, reason being how these are going to affect film properties. However, a few studies included poly lactic acid (PLA) films which were blended with BHT, PG and α-tocopherol and which showed improved antioxidant properties of the films

(Byun et al. 2010; Sahraee et al. 2019). While in another study, cooked turkeys were packed in zein protein-based films blended with BHA (Herald et al. 1996; Sahraee et al. 2019). But in recent past, studies regarding nature-based antioxidants films have gained popularity. Examples of such edible films are discussed in the following section.

28.4.1 Application of Antioxidant Edible Films and Coatings

Antioxidants have been used in food as food additives from dates back. However, antioxidant-based packaging is the latest interest. No doubt, antioxidant-enriched packaging is used for fatty foods like fish, meats, oils, oil seeds and nuts. But this section specifically focuses on edible packaging enriched with antioxidants.

Researchers tried to induce oxidation inhibitory properties in gelatin film by adding β-cyclodextrin-encapsulated curcumin that was used for packaging of red Fuji (*Malus pumila mill*) apple juice to extends its storage life (Sahraee et al. 2019). The investigation revealed that films containing 2.5 mg curcumin exhibited increased antioxidant activity of the films with less degradation of total phenolic compounds compared to pure gelatin films (Wu et al. 2018).

Similarly, in an extended study, the oxidation inhibitory properties of chitosan nanoparticle, polyphenol tea extract-enriched gelatin films were investigated for fish oil. The investigation revealed that films decreased tensile strength and increased water vapour permeability (WVP). Interestingly, oxygen permeability (OP) of films and peroxide value (PV) of oil were found to decrease with increasing nanoparticles concentration as studied by 2,2′-diphenyl-1-picrylhydrazyl (DPPH) radical scavenging activity of films (Bao et al. 2009; Sahraee et al. 2019).

Literature also cites that natural antioxidants like vitamin C and E once incorporated in edible films and coatings will play a major role in limiting lipid oxidation of meat (Kenawi et al. 2011). In contrast, vitamin C content of fruits and vegetables were used to preserve during cold storage using gellan (polysaccharide by *Sphingomonas elodea*) or alginate-based coatings (Robles-Sánchez et al. 2013; Sahraee et al. 2019).

While in another study, peanuts coated with whey protein containing α-tocopherol and ascorbic acid palmitate mixture indicated improved oxygen inhibitory properties of the films (Han et al. 2008; Sahraee et al. 2019).

Encapsulation of active food ingredients has become a trend these days because of enhanced targeted delivery and more bioavailability of sensitive or active ingredient entrapped like α-tocopherol, as it specifically talks of antioxidants. In view of this, researchers studied nano-encapsulated α-tocopherol with methylcellulose films and found that light-barrier properties and antioxidant activity of films were increased with incorporation of nano-encapsulated antioxidant (Noronha et al. 2014; Sahraee et al. 2019).

There are a lot of studies where edible coatings and films have been formed by adding antioxidants for their functional, nutritional, physical and antioxidant retention properties, some of which are compiled and presented in the form of Tables 28.1,

Table 28.1 Antioxidant-based films containing pure compounds

Composite film	Antioxidant used	Properties of film formed	References
CMC	Resveratrol	Film shows antioxidant activity proportional to employed resveratrol concentration No antioxidant activity loss noted during film and coating formation	Pastor et al. (2013)
Methyl cellulose (MC)	Nanoparticles of poly-ε-caprolactone and β- carotene	Showed significant higher antioxidant capacity with 70% β-carotene compared to control MC	Lino (2012)
Low methoxyl pectin	Ascorbic acid	Less degradation of ascorbic acid under high RH storage	De'Nobili et al. (2013)
Corn starch	Oleic acid Alpha-tocopherol	Antioxidant capacity decreases with incorporation of oleic acid Net tocopherol didn't give satisfactorily results	Jiménez et al. (2013)
Sodium caseinates Caseins	Tannic acid and catechins	Surface radical scavenging activity of films containing phenolic compounds increases with storage	Helal et al. (2012)

Table 28.2 Antioxidant-based coatings containing pure compounds

Coatings	Antioxidant additive	Food	Result	References
Alginate	Ascorbic acid and citric acid	Mango	Colour retention and antioxidant potential increased Shelf life extended to 12 days at 4 °C	Robles-Sánchez et al. (2013)
Cassava starch	Citric acid	Fresh cut mango	Delayed deterioration, metabolic activity related to ripening inhibited Better colour preservation and mechanical properties	Chiumarelli et al. (2010)
Alginate pectin	N-acetylcysteine	Pears	Vitamin C loss reduced observed for a week	Oms-Oliu et al. (2008)

28.2, 28.3, 28.4, 28.5 and 28.6 which are distinguished on the basis of antioxidant major class added like pure, EOs and extracts.

For ease of understanding, the major highlights of the tables are mentioned below:

- The antioxidant capacities of the films are dependent on the concentration of active ingredient encased (Noronha 2012; Pastor et al. 2011).
- Addition of these compounds into film formation resulted in variation in mechanical, barrier and optical properties, further oxygen inhibitory capabilities of the film are verified (da Silva Bastos et al. 2009; De'Nobili et al. 2013; Ganiari et al. 2017; Jiménez et al. 2013).

Table 28.3 Antioxidants-based films containing essential oils

Composition film	Essential oils added	Results	References
Quince seed mucilage	Oregano essential oil	As oregano essential oil concentration increased, DPPH scavenging activity also increased	Jouki et al. (2014)
Chitosan	Cinnamon leaf essential oil + oleic acid	Shows high antioxidant activity even incorporation of oleic acid didn't affect it	Perdones et al. (2014)
Starch and chitosan	Basil + thyme	Thyme gave more positive result in case of antioxidant activity Antioxidant capacity decrease during film formation because of volatilizing during drying	Bonilla et al. (2013)
Hake protein	Citronella+ coriander+ tarragon + thyme essential oil	Antioxidant activity improved on essential oil addition Films with coriander and citronella shows increased DPPH radical scavenging capacity	Pires et al. (2013)
Fish skin gelatin	Root essential oils of ginger, turmeric and plai	Plai and turmeric-enriched film shows higher antioxidant activity compared to ginger proven by DPPH and ABTS assays	Tongnuanchan et al. (2013)

Table 28.4 Antioxidants-based coatings containing essential oils

Coatings	Essential oil	Food	Result	References
Pectin	Cinnamon leaf oil	Peach	Radical scavenging activity increased significantly	Ayala-Zavala et al. (2013)
Chitosan	Cinnamon oil	Sweet pepper	Good sensory acceptability, high antioxidant scavenging activity especially against enzymes like catalase and peroxidase	Xing et al. (2011)
Pectin	Cinnamon leaf oil	Grapes	Antioxidant capacity increases significantly	Melgarejo-Flores et al. (2013)
Rice starch	Coconut oil	Tomatoes	Prevents tomatoes ripening	Das et al. (2013)
Cassava starch	Cinnamon bark essential oil + fennel oil	Fuji apple slices	High antioxidant activity with the formulations containing cinnamon essential oils	Oriani et al. (2014)

- Ascorbic and citric acids incorporation verified reduction in oxygen permeability (OP) in films, ultimately ensuring product protection (Atarés et al. 2011; Fabra et al. 2011; Ganiari et al. 2017; Han and Krochta 2007).

Table 28.5 Antioxidants-based films containing extracts

Composite Film	Extract added	Results	References
Gelatin	Green tea extract+ ginger extract+ Gingko leaf extract+ grape seed extract	Gingko leaf extract made film better scavenger as per DPPH assay Similar results were obtained for others too in terms of antioxidant property	Li et al. (2014)
Gelatin	Curcuma ethanol extract	Increased antioxidant activity with increase in concentration of extract	Bitencourt (2013)
Alginate	Ginseng extract	Excellent retention of antioxidant activities	Norajit et al. (2010)

Table 28.6 Antioxidants-based coatings containing extracts

Coating	Extract added	Food	Result	References
Rice starch	Coconut oil + green tea extract	Tomatoes	Phenolic substances make films to exhibit good antioxidant activity even after 20 days	Das et al. (2013)
Konjac glucomann	Pineapple fruit extracts from peel, pulp and core	Rose apple	Core extract more effectively retarded browning Polyphenol oxidase and peroxidase activity reduced	Supapvanich et al. (2012)
Hydroxy propyl methyl cellulose (HPMC)	Propolis ethanolic extract	Grapes	Weight loss and browning prevented during cold storage while gloss and microbial safety improvised	Pastor et al. (2011)

These highlights are major findings of the studies containing pure compounds as edible coating manufacturing base, summarized in Tables 28.1 and 28.2.

Similarly, studies conducted till date on incorporation of essential oils in films and coatings presented in Table 28.3 and 28.4 highlighted that:

- The inhibitory activity depends on the kind of essential oils utilized in film formation.
- Antioxidant activity depends on temperature, concentration, light, type of substrate, physical state of the system and pro-oxidants or synergists.
- High antioxidant capacity of EOs facilitates the improvement in water vapour permeability (WVP) of the film (Atarés et al. 2010).
- Antioxidant power of film depends on concentration of EOs added (Gómez-Estaca et al. 2009; Jouki et al. 2014; Moradi et al. 2012; Shojaee-Aliabadi et al. 2013; Tongnuanchan et al. 2013).

The studies based on antioxidant-rich edible packaging summarized in Tables 28.5 and 28.6 highlights that:

- Phenolic compounds and their antagonistic, synergistic and additive effects results in antioxidant activity (Ganiari et al. 2017; Krochta and Mulder-Johnston 1997).
- Though nature-based antioxidants show strong scavenging and shielding activity, they are found to be less prevalent than synthetic antioxidants.
- Films with added natural extracts exhibited good colour preservation, shielding effect against photo-oxidation and other antioxidant properties (Li et al. 2014; Norajit et al. 2010; Pastor et al. 2013).
- In some cases, addition of extracts doesn't alter the physical properties of the film like moisture content and water solubility.
- Moreover, besides antioxidant activity, many authors have looked into many other functional and physical properties of antioxidant extracts enriched films.

In a nutshell, all the above-mentioned studies presented the following outcomes:

- Functional edible films can be used for packaging of oxidation sensitive food products to extend their shelf life.
- To envelop active compounds, functional and physico-chemical properties need to be worked on.
- Many existing biopolymers are temperature intolerant, brittle, have poor barrier properties and are not cheap (Ali et al. 2013; Sahraee et al. 2019).

Oxygen-barrier properties of degradable polymers are very vital to be studied to delay the oxidation-based deteriorative changes and extending product life. In this context, some significant points need to be pondered on like nature of polymer matrix, concentration of plasticizer and the interaction between plasticizer, cross-linking agents, filler and polymer. The literature has evidences of improved film properties due to cross linking between the active compound and the polymer. Even studies have also revealed the potency of plasticizers in not only improving the structure of the film but also in increasing chain interactions and decreasing polymer-free volume (Jouki et al. 2013; Sahraee et al. 2019).

Additionally, the recent trends of manufacturing edible films with added nanoparticles, nanofibres (Azeredo et al. 2017) and nanoclays are found to improve barrier properties as they make the gas diffusion pathway more complicated The examples of nanocomposite-based nonedible packaging includes poly lactic acid (PLA), polyhydroxy-butyrate, polybutylene succinate and aliphatic polyester (Lee et al. 2015). However, starch and its derivatives are solely edible packaging studied for nanocomposites as per the literature.

The potential applications of antioxidant-based consumable packaging from past case studies mentioned can be compiled as meat, nut, cheese cubes, fruits and vegetables, etc.

28.4.2 Advantages and Disadvantages of Antioxidant-Rich Edible Packaging

As every action comes with associated boon and bane, similarly antioxidant-based edible packaging being most versatile has both advantages and disadvantages attached with it.

Talking of advantages, then it includes:

- No specialized equipment and environment required for its application and storage.
- Not only oxidation-prone food product but other can also be easily packaged.
- Most importantly, it is much more convenient than other type of active packaging that includes oxygen absorber sachets.

However, the sole limitation related to its application is associated with usage of essential oils as radical scavengers because it may lead to organoleptic changes. However, it may be avoided or reduced, once food balances the scents provided by the packaging.

This concern, however, makes food industry and researchers to review the shelf-life of active antioxidant constituents used for at least 2 or 3 months. Well nowadays, there are materials in market with a year stability. Moreover, although EOs contains complex compounds, no other degrading compounds have been known till date.

Further, the edible packaging not only curbs environmental pollution issues but conjointly increases the shelf life of the food products. If these are added with natural extracts which boosts antioxidant activity of the package, then it can solve one more major concern of consumers regarding synthetic food additives and their associated safety and health concerns. However, the effect of ultraviolet (UV) light especially during storage has shown some concern regarding product deterioration but there are researches in past which well explained that antioxidant-rich edible packages can improve the weight loss and respiration rates, conjointly longing storage time when compared to other packaging materials.

28.5 Future Trends

Today's food packaging industry is flooded with plastic packaging and it being hazardous to health and environment have paved way for alternative packaging materials. Though plastic is the cheapest packaging material, these companies are working towards advancing the barrier properties of these traditional polymers using nanoparticles. No matter how striking these materials are, this technique is not commercial at large scale. But in near future, there will be advancement in nano-technology studies and regulations making way for nature-based antioxidant edible packaging with added nanoparticles with more advantages. This technique is gaining recognition among consumers with some fears associated with it as there are no well-established global regulations for its acceptance which is one of the major reason behinds its limited commercialization. However, some or if any of individual state or global regulation are available for edible packaging, it has been presented in next section.

28.6 Regulatory Aspects

Living in global economy requires global regulations which is perhaps the biggest challenge in front of food industries because it is hard to comply with legislations of each country. Therefore, aiming at global legislation, FAO and WHO establishes their own guidelines which are followed globally as uniform regulations and international standards named as Codex Alimentarius.

There are some regulations which are mentioned for usage of edible coatings materials which are as follows:

- First and foremost, regulation is the preparatory material must be **GRAS**, that is, generally regarded as safe.
- Edible packaging regulations should comply with the country's legislation, where it is going to be sold to consumers and consumer safety is the priority among them; therefore, a clear labelling of all the constituents used, processing techniques and functional property food product delivering should be mentioned so as to ensure total transparency with the consumers.
- Edible packaging ingredients being edible, hence must be mentioned as food additives.

In Europe, ingredients used in edible package are included in the regulation EC1331/2008 and EU234/2011 for food additives, enzymes and flavourings (Angelo et al. 2016). It has also been specified that only mentioned ingredients can be used for coating manufacturing.

In US, FDA present the list of additives to be used in food products, however, some of these ingredients have also been allowed specifically for being used as coatings

1. The particular section "Indirect Food Additives: Adhesives and components of coatings" present specific ingredients that can be used in edible packaging production, for example, carnauba wax and bees wax.
2. Also in US, coated fresh cut fruits and vegetables need to be labelled for the consumers. This is just to win consumer trust.

Note: Legislations are available till date regarding ingredients and materials that can be eaten or material approved as food contact material. So, these guidelines must be followed as thumb rule. Being case-specific, the packaging which is to be eaten must follow food additives guidelines and wraps/films not to be eaten must follow food contact material guidelines.

28.7 Conclusion

Antioxidant-rich edible films are nature-based, consumer acceptable, non-additive and biodegradable alternative to artificial or chemical preservatives. These act better in terms of barrier properties and shelf-life extension of food products by inhibiting lipid oxidation. Additionally, films and coatings with antioxidant compounds may result in maintaining food quality by impeding deteriorative changes in food products. Further, nutritional quality especially with the case of encapsulated food material enriched with active agents can be preserved by applying antioxidant-rich edible coatings or films. It has been well established by the researches till date that edible films with added antioxidant agent results in reduced oxygen permeability, water vapour permeability and increased antioxidant activity which is beneficial for slowing down the fruits and vegetables respiration rate. Further, antioxidant-rich edible packaging may delay the rancidity, degradation and discolouration of the product. Natural antioxidants like plants and herb extracts, spice extracts, oleoresins and essential oils, etc., have established a benchmark in enhancing antioxidant retention and activity of the edible coatings.

Additionally, it has been suggested that properties of film can be upgraded by incorporating films with nanomaterials, cross-linking agents and by adding plasticizers, and films with added nanocarriers will be the future and will be more potent in food preservation as a packaging material.

However, both nanotechnology and edible packaging are emerging fields and have a long way to go because these field lack, in part, well-established global or local laws which are necessary to win full customer trust but as consumer perception towards food packaging has changed from plastic-based to biodegradable to edible packaging (the need of hour); it will make statutory bodies too to frame laws which could be used globally and make this packaging flourished more around the globe and contribute to a sustainable development by reducing carbon foot printing as antioxidant involved in packaging material will slow down product deterioration and hence landfilling issues and global environmental issues could be curbed somewhat.

Another area of major concern for antioxidant-enriched edible packaging that requires special attention is the studies related to stability of the antioxidants during storage, their release and the influence on the sensorial properties of food.

References

Akhtar MJ, Jacquot M, Jasniewski J et al (2012) Antioxidant capacity and light-aging study of HPMC films functionalized with natural plant extract. Carbohydr Polym 89:1150–1158. https://doi.org/10.1016/j.carbpol.2012.03.088

Ali A, Maqbool M, Alderson PG, Zahid N (2013) Effect of gum arabic as an edible coating on antioxidant capacity of tomato (Solanum lycopersicum L.) fruit during storage. Postharvest Biol Technol 76:119–124. https://doi.org/10.1016/j.postharvbio.2012.09.011

Angelo M, Cerqueira PR, Couto Teixeira JA, Augusto Vicente A (2016) Edible packaging today. Edible Food Packaging. CRC Press, pp 1–8

Atarés L, Bonilla J, Chiralt A (2010) Characterization of sodium caseinate-based edible films incorporated with cinnamon or ginger essential oils. J Food Eng 100:678–687. https://doi.org/10.1016/j.jfoodeng.2010.05.018

Atarés L, Pérez-Masiá R, Chiralt A (2011) The role of some antioxidants in the HPMC film properties and lipid protection in coated toasted almonds. J Food Eng 104:649–656. https://doi.org/10.1016/j.jfoodeng.2011.02.005

Ayala-Zavala JF, Silva-Espinoza BA, Cruz-Valenzuela MR et al (2013) Pectin-cinnamon leaf oil coatings add antioxidant and antibacterial properties to fresh-cut peach. Flavour Fragr J 28:39–45. https://doi.org/10.1002/ffj.3125

Azeredo HMC, Rosa MF, Mattoso LHC (2017) Nanocellulose in bio-based food packaging applications. Ind Crop Prod 97:664–671. https://doi.org/10.1016/j.indcrop.2016.03.013

Bao S, Xu S, Wang Z (2009) Antioxidant activity and properties of gelatin films incorporated with tea polyphenol-loaded chitosan nanoparticles. J Sci Food Agric 89:2692–2700. https://doi.org/10.1002/jsfa.3775

Benbettaïeb N, Debeaufort F, Karbowiak T (2019) Bioactive edible films for food applications: mechanisms of antimicrobial and antioxidant activity. Crit Rev Food Sci Nutr 59:3431–3455. https://doi.org/10.1080/10408398.2018.1494132

Bhardwaj A, Alam T, Talwar N (2019) Recent advances in active packaging of agri-food products: a review. J Postharvest Technol 7:33–62

Bitencourt CM (2013) Desenvolvimento e aplicação de filmes à base de gelatina aditivados com extrato etanólico de cúrcuma (Cúrcuma longa L.). Biblioteca Digital de Teses e Dissertações da Universidade de São Paulo

Bonilla J, Talón E, Atarés L et al (2013) Effect of the incorporation of antioxidants on physicochemical and antioxidant properties of wheat starch-chitosan films. J Food Eng 118:271–278. https://doi.org/10.1016/j.jfoodeng.2013.04.008

Byun Y, Kim YT, Whiteside S (2010) Characterization of an antioxidant polylactic acid (PLA) film prepared with α-tocopherol, BHT and polyethylene glycol using film cast extruder. J Food Eng 100:239–244. https://doi.org/10.1016/j.jfoodeng.2010.04.005

Chiumarelli M, Pereira LM, Ferrari CC et al (2010) Cassava starch coating and citric acid to preserve quality parameters of fresh-cut "Tommy Atkins" mango. J Food Sci 75:E297–E304. https://doi.org/10.1111/j.1750-3841.2010.01636.x

da Silva Bastos D, de Lima Araújo KG, da Rocha Leão MH (2009) Ascorbic acid retaining using a new calcium alginate-Capsul based edible film. J Microencapsul 26:97–103. https://doi.org/10.1080/02652040802175805

Das DK, Dutta H, Mahanta CL (2013) Development of a rice starch-based coating with antioxidant and microbe-barrier properties and study of its effect on tomatoes stored at room temperature. LWT Food Sci Technol 50:272–278. https://doi.org/10.1016/j.lwt.2012.05.018

De'Nobili MD, Pérez CD, Navarro DA et al (2013) Hydrolytic stability of l-(+)-ascorbic acid in low methoxyl pectin films with potential antioxidant activity at food interfaces. Food Bioprocess Technol 6:186–197. https://doi.org/10.1007/s11947-011-0684-6

Fabra MJ, Hambleton A, Talens P et al (2011) Effect of ferulic acid and α-tocopherol antioxidants on properties of sodium caseinate edible films. Food Hydrocoll 25:1441–1447. https://doi.org/10.1016/j.foodhyd.2011.01.012

Ganiari S, Choulitoudi E, Oreopoulou V (2017) Edible and active films and coatings as carriers of natural antioxidants for lipid food. Trends Food Sci Technol 68:70–82

Gómez-Estaca J, Montero P, Giménez B, Gómez-Guillén MC (2007) Effect of functional edible films and high pressure processing on microbial and oxidative spoilage in cold-smoked sardine (Sardina pilchardus). Food Chem 105:511–520. https://doi.org/10.1016/j.foodchem.2007.04.006

Gómez-Estaca J, Bravo L, Gómez-Guillén MC et al (2009) Antioxidant properties of tuna-skin and bovine-hide gelatin films induced by the addition of oregano and rosemary extracts. Food Chem 112:18–25. https://doi.org/10.1016/j.foodchem.2008.05.034

Halim NRA, Yusof HM, Sarbon NM (2016) Functional and bioactive properties of fish protein hydolysates and peptides: a comprehensive review. Trends Food Sci Technol 51:24–33

Han JH, Krochta JM (2007) Physical properties of whey protein coating solutions and films containing antioxidants. J Food Sci 72:E308–E314. https://doi.org/10.1111/j.1750-3841.2007.00358.x

Han JH, Hwang HM, Min S, Krochta JM (2008) Coating of peanuts with edible whey protein film containing α-tocopherol and ascorbyl palmitate. J Food Sci 73:E349–E355. https://doi.org/10.1111/j.1750-3841.2008.00910.x

Helal A, Tagliazucchi D, Conte A, Desobry S (2012) Antioxidant properties of polyphenols incorporated in casein/sodium caseinate films. Int Dairy J 25:10–15. https://doi.org/10.1016/j.idairyj.2011.12.002

Herald TJ, Hachmeister KA, Huang S, Bowers JR (1996) Corn zein packaging materials for cooked Turkey. J Food Sci 61:415–418. https://doi.org/10.1111/j.1365-2621.1996.tb14206.x

Jamróz E, Kulawik P, Tkaczewska J et al (2021) The effects of active double-layered furcellaran/gelatin hydrolysate film system with Ala-Tyr peptide on fresh Atlantic mackerel stored at −18 °C. Food Chem 338:127867. https://doi.org/10.1016/j.foodchem.2020.127867

Jiménez A, Fabra MJ, Talens P, Chiralt A (2013) Physical properties and antioxidant capacity of starch-sodium caseinate films containing lipids. J Food Eng 116:695–702. https://doi.org/10.1016/j.jfoodeng.2013.01.010

Jouki M, Tabatabaei Yazdi F, Mortazavi SA, Koocheki A (2013) Physical, barrier and antioxidant properties of a novel plasticized edible film from quince seed mucilage. Int J Biol Macromol 62:500–507. https://doi.org/10.1016/j.ijbiomac.2013.09.031

Jouki M, Yazdi FT, Mortazavi SA, Koocheki A (2014) Quince seed mucilage films incorporated with oregano essential oil: physical, thermal, barrier, antioxidant and antibacterial properties. Food Hydrocoll 36:9–19. https://doi.org/10.1016/j.foodhyd.2013.08.030

Kenawi MA, Zaghlul MMA, Abdel-Salam RR (2011) Effect of two natural antioxidants in combination with edible packaging on stability of low fat beef product stored under frozen condition. Biotechnol Anim Husb 27:345–356

Krochta JM, Mulder-Johnston CD (1997) Edible and biodegradable polymer films: challenges and opportunities. Food Technol 51:61–74

Lee SY, Lee SJ, Choi DS, Hur SJ (2015) Current topics in active and intelligent food packaging for preservation of fresh foods. J Sci Food Agric 95:2799–2810. https://doi.org/10.1002/jsfa.7218

León PG, Rojas AM (2007) Gellan gum films as carriers of l-(+)-ascorbic acid. Food Res Int 40:565–575. https://doi.org/10.1016/j.foodres.2006.10.021

Li JH, Miao J, Wu JL et al (2014) Preparation and characterization of active gelatin-based films incorporated with natural antioxidants. Food Hydrocoll 37:166–173. https://doi.org/10.1016/j.foodhyd.2013.10.015

Lino RC (2012) Desenvolvimento de filmes de metilcelulose incorporados por nanopartículas de poli-ε-Caprolactona/β-Caroteno

Malherbi NM, Schmitz AC, Grando RC et al (2019) Corn starch and gelatin-based films added with guabiroba pulp for application in food packaging. Food Packag Shelf Life 19:140–146. https://doi.org/10.1016/j.fpsl.2018.12.008

Melgarejo-Flores BG, Ortega-Ramírez LA, Silva-Espinoza BA et al (2013) Antifungal protection and antioxidant enhancement of table grapes treated with emulsions, vapors, and coatings of cinnamon leaf oil. Postharvest Biol Technol 86:321–328. https://doi.org/10.1016/j.postharvbio.2013.07.027

Moradi M, Tajik H, Razavi Rohani SM et al (2012) Characterization of antioxidant chitosan film incorporated with Zataria multiflora Boiss essential oil and grape seed extract. LWT Food Sci Technol 46:477–484. https://doi.org/10.1016/j.lwt.2011.11.020

Murcia MA, Martínez-Tomé M (2001) Antioxidant activity of resveratrol compared with common food additives. J Food Prot 64:379–384. https://doi.org/10.4315/0362-028X-64.3.379

Nerín C (2010) Antioxidant active food packaging and antioxidant edible films. Oxid Foods Beverages Antioxid Appl:496–515. https://doi.org/10.1533/9780857090331.3.496

Norajit K, Kim KM, Ryu GH (2010) Comparative studies on the characterization and antioxidant properties of biodegradable alginate films containing ginseng extract. J Food Eng 98:377–384. https://doi.org/10.1016/j.jfoodeng.2010.01.015

Noronha CM (2012) Incorporação de nanocápsulas de poli(ε-caprolactona) contendo α-tocoferol EM biofilmes de metilcelulose

Noronha CM, De Carvalho SM, Lino RC, Barreto PLM (2014) Characterization of antioxidant methylcellulose film incorporated with α-tocopherol nanocapsules. Food Chem 159:529–535. https://doi.org/10.1016/j.foodchem.2014.02.159

Oms-Oliu G, Soliva-Fortuny R, Martín-Belloso O (2008) Edible coatings with antibrowning agents to maintain sensory quality and antioxidant properties of fresh-cut pears. Postharvest Biol Technol 50:87–94. https://doi.org/10.1016/j.postharvbio.2008.03.005

Oriani VB, Molina G, Chiumarelli M et al (2014) Properties of cassava starch-based edible coating containing essential oils. J Food Sci 79:E189–E194. https://doi.org/10.1111/1750-3841.12332

Pastor C, Sanchez-Gonzalez L, Marcilla A et al (2011) Quality and safety of table grapes coated with hydroxypropylmethylcellulose edible coatings containing propolis extract. Postharvest Biol Technol 60:64–70. https://doi.org/10.1016/j.postharvbio.2010.11.003

Pastor C, Sánchez-González L, Chiralt A et al (2013) Physical and antioxidant properties of chitosan and methylcellulose based films containing resveratrol. Food Hydrocoll 30:272–280. https://doi.org/10.1016/j.foodhyd.2012.05.026

Perdones Á, Vargas M, Atarés L, Chiralt A (2014) Physical, antioxidant and antimicrobial properties of chitosan-cinnamon leaf oil films as affected by oleic acid. Food Hydrocoll 36: 256–264. https://doi.org/10.1016/j.foodhyd.2013.10.003

Pires C, Ramos C, Teixeira B et al (2013) Hake proteins edible films incorporated with essential oils: physical, mechanical, antioxidant and antibacterial properties. Food Hydrocoll 30:224–231. https://doi.org/10.1016/j.foodhyd.2012.05.019

Rambabu K, Bharath G, Banat F et al (2019) Mango leaf extract incorporated chitosan antioxidant film for active food packaging. Int J Biol Macromol 126:1234–1243. https://doi.org/10.1016/j.ijbiomac.2018.12.196

Riveros CG, Nepote V, Grosso NR (2016) Thyme and basil essential oils included in edible coatings as a natural preserving method of oilseed kernels. J Sci Food Agric 96:183–191. https://doi.org/10.1002/jsfa.7080

Robles-Sánchez RM, Rojas-Graü MA, Odriozola-Serrano I et al (2013) Influence of alginate-based edible coating as carrier of antibrowning agents on bioactive compounds and antioxidant activity in fresh-cut Kent mangoes. LWT Food Sci Technol 50:240–246. https://doi.org/10.1016/j.lwt.2012.05.021

Sahraee S, Milani JM, Regenstein JM, Kafil HS (2019) Protection of foods against oxidative deterioration using edible films and coatings: a review. Food Biosci 32:100451

Salgado PR, Ortiz CM, Musso YS et al (2015) Edible films and coatings containing bioactives. Curr Opin Food Sci 5:86–92

Shojaee-Aliabadi S, Hosseini H, Mohammadifar MA et al (2013) Characterization of antioxidant-antimicrobial κ-carrageenan films containing Satureja hortensis essential oil. Int J Biol Macromol 52:116–124. https://doi.org/10.1016/j.ijbiomac.2012.08.026

Skurtys O (2014) Food hydrocolloid edible films and coatings

Supapvanich S, Prathaan P, Tepsorn R (2012) Browning inhibition in fresh-cut rose apple fruit cv. Taaptimjaan using konjac glucomannan coating incorporated with pineapple fruit extract. Postharvest Biol Technol 73:46–49. https://doi.org/10.1016/j.postharvbio.2012.05.013

Tkaczewska J (2020) Peptides and protein hydrolysates as food preservatives and bioactive components of edible films and coatings - a review. Trends Food Sci Technol 106:298–311

Tongnuanchan P, Benjakul S, Prodpran T (2013) Physico-chemical properties, morphology and antioxidant activity of film from fish skin gelatin incorporated with root essential oils. J Food Eng 117:350–360. https://doi.org/10.1016/j.jfoodeng.2013.03.005

Wu J, Sun X, Guo X et al (2018) Physicochemical, antioxidant, in vitro release, and heat sealing properties of fish gelatin films incorporated with β-cyclodextrin/curcumin complexes for apple juice preservation. Food Bioprocess Technol 11:447–461. https://doi.org/10.1007/s11947-017-2021-1

Xing Y, Li X, Xu Q et al (2011) Effects of chitosan coating enriched with cinnamon oil on qualitative properties of sweet pepper (Capsicum annuum L.). Food Chem 124:1443–1450. https://doi.org/10.1016/j.foodchem.2010.07.105

The manufacturer's authorised representative in the EU is Springer Nature Customer Service Centre GmbH, Europaplatz 3, 69115 Heidelberg, Germany. If you have any concerns regarding our products, please contact ProductSafety@springernature.com

Printed and bound by CPI Group (UK) Ltd, Croydon, CR0 4YY
30/09/2026
02225324-0020